ADVANCES IN SPACE ENVIRONMENT RESEARCH – VOLUME I

ADVANCES IN SPACE ENVIRONMENT RESEARCH – VOLUME I

Proceedings of the WISER Workshops on

World Space Environment Forum (WSEF2002)
and
High Performance Computing in
Space Environment Research (HPC2002)

Adelaide, Australia   22 July to 2 August 2002

Chief-Editor:

A.C.-L. CHIAN
World Institute for Space Environment Research – WISER,
University of Adelaide, Australia and INPE, Brazil

Co-Editors:
I.H. CAIRNS (*University of Sydney, Australia*)
S.B. GABRIEL (*University of Southampton, UK*)
J.P. GOEDBLOED (*FOM-Institute for Plasma Physics, The Netherlands*)
T. HADA (*Kyushu University, Japan*)
M. LEUBNER (*Space Research Institute, Austria*)
L. NOCERA (*Institute for Chemical and Physical Processes/CNR, Italy*)
R. STENING (*University of New South Wales, Australia*)
F. TOFFOLETTO (*Rice University, USA*)
C. UBEROI (*Indian Institute of Science, India*)
J.A. VALDIVIA (*University of Chile, Chile*)
U. VILLANTE (*University of L'Aquila, Italy*)
C.-C. WU (*University of California at Los Angeles, USA*)
Y. YAN (*National Astronomical Observatories*)

Reprinted from *Space Science Reviews*, Volume 107, Nos. 1–2, 2003

INTERNATIONAL
SPACE
SCIENCE
INSTITUTE

Springer-Science+Business Media, B.V.

Space Sciences Series of ISSI

Library of Congress Cataloging-in-Publication Data

Advances in space environment research / edited by A.C.-L. Chian.
   p. cm.
   ISBN 978-94-010-3781-5   ISBN 978-94-007-1069-6 (eBook)
   DOI 10.1007/978-94-007-1069-6

1. Space environment. I. Chian, Abraham C.-L.   2003051317

Printed on acid-free paper

SPACE SCIENCE REVIEWS / *Vol. 107   Nos. 1–2   2003*

TABLE OF CONTENTS

II: MAGNETOSPHERE/BOW SHOCK

III: IONOSPHERE/ATMOSPHERE

IV: SPACE WEATHER/SPACE CLIMATE

V: SPACE PLASMA PHYSICS/ASTROPHYSICS

VI: COMPLEX/INTELLIGENT SYSTEMS

List of Contributors

I.I. Alexeev
Scobeltsyn Institute of Nuclear Physics, Moscow State University
Moscow 119992, Russia
alexeev@dec1.sinp.msu.ru

I.H. Cairns
School of Physics, University of Sydney
Sydney, NSW 2006, Australia
cairns@physics.usyd.edu.au

A.H. Cerqueira
Department of Exact Sciences and Technologies, State University of Santa Cruz
Ilheus, Bahia CEP 45650-000, Brazil
hoth@usesc.br

T. Chang
Center for Space Research, MIT
Cambridge, MA 02139, USA
tsc@space.mit.edu

A.C.-L. Chian
World Institute for Space Environment Research (WISER)
WISER/NITP, University of Adelaide
Adelaide, SA 5005, Australia
achian@physics.adelaide.edu.au
&
National Institute for Space Research (INPE)
P.O. Box 515, Sao Jose dos Campos-SP, CEP 12227-010, Brazil
achian@dge.inpe.br

D.-Y. Chou
Institute of Astronomy, Tsing Hua University
Hsinchu 30043, Taiwan, R.O.C.
chou@phys.nthu.edu.tw

R. Clay
Physics Department, University of Adelaide
Adelaide, SA 5005, Australia
rclay@physics.adelaide.edu.au

D. Cole
IPS Radio and Space Services
P.O. Box 1386 Haymarket, Sydney, NSW 1240, Australia
david@ips.gov.au

B. Dawson
Physics Department, University of Adelaide
Adelaide, SA 5005, Australia
bruce.dawson@adelaide.edu.au

A. De Assis
Fluminense Federal University
Caixa Postal 100294, Niteroi – RJ, CEP 24001-970, Brazil
gmaasda@vm.uff.br

R. Dewar
Department of Theoretical Physics, Australian National University
Canberra, ACT 0200, Australia
robert.dewar@anu.edu.au

A.-C. Donea
Physics Department, University of Adelaide
Adelaide, SA 5005, Australia
adonea@physics.adelaide.edu.au

X. Feng
Laboratory for Space Weather, Center for Space Science and Applied Research
Beijing 100080, China
casfeng@yahoo.com

B.J. Fraser
School of Mathematical and Physical Sciences, University of Newcastle
Callaghan, NSW 2308, Australia
phbjf@cc.newcastle.edu.au

S.B. Gabriel
Aeronautics Research Group, University of Southampton
Southampton, U.K. SO17 1 BJ
sbg2@soton.ac.uk

Goedbloed, J.P.
FOM-Institute for Plasma Physics
P.O. Box 1207, 3430 BE Nieuwegein, The Netherlands
goedbloed@rijnh.nl

J.-S. Guo
Laboratory for Space Weather, Center for Space Science and Applied Research
P.O. Box 8701, Beijing 100080, China
guojs@center.cssar.ac.cn

M.M. Guzzo,
Institute of Physics, State University of Campinas – UNICAMP
Campinas – SP, CEP 13083-970, Brazil
guzzo@ifi.unicamp.br

T. Hada
E.S.S.T., Kyushu University
Fukuoka 816-8580, Japan
hada@esst.kyushu-u.ac.jp

K. He
Institute of Low Energy Nuclear Physics, Beijing Normal University
Beijing 100875, China
kfhe@bnu.edu.cn

V.K. Jordanova
Space Science Center, University of New Hampshire
Durham, NH 03824, USA
vania.jordanova@unh.edu

Y. Kamide
Solar-Terretrial Environment Laboratory, Nagoya University
Toyokawa 442-8507, Japan
kamide@stelab.nagoya-u.ac.jp

M. Karlicky
Astronomical Institute, Academy of Sciences of the Czech Republic
Ondrejov CZ-25165, Czech Republic
karlicky@asu.cas.cz

D. Koga
E.S.S.T., Kyushu University
Fukuoka 816-8580, Japan
daiki@esst.Kyushu-u.ac.jp

G. Lapenta
Plasma Theory Group, Loss Alamos National Laboratory
Los Alamos, NM 87545, USA
lapenta@lanl.gov

M. Leubner
Space Research Institute, Austrian Academy of Sciences
Graz A-8042, Austria
Manfred.leubner@uibk.ac.at

N. Marsh
Danish Space Research Institute
Copenhagen DK-2100, Denmark
ndm@dsri.dk

F. Otsuka
E.S.S.T., Kyushu University
Fukuoka 816-8580, Japan
otsuka@esst.kyushu-u.ac.jp

R. Protheroe
Physics Department, University of Adelaide
Adelaide, SA 5005, Australia
rprother@physics.adelaide.edu.au

Reggiani, N.
Mathematics Faculty, Catholic University of Campinas
Campinas-SP, CEP 13086-900, Brazil
universo@puc-campinas.edu.br

E.L. Rempel
National Institute for Space Research – INPE
P.O. Box 515, Sao Jose dos Campos-SP, CEP 12227-010, Brazil
erico@lac.inpe.br

F.B. Rizzato
Institute of Physics, Federal University of Rio Grande do Sul
Caixa Postal 15051, Porto Alegre-RS, CEP 91500, Brazil
rizzato@if.ufrgs.br

Schunker, H.
Physics Department, University of Adelaide
Adelaide, SA 5005, Australia
hschunke@physics.adelaide.edu.au

R. Sekar
Physical Research Laboratory
Ahmedabad 380 009, India
rsekar@prl.ernet.in

Sen, S.
Research School of Physical Sciences and Eng., Australian National University
Canberra, ACT 0200, Australia
sudip.sen@anu.edu.au

R. Stening
School of Physics, University of New South Wales
Sydney, NSW 2052, Australia
r.stening@unsw.edu.au

M.-T. Sun
Department of Mechanical Engineering, Chang-Gung University
Tao-Yuan 33333, Taiwan, R.O.C.
mtsun@mail.cgu.edu.tw

F. Toffoletto
Department of Physics and Astronomy, Rice University
Houston, TX 77005, USA
toffo@rice.edu

C. Uberoi
Department of Mathematics, Indian Institute of Science
Bangalore 560 012, India
cuberoi@math.iisc.ernet.in

J.A. Valdivia
Department of Physics, University of Chile
Santiago, Chile
alejo@fisica.ciencias.uchile.cl

M.J. Vasconcelos
Department of Exact Sciences and Technologies, State University of Santa Cruz
Ilheus, Bahia, CEP 45650-000, Brazil
mjvasc@uesc.br

U. Vilante
Department of Physics, University of L'Aquila
L'Aquila, Italy
umberto.villante@aquila.infn.it

Y. Voitenko
Center for Plasma Astrophysics, Catholic University of Leuven
Leuven B-3001, Belgium
yuriy.voitenko@wis.kuleuven.ac.be

F. Wei
Laboratory for Space Weather, Center for Space Science and Applied Research
P.O. Box 8701, Beijing 100080, China
weifs@ns.lhp.ac.cn

L. Winton
School of Physics, University of Melbourne
Melbourne, VC 3010, Austria
winton@physics.unimelb.edu.au

C.-C. Wu
Institute of Geophysics and Planetary Phys., University of California at Los Angeles
Los Angeles, CA 90095, USA
wu@physics.ucla.edu

Y. Yan
National Astronomical Observatories, Chinese Academy of Sciences
Beijing 100012, China
yyh@bao.ac.nc

FOREWORD: ADVANCES IN SPACE ENVIRONMENT RESEARCH

ABRAHAM C.-L. CHIAN[1,2] and the WISER team

[1] World Institute for Space Environment Research, WISER/NITP, University of Adelaide,
Adelaide, SA 5005, Australia
[2] National Institute for Space Research (INPE), P.O. Box 515,
12227-010 Sao Jose dos Campos – SP, Brazil

Space environment research is devoted to the study of physical processes of the upper atmospheres of planets, Sun-Earth connections, interplanetary medium, and interactions of interstellar medium and galactic/extragalactic cosmic rays with the heliosphere. This field of research is multi-disciplinary encompassing space physics, astrophysics, computational science, applied mathematics and engineering.

As the world enters the twenty-first century, our society is becoming increasingly dependent on technology which is vulnerable to the physical conditions in the space environment. *Space weather* is influenced by disturbances in the space environment caused by solar flares, coronal mass ejections, magnetic storms and cosmic rays, which can affect the performance and reliability of space-borne and ground-based technological systems such as satellites, precise positioning systems, telecommunications, high speed data and imaging, geologic prospecting, power distribution, gas and oil pipelines, aviation, climate, defense, as well as human health. As society comes to rely more heavily on systems operating in space, the knowledge of space weather will become vital for the maintenance of those systems and the industrial base that they supply. Future economic and social infrastructure will depend upon the nations' capability to predict the potential space storms in advance – *space weather forecasting*, to provide timely, accurate, and reliable space environment observations, specifications and forecasts to alleviate damage to systems and human life.

Motivated by the growing worldwide interest in the study of space environment, the World Institute for Space Environment Research (WISER) was established officially through the collaboration with the National Institute for Theoretical Physics (NITP) at the University of Adelaide in Australia in December 2001. The aim of WISER is to coordinate an international network of research and training centers of excellence dedicated to promoting cooperation in cutting-edge space environment research and training of first-rate space scientists, with emphasis on the application of information technology to theoretical and computational studies of space plasmas and atmospheres, space data analysis, space weather forecasting, and monitoring the impact of space weather on the Earth's climate, environment and technology. In contrast to other space weather programs which focus on national or regional interests, WISER aims to address problems of global concern through collaboration involving all nations, with a special emphasis on the complex/intelligent systems approach to study the dynamics and structures of the space environment.

Space plasmas and planetary atmospheres are complex systems whose dynamics depend on the interactions involving a large number of sub-systems. The Sun-Earth connections, which determine space weather, is the result of a complex chain of spatiotemporal interactions involving the solar interior-solar atmosphere-solar wind-magnetosphere-ionosphere-atmosphere coupling. A distinctive property of fluid motions such as space plasmas and the Earth's atmosphere is the inverse energy cascade, whereby energy transferring into large-scale vortices, coherent structures, or sheared flows gives a remarkable propensity for self-organising behaviour. Solar active regions, interplanetary medium and the Earth's magnetosphere are dominated by waves, instabilities and turbulence, often exhibiting characteristics typical of complex systems.

The WISER scientific program covers the following topics: (1) Waves, instabilities and turbulence in space plasmas, (2) Interstellar medium-heliosphere coupling, (3) Solar wind-magnetosphere-ionosphere-atmosphere coupling, (4) Solar flares and coronal mass ejections, (5) Geomagnetic storms, (6) Magnetospheric radiation belts, (7) Ionospheric irregularities, (8) Dynamics and structures of the Earth's atmosphere, (9) Solar modulation of cosmic rays, (10) Cosmic rays-climate connection, (11) Space weather forecasting and (12) Impact of space weather on terrestrial systems and human health. This scientific program is carried out by the following 6 WISER Workgroups: (1) Sun/Heliosphere, (2) Magnetosphere/Bow Shock, (3) Ionosphere/Atmosphere, (4) Space Weather/Space Climate, (5) Space Plasma Physics/Astrophysics and (6) Complex/Intelligent Systems.

Presently, the following 3 institutions are hosting the WISER Regional Centers of Excellence which coordinate the research and training activities in each region: 1) Asia-Pacific Region: National Institute for Theoretical Physics (NITP), University of Adelaide, Australia; 2) Pan-American Region: National Institute for Space Research (INPE), Brazil; 3) Afro-European Region: International Centre for Theoretical Physics (ICTP), Italy.

In 2002, three international conferences were organized by WISER. In collaboration with the Institute for Chemical and Physical Processes (IPCP) of the Italian National Research Council (CNR), a WISER Workshop on Space Environment Turbulence (ALFVEN2002) was held in Pisa, Italy on 22 June 2002. The aim of this workshop was to review recent advances in the theory of waves, instabilities and turbulence in space plasmas and atmospheres. In collaboration with the National Institute for Theoretical Physics (NITP) of the University of Adelaide, WISER organized the World Space Environment Forum (WSEF2002) from 22 to 25 July 2002 and the WISER Workshop on High Performance Computing in Space Environment Research (HPC2002) from 29 July to 2 August 2002, respectively, in Adelaide, Australia. The aim of the WSEF2002 was to review the state-of-the-art of space environment research and identify the key problems in solar-terrestrial connection to be addressed by the international space science and plasma physics community in the coming years. It also provided a forum for the elaboration of international collaboration programs on research and training in space environment. The aim of the HPC2002 was to review the recent advances on high performance computing and computer modeling in space environment research.

About 80 leading scientists from over 20 countries/regions are taking part in the WISER research and training activities. Other scientists and students from all nations are welcome to join forces with WISER to link nations for the peaceful use of the space environment. Sixty-one delegates from 20 countries/regions who attended the World Space Environment Forum (WSEF2002) approved unanimously the establishment of WISER and recommended strongly that WISER should seek sponsorship through United Nations organisations such as UNESCO, the United Nations University (UNU) and the UN Office for Outer Space Affairs. In addition, the delegates to WSEF2002 approved the organization of three (WSEF, ALFVEN and HPC) series of WISER conferences/workshops and International Advanced Schools on a regular basis.

This volume contains fifty-three papers presented at the WSEF2002 and HPC2002. All contributions have been independently refereed. The papers are grouped in alphabetical order, under six topical areas covered by the six WISER Workgroups.

The hospitality of and financial support from the National Institute for Theoretical Physics (NITP) and the ARC Special Research Centre for the Subatomic Structure of Matter (CSSM) of the University of Adelaide, the Institute for Chemical and Physical Processes (IPCP) of the Italian National Research Council (CNR), and the Air Force Office of Scientific Research (AFOSR) are greatly appreciated. In particular, we acknowledge the strong support of Professor A. W. Thomas, Ms. R. Adorjan, Mrs. Sara Boffa and Mrs. S. Johnson of NITP and CSSM, Dr. M. Martinelli of IPCP/CNR, and Drs. C. Rhoades and P. Bellaire of AFOSR, as well as the valuable assistance in paper review by R. Clay (U. of Adelaide), P. Coddington (U. of Adelaide), E. M. de Gouveia Dal Pino (U. of Sao Paulo), V. Jatenco-Pereira (U. of Sao Paulo), V. Krishan (Indian Inst. Astrophysics), D. Moudry (U. of Alaska), R. Sekar (Physical Res. Lab.) and G. Tanco (U. of Sao Paulo).

I: SUN/HELIOSPHERE

MODELLING OF THE ELECTROMAGNETIC FIELD IN THE INTERPLANETARY SPACE AND IN THE EARTH'S MAGNETOSPHERE

IGOR I. ALEXEEV, ELENA S. BELENKAYA, SERGEY YU. BOBROVNIKOV and
VLADIMIR V. KALEGAEV

Scobeltsyn Institute of Nuclear Physics, Moscow State University, Moscow, Russia

Abstract. A magnetohydrodynamic model of the solar wind flow is constructed using a kinematic approach. It is shown that a phenomenological conductivity of the solar wind plasma plays a key role in the forming of the interplanetary magnetic field (IMF) component normal to the ecliptic plane. This component is mostly important for the magnetospheric dynamics which is controlled by the solar wind electric field. A simple analytical solution for the problem of the solar wind flow past the magnetosphere is presented. In this approach the magnetopause and the Earth's bow shock are approximated by the paraboloids of revolution. Superposition of the effects of the bulk solar wind plasma motion and the magnetic field diffusion results in an incomplete screening of the IMF by the magnetopause. It is shown that the normal to the magnetopause component of the solar wind magnetic field and the tangential component of the electric field penetrated into the magnetosphere are determined by the quarter square of the magnetic Reynolds number. In final, a dynamic model of the magnetospheric magnetic field is constructed. This model can describe the magnetosphere in the course of the severe magnetic storm. The conditions under which the magnetospheric magnetic flux structure is unstable and can drive the magnetospheric substorm are discussed. The model calculations are compared with the observational data for September 24–26, 1998 magnetic storm ($Dst_{min} = -205$ nT) and substorm occurred at 02:30 UT on January 10, 1997.

Key words: magnetic storm, magnetosphere, magnetospheric model, solar wind, substorm

1. Introduction

The investigation of the solar wind origin, the solar wind plasma flow and interplanetary magnetic field structure are the most important problems in the Solar-Terrestrial physics. During the several solar cycles the observations of *Helios, Pioneer, Voyager, Ulysses, Galileo, and Cassini* showed us the large scale solar wind magnetic field structure at the heliocentric distances of 0.3–60.0 a.u. It has been determined that the magnetic field in the heliosphere is organized into two open magnetic flux regions, each with uniform polarity, separated by a reasonably well-defined current sheet (McComas *et al.*, 2000; Balogh and Smith, 2001). Good correlation between the photospheric magnetic field and the sector structure of the IMF is observed. Recent *Ulysses* observations suggest that a single current sheet is conserved even in periods of high solar activity when several magnetic field sectors have been recognized at the Earth's orbit (Balogh and Smith, 2001).

Space Science Reviews **107**: 7–26, 2003.
© 2003 *Kluwer Academic Publishers.*

Due to the solar wind Earth's magnetosphere coupling the magnetospheric state and the magnetospheric dynamics are determined by the mass, momentum, and energy transport across the magnetopause. Properties of the magnetosheath region adjacent to the magnetopause depend strongly on the magnetic shear across the magnetopause: an angle between the magnetosheath and magnetospheric magnetic field directions. The analysis of 68 passes from 1984 to 1986 during the AMPTE/IRM mission (Phan *et al.*, 1994) shows that for the high magnetic shear across the magnetopause, the magnetosheath near the magnetopause is more disturbed. In general, the magnetic field rises when we move from the bow shock to the magnetopause, but for the high magnetic shear this trend often reverses in the vicinity of the magnetopause: the magnetic field decreases, indicating violation of the frozen-in magnetic field condition in this region. The plasma density is relatively constant in the dayside magnetosheath for high shear. However, plasma depletion layer with decreased values of density near the low shear magnetopause is actually detected (Phan *et al.*, 1979; Phan *et al.*, 1994).

The magnetohydrodynamic (MHD) approach is one of the most fruitful tools to investigate the mentioned above phenomena. Although the dimension scales of the Sun, heliosphere and magnetosphere are very different we will try to use a universal physical approach as long as possible.

Based on the MHD approach the equations for determination of the large scale interplanetary magnetic and electric fields are formulated in this paper. A solution for the MHD problem of the solar wind expansion is described. A quasistationary model of the interplanetary magnetic and electric fields which phenomenologically includes the dissipative processes in the solar wind plasma is constructed. Taking into account a finite plasma conductivity a problem of the solar wind magnetized plasma flowing past the magnetosphere is solved. Using a correct analytical solution for the external problem of the flowing past a blunt body, the IMF component normal to the magnetopause, B_n, is calculated. The solution dependence on the magnetic Reinolds number Re_m is examined. In particular, at $Re_m \rightarrow \infty$ (the frozen-in condition) asymptotic behavior of B_n is obtained. 'Shielding' of the magnetospheric magnetic field is described and the observed weakening (by about 10 times) of the solar wind electric field in the process of its penetration the magnetosphere is explained.

Based on the outer problem solution, the key parameters of the magnetospheric dynamic model and their relations to the measured parameters of the interplanetary medium and geomagnetic activity are determined. The created model was used to describe the disturbed states of the magnetosphere in the course of the substorms and magnetic storms. The analysis of the experimental data and the model calculations show that the magnetotail current system and the ring current contributions to the observed depression of the geomagnetic field at the Earth's equator are of the same order. Rapid variations of the geomagnetic field at the equator during the recovery phase of the storm are mainly associated with the magnetotail currents.

The model calculations are compared with the onground and satellite observations in the course of GEM storm on September 24–26, 1998.

2. Solar Wind Flow and Interplanetary Magnetic Field in the Heliosphere

Below we describe the kinematic MHD model (Alexeev *et al.*, 1982) which results coincide with measurements (Balogh and Smith, 2001). We use a simplest kinematic approach because the Mach-Alfven number $Ma^2 \gg 1$ for the solar wind flow, here $Ma^2 = \mu_0 \rho V_0^2 / B^2$, ρ is the plasma density, V_0 is the solar wind speed, and B is the interplanetary magnetic field.

Inside the Sun (a sphere with radius $R_\odot$) the plasma rigid rotation is proposed. In the spherical coordinate system for $R < R_\odot$ the flow velocity is $\vec{V} = \{0, 0, \Omega \cdot R \cdot \sin\theta\}$ where Ω is the Sun's angular velocity.

For $R > R_\odot$, outside the Sun a radial plasma flow is proposed to be $\vec{V} = \{V_0, 0, 0\}$. If inside the Sun magnetic field $\vec{B}$ sources are given, then we can define $\vec{E}$ and $\vec{B}$ from the boundary conditions at $R = R_\odot$ (B_n and E_t are continuous) as well as from the Maxwell's equations and the Ohm's law:

$$\text{rot}\vec{B} = \mu_\circ \vec{J}, \ \text{rot}\vec{E} = \text{div}\vec{B} = 0, \ \text{div}\vec{E} = \frac{\rho_{ch}}{\varepsilon_\circ}, \ \vec{E} + \left[\vec{V} \times \vec{B}\right] = \frac{\vec{J}}{\sigma}. \tag{1}$$

If $\partial/\partial\varphi \equiv 0$, Equations (1) are separated on the following equation arrays:

$$\frac{\partial(RB_\varphi)}{\partial R} - \frac{Re_m}{R_\odot} \cdot RB_\varphi - \mu_\circ\sigma\frac{\partial\Phi}{\partial\theta} = 0,$$

$$\frac{1}{R\sin\theta}\frac{\partial}{\partial\theta}\left(\sin\theta \cdot B_\varphi\right) + \mu_\circ\sigma\frac{\partial\Phi}{\partial R} = 0, \tag{2}$$

$$\frac{\partial(RB_\theta)}{\partial R} - \frac{\partial B_R}{\partial\theta} - \frac{Re_m}{R_\odot} \cdot RB_\theta = 0,$$

$$\frac{1}{R}\frac{\partial(R^2 B_R)}{\partial R} + \frac{1}{\sin\theta} \cdot \frac{\partial}{\partial\theta}(\sin\theta \cdot B_\theta) = 0. \tag{3}$$

If conductivity σ =constant, then a solution of Equations (2,3) depends on the magnetic Reynolds number, $Re_m = \mu_\circ\sigma V_\circ R_\odot$. In Equation (2) $\Phi(R, \theta, \varphi)$ is a potential of the interplanetary electric field, $\vec{E} = -\nabla\Phi$. Equation $\rho_{ch} = \varepsilon_\circ \text{div}\vec{E}$ defines a charge density ρ_{ch}.

Let us describe a simple case when $\vec{B}$ is constant inside the Sun. In this case the electric currents $\vec{j} \equiv 0$ inside the sphere. Introducing the electromotive force of the unipolar inductor $\Phi_\circ = \Omega B_\circ R_\odot^2/2$ and a dimensionless distance $r = R/R_\odot$ inside the Sun (for $R \leq R_\odot$ or $r \leq 1$) we have:

$$\Phi = \Phi_\circ\left(r^2\sin^2\theta - \frac{2}{3}\right), \ E_R = -2\frac{\Phi_\circ}{R_\odot}r\sin^2\theta, \ E_\theta = E_R\frac{\cos\theta}{\sin\theta},$$

$$E_\varphi = B_\varphi = \vec{J} = 0, \ \rho = \frac{\Phi_\circ}{\pi R_\odot^2}, \ B_R = B_\circ \cdot \cos\theta, \ B_\theta = -B_\circ \cdot \sin\theta. \tag{4}$$

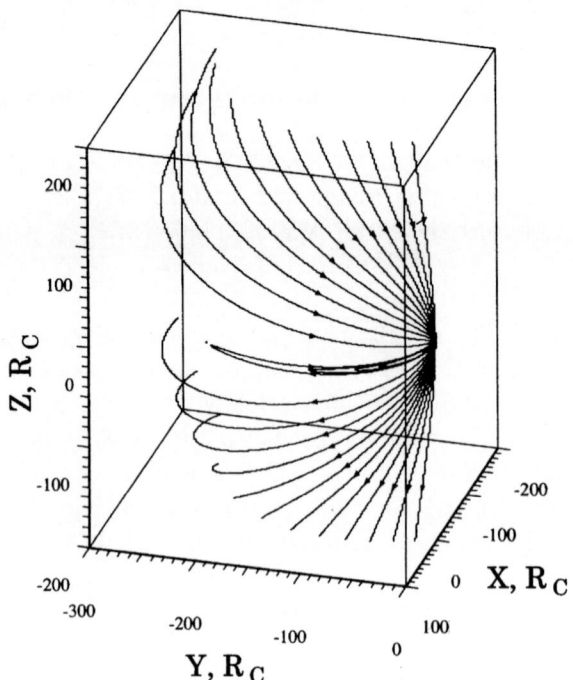

Figure 1. Interplanetary magnetic field lines which attached to the Sun at the heliolongitude $\varphi = 0$.

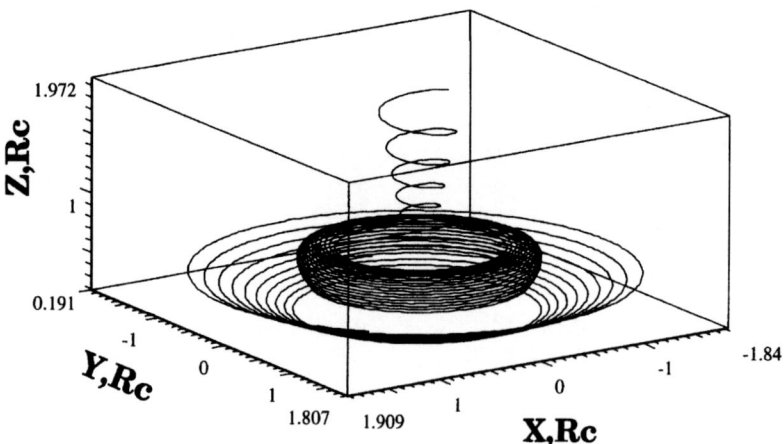

Figure 2. Three–dimensional electric current patterns in the inner heliosphere.

Outside the Sun (for $R > R_\odot$):

$$\Phi = -\frac{\Phi_\infty}{2}Q_2(t) \cdot \left(\cos 2\theta + \frac{1}{3}\right), \qquad E_R = 6\frac{Re_m}{R_\odot}\frac{F_2(t)}{Q_2(t)} \cdot \Phi,$$

$$E_\theta = -\frac{\Phi_\infty Re_m}{R_\odot}Q_2(t)\sin 2\theta, \qquad E_\varphi = 0,$$

$$B_R = B_\circ\left(\frac{Re_m}{t}\right)^3\frac{t+2}{Re_m+2}\cos\theta, \qquad B_\theta = \frac{B_R}{t+2}\frac{\sin\theta}{\cos\theta}, \qquad (5)$$

$$B_\varphi = -B_\circ\frac{Re_m\Omega R_\odot F_2(t)}{2V_0 Q_2(Re_m)}\sin 2\theta, \quad J_R = \sigma \cdot E_R, \quad J_\varphi = \sigma V B_\theta,$$

$$J_\theta = -\frac{\sigma Re_m\Phi_\infty}{R_\odot^3}\left(1+\frac{4}{t}\right)\sin 2\theta, \qquad \rho = -\frac{E_R}{4\pi R_\odot t}.$$

Here $\Phi_\infty = \Phi_\circ/Q_2(Re_m)$ is a potential drop between the pole and equator for $R \gg R_\odot$, and $t = Re_m \cdot r$. The polynomials $F_2(1/t) = t(1 + 6t + 12t^2)$ and $Q_2(1/t) = 1 + 6t + 18t^2 + 24t^3$.

On the surface $R_\odot$ we have the surface current $\vec{I}$ and the surface charge Σ:

$$\mu_\circ I_\varphi = B_\circ\frac{Re_m+3}{Re_m+2}\sin\theta, \quad \mu_\circ I_\theta = \xi B_\circ\frac{Re_m F_2(Re_m)}{2Q_2(Re_m)}\sin 2\theta,$$

$$\Sigma(\theta) = \frac{\Phi_\circ}{4\pi R_\odot}\left[2\sin^2\theta - \frac{3F_2(Re_m)}{Q_2(Re_m)}\left(\cos 2\theta + \frac{1}{3}\right)\right]. \qquad (6)$$

The solutions of the magnetic field line equations are

$$\frac{t+2}{t}\sin^2\theta = \frac{Re_m+2}{Re_m}\sin^2\theta_\circ, \qquad (7)$$

here $\theta_\circ$ – is the latitude of the field line Sun's foot (at the $R = R_\odot$). If $\sin^2\theta_\circ > (1 + 2/Re_m)^{-1}$, the field lines are closed. In the opposite case the field lines are open.

In Figure 1 the interplanetary magnetic field lines are presented. A dot near the closed field line at the heliocentric distance about 215 $R_\odot$ in the ecliptic plane is the Earth's position. In Figure 2 the three-dimensional heliospheric electric current lines are shown. Electric currents enter the polar Sun chromosphere at heliolatitudes more than $\sim 35°$ and go out from the Sun at the lower heliolatitudes. Electric currents have two distinct components: open current lines attached to the Sun and dragged outward into the heliosphere, and closed electric currents attached to the Sun at the both (start and finish) points.

3. Magnetic Field Structure in the Magnetosheath

In this section we will investigate the magnetosheath magnetic field and plasma properties. Four regions in the Earth's environment with the different kinds of magnetic field will be considered: the region of the supersonic solar wind flow, the magnetosheath, the magnetopause, and the magnetosphere. The inner boundary of the magnetopause is a surface being flown past, where $V_n = 0$. The magnetopause is assumed to be a thin dissipative boundary layer between the magnetosheath proper and the magnetosphere where the finite solar wind plasma conductivity is taken into account. As follows from the boundary layer theory (Van Dyke, 1995), the thickness of the boundary layer is equal to $R_1/\sqrt{Re_m}$, here R_1 is the magnetopause subsolar distance, and $Re_m = \mu_o \sigma V_o R_1$. The kinematic approximation allowing to calculate the magnetic field for the known velocity field will be used (see Alexeev and Kalegaev (1995)).

To utilize the kinematic approximation, we should obtain the plasma flow structure in the magnetosheath. Based on solution (5) at the distance about 200 $R_\oplus$ we have in the supersonic flow region upstream the bow shock: $\rho = \rho_\infty$, $\vec{V} = \vec{V}_\infty = (-V_\infty, 0, 0)$, and $\vec{B} = \vec{B}_\infty = (-B_\infty^\parallel, 0, -B_\infty^\perp)$, where ρ is plasma density. Here the Cartesian coordinates (X, Y, Z) are used, where the X axis is directed along the Sun–Earth line and the IMF lies in XZ plane.

Let us represent the magnetospheric surface as a paraboloid of revolution. The parabolic dimensionless coordinates (α, β, and φ) are

$$2x/R_1 = \beta^2 - \alpha^2 + 1, \quad y/R_1 = \alpha\beta \sin\varphi, \quad z/R_1 = \alpha\beta \cos\varphi. \tag{8}$$

The magnetospheric surface and the bow shock are described by the $\beta = 1$ and $\beta = \beta_{bs}$ relations. In the supersonic region in parabolic coordinates:

$$\rho_1 = \rho_\infty, \quad \vec{V}_1 = \vec{V}_\infty = V_\infty(-\alpha/h; \beta/h; 0), , \quad h = \sqrt{\alpha^2 + \beta^2}. \tag{9}$$

In the magnetosheath div $\rho_2\vec{V}_2 = 0$, $\vec{V}_2 = -\nabla U_2$, here $\rho_2 = \kappa\rho_1$ is the magnetosheath plasma density downstream the bow shock and κ determines the plasma compression at the bow shock. The boundary conditions

$$V_n|_{\beta=1} = 0, \quad \{V_\tau\}|_{\beta=\beta_{bs}} = 0, \quad \{\rho V_n\}|_{\beta=\beta_{bs}} = 0 \tag{10}$$

at the magnetopause and at the bow shock will be taken into account. The partial solution inside the magnetosheath exists:

$$U_2 = -V_\infty R_1[0.5(\beta^2 - \alpha^2) - \ln\beta], \quad \frac{\rho_2}{\rho_1} = \kappa = (1 - \beta_{bs}^{-2})^{-1}. \tag{11}$$

The magnetic field in the supersonic region is uniform. The orthogonal and parallel to the solar wind velocity components read:

$$\vec{B}^\perp = -B_\infty^\perp \left(\frac{\beta}{h}\cos\varphi; \frac{\alpha}{h}\cos\varphi; -\sin\varphi\right), \tag{12}$$

$$\vec{B}^{\parallel} = -B_{\infty}^{\parallel} \left(-\frac{\alpha}{h}; \quad \frac{\beta}{h}; \quad 0 \right). \tag{13}$$

The magnetic field in the magnetosheath proper besides the magnetopause boundary layer will be assumed frozen in the plasma flow. Inside the magnetopause boundary layer the solar wind plasma will be considered as an ohmic conducting fluid. In both these regions $\vec{V}$ is presented by (11), and a solution of Equations (1) can be received if the boundary conditions are defined (Moffat, 1978; Alexeev and Kalegaev, 1995).

Inside the magnetosphere the magnetic field is described by the paraboloid model (Alexeev and Feldstein, 2001; Alexeev *et al.*, 1996))

$$\vec{B}_m = \vec{B}_d + \vec{B}_r + \vec{B}_t + \vec{B}_{cf} + k_{mp}^{\perp} \cdot \vec{B}_{\infty}^{\perp} + k_{mp}^{\parallel} \cdot \vec{B}_{\infty}^{\parallel} \tag{14}$$

as a sum of magnetic fields of the geomagnetic dipole, the ring current, the geotail current system, the Chapman-Ferraro currents, and also some parts of the IMF parallel and orthogonal components penetrating the magnetosphere. Reconnection efficiency coefficients $k_{mp}^{\perp}$ and $k_{mp}^{\parallel}$ are yet undefined and will be determined from the solution of the Equations (1).

We will use the boundary conditions at the magnetopause inner boundary, $\{\vec{B}\}|_{\beta=1} = 0$, at the bow shock, $\{B_n\}|_{\beta=\beta_{bs}} = 0$, and the special asymptotic condition on the magnetopause external surface which is often used in the boundary layer theory (van Dyke, 1995:

$$\lim_{\beta \to 1} \vec{B}_{msh} = \lim_{\beta \cdot Re_m^{1/2} \to \infty} \vec{B}_{mp}. \tag{15}$$

In the 'frozen' magnetosheath's region besides the magnetopause the orthogonal to the solar wind velocity component of the magnetic field reads (Kalegaev, 2000):

$$\vec{B}^{\perp} = -\frac{\sqrt{\kappa} B_{\infty}^{\perp}}{\sqrt{\beta^2 - 1}} \left(\frac{\beta^2}{h} \cos\varphi; \quad \frac{\alpha(\beta^2 - 1)}{h\beta} \cos\varphi; \quad -\beta \sin\varphi \right). \tag{16}$$

The 'parallel' component reads:

$$\vec{B}^{\parallel} = -\kappa B_{\infty}^{\parallel} (-\alpha/h; \quad (\beta - \beta^{-1})/h; \quad 0). \tag{17}$$

Magnetic field compression in the magnetosheath depends on the plasma compression coefficient κ at the bow shock and is defined by the boundary condition for the normal component of the magnetic field.

Inside the magnetopause boundary layer the dissipative solution is a sum of two terms, originated from the IMF and the magnetospheric magnetic field. The latter term describing the magnetospheric magnetic field diffusion into the magnetosheath is found by Alexeev and Kalegaev (1995) as expansion in confluent hypergeometrical functions. This solution decreases exponentially when we move from the magnetopause to the bow shock.

The former term for the orthogonal component describes the IMF penetration into the magnetosphere (Kalegaev, 2000):

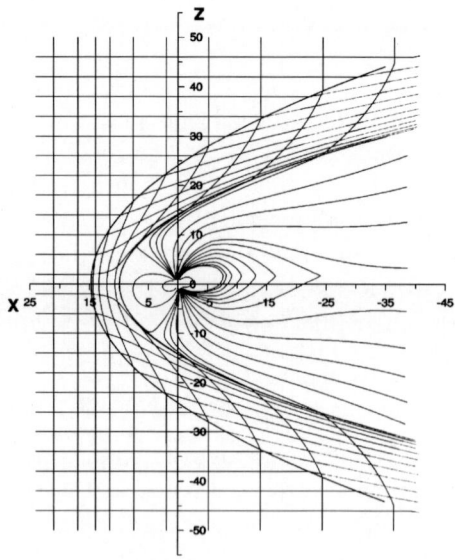

Figure 3. Magnetic field lines (vertical) and flow lines (horizontal) lying in the noon-midnight plane.

$$\vec{B}^{\perp} = -B_{\infty}^{\perp} f^{\perp'} \left(\frac{\beta}{h} \cos\varphi; \; \frac{\alpha f^{\perp}}{h\beta^2 f^{\perp'}} \cos\varphi; \; -\sin\varphi \right). \tag{18}$$

The 'parallel' component inside the magnetopause reads:

$$\vec{B}^{\parallel} = f^{\parallel} \cdot \vec{V} - (R_1 V_{\infty}/Re_m)\nabla f^{\parallel}. \tag{19}$$

Here $f^{\perp} = \beta^{1+2a}[d_1 M(a; b; -Re_m\beta^2/2) + d_2 U(a; b; -Re_m\beta^2/2)]$, $f^{\parallel} = B_{\infty}^{\parallel}(c_1 + c_2\Gamma(Re_m/2; -Re_m\beta^2/2))/V_{\infty}$, M and U are confluent hypergeometrical functions; $2b = 2 + \sqrt{Re_m^2 + 4}$; $a = b/2 + Re_m/4 - 1$ (see Alexeev and Kalegaev, 1995). The constants d_1, d_2, c_1, c_2 are determined from the boundary conditions (see Kalegeav, 2000). For the large Re_m the magnetopause reconnection efficiency coefficients are $k_{mp}^{\perp} \approx 0.9\sqrt{\kappa}\,Re_m^{-1/4}$, $k_{mp}^{\parallel} \approx \dfrac{2\kappa}{\sqrt{\pi\,Re_m}}$. They depend on Re_m and the plasma compression coefficient κ.

Figure 3 shows magnetic field lines and flow lines in the noon-midnight plane for the IMF $\vec{B}_{\infty} = (0,0,-10 \text{ nT})$, $Re_m = 20$. The paraboloid model of the Earth's magnetosphere (Alexeev and Feldstein, 2001) was used in calculation of the magnetospheric magnetic field under condition taken at 14:00 UT on August 29, 1984. The magnetic field outside the magnetosheath is uniform. Inside the magnetosheath it has the same topology as in the case of the subsonic flow. However, the magnetic field intensity is increased. One can obtain that the magnetic field in the magnetosheath increases by a factor of $\sqrt{\kappa}$ times compared with the magnetic field for the subsonic flow found by Alexeev and Kalegaev (1995).

The obtained solution is only the first approximation to the real magnetic field in the magnetosheath. Firstly, we can use it only in the dayside magnetosheath

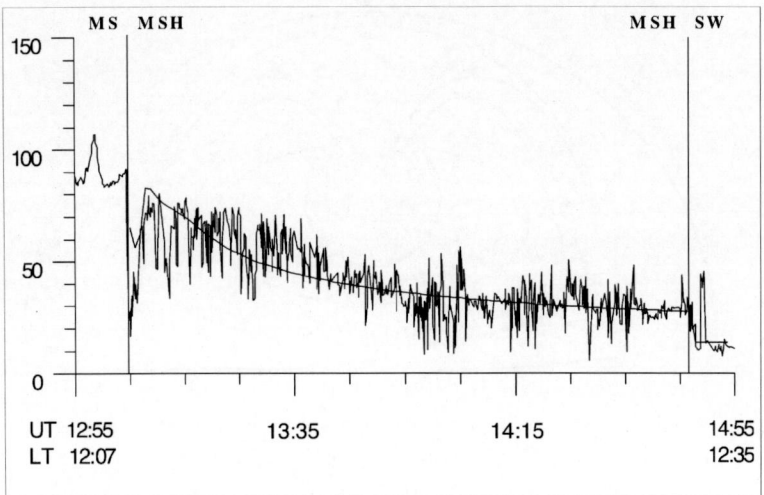

Figure 4. The magnetic field structure during the AMPTE/IRM magnetosheath crossing on August 28, 1984 (solid line) and calculated magnetic field (heavy solid line).

and rather for the high magnetic shear magnetopause, when the plasma density is about constant. Secondly, the flow properties near the magnetopause should be calculated depending on the magnetic field. The kinematic approach is valid only under $\sqrt{Re_m}/Ma^2 \ll 1$ condition, when one can neglect the Ampere force near the magnetopause (see Alexeev and Kalegaev, 1995).

Figure 4 presents the magnetic field measurements during the AMPTE/IRM magnetosheath crossing on August 28, 1984. The heavy solid line is for the calculation results. The IMF and solar wind hourly averaged data measured by the IMP-8 were used in calculations. We can see that on average the measured magnetic field is in good agreement with our results. The magnetic field behavior near the magnetopause is well explained by the coupling between IMF and the diffused magnetospheric magnetic field. Strong plasma compression at the bow shock is responsible for the high values of magnetic field near the magnetopause.

4. Modelling of the Earth's Magnetosphere

The continuity equations for the magnetic field and electric current density:

$$\operatorname{div}\vec{B} = 0 \quad \text{and} \quad \operatorname{div}\vec{j} = 0$$

are valid for all model calculations. All magnetospheric current systems are closed and one can see the typical magnetospheric closed currents' loops in Figure 5. There are:
1. Ring current;
2. Tail current and closure magnetopause current;

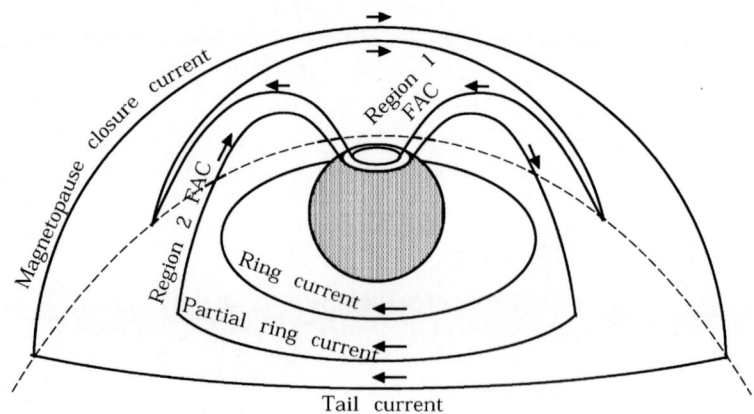

Figure 5. Current systems used for calculation of the magnetic field in the magnetosphere.

3. Partial ring current and Region 2 field-aligned currents;
4. Region 1 field-aligned currents closed by the magnetopause current on the day-side.

The Region 1 and 2 field-aligned currents are located at the polarward and equator-ward boundaries of the auroral oval, correspondingly (Iijima and Potemra, 1976).

MAGNETIC STORM STUDY: SEPTEMBER 24–26, 1998

Investigation of the magnetospheric storms is of great interest as it enables reveal-ing relative contributions of the magnetospheric current systems to the magnetic disturbances at the Earth's surface and onboard spacecrafts. A correct taking into account the contribution of each current system to the magnetic field allows predic-tion of magnetic situation in the Earth's environment and the character and value of magnetic disturbance onground. These predictions will enable avoiding of the unexpected situations of braking onground equipment and spacecrafts devices.

Recent publications dealing with one of the most interesting events, the mag-netic storm on 9–12 January, 1997, show lacking of consensus around the problem of a relative contribution by various current systems to the magnetospheric mag-netic field. The most contradictory item is taking into account the influence of the magnetotail current system. So, in the paper by Turner *et al.* (2000) the contribution of the magnetotail current system to D_{st} is estimated to be $\sim 25\%$. Meanwhile, in the paper by Alexeev *et al.* (2001) the contribution of the magnetotail current system in the course of the overall storm is estimated to be $\sim 50\%$, prevailing 60% in the main phase. This paper note a significant difference between models of the magnetospheric magnetic field, especially when determine a quiet contribution to D_{st} by all the magnetospheric current systems.

The magnetic storm on 24–27 September, 1998 is also one of the most interest-ing events (see Figure 6). An arrival of a dense cloud of the solar wind plasma at 23:45 on 24 September was accompanied with a northward turning of the IMF,

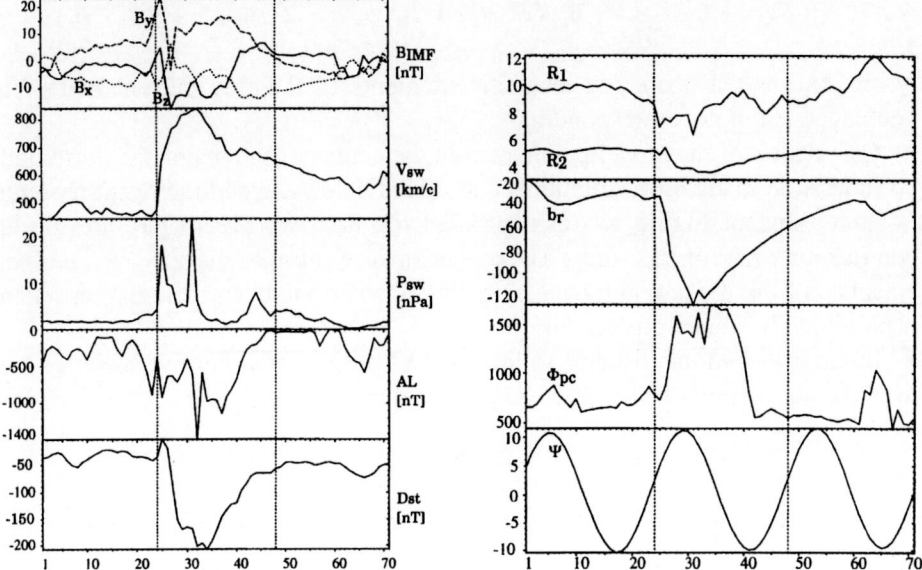

Figure 6. Solar wind data, D_{st}, AL (left panel), and the model parameters R_1, R_2, both in R_E, b_r [nT], $\Phi_{pc} = \Phi_\infty$ [MWb], and Ψ [°] (right).

this direction remains for 3 hours. This leads to the interesting phenomena in the Earth's magnetosphere, e.g., to a significant decrease of the polar cap during 30 minutes (Clauer *et al.*, 2001). After that the IMF had a strong negative north-south component. Moreover, in 6 hours the second shock wave of the solar wind dense plasma encounters the magnetosphere. At almost the same time a significant substorm activity has been detected (AL index). The both events, however, weakly influenced the dynamics of the D_{st} variation. This behavior of the D_{st} index again attracts attention to the question about dynamics of the magnetospheric current systems and their relative contributions to the magnetic field at the Earth's surface.

In the paraboloid model the magnetic field and D_{st} index are calculated in two stages. Various current systems depend on the concrete parameters unambiguously and a fixed set of these parameters unambiguously defines the magnetic field over the entire magnetosphere. The parameter changes is determined by experimental data variations. So, the magnetospheric field dynamics is determined by empirical data. The dipole magnetic field, B_d, depends on only one parameter – the dipole tilt angle ψ. The magnetic field of shielding currents on the magnetopause, B_{cf}, depends on the dipole tilt angle ψ and the distance R_1 to the subsolar point. The tail current magnetic field, B_t, depends on ψ, R_1, on the distance R_2 to the inner edge of the tail current sheet, and on the magnetic flux in the magnetotail lobe Φ_∞. The ring current magnetic field, B_r, depends on ψ and on the magnetic field strength at the Earth's center b_r. Thus, the magnetic field in the paraboloid model is calculated in the form (Alexeev and Feldstein, 2001):

$$\vec{B}_m = \vec{B}_d(\psi) + \left(1 + \frac{M_{rc}}{M_E}\right) \vec{B}_{cf}(\psi, R_1) + \vec{B}_t(\psi, R_1, R_2, \Phi_\infty) + \vec{B}_r(\psi, b_r),$$

where M_{rc} and M_E are the magnetic moments of the ring current and of the geomagnetic dipole, correspondingly.

Calculation of the model parameters in the course of the magnetic storm under consideration made using submodels described below yielded the result presented in right panel of Figure 6. Calculation of the first two parameters are the less contradictory part of this study. The magnetopause subsolar distance, R_1, has been calculated from the pressure balance at the subsolar point. The calculation scheme is based on the iterative procedure.

In order to find the distance to the inner edge of the tail current sheet we have used the results by Feldstein *et al.* (1999) study:

$$R_2 = 1/\cos^2 \varphi_n, \quad \varphi_n = 64.9° + (|D_{st}|[\text{nT}]/31)°. \tag{20}$$

Submodels: Ring Current

The calculation of the ring current contribution to the D_{st} index is based on the results of Burton *et al.* (1975), O'Brien and McPheron (2000) and Alexeev *et al.* (2001). Starting from the Dessler-Parker-Sckopke relation between b_r and the ring current particle energy ε_r (Dessler and Parker, 1959) as well as from the injection equation for the dependence of the energy, ε_r, on the time:

$$b_r = -\frac{2}{3} B_0 \frac{\varepsilon_r}{\varepsilon_d}, \quad \text{here } \varepsilon_d = \frac{1}{3} B_0 M_E, \quad \text{and } \frac{d\varepsilon_r}{dt} = U - \frac{\varepsilon_r}{\tau}, \tag{21}$$

we received equation for the ring current field:

$$\frac{db_r}{dt} = F(E) - \frac{b_r}{\tau}, \quad \tau\,(hours) = 2.37 e^{9.74/(4.78 + E_y)}. \tag{22}$$

Here injection function, $F(E)$, is determined by the dawn-dusk solar wind electric field E_y:

$$F(E) = f_{pr}(p_{sw}) f_{ar}(AL) \begin{cases} d \cdot (E_y - 0.5) & E_y > 0.5 mV/m. \\ 0 & E_y < 0.5 mV/m \end{cases} \tag{23}$$

The additional factors $f_{pr}(p_{sw})$ and $f_{ar}(AL)$ makes it feasible to take into account the influence on the injection of the solar wind dynamic pressure and substorm activity. Large values of the solar wind dynamical pressure increase the effectiveness of injection to the ring current due to increasing of the solar wind plasma transport across magnetopause and into the plasma sheet. AL index shows a fraction of energy flowing directly to the ionosphere. Large values of the AL index or, on the other words, increasing of substorm activity show, from our point of view, that the most part of the energy accumulated in the magnetotail is directly transferred to the ionosphere. That's why the injection to the ring current decreases. These effects are described by factors $f_{pr}(p_{sw})$, $f_{ar}(AL)$ in Equation (23).

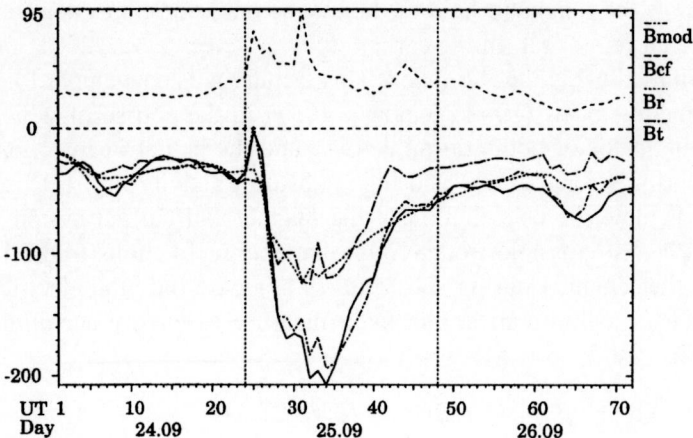

Figure 7. Model magnetic field (dot-dot-dashed curve) vs D_{st} (solid curve). Contributions of magnetopause currents (dashed curve), ring current (dotted curve), and tail current (dot-dashed curve) are demonstrated. One vertical step is 10 nT.

Submodels: Tail Current System

Using an enormous experience of modelling the magnetic storms and an ideas developed in the present paper we elaborated new submodel describing the dependence of the magnetotail current system on the measurement data. As in the earlier papers (see, e.g., Alexeev *et al.* (2001)), the magnetic flux in the magnetotail lobes is the key parameter for both the tail current system and the magnetosphere as a whole, being calculated as a sum of two terms:

$$\Phi_\infty (t) = \Phi_0 (t) + \Phi_s (t). \tag{24}$$

Here $\Phi_0 (t)$ is the quiet time value of the magnetic flux, $\Phi_s (t)$ is the magnetic flux value associated with the electric field enhancement in the solar wind. To calculate the quiet value, use is made here of our assumption of the basic state of the magnetosphere.

For each value of the solar wind dynamical pressure, the current system parameters for which the energy of the magnetosphere interaction with the solar wind plasma has minimum can be found. At the same time, the characteristic time scale of the current system response to the changes in the interplanetary medium is about one hour. Thus, at the present moment the magnetotail current system tends to realize a state corresponding to the dynamical pressure value of one hour before:

$$\Phi_0 (t) = 400 \cdot (p_{sw} (t - 1))^{1/6.6}. \tag{25}$$

When calculating the dynamical contribution, use was made of the reliable formula (Alexeev *et al.*, 2001):

$$\Phi_s (t) = b_t (t) \frac{\pi R_1^2}{2} \sqrt{\frac{2R_2}{R_1} + 1}, \tag{26}$$

where $b_t(t)$ is the magnetic field of the magnetotail current system near the inner edge of the current sheet. In the earlier study (Alexeev *et al.*, 2001) this value has been determined using the AL index. This technique is convenient for comparably weak magnetic storms (10–12 January, 1997) in the course of which AL index is a good indicator of both auroral activity and the extent of injection to the ring current. During strong storms and high substorm activity a direct use of the AL index to calculate the contribution of the magnetotail current system is not quite correct, as it is impossible to determine the energy fractions transferred through one or another channel during substorm. In this case only the electric field in the interplanetary medium can be the main measure of energy accumulation in the magnetotail:

$$b_t(t) = f_{vb}\left(V, B_{imf}, t\right) f_{pb}\left(p_{sw}\right) f_{ab}\left(AL\right) \tag{27}$$

In the above expression the terms f_{ab} and f_{pb} indicate the influence of substorm activity and solar wind dynamical pressure on the tail current. The most essential magnetic field dependence on the electric field in the solar wind is represented by f_{vb} which is determined by the following expression:

$$f_{vb}\left(V, B_{imf}, t\right) = \begin{cases} 0 & b_z \geq 0 \\ \frac{V(t-1)}{300}(b_z(t-1) - & b_z < 0 \\ \left|b_y(t-1)\right| f_{pb}\left(p_{sw}\right)) \end{cases} \tag{28}$$

In order to explain this formula we should kept in mind that the current in the current sheet depends mainly on the term $V \cdot b_z$, and one hour delay should be taken into acount. Here one hour was obtained for two reasons. Firstly, this is the minimal time step used when studying magnetic storm and modeling the D_{st} variation. Secondly, an average duration of substorm is also about one hour. Thus, minute time scale would be an overestimation of accuracy.

Figure 7 represents the measured D_{st}-index (see the solid curve) and the magnetic field at the Earth's surface given by the Paraboloid model (see the dot-dot-dashed curve). One can see good agreement of the model calculations with the real D_{st}. The root mean square deviation is about 10 nT. Figure 7 also represents a relative contribution of the large scale magnetospheric current systems to D_{st} (see the figure legend).

SUBSTORM STUDY

The most important feature of substorms is the onset. Thus, our model must represent the substorm onset in terms of its parameters. Lets consider three major parameters of the Paraboloid model: R_1-distance to the subsolar point, R_2-distance to the Earth-ward edge of the tail current sheet, and Φ_{pc}-magnetic flux in the polar cap; and three sources of the magnetospheric magnetic field: the dipole, the magnetopause currents screening the dipole and the tail current system. Playing with

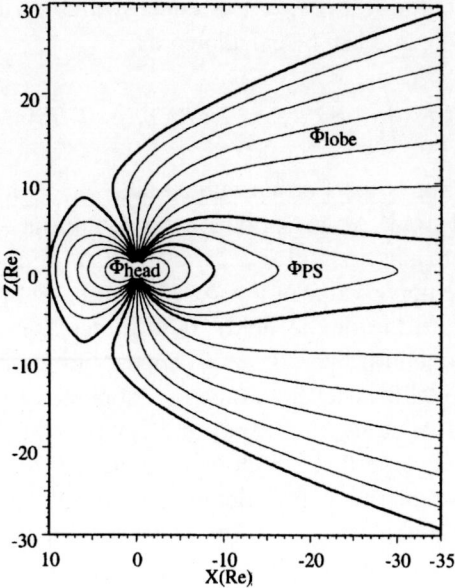

Figure 8. Magnetic field lines in Paraboloid model (noon-midnight cross-section), and division on three major magnetospheric domains.

these parameters of the magnetic field we have found that at the appropriate values of these parameters the point with $B_z = 0$ appears at the inner edge of the tail current sheet. This can be related to the neutral line formation in the plasma sheet at the onset time. However, the question remains to be answered: Does it really have physical meaning?

The neutral point appears at relatively high values of the magnetic flux in the polar cap (in the tail lobe). This high value of Φ_{pc} corresponds to the enhanced cross-tail current which is observed during substorms. Thus, the neutral point in the Paraboloid model is formed at the time of the highly enhanced cross-tail current and this is very similar to the substorm onset conditions.

Figure 8 represents the model field lines at the noon-midnight meridian plane. Based on this Figure we can define three distinct magnetospheric magnetic fluxes components:

− Φ_{head} is the magnetic flux in the inner magnetosphere;
− Φ_{lobe} is the magnetic flux in the tail lobe equal to the polar cap magnetic flux (Φ_{pc}), and equal to the model parameter Φ_{∞};
− Φ_{ps} is the magnetic flux through the plasma sheet (from the inner edge of the current sheet and up to the infinity). This magnetic flux forms the auroral oval.

The field lines forming all these fluxes attached to the Earth. Then, the main condition for stability of the magnetosphere is the following equation:

$$\Phi_{dipole} = \Phi_{head} + \Phi_{lobe} + \Phi_{ps}. \tag{29}$$

Left side of the Equation (29), Φ_{dipole} is equal to the Earth dipole magnetic flux across equatorial plane and can be calculated as

$$\Phi_{dipole} = \int_0^{2\pi} R d\varphi \int_{R_E}^{\infty} B_{dz} dR = 2\pi R_E^2 B_0 \approx 7,700 \text{ MWb} .\tag{30}$$

Here $R_E = 6,400$ km is the Earth's radius, and $B_0 = 30,192$ nT is the dipole equatorial onground field. At any point of the equatorial plane the normal component of the magnetospheric field B_{mz} is positive. As consequence $\Phi_{ps} \geq 0$. This inequality limited the upper value of the Φ_{∞}. The tail current magnetic field directs southward (opposite to the dipole field). If Φ_{∞} increase for the some threshold value $\Phi_{\infty} = \Phi_{thr}$, a neutral line ($B_m = 0$) appears in the equatorial plane. In the framework of the model for each set of the model parameters R_1, R_2 we have calculated the threshold value Φ_{thr} (Alexeev and Bobrovnikov, 2002). If $\Phi_{\infty} > \Phi_{thr}$ then an additional magnetic flux which is not attached to the Earth appears in the tail. We suppose that this inequality determine the substorm onset conditions in terms of the model parameters.

It is most important for our research that with an increase of the cross tail current the B_z component of the magnetic field goes to zero somewhere in the magnetosphere. It corresponds to conditions when additional magnetic flux is needed to transform the magnetospheric configuration. This is what we call the global instability of the magnetosphere. The neutral point formation is treated as an incapability of the magnetosphere to redistribute the magnetic flux in accordance with new the external or internal conditions.

One can obtain from Equation (29) that the increase of the magnetic flux in one of the defined domains leads to a decrease of fluxes in other ones. After the IMF northward turning an inner magnetospheric domain grows and Φ_{head} increases. At the same time tail lobe magnetic flux Φ_{lobe} also increases (or remains unchanged). In fact, the both of these processes together cut the magnetic flux in the plasma sheet.

The Scenario of Substorm Triggered by the IMF Northward Turning

The scenario of substorms triggered by the IMF northward turning can be described as follows. When the IMF turns southward and energy transport into the magnetosphere increases, above mentioned distances (R_1 and R_2) decrease and an increasing of the tail flux (Φ_{lobe}) begins. This process has low a probability to reach the critical surface and thus low probability of substorm onset. The tail lobe magnetic flux approaches very close to the 'critical' value, but the behavior of the model parameters (R_1 and R_2) does not permit Φ_{lobe} to exceed it. When the IMF turns northward, a negative contribution of Region 1 field-aligned currents to the magnetic field in the day side magnetosphere disappears, magnetic field pressure in dayside magnetosphere increases and the magnetopause moves away from the Earth. Moreover, a positive contribution of field-aligned currents to the magnetic field in the nightside magnetosphere also disappears and the Ampere

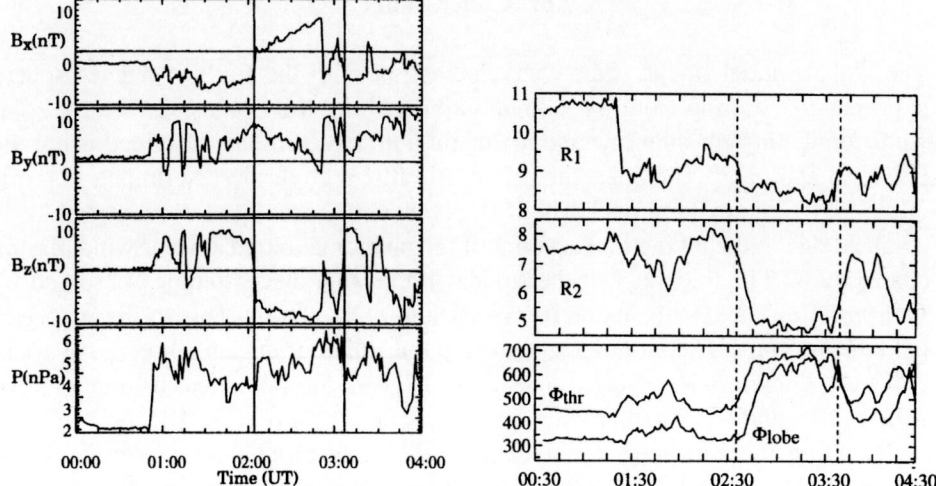

Figure 9. Wind data (left panel), and a dynamics of the model parameters R_1, R_2, both in R_E, and Φ_{lobe}, Φ_{ths} [MWb] (right panel) during substorm of January 10, 1997.

force decreases, the tail current sheet moves tailward. It is important to note that changes in the IMF which first appear at the subsolar point can change the current in plasma sheet with some delay. This delay can be estimated as 5–15 minutes. During this period the following process will take place.

When IMF turns northward the Region 1 field-aligned currents disappear and the magnetic pressure at the day side magnetosphere (near the subsolar point) increases. The magnetopause moves away from the Earth (R_1-increases). The same processes begins in the tail – the current sheet moves tail-ward (R_2-increases). During this period of time the tail lobe magnetic flux Φ_{lobe} will not decrease, but the threshold value will. If during the substorm growth phase the sufficient amount of energy has been stored in the magnetospheric tail, the tail lobe magnetic flux becomes very close to the 'threshold' value (metastable state) and it is very possible that the Φ_{lobe} magnetic flux will reach and exceed the threshold value.

It is important to note once more that the tail will react to the changes in the IMF with some delay but the Region 1 field-aligned currents will react to those changes almost at the same time. This is the core of the substorm triggering by the IMF northward turning.

Based on the solar wind data which are presented in the left panel of Figure 9 and proposed scenario we have calculated the dynamics of the model parameters (see Figure 9).

At the time marked by the second vertical dashed line the magnetic flux in the tail lobe exceeds its threshold value. And this time corresponds well to the auroral intensifications observed by the polar spacecraft and ground magnetometers.

5. Conclusions

The global model of the solar wind interaction with the Earth's magnetosphere is presented. Starting from the Sun and approaching the Earth's surface the solar wind conducting plasma flow and magnetic field structure are described using the universal approach. The open solar magnetic flux regions are determined in terms of the MHD kinematic model. The IMF component normal to the ecliptic plane is obtained due to the taking into account of the phenomenological solar wind plasma conductivity. The thickness of the solar wind plasma discontinuities observed by *Interball* 1 and *IMP* 8 is about 10^5 km (Dalin *et al.*, 2002). Due to the magnetic field diffusion the discontinuities created in the solar corona and observed at 1 a.u. will have thickness $d = \sqrt{Re_m R_\odot}$ 1 a.u.. It gives the numerical estimations for Re_m (10^4) and for the solar wind plasma conductivity, $\sigma = 3 \cdot 10^{-5}$ S.

Based on the heliospheric magnetic field structure obtained in Section 2, the magnetic field in the dayside magnetosheath is calculated. The magnetic field component penetrating into the magnetosphere is determined by the solar wind conductivity and plasma compression at the Earth's bow shock. The IMF components parallel and perpendicular to the solar wind flow penetrate inside the magnetosphere by different manner: the first one is much more weakened inside the magnetosphere. For $Re_m = 10^4$ and bow shock plasma compression coefficient $\kappa = 2$ we received the reconnection efficiency coefficient $k_{mp}^\perp = 0.12$. In good accordance with observations we have dayside magnetopause thickness $0.1 R_E$ and polar cap potential drop ~ 61 kV for quiet solar wind conditions.

The obtained information about the IMF structure give us possibility to describe a magnetospheric dynamics in the course of magnetic storm and substorm. During disturbed interval it was shown by the model calculations that the tail current magnetic field contribution to D_{st} has approximately the same magnitude as the ring current one. Model results can be used for forecast the magnetospheric dynamics based on the solar wind observations. Checking the model calculations by comparison with observed D_{st}, give us the discrepancy about 10 nT for severe magnetic storm.

The proposed model is the next step in the development of the magnetospheric dynamic model. One of the main demands to the dynamic model is the possibility to relate the solar wind and magnetospheric parameters. The model covers the whole region of the solar wind-magnetosphere coupling including the Sun environment, Earth's bow shock, magnetosheath and magnetosphere. Based on the proposed approach we can see that the magnetospheric magnetic field is determined by input parameters, depending on factors, which have both magnetospheric and interplanetary origin. The IMF penetrating into the magnetosphere directly controls the polar cap dimensions, potential difference across the polar cap, energy input rate into the magnetosphere. The proposed model relates the reconnection efficiency with the solar wind parameters: velocity, density, and conductivity. So, for the known solar wind parameters, we can calculate the state of the magnetosphere. The evolution of the magnetosphere can be presented as a sequence of such states. The constructed models can be used for the real-time forecast of the magnetospheric disturbances based on the Sun and the solar wind observations.

Acknowledgements

This research was supported by Russian Foundation for Basic Research Grants 01-05-65003, 00-15-96623, 01-07-90117 and 02-05-74643. IA is grateful to the Organizing Committee of the WISER HPC2002 chaired by A. C.-I. Chian for the financial support. Authors thank the referee for the useful comments.

References

Abramowitz, M. and Stegun, I. A.: 1972, *Handbook of Mathematical Functions*, 8th ed., Dover, Mineola New York, pp. 832.

Alexeev, I. I. and Bobrovnikov, S. Yu.: 2002, 'Substorm Study on the Basis of Paraboloid Model of the Magnetosphere', *Geomagnetizm and Aeron. Int.* **3N 1**, 35–44.

Alexeev, I. I. and Feldstein, Y. I.: 2001, 'Modeling of Geomagnetic Field During Magnetic Storms and Comparison with Observations', *J. Atmos. Sol. Terr. Phys.* **63**, 331–340.

Alexeev, I. I. and Kalegaev, V. V.: 1995, 'Magnetic Field and the Plasma Flow Structure Near the Magnetopause', *J. Geophys. Res.* **100**, 19,267–19,276.

Alexeev, I. I. *et al.*: 1982, 'On Interplanetary Electric and Magnetic Fields', *Solar Physics* **79**, 385–397.

Alexeev, I. I. *et al.*: 1996, 'Magnetic Storms and Magnetotail Currents', *J. Geophys. Res.* **101**, 7737–7748.

Alexeev, I. I. *et al.*: 2001, 'Dynamic Model of the Magnetosphere: Case Study for January 9–12, 1997', *J. Geophys. Res.* **106**, 25,683–25,694.

Balogh, A. and Smith, E. J.: 2001, 'The Heliospheric Magnetic Field at Solar Maximum: Ulysses Observations', *Space Sci. Rev.* **97**, N 1-4, 148–160.

Burton, R. K. *et al.*: 1975, 'An Empirical Relationship Between Interplanetary Conditions and *Dst*', *J. Geophys. Res.* **80**, 4204–4213.

Clauer C. R. Jr. *et al.*: 2001, 'Special Features of the September 24–27, 1998 Storm During High Solar Wind Dynamic Pressure and Northward Interplanetary Magnetic Field', *J. Geophys. Res.* **106**, 25,695–25,712.

Dalin, P. *et al.*: 2002, 'A Survey of Large, Rapid Solar Wind Dynamic Pressure Changes Observed by Interball 1 and IMP 8', *Ann. Geophys.* **20**, 293–299.

Dessler, A. J. and Parker, E. N.: 1959, 'Hydromagnetic Theory of Geomagnetic Storms', *J. Geophys. Res.* **64**, 2239–2252.

Feldstein, Y. I. *et al.*: 1999, 'Dynamics of the Auroral Elecrtojets and Their Mapping to the Magnetosphere', *Ragiation Measurements* **30**, N 5, 579–587.

Iijima, T. and Potemra, T. A.: 1976, 'The Amplitude Distribution of Field-Aligned Currents at Northern High Latitudes Observed by Triad', *J. Geophys. Res.* **81**, 2165–2174.

Kalegaev, V.: 2000, 'Magnetosheath Conditions and Magnetopause Structure for High Magnetic Shear', *Phys. and Chem. of the Earth* **25**, 173–176.

McComas, D. J. *et al.*: 2000, 'Solar Wind Observations over Ulysses' First Full Polar Orbit', *J. Geophys. Res.* **105**, 10,419–10,433.

Moffatt, H. K.: 1978, *Magnetic Field Generation in Electrically Conducting Fluids*, Cambridge Univ. Press, New York, pp. 244.

O'Brien, T. P. and McPherron, R. L.: 2000, 'An Empirical Phase Space Analysis of Ring Current Dynamics: Solar Wind Control of Injection and Decay', *J. Geophys. Res.* **105**, 7707–7719.

Paschmann, G. B. *et al.*: 1979, 'Plasma Acceleration at the Earth's Magnetopause: Evidence for Magnetic Reconnection', *Nature* **282**, 243–246.

Phan, T.-D. *et al.*: 1994, 'The Magnetosheath Region Adjacent to the Dayside Magnetopause: AMPTE/IRM Observations', *J. Geophys. Res.* **99**, 121–141.

Phan, T.-D. *et al.*: 1997, 'Low-Latitude Dusk Flank Magnetosheath, Magnetopause, and Boundary Layer for Low Magnetic Shear: Wind Observations', *J. Geophys. Res.* **102**, 19,883–19,895.

Turner, N. E. *et al.*: 2000, 'Evaluation of the Tail Current Contribution to *Dst*', *J. Geophys. Res.* **105**, No. A3, 5431–5440.

Van Dyke, M.: 1965, *Perturbation Methods in Fluid Mechanics*, Academic Press, San Diego, California, pp. 1–229.

TYPE II SOLAR RADIO BURSTS: THEORY AND SPACE WEATHER IMPLICATIONS

IVER H. CAIRNS*, S. A. KNOCK, P. A. ROBINSON and Z. KUNCIC

School of Physics, University of Sydney, Sydney, NSW 2006, Australia
*(*author for correspondence, e-mail; cairns@physics.usyd.edu.au)*

Abstract. Recent data and theory for type II solar radio bursts are reviewed, focusing on a recent analytic quantitative theory for interplanetary type II bursts. The theory addresses electron reflection and acceleration at the type II shock, formation of electron beams in the foreshock, and generation of Langmuir waves and the type II radiation there. The theory's predictions as functions of the shock and plasma parameters are summarized and discussed in terms of space weather events. The theory is consistent with available data, has explanations for radio-loud/quiet coronal mass ejections (CMEs) and why type IIs are bursty, and can account for empirical correlations between type IIs, CMEs, and interplanetary disturbances.

Key words: coronal mass esjections, electron reflection, plasma waves, radiation, shocks, solar radio emission, type II bursts

1. Introduction

Type II solar radio bursts were discovered in dynamic spectra as slowly drifting bands, often in pairs differing in frequency by a factor ≈ 2 (Wild *et al.*, 1954). They were quickly interpreted in terms of a coronal shock wave accelerating electrons, driving Langmuir waves near the electron plasma frequency f_p, and producing radio emission near f_p and $2f_p$ (Wild *et al.*, 1954, Nelson and Melrose, 1985). Interplanetary type II bursts were discovered in spacecraft data (Cane *et al.*, 1982) and definitively associated with coronal mass ejections (CMEs), traveling shock waves, and radiation near f_p and $2f_p$ (Cane and Stone, 1984; Lengyel-Frey, 1992; Reiner *et al.*, 1998a). The basic model for coronal type IIs is strongly supported by *in situ* observations of an interplanetary type II source: Bale *et al.* (1999) observed the shock wave, reflected electrons, Langmuir waves, and radiation generated near the local f_p and $2f_p$.

The primary purpose of this paper is to summarize and review our recent theory for type II bursts (Knock *et al.*, 2001, 2003) and related work (Kuncic *et al.*, 2002a, 2002b). Earlier, qualitative theories are reviewed elsewhere (Nelson and Melrose, 1985; Robinson and Cairns, 2000). The secondary purpose is to relate the theory to space weather physics, resulting in explanations for known empirical connections between type IIs and space weather events like CMEs, as well as new predictions. The theory is an analytic, quantitative description of type IIs with four stages:

Space Science Reviews **107**: 27–34, 2003.
© 2003 *Kluwer Academic Publishers.*

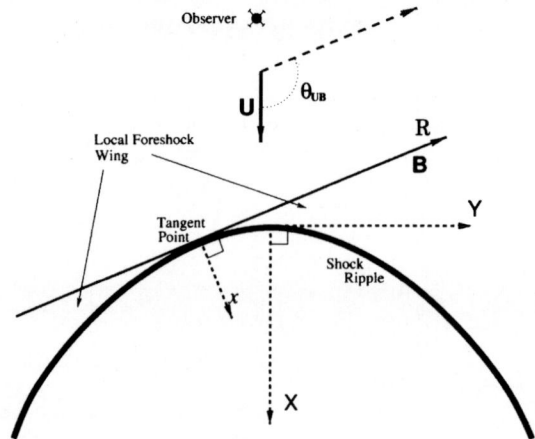

Figure 1. Foreshock geometry, shock and plasma variables, and coordinate systems.

(i) electron reflection and acceleration at the shock, including the shock's magnetic mirror and electrostatic cross-shock potential (Kuncic *et al.*, 2002a), (ii) formation of electron beams in the upstream foreshock region by time-of-flight effects (Filbert and Kellogg, 1979; Cairns, 1986), (iii) energy flow into Langmuir waves from electron beams, and (iv) specific nonlinear Langmuir processes for f_p and $2f_p$ radiation (Robinson and Cairns, 2000). Section 2 summarizes the basic physics and results while Section 3 contains predicted trends for type II emission. Comparisons between observation and theory, connections to space weather events, and future work are discussed in Section 4.

2. Basic Theory

Consider the global shock or a localized ripple thereon (cf. Bale *et al.*, 1999). Important shock/ripple parameters are (Figure 1): $U = |\mathbf{V}_{sh} - \mathbf{v}_{sw}|$ is the shock's speed relative to the plasma, where $\mathbf{V}_{sh}$ and $\mathbf{v}_{sw}$ are the (aligned) shock velocity and solar wind velocity, respectively, $\mathbf{B}_1$ is the upstream magnetic field vector, θ_{bu} the angle between $\mathbf{V}_{sh}$ and $\mathbf{B}_1$, and b is the shock's curvature parameter defined by $X = bY^2$. Then $b = R_c^{-1}$, where R_c is the radius of curvature at the shock's nose. The important plasma properties are the electron and ion temperatures T_e and T_i and the parameter κ of the incoming, gyrotropic, generalized-Lorentzian electron distribution, defined by

$$f_\kappa(v_\parallel, v_\perp) = \frac{\Gamma(\kappa+1)}{\Gamma(\kappa)} \pi^{-3/2} V_e^{-3} \left(1 + \frac{v_\parallel^2 + v_\perp^2}{V_e^2} \right)^{-(\kappa+1)} \tag{1}$$

where $V_e = \sqrt{k_B T_e/m_e}$ is the electron thermal speed. The reduced distribution $F_\kappa(v_\parallel)$ formed by integrating (1) over $v_\perp$ varies asymptotically as $v_\parallel^{-2(\kappa+1)}$: lower κ corresponds to more high-velocity particles.

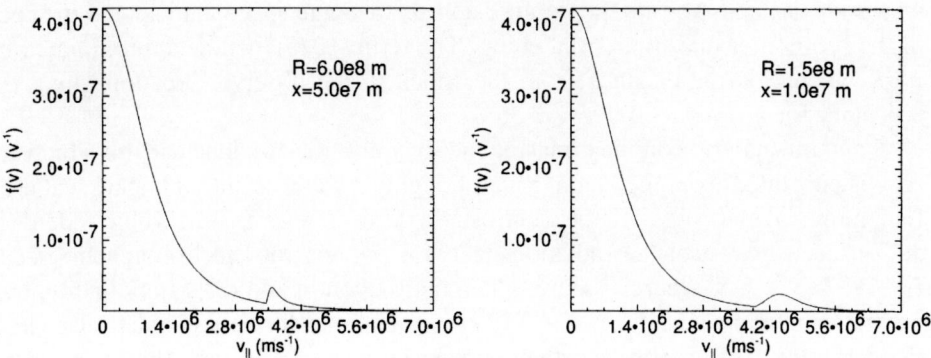

Figure 2. Beams in $F_\kappa(v_\parallel)$ at two foreshock locations (Knock *et al.*, 2001).

Electron reflection is best described in the de Hoffman-Teller frame, where the convection electric field vanishes, due to conservation of magnetic moment and energy, the latter subject to the electrostatic cross-shock potential ϕ_{cs}. The magnetic mirror ratio B_2/B_1 is predicted by the Rankine-Hugoniot conditions, which depend on the plasma's normal flow speed relative to the shock $U_n = (\mathbf{V}_{sh} - \mathbf{v}_{sw})_n$, the angle θ_{Bn} between $\mathbf{B}_1$ and the local normal, the Alfven speed V_A and the sound speed c_S. Similar dependences exist for ϕ_{cs}, predictable analytically (Kuncic *et al.*, 2002a), which modifies the shock's loss cone at low $v_\parallel$ and makes it more difficult to reflect low $v_\parallel$ electrons. Reflection by the shock's magnetic mirror leads to shock-drift acceleration (SDA) in the plasma rest frame, similar to a ping-pong bat accelerating a ball: the reflected particle speed $v_\parallel^r$ is related to the initial speed $v_\parallel^i$ by

$$v_\parallel^r \approx 2v_d \tan\theta_{Bn} - v_\parallel^i \,, \tag{2}$$

where v_d is the component of $\mathbf{U}$ perpendicular to $\mathbf{B}_1$.

Liouville's Theorem is used to predict the reduced distribution $F_\kappa(v_\parallel)$ throughout the foreshock, by tracing particle paths back to the shock (with B_2/B_1 and ϕ_{cs} varying with position), unfolding the effects of SDA taking into account (2), equating $f(v_\parallel^r, v_\perp)$ to $f_\kappa(v_\parallel^i, v_\perp)$, and then integrating over $v_\perp$. Figure 2 shows that a beam develops by time-of-flight effects (Cairns, 1986; Knock *et al.*, 2001), as for Earth's bow shock (Filbert and Kellogg, 1979; Cairns, 1987; Fitzenreiter *et al.*, 1990). This beam is unstable to growth of Langmuir waves, with the available free energy varying with position (Knock *et al.*, 2001). Quasilinear relaxation relates the wave energy density at saturation to the available free energy. At marginal stability, as predicted by stochastic growth theory, the power flow into Langmuir waves equals the total time derivative of the available free energy, yielding (in steady-state)

$$\frac{d}{dt}W_L = \mathbf{v} \cdot \frac{\partial}{\partial \mathbf{r}}\left(\frac{N_b v_b \Delta v_b}{3}\right), \tag{3}$$

where N_b, v_b, and Δv_b are the number density, average speed, and spread in speed of the beam after quasilinear flattening. The term $\mathbf{v}.\partial/\partial\mathbf{r}$ is now approximated by v_b/l, where l is the distance from the shock to the observer location along the trajectory for $v_\| = v_b$.

Standard analytic nonlinear plasma theory yields the efficiencies with which energy is converted from the beam-driven Langmuir waves L into: (1) backscattered Langmuir waves L', $\phi_{L'}$, via the electrostatic decay $L \longrightarrow L'+S$, where S denotes an ion acoustic wave; (2) radiation near f_p, ϕ_F, via the electromagnetic decay $L \longrightarrow T(f_p) + S'$ where T represents a radio photon; and (3) $2f_p$ radiation, ϕ_H, via the coalescence $L + L' \longrightarrow T(2f_p)$. Functional forms are stated elsewhere (Knock *et al.*, 2001); although they depend on v_b, Δv_b, V_e, and the ion acoustic speed V_S, characteristic values are $\phi_{L'} \approx 10^{-3}$ and $\phi_F \approx \phi_H \approx 10^{-6}$. Combined with (3), these efficiencies yield the volume emissivities of radiation (power output per unit volume and solid angle) throughout the foreshock, with

$$j_M = \frac{\phi_M}{\Delta\Omega_M} \frac{N_b m_e v_b^3}{3l} \frac{\Delta v_b}{v_b}, \tag{4}$$

where $M = F$ or H, and $\Delta\Omega_F \approx 2\pi$ and $\Delta\Omega_H \approx 4\pi$ are the solid angles into which the radiation is produced.

Figure 3 shows j_F and j_H as functions of foreshock position (Knock *et al.*, 2001) for parameters appropriate to Bale *et al.*'s (1999) type II shock: $U = 744$ km s^{-1}, $\kappa = 2.5$, $b = 10^{-9}$ m^{-1}, $T_e = 3T_i = 1.5 \times 10^5$ K, $N_e = 7 \times 10^6$ m^{-3}, and $B_1 = 6$ nT. Fundamental radiation is predominantly produced where v_b is large, near the tangent field line, due to strong dependences of ϕ_F on v_b. Harmonic radiation, however, is produced over a greater area but with smaller peak magnitude. The peak values and characteristic ranges of j_F and j_H are comparable to those observed and predicted (Robinson and Cairns, 2000) for interplanetary type III bursts near 1 AU. Integrating the volume emissivities over the foreshock yields the flux $\int j_M/D^2 \, d^3V$ of radiation, where D is the distance between the 3-D source element and observer.

3. Theoretical Trends for type II Emission

Figure 4 shows the fluxes of f_p and $2f_p$ radiation predicted as functions of U and κ for an observer at $(X, Y) = (-10^9$ m, 0) upstream of a single 3-D ripple (Knock *et al.*, 2003). Other parameters are as for Figure 3. The fluxes increase with increasing U and decreasing κ. Below ≈ 200 km s^{-1}, decreasing U at constant κ causes a rapid, approximately logarithmic drop-off in the fluxes. Similarly, increasing κ by 1 at constant U causes the harmonic flux to decrease by about 1 order of magnitude, while the fundamental flux varies by closer to 2 orders of magnitude between $\kappa = 2$ and 3. Accordingly, relatively small changes in U near 150 km s^{-1} and κ near 3 can change the predicted flux by orders of magnitude, thereby appearing to turn the emission on and off.

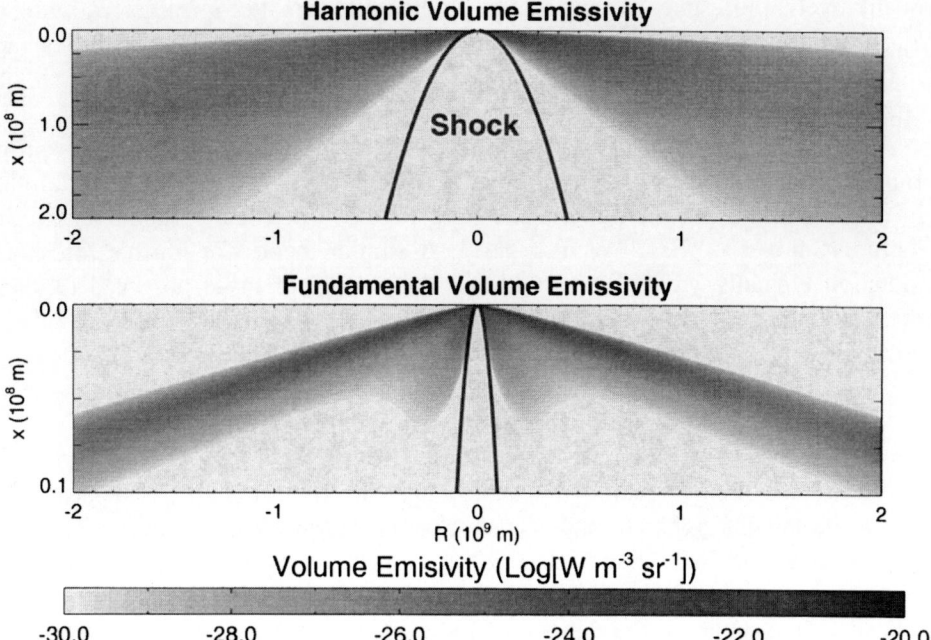

Figure 3. Volume emissivities j_F and j_H in the foreshock (cf., Knock *et al.*, 2001) for the shock and plasma parameters described in the text.

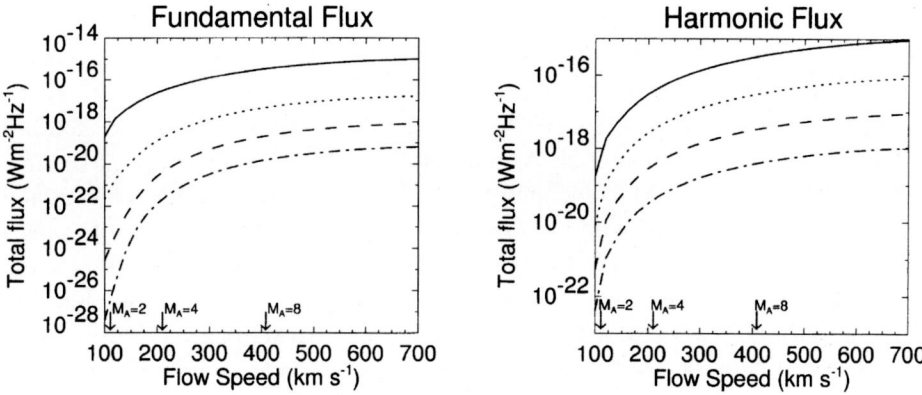

Figure 4. Variations of the predicted fundamental and harmonic flux with U and κ (Knock *et al.*, 2003): κ decreases from 2 to 5 from the top to bottom curve.

Figure 4 predicts that radio-loud shocks should be faster and move through solar wind regions with smaller v_{sw} and κ. The dependence on U follows from (i) SDA producing more fast electrons via (2), (ii) the shock's increased mirror ratio, and (iii) the increased foreshock volume at large $v_\parallel$. These effects increase the number or speeds of fast electrons, as does decreasing κ, leading to increased emission. Other trends as functions of b, T_e, θ_{UB}, N_e and B_1 exist (Knock *et al.*, 2003):

qualitatively, more intense radiation is predicted for smaller b (larger R_c), higher T_e and N_e, θ_{UB} closer to 90 deg, and lower B_1 and T_i. Since shock properties vary in the inhomogeneous solar wind, these dependences suggest type II bursts should be bursty and time-varying.

Preliminary comparisons with the fluxes $\approx 10^{-18}$ W m^{-2} Hz^{-1} observed for Bale *et al.*'s (1999) type II are encouraging (Knock *et al.*, 2001) given uncertainties in the shock and plasma parameters, with Figure 4 predicting fundamental and harmonic fluxes $\approx 10^{-17}$ W m^{-2} Hz^{-1}. A similar theory for Earth's foreshock radiation typically yields fluxes within a factor of 2 of those observed (Kuncic *et al.*, 2002b).

4. Relationships to Space Weather Events

The preceding theory predicts that radio-loud type II's should be associated with faster and larger (higher U and R_c) shocks and coronal mass ejections (CMEs). This is consistent with known space weather correlations (Cane and Stone, 1984; Gopalswamy *et al.*, 2001a). Since it is U and κ rather than V_{sh} alone that are relevant, the scatter in correlations of radio-loud or radio-quiet shocks with V_{sh} may be due to variations in v_{sw}, κ, and b. Radio-quiet CMEs are expected to have low U, large b and κ, quasi-radial $\mathbf{B}_1$, and small T_e and N_e.

The theory predicts 'hot spots' or localized bursts of type II radiation when the shock moves into solar wind regions with more high-v particles (lower κ), higher N_e or higher T_e. Examples are corotating interaction regions (CIRs), which increase the number of high-v electrons, and prominences. The theory is thus consistent with Reiner *et al.*'s (1998b) observations of hot spots associated with a CIR and prominence material. A more global example is of a fast CME shock overtaking an earlier CME, encountering the enhanced fast particles and heated, denser plasma associated with the slower CME's shock. Qualitatively one expects the second CME to produce a stronger type II burst than the first CME (if upstream parameters are otherwise identical), sometimes with neither CME being radio-loud until the second shock interacts with the first's downstream plasma. These expectations appear consistent with 'colliding CME' events recently seen (Gopalswamy *et al.*, 2001b).

These results point to the theory having significant future potential as a predictor of space weather. Predictions for multiple ripples, dynamic spectra, and coronal type IIs are still being developed. These indicate that IMF direction is important, with possible space weather implications, and that simulations of rippled shocks from the corona to 1 AU are necessary. Finally, we caution that the flux, shock, and plasma parameters of interplanetary type IIs remain poorly known and that detailed data-theory comparisons must still be performed.

5. Conclusions

An analytic, quantitative theory for type II solar radio bursts now exists (Knock *et al.*, 2001, 2003), treating electron energization at the shock, formation of electron beams, and transfer of electron energy into Langmuir waves and radio emission in a 3-D source volume. Predictions suitable for observational testing exist and preliminary comparisons are encouraging. Theory predicts that the radio flux depends sensitively on properties of the shock and solar wind plasma, so that type IIs should often be bursty and irregular. Radio-loud shocks and CMEs are predicted to be faster, larger, and to move through plasmas with lower v_{sw} and κ, with the first two dependences being consistent with known correlations and the others as yet untested. Explanations exist, and are qualitatively consistent with observations, for intensifications of type IIs associated with CIRs, prominences and colliding CMEs. While the theory has significant potential for space weather prediction, it must first be applied to coronal type IIs and tested in detail.

References

Bale, S. D., Reiner, M. J., Bougeret, J. L., Kaiser, M. L., Krucker, S., Larson, D. E. and Lin, R. P.: 1999, 'The Source Region of an Interplanetary Type II Radio Burst', *Geophys. Res. Lett.* **26**, 1573–1576.

Cairns, I. H.: 1986, 'Source of Free Energy for Type II Solar Radio Bursts', *Proc. Astron. Soc. Austr.* **6**, 444–446.

Cairns, I. H.: 1987, 'The Electron Distribution Function Upstream from the Earth's Bow Shock', *J. Geophys. Res.* **92**, 2315–2327.

Cane, H. V., Stone, R. G., Fainberg, J., Steinberg, J. -L. and Hoang, S.: 1982, 'Type II Solar Radio Events Observed in the Interplanetary Medium. 1 General Characteristics', *Solar Phys.* **78**, 187–198.

Cane, H. V. and Stone, R. G.: 1984, 'Type II Solar Radio Bursts, Interplanetary Shocks, and Energetic Particle Events', *Astrophys. J.* **282**, 339–344.

Filbert, P. C. and Kellogg, P. J.: 1979, 'Electrostatic Noise at the Plasma Frequency Beyond the Earth's Bow Shock', *J. Geophys. Res.* **84**, 1369–1381.

Fitzenreiter, R. J., Klimas, A. J. and Scudder, J. D.: 1990, 'Three-Dimensional Model for the Spatial Variation of the Foreshock Electron Distribution: Systematics and Comparison with Data', *J. Geophys. Res.* **95**, 4155–4173.

Gopalswamy, N., Yashiro, S., Kaiser, M. L., Howard, R. A. and Bougeret, J. -L.: 2001a, 'Characteristics of Coronal Mass Ejectrions Associated with Long-Wavelength Type II Radio Bursts', *J. Geophys. Res.* **106**, 29219–29229.

Gopalswamy, N., Yashiro, S., Kaiser, M. L., Howard, R. A. and Bougeret, J. -L.: 2001b, 'Radio Signatures of Coronal Mass Ejection Interaction: Coronal Mass Ejection Cannibalism?', *Astrophys. J.* **548**, L91–94.

Knock, S. A., Cairns, I. H., Robinson, P. A. and Kuncic, Z.: 2001, 'Theory of Type II Radio Emission from the Foreshock of an Interplanetary Shock', *J. Geophys. Res.* **106**, 25041–25051.

Knock, S. A., Cairns, I. H., Robinson, P. A. and Kuncic, Z.: 2003, 'Theoretically predicted properties of Type II radio Emission from an Interplanetary Foreshock', *J. Geophys. Res.* **108**, 1126–1137.

Kuncic, Z., Cairns, I. H. and Knock, S. A.: 2000a, 'Analytic Model for the Electrostatic Potential Jump Across Collisionless Shocks, with Application to Earth's Bow Shock', *J. Geophys. Res.* **107**(A8), 10.1029/2001JA000250.

Kuncic, Z., Cairns, I. H. and Knock, S. A.: 2000b, 'Earth's Foreshock Radio Emission', *Geophys. Res. Lett.* **29**(8), 47(1)–47(4).

Lengyel-Frey, D.: 1992, 'Location of the Radio Emitting Regions of Interplanetary Shocks', *J. Geophys. Res.* **97**, 1609–1617.

Nelson, G. S. and Melrose, D. B.: 1985, 'Type II Bursts', in N. Labrum and D. J. McLean (eds.), *Solar Radiophysics*, Cambridge, Cambridge University Press, pp. 333–359.

Reiner, M. J., Kaiser, M. L., Fainberg, J. and Stone, R. G.: 1998a, 'A New Method for Studying Remote Type II Radio Emissions from Coronal Mass Ejection-Driven Shocks', *J. Geophys. Res.* **103**, 29651–29664.

Reiner, M. J., Kaiser, M. L., Fainberg, J., Bougeret, J. -L. and Stone, R. G.: 1998b, 'On the Origin of Radio Emission Associated with the January 6–11, 1997, CME', *Geophys. Res. Lett.* **25**, 2493–2496.

Robinson, P. A. and Cairns, I. H.: 2000, 'Theory of Type III and Type II Solar Radio Emissions, in R. G. Stone *et al.*(eds), *Radio Astronomy at Long Wavelengths*, American Geophysical Union, Washington, pp. 37–45.

Wild, J. P., Murray, J. D. and Rowe, W. C.: 1954, 'Harmonics in Spectra of Solar Radio Disturbances', *Austr. J. Sci. Res.* **A7**, 439–459.

PROBING SOLAR SUBSURFACE MAGNETIC FIELDS

DEAN-YI CHOU[1], ALEXANDER SEREBRYANSKIY[1] and MING-TSUNG SUN[2]

[1]*Institute of Astronomy and Department of Physics,*
Tsing Hua University, Hsinchu, 30043, Taiwan, R.O.C.
[2]*Department of Mechanical Engineering,*
Chang-Gung University, Tao-Yuan, 33333, Taiwan, R.O.C.

Abstract. Helioseismology uses solar p-mode oscillations to probe the structure of the solar interior. The modifications of p-mode properties due to the presence of solar magnetic fields provide information on the magnetic fields in the solar interior. Here we review some of results in helioseismology on the magnetic fields in the solar convection zone. We will also discuss a recent result on the magnetic fields at the base of the convection zone.

Key words: helioseismology, solar interior, solar magnetic fields

1. Introduction

Observations indicate that the magnetic fields on the Sun emerge from below. It has been suggested that the solar magnetic fields are generated at the base of the convection zone (CZ) for two reasons (Spiegel and Weiss, 1980). First, the radial and latitudinal differential rotation in this region could generate magnetic fields through a dynamo mechanism. Second, magnetic fields could be stored in this sub-adiabatic region for an extended period of time for the dynamo to operate. The magnetic fields rise to the surface appearing as active regions due to magnetic buoyancy. Helioseismology uses the solar p-mode waves to probe the subsurface magnetic fields. The presence of subsurface magnetic fields changes the physical conditions of the solar interior; it in turn changes the properties of the solar p-mode waves. Studying the influence of these changes on p-mode waves could provide information on the distribution of the subsurface magnetic fields. The influence of magnetic fields on p-modes decreases rapidly with depth because the ratio of magnetic pressure to gas pressure decreases rapidly with depth. Thus probing the magnetic field deep in the CZ is difficult. Here we review some methods used in helioseismology to probe the subsurface magnetic fields.

Space Science Reviews **107**: 35–42, 2003.
© 2003 *Kluwer Academic Publishers.*

2. Near-surface Magnetic Fields

2.1. NORMAL MODE ANALYSIS (FREQUENCY ANALYSIS)

The change in physical conditions of a medium due to the presence of magnetic fields results in a change in the frequency of the resonant p-modes. It has been found that the frequencies vary with solar cycle (Libbrecht and Woodard, 1990). The magnitude of change in frequency is well correlated with solar activity (Howe *et al.*, 1999). It has been argued that measured solar-cycle variations of mode frequencies are due to the near-surface magnetic fields (Goldreich *et al.*, 1991; Dziembowski *et al.*, 2000).

The normal mode decomposition can be also applied to a local area to study the structure of active regions, such as ring-diagram analysis (Hill, 1988) and Fourier-Hankel decomposition (Braun *et al.*, 1997). Chen *et al.* (1997) inverted the measured absorption coefficients of sunspots from Fourier-Hankel decomposition to obtain the depth distribution of the absorption region, due to magnetic fields, in the range of 0–30 Mm below the surface.

2.2. TIME-DISTANCE ANALYSIS

Duvall *et al.* (1993) first applied time-distance analysis to helioseismology. The modes with the same angular phase velocity have approximately the same ray path and form a wave packet. The relation between the travel time and travel distance of a wave packet can be measured with the temporal cross-correlation between the time series at two points. If the physical conditions of material along the ray path of the wave packet change, the travel time would change. Thus travel time provides information on the physical conditions of the solar interior. For example, the difference between the travel times of opposite directions is sensitive to the flow along the ray path (Duvall *et al.*, 1996). The mean of travel time of opposite directions provides information on the change in wave speed (Kosovichev, 1996). Since the measured mean and difference of travel times are integral effects over the ray path, it needs inversion to obtain the spatial distribution of the flow and wave-speed perturbation below the surface (Kosovichev *et al.*, 2000). Time-distance analysis can probe the *local* effects of subsurface magnetic fields in active regions down to about 30 Mm.

2.3. ACOUSTIC IMAGING (ACOUSTIC HOLOGRAPHY)

In 1997 a method, acoustic imaging, was developed to construct the acoustic signals below the surface (Chang *et al.*, 1997) using the helioseismic data taken with the Taiwan Oscillation Network (TON). The acoustic signals measured at the surface are coherently added, based on the time-distance relation, to construct the signal at a target point and a target time (Lindsey and Braun, 1998; Chou *et al.*, 1999; Chou, 2000). For a target point on the surface, one can use the measured time-distance relation. For a target point located below the surface, one has to use the time-distance relation computed with a standard solar model and the ray approximation. Since each point on the time-distance curve corresponds to a wave-packet formed by the modes with the same ω/l, the spatial range of data used in coherent sum determines the modes used to construct the signal, which determines the spatial resolution of acoustic imaging.

The constructed signal contains information on intensity and phase. The intensity at the target point can be computed by summing the square of the amplitude of the constructed signal over time. The constructed intensity in magnetic regions is smaller than that in the surrounding quiet Sun. The phase of constructed signals can be studied by the cross-correlation function between the signals constructed with outgoing waves and ingoing waves (Chen *et al.*, 1998). A review of the use of acoustic imaging to probe the subsurface magnetic fields is given in Chou (2000). The inversion problem of acoustic imaging is discussed in Sun and Chou (2002a, 2002b). Similar to time-distance analysis, the sensitivity of acoustic imaging allows probing of the *local* effects of subsurface magnetic fields down to about 30 Mm below the photosphere.

3. Magnetic Fields in the Convection Zone

Although local time-distance analysis is sensitive only to the near-surface magnetic fields, its sensitivity increases if the cross-correlation function is averaged over longitude. With longitudinal averaging, the meridional flows deep in the solar CZ has been detected (Giles *et al.*, 1997). It has been shown that the meridional flows penetrate into the entire CZ. Recently, it was shown using the TON data that the surface meridional flow varies with solar cycle (Chou and Dai, 2001): a new divergent flow was created at the active latitudes in each hemisphere at solar maximum. The new component is superposed on the original flow to transform the flows from the one-cell pattern into the two-cell pattern in each hemisphere. This new flow penetrates at least down to 70 Mm (0.1 $R_\odot$). The fact that the center of this new flow coincides with the centroid of the active longitudes along the solar cycle suggests that the new flow is created by subsurface magnetic fields. If so, the solar-cycle variations of meridional flow might be used as an indirect probe of the magnetic fields deep in the CZ. The equipartition field strength corresponding the velocity of new divergent flow, about 20 m/s, at 0.9 $R_\odot$ is about 1600 gauss. The existence of this new divergent flow is confirmed by Beck *et al.* (2002) using Michelson Doppler Imager (MDI) data (Scherrer *et al.*, 1995).

4. Magnetic Fields at the Base of the Convection Zone

How and where solar magnetic fields are generated is a long standing unanswered question in astronomy (Cowling, 1934; Parker, 1955; Babcock, 1961). It is believed that magnetic fields are generated at the boundary between the radiative zone and the CZ by a dynamo mechanism. Many attempts have been made to detect the magnetic fields in this region (Gough et al., 1996; Basu, 1997; Howe et al., 1999; Basu and Antia, 2001; Eff-Darwich et al., 2001; Antia et al., 2001). Until now no clear evidence of the magnetic field in this region has been found. Recently, Chou and Serebryanskiy (2002) measured solar-cycle variations of the travel time of acoustic waves with different ray paths to probe the magnetic fields at the base of the convection zone (BCZ). The modes with the similar phase velocity form a wave packet. Different wave packets penetrate into different depths: the wave packet with a larger phase velocity penetrates into a greater depth as shown in Figure 1. If a magnetic field is present at the BCZ, it would affect the wave packets penetrating into the BCZ, while leave other wave packets intact. If the magnetic fields at the BCZ vary with the solar cycle, travel time is expected to vary with the solar cycle as well. However, the change in travel time due to the magnetic fields at the BCZ is small because the ratio of magnetic pressure to gas pressure is small. To improve the S/N, one can measure multiple-bounce travel time because the change in travel time is linearly proportional to the number of bounces. Chou and Serebry-anskiy (2002) measured the time for a wave packet which takes N bounces to travel around the Sun to come back to the same spatial point. Thus the problem becomes measuring solar cycle variations of travel time with the auto-correlation function of the time series at the same spatial point. Chou and Serebryanskiy (2002) use two different approaches: 1) the direct computation of auto-correlation function, and 2) the power spectrum simulation analysis, to measure the travel time of wave packets.

In the first approach, the Doppler images are filtered with a phase-velocity filter to isolate the signals in a range of the phase velocity. The center of filter is chosen such that the one-bounce travel distance is $360°/N$, where N is an integer. The auto-correlation function of time series at each spatial point is computed. The auto-correlation functions are then averaged over the solar disk. The phase travel time τ_N is determined from the averaged auto-correlation function.

The correlation function is the inverse Fourier transform of the power spectrum of p-modes. Thus the signal corresponding to the travel time perturbation detected in the first approach should also exist in the mode frequencies which are determined from the power spectrum. In the second approach, the measured mode frequencies are used to construct the power spectrum in (l, ω) with assumed line widths and relative mode amplitudes. The constructed power spectrum is filtered with the same phase velocity filter as in the first approach, prior to computing the cross-correlation function with the inverse Fourier transform. The cross-correlation function at zero travel distance corresponds to the auto-correlation function aver-

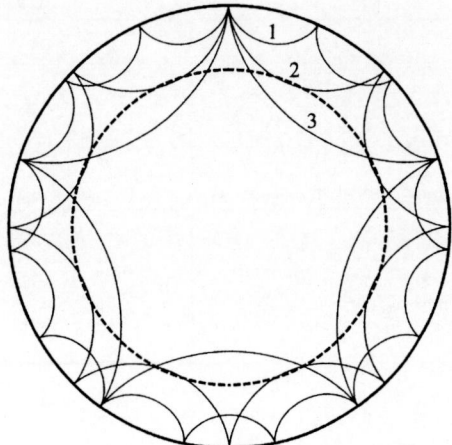

Figure 1. Diagram showing ray paths of three different wave packets computed from a standard solar model (Scherrer *et al.*, 1996) with the ray approximation. The thick solid line is the solar surface, and the dashed line is the BCZ at 0.713 $R_\odot$ (Basu, 1997). Ray 1, which takes 15 bounces to go around the Sun, is not affected by the magnetic fields at the BCZ. Ray 2, taking 8 bounces to go around the Sun, has the lower turning point very close to the base of the CZ. Ray 3, taking 5 bounces to go around the Sun, can penetrate into the radiative zone.

aged over all spatial points, which is used to determine the travel time of the wave packet.

To study solar cycle variations, in the first approach, Chou and Serebryanskiy (2002) have used the MDI Doppler images in the periods of solar minimum and maximum. The travel time at solar maximum is shorter than that at minimum. The difference is defined as $\delta\tau_N$. The change in one-bounce travel time is equal to $\delta\tau_N/N$. The value of $\delta\tau_N/N$ versus N is shown by the open circles in the left panel of Figure 2.

In the second approach, the mode frequencies measured with the data taken with MDI and the Global Oscillation Network Group (GONG) have been used. The change in travel time relative to solar minimum for various periods along the solar cycle, $\delta\tau_N/N$, is shown by the dots in Figure 2. The results of two different approaches are consistent. Both show that the value of $\delta\tau_N/N$ is approximately the same for all N's, except a small drop at $N = 8$. This approximately constant change is caused by the magnetic fields near the surface because the rays paths of different wave packets near the surface are approximately the same. The interesting phenomenon in Figure 2 is the additional decrease in travel time at $N = 8$, which increases with the solar activity. The additional changes in travel time are different in two analysis. This may be caused by the differences in data analysis: for example, the temporal filters are different in two analysis, and the second approach does not use information on azimuthal degree m.

Simulation tests have been carried to show that the anomaly at $N = 8$ is unlikely to be caused by the analysis procedure (Chou and Serebryanskiy, 2002). It

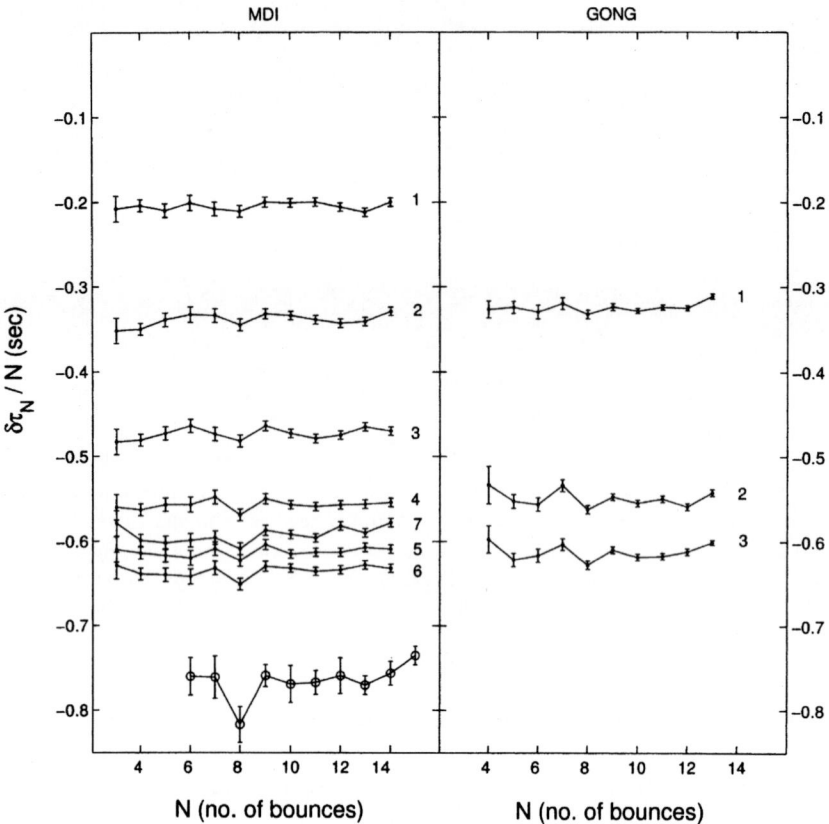

Figure 2. Change in one-bounce travel time relative to solar minimum versus number of bounces N, which corresponds to different wave packets (from Chou and Serebryanskiy, 2002). The result from the direct computation (first approach) is denoted by the open circle in the left panel. The error bar is an estimate of fluctuation from averaging the travel time of different time series over solar minimum or maximum. The result from the power spectrum simulation analysis (second approach) is denoted by the dot. The left panel is computed from the MDI mode frequencies, and the right panel from the GONG mode frequencies. The sequence of the averaging periods along the solar cycle (from minimum to maximum) is indicated by the number associated with each curve. It is noted that averaging periods for two analysis are different.

has also been argued that the anomaly at $N = 8$ is not caused by the organized spatial distribution of the near-surface magnetic fields (Chou and Serebryanskiy, 2002; Chou *et al.*, 2002). The fact that the ray path of the wave packet of $N = 8$ has the lower turning point at the BCZ as shown in Figure 1 suggests that the additional decrease in travel time at $N = 8$ may be caused by the solar-cycle varying perturbations at the BCZ. The magnetic field in solar interior can change the travel time in a complicated way. The magnetic field has a direct influence on the dispersion relation and causes a change in phase velocity. The magnetic field can also change the thermal structure; it in turn causes the change in wave speed. The change in dispersion relation alone has difficulty to explain the phenomenon that the anomaly in travel time occurs only at $N = 8$. The combination of changes in dispersion relation and thermal structure might be able to explain the observed results. If the anomaly at $N = 8$ is caused solely by the change in dispersion relation due to the magnetic fields at the BCZ, the magnitude of the additional decrease can be used to estimate the field strength at the BCZ, $B \sim 4 - 7 \times 10^5$ gauss for a filling factor of unity (Chou and Serebryanskiy, 2002).

5. Conclusion

Among methods discussed here, the method discussed in Section 4 is the most sensitive way to probe the magnetic fields at the BCZ. This method can be further developed, such as the study of m-dependence and frequency-dependence of the travel time variations and the inversion of the travel time variations, to investigate the spatial distribution of the wave-speed perturbations.

Acknowledgement

The authors were supported by NSC of ROC under grants NSC-90-2112-M-007-036 and NSC-90-2112-M-182-001.

References

Antia, H. M. *et al.*: 2001, *Mon. Not. R, Astron. Soc.* **327**, 1029.

Babcock, H. W.: 1961, *Astrophys. J.* **133**, 572.

Basu, S.: 1997, *Mon. Not. R, Astron. Soc.* **288**, 572.

Basu, S. and Antia, H. M.: 2001, in A. Wilson (ed.), Proc. SOHO 10/GONG 2000 Workshop: Helio- and Astero-seismology at the Dawn of the Millennium, ESA SP-464, ESA, Noordwijk, p. 297.

Beck, J. G., Gizon, L. and Duvall, T. L. Jr.: 2002, *Astrophys. J.* **575**, L47.

Braun, D. C., Duvall, T. L. Jr. and LaBonte, B. J.: 1997, *Astrophs. J.* **319**, L27.

Chang, H. -K., Chou, D. -Y., LaBonte, B. and the TON Team: 1997, *Nature* **389**, 825.

Chen, H. -R., Chou, D. -Y. and the TON Team: 1997, *Astrophys. J.* **490**, 452.

Chen, H. -R., Chou, D. -Y., Chang, H. -K., Sun, M. -T., Yeh, S. -J, LaBonte, B. and the TON Team: 1998, *Astrophys. J.* **501**, L139.

Chou, D. -Y., Chang, H. -K., Sun, M. -T., LaBonte, B., Chen, H. -R., Yeh, S. -J. and the TON Team: 1999, *Astrophys. J.* **514**, 979.

Chou, D. -Y.: 2000, *Solar Phys.* **192**, 241.

Chou, D. -Y. and Dai, D. -C.: 2001, *Astrophys. J.* **559**, L175.

Chou, D. -Y. and Serebryanskiy, A.: 2002, *Astrophys. J.* **578**, L157.

Chou, D. -Y., Serebryanskiy, A. and Sun, M. -T.: 2002, in Proc. SOHO 12/GONG 2002 Workshop, ESA, Noordwijk, in press.

Christensen-Dalsgaard, J. *et al.*: 1996, *Science* **272**, 1286.

Cowling, T. G.: 1934, *Mon. Not. R, Astron. Soc.* **94**, 39.

Duvall, T. L. Jr., Jefferies, S. M., Harvey, J. W. and Pomerantz, M. A.: 1993, *Nature* **362**, 430.

Duvall, T. L. Jr., D'Silva, S., Jefferies, S. M., Harvey, J. W. and Schou, J.: 1996, *Nature* **379**, 235.

Dziembowski, W. A., Goode, P. R., Kosovichev, A. G. and Schou, J.: 2000, *Astrophys. J.* **537**, 1026.

Eff-Darwich, A., Perez Hernandez, F. and Korzennik, S. G.: 2001, in A. Wilson (ed.), Proc. SOHO 10/GONG 2000 Workshop: Helio- and Astero-seismology at the Dawn of the Millennium, ESA SP-464, ESA, Noordwijk, p. 289.

Giles, P. M., Duvall, T. L. Jr., Scherrer, P. H. and Bogart, R. S.: 1997, *Nature* **390**, 52.

Goldreich, P., Murray, N., Willette, G. and Kumar, P.: 1991, *Astrophys. J.* **370**, 752.

Gough, D. O. *et al.*: 1996, *Science* **272**, 1296.

Hill, F.: 1988, *Astrophys. J.* **333**, 996.

Howe, R., Komm, R. and Hill, F.: 1999, *Astrophys. J.* **524**, 1084.

Kosovichev, A. G.: 1996, *Astrophys. J.* **461**, L55.

Kosovichev, A. G., Duvall, T. L. Jr. and Scherrer, P. H.: 2000, *Solar Phys.* **192**, 159.

Libbrecht, K. G. and Woodard, M. F.: 1990, *Nature* **345**, 779.

Lindsey, C. and Braun, D. C.: 1998, *Astrophys. J.* **499**, L99.

Parker, E. N.: 1955, *Astrophys. J.* **122**, 293.

Scherrer, P. H. *et al.*: 1995, *Solar Phys.* **160**, 237.

Spiegel, E. A. and Weiss, N. O.: 1980, *Nature* **287**, 616.

Sun, M. -T., Chou, D. -Y.: 2002a, and the TON Team, *Solar Phys.* **209**, 5.

Sun, M. -T., Chou, D. -Y.: 2002b, and the TON Team, *in this proceedings*.

A CLASS OF TVD TYPE COMBINED NUMERICAL SCHEME FOR MHD EQUATIONS WITH A SURVEY ABOUT NUMERICAL METHODS IN SOLAR WIND SIMULATIONS

XUESHANG FENG[1], S. T. WU[2], FENGSI WEI[1] and QUANLIN FAN[1]

[1]*Laboratory for Space Weather, Center for Space Science and Applied Research, Beijing 100080,*
[2]*Center for Space Physics and Aeronomic Research, The University of Alabama in Huntsville,*
AL 35899 U.S.A.

Abstract. It has been believed that three-dimensional, numerical, magnetohydrodynamic (MHD) modelling must play a crucial role in a seamless forecasting system. This system refers to space weather originating on the sun; propagation of disturbances through the solar wind and interplanetary magnetic field (IMF), and thence, transmission into the magnetosphere, ionosphere, and thermosphere. This role comes as no surprise to numerical modelers that participate in the numerical modelling of atmospheric environments as well as the meteorological conditions at Earth. Space scientists have paid great attention to operational numerical space weather prediction models. To this purpose practical progress has been made in the past years. Here first is reviewed the progress of the numerical methods in solar wind modelling. Then, based on our discussion, a new numerical scheme of total variation diminishing (TVD) type for magnetohydrodynamic equations in spherical coordinates is proposed by taking into account convergence, stability and resolution. This new MHD model is established by solving the fluid equations of MHD system with a modified Lax–Friedrichs scheme and the magnetic induction equations with MacCormack II scheme for the purpose of developing a combined scheme of quick convergence as well as of TVD property. To verify the validation of the scheme, the propagation of one-dimensional MHD fast and slow shock problem is discussed with the numerical results conforming to the existing results obtained by the piece-wise parabolic method (PPM). Finally, some conclusions are made.

Key words: combined numerical scheme of TVD type, numerical methods, solar wind simulations

1. Introduction

Numerical MHD studies have been commonly used over the last three decades to simulate solar-terrestrial physical phenomena. Both theoretical analysis and numerical modelling of physical events have undergone a revolution with a tendency of developing numerical prediction models for adverse space weather events.

There are many works on space weather events. To say a few, Wu and his co-workers selected three CME events to numerically reproduce the observations by LASCO/SOHO by introducing sound mechanisms of the initiation processes in a two-dimensional MHD simulation (Wu, Guo and Dryer, 1997; Wang *et al.*, 1998; Wu *et al.*, 2000a, 2000b). Linker, Mikic and their co-workers simulated the global solar corona by using observed photospheric magnetic fields as a boundary

Space Science Reviews **107**: 43–53, 2003.

condition and interpreting some solar observations including: eclipse images of the corona, Ulysses spacecraft measurements of the interplanetary magnetic field, and coronal hole boundaries from Kitt Peak He 10830 Å maps and extreme ultraviolet images from the Solar Heliospheric Observatory (SOHO) (Linker *et al.*, 1999; Mikic *et al.*, 1999; Riley *et al.*, 1997, 1999, 2001).

In space weather event studies, the prediction of IMF B_z is very important since the long duration southward interplanetary magnetic field (IMF) in solar magneto-spheric coordinate system (GSM), usually called $-B_z$ or B_s plays a crucial role in determining the amount of solar wind energy to be transferred to the magneto-sphere. Thus, understanding the causes of and predicting the length and strength of the large southward IMF B_z are key goals in the study of the occurrence of intense geomagnetic storms. Numerical study of the southern IMF component B_z is made with 3D-time dependent MHD equations with McCormack difference scheme with near real initial-boundary conditions that are based on the source surface magnetic field observation. As an example, the propagation and evolution of the January 1997 interplanetary CME are numerically studied using this 3-D MHD model. The numerical results show that the parameters obtained near the earth are in agreement with observations of WIND satellite, especially the temporal behavior B_z near the earth (Shi, Wei and Feng, 2001). Other quantitative studies of southward IFM B_z are made by Wu *et al.* (1996a), Wu and Dryer (1996b) and Wu and Dryer (1997). A recent review of previous attempts for predicting large IFM B_z based on some numerical methods and the Hakamada–Akasofu scheme is given by Chao and Chen (2001).

Models describing the process of CME formation, interplanetary propagation and interaction with the ionospheric-magnetospheric system have also made progress. Interested readers may refer to the recent work by Groth *et al.* (1999, 2000a, 2000b), Gombosi *et al.* (2000a, 2000b, 2001), Song *et al.* (1999; 2000) and references therein. Three-dimensional propagation of CMEs in a structured solar wind has also got attention (Odstrcil and Pizzo, 1999a, 1999b, 1999c; Riley *et al.*, 2002; Vandas *et al.*, 2002). For more complete discussion of the observational properties and theory of CMES, we can refer to the following papers: Dryer (1998), Wu, Andrews and Plunkett (2001), St Cyr (2000), Webb *et al.* (2001), Klimchuk (2001) and Burlaga *et al.* (2002).

In what follows we first briefly mention some current used numerical methods in solar wind simulation. Then, based on our analysis, a new combined numerical scheme of TVD type for magnetohydrodynamic equations is introduced. Finally, some conclusions about the properties a high quality numerical scheme of solar wind simulation should have is given.

2. Numerical Methods in Solar Wind Modelling

In the past three decades, solar-terrestrial physicists have introduced many kinds of numerical schemes used in computational fluid mechanics to the magnetohydro-dynamic system in order to simulate various phenomena of solar-terrestrial physics. The numerical study of solar wind has undergone a transit from its early simulation problems in supersonic and superAlfvenic domain to the recent works from the solar surface, the interplanetary space, and the interaction between the solar wind and the earth's magnetosphere.

Nakagawa (1987) developed a method to solve MHD initial boundary value problems by the method of characteristics. The theoretical method was used to generate a numerical scheme named Full-Implicit-Continuous-Eulerian (FICE) to solve multidimensional transient MHD flows (Hu and Wu, 1984; Wu et al., 1986). A three-dimensional extended coronal model was presented by Wu and Wang (1991), but the scheme did not efficiently handle very small values of the plasma β, where $\beta = \frac{8\pi}{|\mathbf{B}|^2}$ is the thermal pressure and the magnetic field strength. To extend the simulation to low β plasma flow, an improved numerical scheme the so called combined difference scheme, was developed (Guo and Wu, 1998; Wu, Guo and Dryer, 1997). In the method, the mass and energy equations are solved by the upwind scheme and the momentum and magnetic induction equations are solved by the Lax-Wendroff scheme.

Tanaka (1994) developed a finite volume TVD scheme of Roe type on an un-structured grid system for three-dimensional MHD simulation of inhomogeneous systems, including strong background potential fields, to simulate magnetospheric problems by decomposing the magnetic field into a potential part and an non-potential one, in which the potential part is time-independent and the non-potential one is time-dependent. This decomposition can efficiently simulate magnetohydro-dynamic problem of low β plasma and avoid the unrealistic flow caused by the numerical non-zeroness of $\nabla \cdot \mathbf{B}$. By using this method, Tanaka (1995) also discussed the generation mechanism for magnetosphere-ionosphere current systems deduced from a three-dimensional MHD simulation of the solar wind-magnetosphere-iono-sphere coupling processes.

Panitchob, Wu and Suess (1987) used a grid generation technique to establish an adaptive grid unsteady MHD model to simulate the propagation of a solar-flare-generated shock wave in solar wind flow in the heliographic equatorial plane. A Lax–Wendroff scheme with artificial diffusions (Han, Wu and Dryer, 1988) has been used to model the three-dimensional supersonic and superAlfvenic solar wind flow with a conservative form of the MHD system. This scheme has also been used by Washimi (1990) to simulate the two-dimensional structure of the solar wind and interplanetary magnetic field near the sun.

In a series of works, Steinolfson and his coworkers (see Steinolfson (1994) and references therein) carried out the numerical study of coronal structure and coronal mass ejections by using two-step explicit Lax–Wendroff scheme with a fourth-order damping at each time step, and Steinolfson (1991) showed that this algorithm produces results virtually identical to those from the well-known MacCormack method. Zhang and Wei (2000) used MacCormack II scheme to discuss the interaction between solar wind background and magnetic field in the meridional plane by using the non-reflective projective characteristics method at the inner boundary. Using two-step Lax-Wendroff scheme with flux-corrected transport technique, Zhang and Wang (2000) discussed the coronal mass ejection-shock system in the inner corona.

In recent years, a 3-D time-dependent resistive MHD to investigate the structure of the solar corona has been developed by Mikic, Linker and their co-workers (Linker *et al.*, 1999; Mikic *et al.*, 1999). In this model, the spherical (r, θ, ϕ) grid permits non-uniform spacing of mesh points in both r and θ. At the lower boundary the radial component of the magnetic field B_r is based on the observed line of sight measurements of the photospheric magnetic field. For the density and temperature, characteristic values are used. As usual, an initial estimate of the field and plasma parameters are found from a potential field model and a Parker transonic solar wind solution; then this initial solution is advanced in time by a leapfrog time integration scheme with a semi-implicit method until a steady state is obtained. In the radial r and meridional θ directions, a finite-differencing approach of upwind type is used; in the azimuth θ, the derivatives are calculated pseuospectrally (Lionello, Mikic and Linker, 1998). The time step is chosen according to accuracy constraints; an implicit linear operator is used in the momentum equation to ensure the unconditional stability of wave-like modes. The implicit operator has the ability to modify the dispersion properties of high-k modes with k the spatial wave vector. These modes can not be expressed accurately by the spatial differencing technique. This scheme is particularly efficient for simulating the solar evolution with long wavelengths.

MHD simulations have also been for long used to simulate the global magnetospheric configuration and to investigate the response of magnetosphere ionosphere system to changing solar wind conditions. The first global-scale 3D MHD simulations of the solar wind-magnetosphere system were published in the early 1980s Brecht *et al.*, 1982; Leboeuf *et al.*, 1981; Wu, Walker and Dawson, 1981) by using a Rusanov scheme, a hybrid scheme of flux corrected tansport (FCT) type and a Lax-Wendroff scheme with artificial diffusion. Later, several magnetospheric models have been established (Janhunen, 1996; Lyon, Feddr and Huba, 1986; Ogino, 1986; Powell *et al.*, 1999; Reader, Walker and Ashour-Abdalla, 1998; Ridley *et al.*, 2002; Watanabe and Sato, 1990; Winglee, 1994; Tanaka, 1995; White *et al.*, 1998). Ogino (1986), Watanabe and Sato (1990) and Winglee (1994) used a relatively simple central differencing methods such as two-step Lax–Wendroff scheme. Lyon *et al.* (1986) and White *et al.* (1998) employed an approximate Riemann solver based on the five waves associated with the fluid dynamics system, and treated the electromagnetic effects by the constrained-transport technique (Evans and Hawley, 1988). The numerical method by Raeder *et al.* (1998) is similarly structured, but without the use of a Reimann solver; a finite difference scheme that is conservative for the fluid dynamics system is combined with the constrained-transport technique. Tanaka (1995), Janhunen (1996) and Powell *et al.* (1995) applied approximate Riemann solvers based on the waves associated with the full magneto-hydrodynamic system with a special symmetric formulation of the MHD system suggested by Godunov (1972). The numerical schemes (Lyon, Fedder and Huba, 1986; Powell *et al.*, 1995; Reader, Walker and Ashour-Abdalla, 1998; Tanaka, 1995; White *et al.*, 1998) are, due to the high resolution approach, second-order accurate in smooth regions, and locally first order in discontinuous regions. However, the method used by Janhunen (1996) is just first-order accurate. De Zeeuw *et al.* (2000) developed a parallel adaptive solution method for MHD in Cartesian coordinates to simulate the magnetospheric plasma flows. This scheme adopts a cell-centered upwind finite-volume discretization procedure and uses approximate Riemann solvers, with explicit multi-stage time stepping to solve the MHD equations in divergence form. Recent applications see the study of magnetospheric events. In these simulations, the observed upstream solar wind conditions are used to drive the magnetosphere-ionosphere system and numerical predictions are compared with ground-based or satellite observations (Fedder *et al.*, 1998; Raeder, Walker and Ashour-Abdalla, 1998).

3. A Class of TVD Type Combined Numerical Scheme for MHD Equations

As a basic model of describing space plasma, MHD system is a non-strictly hyperbolic system composed of eight equations coupled by fluid dynamics and magnetic induction. Differing from the usual fluid dynamic equations with the unique sound wave mode, MHD has multiple modes of sound wave, Alfven wave, fast magnetosonic and slow magnetosonic wave, which make the eigenvalues and eigenvectors of MHD system so complex that many modern high resolution numerical schemes used in simulating supersonic flow problems of computational fluid mechanics can not be transplanted directly to the numerical study of supersonic, super-Alfvenic solar wind flow. Lax and Wendroff (1960) has pointed out that the shock-capturing capability of a numerical scheme in conservative form is superior to that of the numerical scheme in non-conservative form with the same accuracy. But, in MHD simulation we are facing another problem about how to keep $\nabla \cdot \mathbf{B} = 0$ numerically in order to avoid the non-physical flow caused by the numerical non-zeroness of $\nabla \cdot \mathbf{B}$. Former experiences have told us that when numerical schemes in conservative form are considered, we have to deal with $\nabla \cdot \mathbf{B}$ numerically (Brackbill and Barnes, 1980; Tanaka, 1994; Powell *et al.*, 1995, 1999; Toth, 2000), which is of course time-consuming. However, the numerical scheme in non-conservative form for MHD system can lead to a tolerable error in keeping $\nabla \cdot \mathbf{B} = 0$ numerically and thus reduce the numerical error. Taking the above consideration, the TVD Lax-Friedrichs scheme is used for the fluid equations of the MHD system and the well-known MacCormack II is employed for the magnetic induction equations. We are here interested in the TVD Lax-Friedrichs scheme (monotonic upwind schemes) proposed in Cartesian coordinates for hyperbolic conservation law system by Lax and Wendroff (1960) and Yee (1989) since it does not involve the use of a Riemann solver, and thus can be implemented for any system of conservation laws without any knowledge of the characteristic waves. Instead of using the Lax–Freidrichs form presented by Yee (1989), we employ the modified form given by Toth and Odstrcil (1996). The modified Lax–Freidrichs scheme consists of two steps of predictor and corrector steps and is of second order accuracy both in time and space. Compared to those of routine TVD and Godunov type, the present scheme need not use the calculation of eigenvectors associated to the Jacobian matrices in every time step and thus reduce the amount of work.

In order to see the shock-capturing ability, correctness and performance, of the proposed combined numerical scheme in 1-D, we shall compare two MHD shock simulations with the PPM results of Dai and Woodward (1994). Let us consider a one-dimensional case with all variables depending only on the time and the space coordinate x. Then, after proper normalization of variables such that the factor of $1/4\pi$ does not appear, the MHD system can be written as follows:

$$\rho_t + (\rho u)_x = 0$$

$$(\rho u)_t + (\rho u^2 + P_0 - B_x^2)_x = 0$$

$$(\rho v)_t + (\rho u v - B_x B_y)_x = 0$$

$$(\rho w)_t + (\rho u w - B_x B_z)_x = 0$$

$$(B_y)_t + (B_y u - B_x v)_x = 0$$

$$(B_z)_t + (B_z u - B_x w)_x = 0$$

$$e_t + \left((e + P_0)u - B_x(B_x u + B_y v + B_z w)\right)_x = 0$$

where $P_0 = p + \frac{|\mathbf{B}|^2}{8\pi}$ is the total pressure, $p = \Re \rho T$ the thermal pressure. Obviously, B_x is constant due to $\nabla \cdot \mathbf{B} = \nabla_x(B_x) = 0$. The initial conditions for Problem I are $(\rho, p, u, v, w, B_y/\sqrt{4\pi}, B_z/\sqrt{4\pi}) = (3.896, 305.9, 0, -0.0058, -0.0028, 3.951, 15.8), x < 0.2$, and $(1, 1, -15.3, 0, 0, 1, 4), x > 0.2, B_x = 5$. This stands for a MHD fast shock with a Mach number of 10. The profiles of variables at $t = 0.05$ and $t = 1$ are given in Figure 1. The initial conditions for Problem II are $(\rho, p, u, v, w, B_y/\sqrt{4\pi}, B_z/\sqrt{4\pi}) = (3.108, 1.4336, 0, 0.2633, 0.2633, 0.1, 0.1), x < 0.2$ and $(\rho, p, u, v, w, B_y/\sqrt{4\pi}, B_z/\sqrt{4\pi}) = (1, 0.1, -0.9225, 0, 0, 1, 1), x > 0.2, B_x = 5$ and the Mach number is 3.43. This represents a slow MHD shock. The profiles of variables at $t = 1.6$ are given in Figure 2.

In our calculation, the Courant number is 0.5. However, we can enlarge the Courant number to be 0.8. It should be mentioned that no artificial viscosity is introduced here. The specific heat ratio $\gamma = 5/3$, $B_x = 5$ in the above two

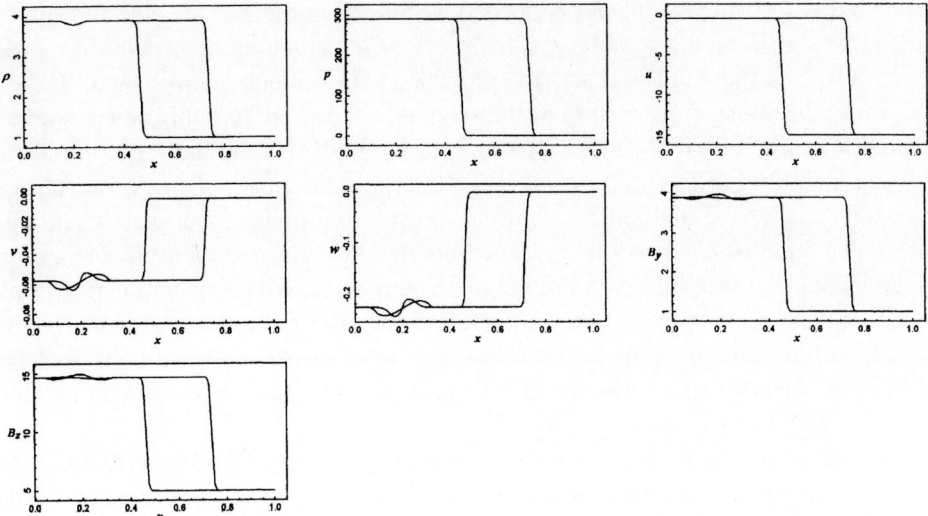

Figure 1. The propagation of a one-dimensional MHD fast shock: The profiles of density, pressure, velocity and magnetic field at $t = 0.05$ and $t = 0.1$ are shown here.

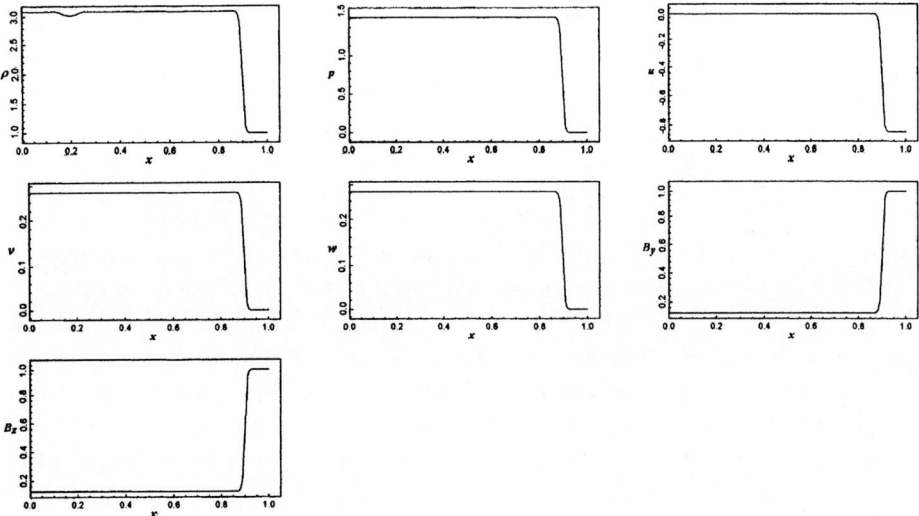

Figure 2. The propagation of a one-dimensional MHD slow shock: The profiles of density, pressure, velocity and magnetic field at $t = 0.16$ are shown here.

examples. The grids are uniform with 200 computational cells between zero and unity for the results. Meanwhile, the resulting numerical values of the magnetic field components are multiplied by the factor $\sqrt{4\pi}$ for sake of their comparison with those of Dai and Woodward (1994).

Our results for the fast and slow MHD shocks in Figures 1 and 2 are in good agreement with counterpart Figures 4 and 5 of Dai and Woodward (1994), which shows that to some extent our proposed combined numerical scheme has almost the same shock-capturing ability as the piece-wise parabolic method of third order accuracy. But, the numerical scheme proposed here is much simpler than PPM.

From the above construction of the scheme, we can see that this new combined numerical scheme is of second order accuracy both in time and space, and has strong stability. Different from other TVD type numerical schemes, our chosen scheme has been simplified by avoiding the splitting of Jacobian matrix and thus the calculation of eigenvectors. Meanwhile, the different treatment of scheme implementation of the fluid part and magnetic part of the MHD system can not only maintain the necessary resolution but also keep the numerical error caused by $\nabla \cdot \mathbf{B} \neq 0$ in tolerable range. The numerical experiments of one-dimensional fast and slow MHD shock waves can give us reliable numerical results as compared to those of Dai and Woodward (1994).

By the way, the application of this combined numerical scheme in spherical coordinates to the numerical study of solar wind ambient with Parker solar wind solution and dipolar magnetic field as initiation can give us satisfactory numerical results under very small plasma β, which will be presented elsewhere. The further application of this scheme to model 3-D solar wind ambient and disturbance propagation in corona and interplanetary space is left for future work.

4. Conclusion

In summary, although both upwind and symmetric TVD schemes in the framework of the shock-capturing approach are thoroughly investigated and applied with great success to a number of complicated multidimensional gasdynamic problems, the extension of these schemes to magnetohydrodynamic (MHD) equations is not a simple task. First, the exact solution of the MHD Riemann problem is too multivariant to be used in regular calculations. On the other hand, the extensions of Roe's approximate Riemann problem solvers for MHD equations in the general case are non-unique (Kulikovskiy and Lyubimov, 1965) and need further investigation. Taking account of the above mentioned remarks we need to construct some simplified approaches, which should (I) satisfy the TVD property by using the conservative form of MHD system in order to exactly catch the discontinuity, (II) keep the $\nabla \cdot \mathbf{B} = 0$ constraint numerically in order to reduce the non-physical numerical flow as much as possible, and (III) be economical and robust (such as to avoid the splitting of associated Jacobian matrices and the calculation of corresponding eigenvalues and eigenvectors), (VI) be able to reproduce the typical characteristics of the solar wind if used in a solar wind simulation.

From the point of view of space weather numerical prediction, an operable numerical MHD model of CMEs should bear the following: 1) reasonable triggering mechanism (How to drive or fuel a CME), 2) input (Initial-Boundary Conditions) based on the global structures of magnetic field and velocity map derived from solar observations, 3) be able to reproduce observation and give the associated interpretation, furthermore predict CMEs solar-terrestrial phenomena, 4) robust global MHD code with high resolution.

It should be noted that our combined numerical scheme of TVD type for the MHD system is developed with the attention of satisfying the above properties. The one-dimensional numerical results of MHD shock waves are just preliminary. As for its potential capability of simulating solar wind and space weather events, we need further verifications.

Acknowledgements

This work was jointly supported by the National Natural Science Foundation of China (NO. 49925412 and NO. 49990450) and the National Basic Research Science Foundation (NO. G2000078405). Dr. Wu is supported by NSF grant (ATM0070385) and AFOSR grant (F46920-00-0-0204). Dr. Feng would like to express his thanks for the hospitality of Center for Space Physics and Aeronomic Research, The University of Alabama in Huntsville, where part of the present work is done.

References

Brackbill, J. U. and Barnes, D. C.: 1980, *J. Comput. Phys.* **35**, 426–430.

Brecht, S. H. *et al.*: 1982, *J. Geophys. Res.* **87**, 6098–6108.

Burlaga, L. F., Plunkett, S. P., St. Cyr, O. C.: 2002, *J. Geophys. Res.* **107**(A10), 1266, doi:10.1029/2001JA000255.

Chao, J. K. and Chen, H. H.: 2001, 'Prediction of Southward IMF B_z', in P. Song, H. J. Singer and G. L. Siscoe (eds), *Space Weather*, Geophysical Monograph 125, AGU Washington, pp. 143–158.

Dai, W. and Woodward, P. R.: 1994, *J. Comput. Phys.* **115**, 485–514.

Dryer, M.: 1998, *AIAA Journal* **36**, 365–390.

De Zeeuw, D. L. *et al.*: 2000, *IEEE Trans. Plasma Sci.* **28**, 1956–1965.

Evans, C. R. and Hawley, J. F.: 1988, *Astrophys. J.* **332**, 659–677.

Fedder, J. A., Linker, S. P. and Lyon, J. G.: 1998, *J. Geophys. Res.* **103A**, 14799–14810.

Godunov S. K.: 1972, 'Symmetric Form of magnetohydrodynamics (in Russian)', in Numerical Methods for Mechanics of Continium Medium, Vol. 1, Siberian Branch of USSR Acad. of Sci., pp. 26–34.

Gombosi, T. I. *et al.*: 2000a, *Adv. Space Res.* **26**(1), 139–149.

Gombosi, T. I. *et al.*: 2000b, *J. Atmos. Solar-Terr. Phys.* **62**(16), 1515–1525.

Gombosi, T. I. *et al.*: 2001, 'From Sun to Earth: Multiscale MHD simulation of space weather', in P. Song, H. J. Singer, and G. L. Siscoe (eds), *Space Weather*, Geophysical Monograph 125, AGU Washington, pp. 169–176.

Groth, C. P. T. *et al.*: 1999, *Space Sci. Rev.* **87**, 193–198.

Groth, C. P. T. *et al.*: 2000a, *Adv. Space Res.* **26**(5), 793–800.

Groth, C. P. T. *et al.*: 2000b, *J. Geophys. Res.* **105**, 25053–25078.

Guo, W. P. and Wu, S. T.: 1998, *Astrophys. J.* **494**, 419–429.

Han, S. M., Wu, S. T. and Dryer, M.: 1988, *Comp. Fluids* **16**, 81–103.

Harned, D. S. and Kerner, W.: 1985, *J. Comput. Phys.* **60**, 62–75.

Hu, Y. Q. and Wu, S. T.: 1984, *J. Comput. Phys.* **55**, 33–64.

Janhunen, P.: 1996, 'GUMICS-3: A Global Ionosphere-Magnetosphere Coupling Simulation with High Ionospheric Resolution', in Proc. ESA 1996 Symposium on Environment Modelling for Space-based Applications, ESA SP-392, pp. 233–239.

Klimchuk, J. A.: 2001, 'Theory of Coronal mass ejections', in P. Song, H. J. Singer and G. L. Siscoe (eds), *Space Weather*, Geophysical Monograph 125, AGU Washington, pp. 143–158.

Kulikovskiy, A. and Lyubimov, G.: 1965, *Magnetohydrodynamics*. Addison-Wesley, Reading, MA.

Lax, P. D. and Wendroff, B.: 1960, *Comm. Pure Appl. Math.* **13**, 217–237.

Leboeuf, J. N., Tajima, T., Kennel, C. F. and Dawson, J. M.: 1981, *Geophys. Res. Lett.* **8**, 257–260.

Linker, J. A. *et al.*: 1999, *J. Geophys. Res.* **104**, 9809–9830.

Lionello, R. Z., Mikic, Z. and Linker, J. A.: 1998, *J. Comput. Phys.* **140**, 172–201.

Lyon, J. G., Fedder, J. A. and Huba, J. G.: 1986, *J. Geophys. Res.* **91**, 8057–8064.

Mikic, Z. *et al.*: 1999, *Phys. Plasmas* **6**, 2217–2224.

Nakagawa, Y., Hu, Y. Q. and Wu, S. T.: 1987, *Astron. Astrophys.* **179**, 354–370.

Odstrcil, D. and Pizzo, V. J.: 1999a, *J. Geophys. Res.* **104**, 483–492.

Odstrcil, D. and Pizzon, V. J.: 1999b, *J. Geophys. Res.* **104**, 493–503.

Odstrcil, D. and Pizzo, V. J.: 1999c, *J. Geophys. Res.* **104**, 28225–28239.

Ogino, T.: 1986, *J. Geophys. Res.* **91**, 6791–6806.

Panitchob, S., Wu, S. T. and Suess, S. T.: 1987, *AIAA Paper*, 87-1218-CP.

Powell, K. G. *et al.*: 1995, *AIAA Paper*, 95-1704-CP.

Powell, K. G. *et al.*: 1999, *J. Comput. Phys.* **154**, 284–309.

Raeder, J., Walker, J. and Ashour-Abdalla, M.: 1998, *J. Geophys. Res.* **103**, 14787–14797.

Ridley, A. J. *et al.*: 2002, *J. Geophys. Res.* **107**(A10), 1290, doi: 10.1029/2001JA000253.

Riley, P. *et al.*: 1997, *Adv. Space Res.* **20**, 15–22.

Riley, P., Gosling, J. T., McComas, D. J. *et al.*: 1999, *J. Geophys. Res.* **104**, 9871–9879.

Riley, P. *et al.*: 2001, 'MHD Modeling of the Solar Corona and Inner Heliosphere: Comparison with Observations', in P. Song, H. J. Singer, and G. L. Siscoe (eds), *Space Weather*, Geophysical Monograph 125, AGU Washington, pp. 159–168.

Riley, P., Linker, J. A. and Mikic, Z.: 2002, *J. Geophys. Res.* **107**(A7), doi: 10.1029/2001JA000299.

Shi, Y., Wei, F. S. and Feng, X. S.: 2001, *Science in China (A)* **30**, 61–64.

Song, P. *et al.*: 1999, *J. Geophys. Res.* **104**, 28,361–28,378.

Song, P. *et al.*: 2000, *Planetary Space Sci.* **48**, 29–39.

St Cyr, O. C. *et al.*: 2000, *J. Geophys. Res.* **105**, 12493–12506.

Steinolfson, R. S.: 1991, *Astrophys. J.* **382**, 677–687.

Steinolfson, R. S.: 1994, *Space Sci. Rev.* **70**, 289–294.

Tanaka, T.: 1994, *J. Comput. Phys.* **111**, 381–389.

Tanaka, T.: 1995, *J. Geophys. Res.* **100**, 12057–12074.

Toth, G. and Odstrcil, D.: 1996, *J. Comput. Phys.* **128**, 82–100.

Toth, G.: 2000, *J. Comput. Phys.* **161**, 605–652.

Vandas, M., Odstrcil, D. and Watari, S.: 2002, *J. Geophys. Res.* **107**(A9), 1236, doi: 10.1029/2001JA005068.

Wang, A. H., Wu, S. T., Suess, S. T. and Poletto, G.: 1998, *J. Geophys. Res.* **103**, 1913–1922.

Washimi, H.: 1990, *Geophys. Res. Lett.* **17**, 33–40.

Watanabe, K. and Sato, T.: 1990, *J. Geophys. Res.* **95**, 13437–13454.

Webb, D. R. *et al.*: 2001, 'The Solar Sources of Geoeffective Structures', in P. Song, H. J. Singer and G. L. Siscoe (eds), *Space Weather*, Geophysical Monograph 125, AGU Washington, pp. 123–143.

White, W. W. *et al.*: 1998, *Geophys. Res. Lett.* **25**, 1605–1608.

Winglee, R. M.: 1994, *J. Geophys. Res.* **99**, 13437–13454.

Wu, C. C., Walker, R. J. and Dawson, J. M.: 1981, *Geophys. Res. Lett.* **8**, 523–526.
Wu, C. C., Dryer, M. *et al.*: 1996a, *J. Atmos. Solar-Terr. Phys.* **58**, 1805–1812.
Wu, C. C. and Dryer, M.: 1996b, *Geophys. Res. Lett.* **23**, 1709–1712.
Wu, C. C. and Dryer, M.: 1997, *Solar Phys.* **173**, 391–408.
Wu, S. T., Andrews, M. D. and Plunkett, S. P.: 2001, *Space Sci. Rev.* **95**, 191–213.
Wu, S. T., Hu, Y. Q., Nakagawa, Y. and Tandberg-Hanssen, E.: 1986, *Astrophys. J.* **306**, 751–761.
Wu, S. T. and Wang, A. H.: 1991, *Adv. Space Res.* **11**, 187–204.
Wu, S. T., Guo, W. P. and Wang, J. F.: 1995, *Solar Phys.* **157**, 325–348.
Wu, S. T., Guo, W. P. *et al.*: 1997, *Solar Phys.* **175**, 719–735.
Wu, S. T., Guo, W. P. and Dryer, M.: 1997, *Solar Phys.* **170**, 265–282.
Wu, S. T., Guo, W. P. *et al.*: 1999, *J. Geophys. Res.* **104**, 14789–14802.
Wu, S. T. *et al.*: 2000a, *Astrophys. J.* **545**, 1101–1115.
Wu, S. T. *et al.*: 2000b, *J. Atmos. Solar-Terr. Phys.* **62**(16), 1489–1498.
Yee, H. C.: 1989, *A Class of High Resolution Explicit and Implicit Shock-Capturing Methods.* NASA TM-101088.
Zhang, B. C. and Wang, J. F.: 2000, *J. Geophys. Res.* **105**, 12593–12603.
Zhang, J. H. and Wei, F. S.: 1993, *Science in China(A)* **23**(4), 427–436.

SOLAR ENERGETIC PARTICLE EVENTS: PHENOMENOLOGY AND PREDICTION

S. B. GABRIEL and G. J. PATRICK

Astronautics Research Group, University of Southampton, U.K.
(e-mails: sbg2@soton.ac.uk; G.J.Patrick@soton.ac.uk)

Abstract. Solar energetic particle events can cause major disruptions to the operation of spacecraft in earth orbit and outside the earth's magnetosphere and have to be considered for EVA and other manned activities. They may also have an effect on radiation doses received by the crew flying in high altitude aircraft over the polar regions. The occurrence of these events has been assumed to be random, but there would appear to be some solar cycle dependency with a higher annual fluence occuring during a 7 year period, 2 years before and 4 years after the year of solar maximum. Little has been done to try to predict these events in real-time with nearly all of the work concentrating on statistical modelling. Currently our understanding of the causes of these events is not good. But what are the prospects for prediction? Can artificial intelligence techniques be used to predict them in the absence of a more complete understanding of the physics involved? The paper examines the phenomenology of the events, briefly reviews the results of neural network prediction techniques and discusses the conjecture that the underlying physical processes might be related to self-organised criticality and turblent MHD flows.

Key words: intermittancy, neural networks, prediction, protons, SOC, solar energetic particles

Abbreviations: CME – Coronal Mass Ejection, pfu – Proton Flux Unit (protons $cm^{-2} sec^{-1} str^{-1}$) SEC – Space Environment Centre, SEPE – Solar Energetic Particle Event, SPE – Solar Proton Event, SOC – Self Organised Criticality

1. Introduction

The effects of solar energetic particle events (SEPEs) on the design and operation of spacecraft are well known and documented (Feynman and Gabriel, 1996). However, on the contrary, our understanding of what causes them is still quite poor. It is more or less generally accepted, that coronal mass ejections(CMEs) play a key role in the acceleration of the particles during SEPEs. So, if we could predict the onset of CMEs then in principle we could predict the occurrence of SEPEs. In practice, the situation is not that simple, because not all CMEs produce large SEPEs at the earth. The reader is referred to other papers for more comprehensive and detailed discussions of the connection between CMEs and SEPEs, for example Kahler *et al.* (1984). Given that our current understanding of the physical mechanisms that cause SEPEs makes their (deterministic) prediction difficult, if not impossible, is there anything else we can do?

Space Science Reviews **107**: 55–62, 2003.
© 2003 *Kluwer Academic Publishers.*

Recently, at the University of Southampton, we have been using artificial intelligence techniques to predict SEPEs. So far we have focused on long lead times of the order of 48 hours and neural networks, with the ratio of the GOES XL and XS x-ray fluxes as the inputs. The overall success rate has been about 65%. After trying many different input combinations, we have concluded that this is probably about the best that we can do with this technique using X-ray fluxes as inputs to the networks. Consequently we have started to look at other potential methods. Firstly, we are looking at alternatives to the X-ray fluxes as inputs, and secondly at some of the more modern signal processing techniques, such as wavelets. Other inputs that have been used include solar radio data from various ground observations and the results of these investigations have been presented elsewhere (Patrick *et al.*, 2002). The paper briefly describes the phenomonology of SEPEs, the data sets that have been assembled, and presents some of the results from the neural network prediction tool. In addition, we have taken a step towards the physics and have begun to look at what type of processes might be governing these events by looking at the distributions of their size (energy), waiting times and durations.

2. A Brief Description of SPE Phenomenology

There are many sources of solar protons ranging in energies from around 1 keV (solar wind) to greater than 500 MeV. Only those particles with energies from about 1 MeV and above, produced in what are called solar energetic particle events (SEPEs) or solar proton events (SPEs) are discussed herein. These events are sporadic in nature, occurring at any time throughout the solar cycle and exhibit a wide range in duration (2–20 days) and fluxes (5–6 orders of magnitude) (Gabriel and Feynman, 1996). Fluences (time integrated flux over event duration) can vary from just above the cosmic ray background to almost 10^{10} p/cm^2 at energies E > 10 MeV, as in the case of the October 1989 event. Event fluences > 1.5×10^9 p/cm^2 (E > 10 MeV) are very rare with there having been only about 14 since 1963. There were no events with E > 10^{10} p/cm^2 at E > 10 MeV between the famous August 1972 and October 1989 events; both had associated geomagnetic storms which caused widespread power outages in Canada and the USA.

The mechanism responsible for proton acceleration and the cause of SPEs are subjects which are widely discussed in the literature, with much controversy in particular over the role of flares (Gosling, 1993; Pudovkin, 1995). However, it appears that the generally accepted view is that particle acceleration, in the largest SPEs, is caused by CME driven shocks in the corona and interplanetary space. There is growing evidence that there are two types of SPEs, associated with impulsive and gradual x-ray flares (Reames *et al.*, 1997). In the latter case, particles have elemental abundances and isotopic compositions characteristic of the corona and apparently arise from regions that have an electron temperature of 1–2 MK, whereas SPEs with impulsive x-ray flares show a marked enhancement of heavy

ions and He3/He4 ratios 2–4 orders of magnitude larger than in the solar atmosphere or solar wind; they are dominated by electrons and the composition suggests
that the ions come from deep within the corona ($T_e \sim 3$–5 MK). It would appear
that the particles associated with these events are directly accelerated in solar flares.
For a description of SPE propagation characteristics, and fluence and flux spectra,
the reader is referred to Gabriel (1998).

There is no clear solar cycle dependency of solar proton event fluences, certainly
in terms of event occurrences, with events taking place at any time throughout a
cycle. Nevertheless, Feynman *et al.* (1990a) found that, by defining the time of
sunspot maximum to the nearest 0.1 years, the annual integrated fluence could be
divided into two periods, a high fluence, active sun period of 7 years and a low
fluence, quiet sun period of 4 years. The active period extends from 2 years before
the year of solar maximum to 4 years after. For sunspot numbers greater than 50
there is no correlation of SPE fluences with sunspot number; the sunspot number at
the time of the great event of November 1960 was < 60 and < 90 when the August
1972 event occurred (Feynman *et al.*, 1990b).

On shorter time scales of the order of the duration of the event, the particle
fluxes rise over a period of 0.5–1 day with a slower decay of about a few days.

3. The Data

A list of SEPEs was taken from the JPL-91 model and extended to 1999 using
GOES data. Only SPEs occurring in solar active years as defined by Feynman *et al.*
(1990a) were considered for the study, based on solar maximum years of 1968.9,
1979.9, 1989.9 and 2001.2 for cycles 20, 21, 22 and 23 respectively.

For each SEPE GOES daily x-ray fluxes were extracted for a time span centered
on the event and extending to 81 days before and afterwards. Higher resolution
hourly data were extracted for a smaller period spanning from 7 days before to
2 days after each event, providing a dataset of flux-time vectors corresponding
to occurrences of SEPEs. Non-event or quiet periods were defined over the same
active years as times at which the > 10 MeV proton flux was at background levels
(approximately 0.1 pfu) for at least 10 consecutive days and similar x-ray data
extractions were performed. The dataset has recently been extended to include
daily values of solar radio fluxes, sunspot number and plage indices.

4. Prediction

Currently there are two models for the real-time prediction of SPEs (Balch, 1999;
Garcia, 1994). Both models assume a relation between SPEs and discrete x-ray
flares, and require a flare before a prediction can be made, inherently limiting lead
times to several hours. Is it possible to predict a SPE further in advance without the
need for a discrete x-ray flare?

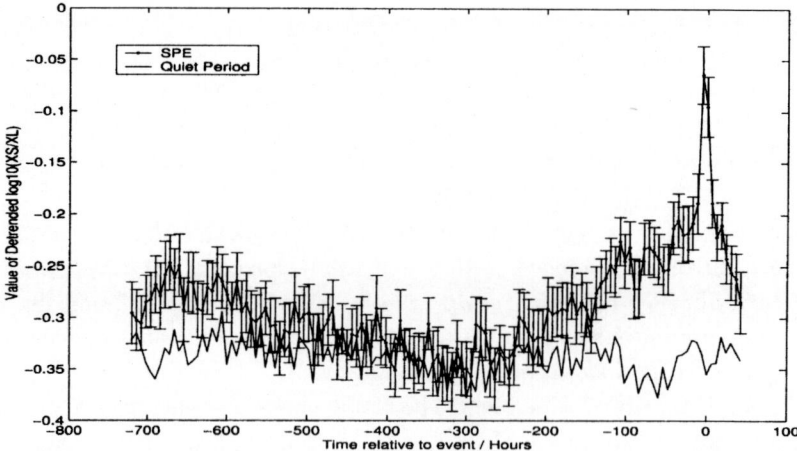

Figure 1. The average value of XS/XL ratio for periods prior to 98 SPEs and 195 randomly selected quiet periods. Error bars denote standard error.

4.1. NEURAL NETWORKS

A neural network model has been created from the dataset described above: it predicts the occurrence of an SPE 48 hours in advance by classifying x-ray inputs as relating to the case of an SEPE or a quiet period. The model takes a detrended ratio of the GOES XS/XL x-ray fluxes as inputs, consisting of 3-hour averages over a 120-hour period, and then projects the input vector onto principal components using coefficients derived from the training dataset. The transformed inputs are then scaled and sent to 10 neural networks. Each network has been trained with subsets of the principal training set using values of 0 and 1 to represent the two possible outcomes of an SEPE and a quiet period. The outputs from the 10 networks give several responses to a single input vector and are used to give a majority decision as to the predicted outcome. Several neural networks operating in this configuration were found to generate superior performance to a single network trained on the entire training set.

The best performance was generated by Radial Basis Function (RBF) networks (Broomhead *et al.*, 1988) which correctly classified 27/37 SPEs and 64/113 quiet periods from previously unseen test data: an overall success rate of 65% with a 48-hour lead time. Multi Layer Perceptron (MLP) and linear models also gave comparable results. An examination of the data shows that on average the XS/XL ratio is higher prior to SPEs than prior to quiet periods, and it is this phenomenon that appears to make SPE prediction possible on larger timescales (Figure 1). This pattern of elevated XS/XL may occur far more frequently than just prior to SPEs though. Recent examination of the XS/XL ratio prior to significant x-ray flares (class M and X) shows, on average, a similar rise over a period of hundreds of hours prior to flare occurrence.

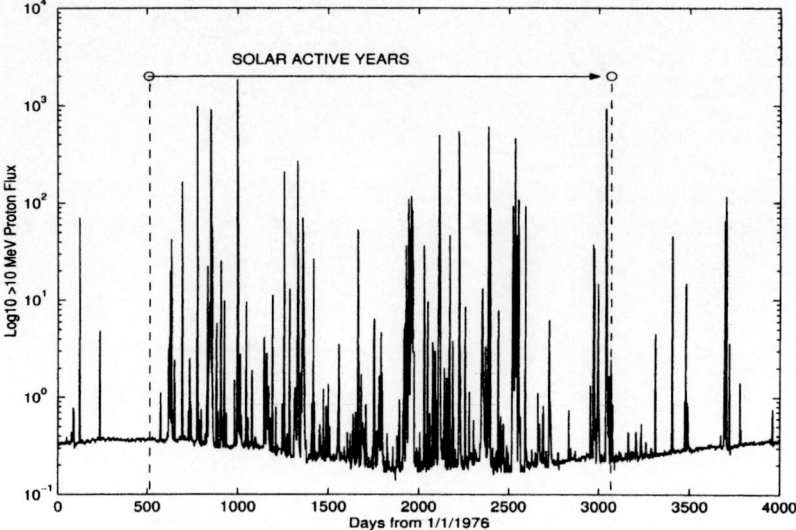

Figure 2. Behaviour of > 10 MeV Flux over cycle 21 (1/1/76 to 31/12/86).

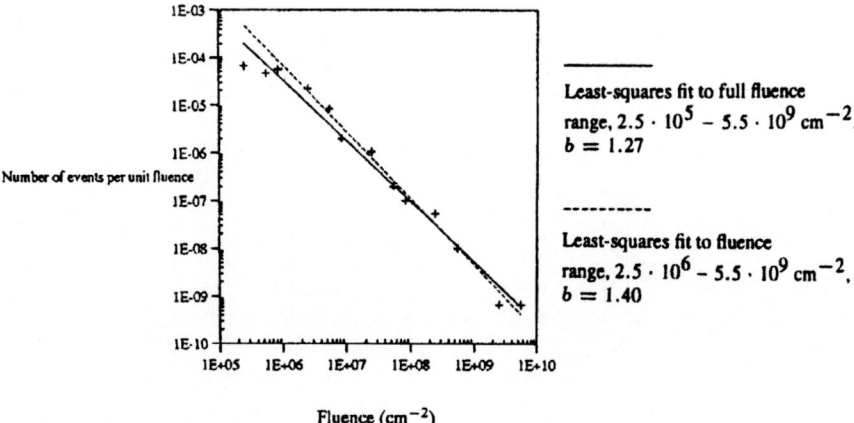

Least-squares fit to full fluence range, $2.5 \cdot 10^5 - 5.5 \cdot 10^9 \, cm^{-2}$, $b = 1.27$

Least-squares fit to fluence range, $2.5 \cdot 10^6 - 5.5 \cdot 10^9 \, cm^{-2}$, $b = 1.40$

Figure 3. Fluence distributions for E > 30 MeV for period 1965 to 1990.

A bank of neural networks has been operating in real time since January 2002 using x-ray data from the GOES-8 satellite to make predictions once per hour. During this period, 12 enhancements of > 1 p.f.u. in the > 10 MeV proton flux have occurred, 7 of which were predicted by the model to within 3 hours, 48 hours in advance. This is a hit-rate of 58% which is comparable to the performance on test data. However, over the trial period 4436 predictions were made, 964 of which were false alarms: a ratio of 21%. This relatively high instance of false alarms could be due to pre-flare variations in the x-ray ratio being interpreted as SPE precursors. Only 1% of all x-ray flares are associated with SPEs.

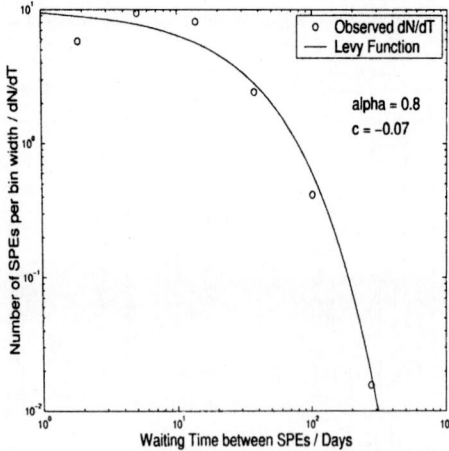

Figure 4. dN/dT as a function of waiting time for SPEs in cycles 20–23.

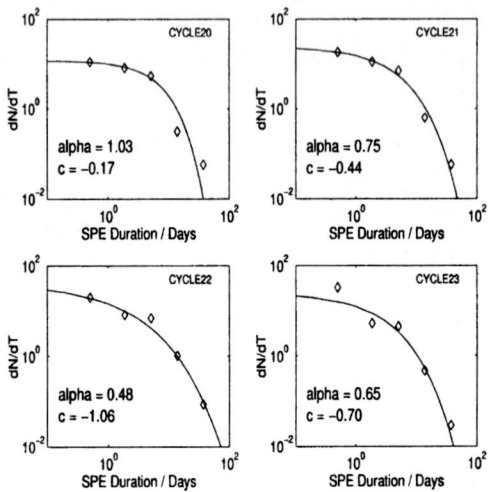

Figure 5. dN/dT as a function of duration for SPEs in cycles 20–23 shown individually.

5. SOC and Intermittency

Figure 2 shows the solar cycle flux for cycle 21. Intermittent or bursty behaviour of the flux is very prominent. Even if we cannot predict these events, can we learn something about the processes involved in their production from examining certain characteristics of their distributions? Firstly, what might the intermittent behaviour in Figure 2 tell us? Carbone *et al.* and others have looked at this type of behaviour in other phenomena, such as X-ray flares and laboratory plasmas, and concluded that it could be indicative of self-organised criticality(SOC) or perhaps turbulence in the MHD equations describing the plasmas. To further examine the applicability to SEPEs, we derived the distributions for 3 of the characteristics of these events, size (or fluence), waiting times and durations. At high fluences the distribution (Figure 3) is well described by a power law (Gabriel and Feynman, 1996), but at lower values flattens out so that the least squares fit is inaccurate. Figures 4 and 5 show that the waiting times and durations are well described by a Levy function, given by:

$$y = A \exp(cx^{\alpha}) \tag{1}$$

where $0 < \alpha < 2$ is an index of the peakedness of the distribution, 'c' is a measure of the skewness and 'A' is a scaling factor. Based on these plots, it seems that the underlying physical processes involved in the production of SEPEs are close to SOC(fractal) or are indicative of turbulent plasma behaviour (cf. magnetic reconnection, multi-fractal), (Carbone, 2002).

6. Conclusions

The 65% success rate achieved using Neural Networks is considered to be very good considering the 48-hour lead time but the technique needs to be improved for operational applications. Considering there is no reported evidence of SPE precursors before the associated x-ray flare it is surprising that the method can predict at all; it is likely that a general association between SPEs and active regions is being detected. The distributions of waiting times and durations (and probably fluences) seem well described by Levy functions. This suggests, along with the intermittency in the fluxes, that SOC or turbulent MHD plasma flows (reconnection?) is involved in the physical processes that give rise to SEPEs. Future work will include conventional time series analysis, spatially resolved images (e.g. SOHO and neural nets) and the use of wavelets. If the process is in a state of SOC, one could conclude that the events are not predictable, at least based on time series analysis (rather than using pre-cursors). However, more work is needed in this area to be able to make this very significant and fundamental conclusion.

References

Balch, C. C.: 1999, 'SEC Proton Prediction Model: Verification and Analysis', *Radiation Measurements* **30**, 231–250.

Broomhead, D. S. and Lowe, D.: 'Radial Basis Functions, Multi-Variable Functional Interpolation and Adaptive Networks', *Royal Signals and Radar Establishment, Malvern*, Memorandum 4148.

Carbone, V.: 'Bursty Behaviour in Complex Systems: Examples From Plasma Physics', *http://www.phys.uit.no/ ashild/Plasmasymp02web/Talks/carbone-norvegia1.pdf*

Feynman, J., Armstrong, T. P., Dao-Gibner L. and Silverman, S. M.: 1990, 'Solar Proton Events During Solar Cycles 19, 20, and 21', *Solar Phys.* **126**, 385.

Feynman, J., Armstrong, T. P., Dao-Gibner L. and Silverman, S. M.: 1990, 'A New Interplanetary Proton Fluence Model', *J. Spacecraft Rockets* **27**, 403.

Feynman, J. and Gabriel, S. B.: 1996, 'High Energy Charged Particles in Space at One Astronomical Unit', *Trans. Nucl. Sci.* **43**, 344.

Gabriel, S. B. and Feynman, J.: 1996, 'Power-Law Distribution for Solar Energetic Proton Events', *Solar Phys.* **165**, 337–346.

Gabriel, S. B.: 1998, *Cosmic Rays and Solar Protons in the Near-Earth Environment and Their Entry into the Magnetosphere*, in Proceedings of ESA Workshop on Space Weather, (Invited Paper), ESA WPP-155, pp. 99–106, November 1998.

Garcia, H. A.: 1994, 'Temperature and Hard X-Ray Signatures for Energetic Proton Events', *Astrophys. J.* **420**, 422–432.

Gosling, J. T.: 1993, 'The Solar Flare Myth', *J. Geophys. Res* **98**, 18937.

Kahler, S. W. *et al.*: 1984, 'Associations Between Coronal Mass Ejections and Solar Energetic Proton Events', *J. Geophys. Res. – Space Phys.* **89**, 9683–9693.

Patrick, G. P., Gabriel, S. B., Rogers, D. and Clucas, S.: 2002, *Neural Network Prediction of Solar Proton Events with Long Lead Times*, in Proc. SOLSPA: The Second Solar Cycle and Space Weather Euroconference, Vico Equense, Italy, 24th–29th September 2001, ESA SP-477, February 2002.

Pudovkin, M. I.: 1995, 'Comment on "The Solar Flare Myth" by J.T. Gosling', *J. Geophys. Res* **100**, 7917.

Reames, D. V. *et al.*: 1997, 'Energy Spectra of Ions Accelerated in Impulsive and Gradual Solar Events', *Astrophys. J.* **483**, 515–522.

COMPUTER SIMULATIONS OF SOLAR PLASMAS

J. P. GOEDBLOED[1,3], R. KEPPENS[1] and S. POEDTS[2]

[1]FOM-Institute for Plasma Physics, P.O. Box 1207, 3430 BE Nieuwegein, the Netherlands
[2]Center for Plasma-Astrophysics, K.U.Leuven, 3001 Heverlee, Belgium
[3]Astronomical Institute, Utrecht University, Utrecht, the Netherlands (goedbloed@rijnh.nl)

Abstract. Plasma dynamics has been investigated intensively for toroidal magnetic confinement in tokamaks with the aim to develop a controlled thermonuclear energy source. On the other hand, it is known that more than 90% of visible matter in the universe consists of plasma, so that the discipline of plasma-astrophysics has an enormous scope. Magnetohydrodynamics (MHD) provides a common theoretical description of these two research areas where the hugely different scales do not play a role. It describes the interaction of electrically conducting fluids with magnetic fields that are, in turn, produced by the dynamics of the plasma itself. Since this theory is scale invariant with respect to lengths, times, and magnetic field strengths, for the nonlinear dynamics it makes no difference whether tokamaks, solar coronal magnetic loops, magnetospheres of neutron stars, or galactic plasmas are described. Important is the magnetic geometry determined by the magnetic field lines lying on magnetic surfaces where also the flows are concentrated.

Yet, transfer of methods and results obtained in tokamak research to solar coronal plasma dynamics immediately runs into severe problems with trans'sonic' (surpassing any one of the three critical MHD speeds) stationary flows. For those flows, the standard paradigm for the analysis of waves and instabilities, viz. a split of the dynamics in equilibrium and perturbations, appears to break down. This problem is resolved by a detailed analysis of the singularities and discontinuities that appear in the trans'sonic' transitions, resulting in a unique characterization of the permissible flow regimes. It then becomes possible to initiate *MHD spectroscopy of axi-symmetric transonic astrophysical plasmas*, like accretion disks or solar magnetic loops, by computing the complete wave and instability spectra by means of the same methods (with unprecedented accuracy) exploited for tokamak plasmas. These large-scale linear programs are executed in tandem with the non-linear (shock-capturing, massively parallel) Versatile Advection Code to describe both the linear and the nonlinear phases of the instabilities.

Key words: MHD waves, transonic plasmas

1. Introduction and Outline

Magnetized plasmas are essentially extended structures because magnetic field lines do not have a beginning or end ($\nabla \cdot \mathbf{B} = 0$). This implies that regions of space are connected that have very different physical properties. For example, solar magnetism arises due to nuclear fusion powering in the extremely dense core of the Sun, radiation transport establishing a convectively stable temperature profile up to the convection zone ($R \sim 0.7\, R_\odot$) where dynamo action by convective instability and differential rotation produces concentrated magnetic field bundles that

Space Science Reviews **107**: 63–80, 2003.

are expelled from the sun proper, giving rise to tremendously complex magnetic field structuring and dynamics in the photosphere and corona (see any SOHO or TRACE web site). Along the magnetic field lines that escape, the solar wind carries a tenuous plasma that is accelerated to transonic speeds and exhibits discontinuous flow (shocks) when the magnetospheres of the planets are encountered and when the heliosphere is finally terminated beyond the solar system.

Obviously, such a complex system (with the huge variety of relevant spatial and temporal scales) cannot be described by a single analytical or computational model. Instead, we here present an approach to some of the plasma dynamical problems encountered in astrophysical plasmas (encompassing solar and space plasmas as well) that is motivated by an attempt to exploit methods that have proved their power for laboratory plasmas. We start by confronting the basic facts of solar magnetism with the main constituents of magnetohydrodynamics (MHD), viz. the description by conservation laws (the most important one being magnetic flux conservation), the occurrence of specific waves and instabilities (in particular, Alfvén wave dynamics), the distinct stationary flow patterns with *trans'sonic' transitions* (apostrophes indicating three, rather than one, critical speeds for magnetized plasmas), and the different types of nonlinear dynamics (e.g. shocks).

Whereas subsonic MHD has been highly developed in the context of laboratory plasma fusion research (where plasmas are basically in static equilibrium and perturbations are controlled to avoid the occurrence of sudden disruptions), in transonic MHD models of astrophysical plasmas the basic equilibrium consists of stationary flows admitting a much larger variety of waves and instabilities whereas sudden transitions by shocks are a rule rather than exception. Evidently, since the construction of the dynamical picture for the much simpler static laboratory plasmas took 40 years of intensive research, a similar description for transonic plasmas is still far from completion. Hence, to appreciate the immense theoretical problems associated with trans'sonic' plasma flows, we first recapitulate the results of the simpler static laboratory plasmas (where spectral analysis yields detailed information about the underlying equilibria: MHD spectroscopy), then generalize this method to plasmas with background equilibrium flows (where rotations and outflows produce new types of waves and instabilities), and then try to generalize the obtained picture to trans'sonic' background flows (i.e. construct two-dimensional equilibrium flow patterns, their waves and instabilities, and find the associated shock solutions).

The first part [2. MHD Modeling; 3. MHD Waves; 4. Spectral Theory; 5. Waves in Tokamaks] recapitulates the basic facts of subsonic MHD based on material in *Principles of magnetohydrodynamics* by J. P. Goedbloed and S. Poedts (Cambridge University Press, to appear). The second part [6. Waves in Astrophysical Objects; 7. Transonic Flow: Singularities; 8. Large-Scale Nonlinear Computing; 9. Waves in Astrophysical Objects Revisited] then formulates effective methods and results for transonic flows that are relevant not only for solar plasmas but for astrophysical plasmas in general.

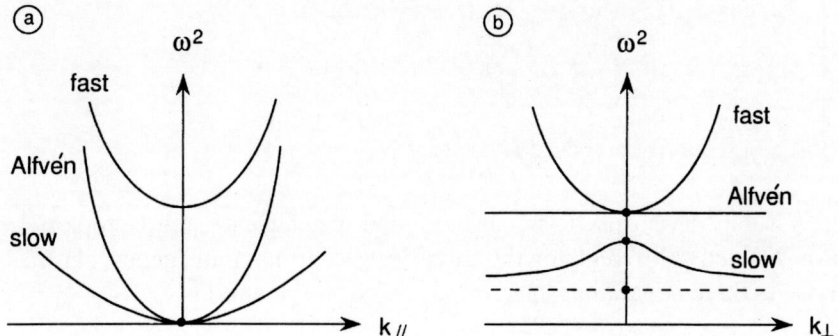

Figure 1. Parallel and perpendicular wavenumber dependence of the frequencies of the three MHD waves.

2. MHD modeling

MHD modeling consists of prescribing: (1) The nonlinear partial differential equations for the motion of *a (perfectly) conducting fluid interacting with a magnetic field* (a perfect transposition of the laws of gas dynamics and electrodynamics); (2) A particular plasma confinement structure, i.e. a generic *magnetic geometry*, fixing the boundary conditions to be imposed. Examples of the latter are the toroidal magnetic confinement geometry of a tokamak (closed in itself), coronal magnetic loops 'closed' onto the photosphere, and magnetic flux bundles emanating from the Sun with 'open' ends associated with the solar wind and the heliosphere.

The strength of the MHD model is that it is *scale invariant*: The MHD equations are unchanged by changing the scales of *length* L_0, *magnetic field* B_0, and density ρ_0, or Alfvén speed $v_A \equiv B_0/\sqrt{\mu_0\rho_0}$, i.e. *time scale* $\tau_A \equiv L_0/v_A$. Thus, MHD is an excellent tool for global analysis of magnetized plasmas on all scales, which justifies the transfer of methods and results from laboratory to astrophysical plasmas.

3. MHD Waves

The three MHD waves (Alfvén, slow, and fast magnetosonic) permit a complete description of the response to arbitrary excitations of a magnetized plasma. However, in the analysis of confined plasmas, the Alfvén waves are the most prominent ones since (1) they may propagate as point disturbances along the magnetic field lines, so that *Alfvén waves 'sample' the magnetic geometry*, (2) their frequency vanishes for $k_\parallel \to 0$ which marks the condition for *marginal stability of tokamaks as well as coronal magnetic flux tubes* (Figure 1(a)). On the other hand, the unique (anisotropic) properties of the three MHD waves are best appreciated by considering the asymptotic dependence of their frequency on the wave number perpendicular to the magnetic field (Figure 1(b)):

$$
\begin{cases}
\partial\omega/\partial k_\perp > 0, & \omega_f^2 \to \infty & \text{(fast)}, \\[2mm]
\partial\omega/\partial k_\perp = 0, & \omega_A^2 \to k_\parallel^2 b^2 & \text{(Alfvén)}, \\[2mm]
\partial\omega/\partial k_\perp < 0, & \omega_s^2 \to k_\parallel^2 \dfrac{b^2 c^2}{b^2 + c^2} & \text{(slow)},
\end{cases}
\tag{1}
$$

where b is the Alfvén speed and c is the sound speed. Hence, the asymptotic spectra behave distinctly different for the three waves. In inhomogeneous plasmas, they give rise to three *continuous spectra*: $\omega_F^2 \equiv \infty$, $\{\omega_A^2\}$, and $\{\omega_S^2\}$.

4. Spectral Theory

Analogous to quantum mechanics, spectral theory of MHD waves and instabilities revolves about the two equivalent view points of *force* and *energy*, respectively leading to a spectral differential equation in terms of the plasma displacement vector field $\boldsymbol{\xi}$ (Bernstein *et al.*, 1958):

$$
\mathbf{F}(\boldsymbol{\xi}) = \rho \frac{\partial^2 \boldsymbol{\xi}}{\partial t^2} = -\rho\omega^2 \boldsymbol{\xi} ,
\tag{2}
$$

and a variational principle for the eigenfrequencies ω^2 of the modes:

$$
\delta(W/I) = 0, \quad W \equiv -\tfrac{1}{2} \int \boldsymbol{\xi}^* \cdot \mathbf{F}(\boldsymbol{\xi}) \, dV , \quad I \equiv \tfrac{1}{2} \int |\boldsymbol{\xi}|^2 \, dV ,
\tag{3}
$$

which involves the quadratic forms $W[\boldsymbol{\xi}]$ for the potential energy and $I[\boldsymbol{\xi}]$ related to the kinetic energy of the perturbations. Whereas quantum mechanical spectral theory has led to a deep understanding of atomic and subatomic structures (occupying much of 20th century physics), the analogous theory for fluids and plasmas is still in its infancy. Yet, the observation of a *classical spectrum* of oscillations and comparison with computed eigenvalues may lead to a firm knowledge of the internal characteristics of fluids and plasmas, which we have called *MHD spectroscopy* (Goedbloed *et al.*, 1993). Relevant examples are helioseismology, sunspot seismology, MHD spectroscopy of tokamaks, and magnetoseismology of accretion disks (Keppens *et al.*, 2002).

In Figure 2, we recall the principle of helioseismology: Comparison of computed frequencies for the p and g modes of a solar model with the observed ones led to the validation of the standard solar model, and may lead to improvements with respect to 2D extensions such as the influence of differential rotation and magnetic fields. The three boxed activities together show what is involved in MHD spectroscopy. For the present purpose, we concentrate on two of them, viz. *analysis* to reveal the structure of spectra and *numerical tools* to compute them. Once these issues are resolved, we will have obtained a very powerful instrument to analyze magnetically confined plasmas.

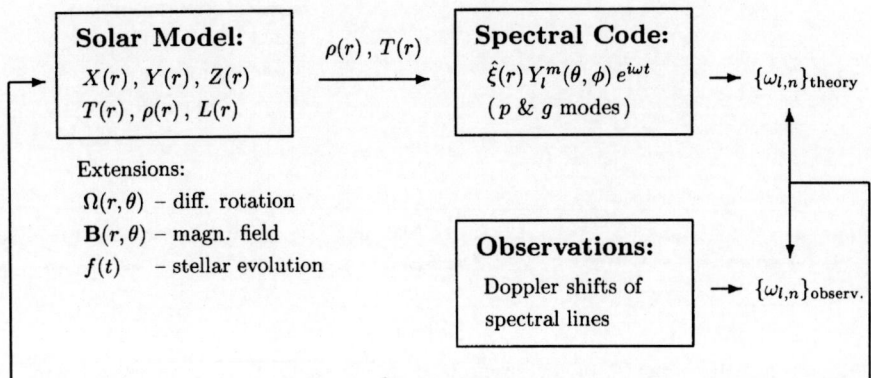

Figure 2. Systematics of helioseismology.

5. Waves in Tokamaks

Let us see how this program is carried out in toroidal fusion experiments of the tokamak type. First, consider the standard case of a static axi-symmetric equilibrium. The basic approach here is to split the problem in a study of the *static equilibrium*, basically described by the force balance equations $\nabla p = \mathbf{j} \times \mathbf{B}$, $\mathbf{j} = \nabla \times \mathbf{B}$, $\nabla \cdot \mathbf{B} = 0$, and the *linear waves and instabilities* described by Equations (2) or (3). The most important property of these equilibria is that they consist of nested *magnetic surfaces* of the magnetic field $\mathbf{B}$ and the current density $\mathbf{j}$, producing confinement of the pressure gradient ∇p through the Lorentz force. We recall from the introduction that the systematic analysis of the spectra (with top priority on the practical issue of improving overall stability for higher values of $\beta \equiv 2\mu_0 p / B^2$) has taken about 40 years of intensive research. This has led to steady increase of confinement, and concomitant understanding of the processes involved, from μseconds in the early days to minutes at present. For our present purpose (a similar effort for astrophysical plasmas with sizeable background flow), this implies: a lot of work ahead and great promise for understanding in the end!

One of the intriguing aspects of wave dynamics in toroidal plasmas is the occurrence of singular perturbations and *continuous spectra* which manifest the preference of the waves and instabilities to localize inside the magnetic surfaces. In Figure 3 we show the schematic structure of the MHD spectrum which clearly demonstrates this (Goedbloed, 1975). Most important: through this singular asymptotics, the three MHD wave spectra maintain the essential features of Equation (1) shown by Figure 1(b) and, thus, make them suitable to be used in MHD spectroscopy. Techniques to accurately compute the static equilibria of tokamaks (Huysmans *et al.*, 1991) and a large-scale spectral code to compute the spectra of these 2D equilibria (Kerner *et al.*, 1998) were developed. Recently, the necessary accurate MHD spectra of ideal and resistive waves in static tokamak equilibria could be com-

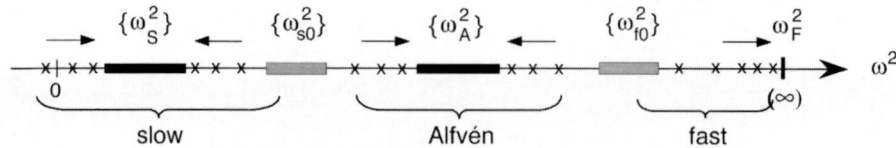

Figure 3. Schematic structure of the spectrum of MHD waves for a static equilibrium. Three sub-spectra of fast, Alfvén, and slow modes concentrate about continua $\omega_F^2 \equiv \infty$, $\{\omega_A^2\}$, and $\{\omega_S^2\}$, separated by regions with non-monotonic discrete modes. The inhomogeneity is chosen to be small so that sub-spectra are well separated.

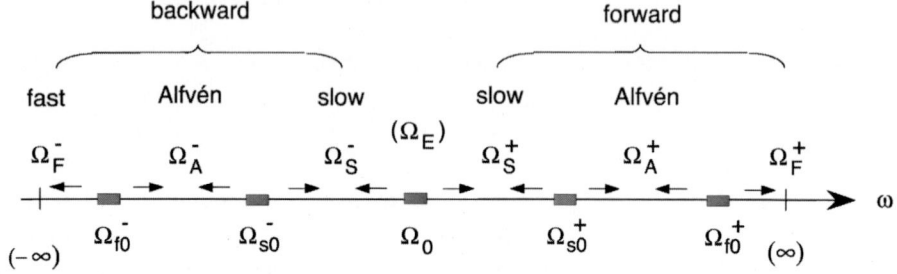

Figure 4. Schematic structure of the spectrum of MHD waves for an equilibrium with flow: The three sub-spectra split into six sub-spectra of forward and backward propagating fast, Alfvén, and slow modes, concentrated about the continua $\Omega_F^\pm \equiv \pm\infty$, $\{\Omega_A^\pm\}$, $\{\Omega_S^\pm\}$, where $\Omega_\ell^\pm \equiv \Omega_0 \pm \omega_l$ ($\ell = F, A, S$) with Doppler shift $\Omega_0 \equiv \mathbf{k} \cdot \mathbf{v}$. The picture should be asymmetric with respect to $\omega = 0$ (not indicated).

puted in full detail (Van der Holst *et al.*, 1999) with the powerful Jacobi–Davidson method (Sleijpen and Van der Vorst, 1996).

An exciting new development in tokamak research is the realization that static equilibria are actually not quite adequate since heating by neutral beams causes sizeable toroidal flows and divertor operation for exhaust removal causes (supersonic!) poloidal flows in the outer layers. Hence, the *paradigm of static equilibrium breaks down*. However, since astrophysical plasmas are all dominated by flow, the good side about this development is that the subject of plasmas with background flow now becomes a common research theme for laboratory and astrophysical plasmas.

In order to enter this common field, all spectral calculations have to be redone with proper incorporation of the background flow of the equilibrium. Again exploiting the standard approach, with a split in equilibrium and perturbations, this first involves construction of a stationary state (where $\mathbf{v} \neq 0$ so that all MHD equations contribute now) and, next, computation of the waves and instabilities by means of a quadratic eigenvalue equation (Frieman and Rotenberg, 1960):

$$\mathbf{F}(\boldsymbol{\xi}) + \nabla \cdot \left[\rho (\mathbf{v} \cdot \nabla \mathbf{v}) \boldsymbol{\xi} - \rho \mathbf{v} \mathbf{v} \cdot \nabla \boldsymbol{\xi} \right] + 2i\rho \omega \mathbf{v} \cdot \nabla \boldsymbol{\xi} + \rho \omega^2 \boldsymbol{\xi} = 0 . \qquad (4)$$

We note in passing that, in contrast to the widely used spectral Equation (2) for static equilibria, the Frieman and Rotenberg spectral Equation (4) has rarely been applied to realistic stationary states. The obvious reason is that the equilibria are much more complicated and that the eigenvalues are complex (admitting overstable modes). Another, even more fundamental, problem will be faced in Section 6.

The schematic spectral structure of stationary equilibria (Figure 4) is again con- centrated about the continuous spectra, which now split into six due to the Doppler shift. [An additional, somewhat esoteric, Eulerian entropy continuum $\{\Omega_E\}$ is not found from the Lagrangian Equation (4), but only when the primitive, Eulerian, variables are exploited.] On the road to a systematic MHD spectroscopy of moving plasmas, with precise input of tokamak equilibria, these continua turn out to con- tain large gaps where new global Alfvén waves driven by the toroidal flow (called TFAEs) were discovered (Van der Holst, 2000). This appears to open up a new chapter in MHD spectroscopy which, obviously, calls for a generalization admit- ting poloidal flows as well. To do this, we constructed the necessary numerical tools FINESSE (Beliën *et al.*, 2002) to compute the stationary axi-symmetric equi- libria and PHOENIX Van der Holst *et al.*, 2003) for the perturbations, and started to apply them to tokamaks. This worked well as long as the poloidal velocities were restricted to sub'sonic' flows. In this manner, we have contributed to MHD spectroscopy as a highly developed tool to investigate the dynamics of plasmas in future fusion machines.

6. Waves in Astrophysical Objects: A Hair in the Soup

So far, so good. But why did we have to restrict the poloidal velocities to sub'sonic' speeds? [Recall our use of apostrophes to indicate the occurrence of three (slow/ Alfvén/ fast), rather than one, critical MHD speeds.] Obviously, such a restriction is prohibitive if we wish to exploit the same tools for astrophysically relevant flows, which are usually trans'sonic'. What happens precisely when the critical MHD speeds are surpassed?

Consider the stationary equilibrium equations for rotating and gravitating mag- netized plasmas:

$$\nabla \cdot (\rho \mathbf{v}) = 0 ,$$

$$\rho \mathbf{v} \cdot \nabla \mathbf{v} + \nabla p = \mathbf{j} \times \mathbf{B} - \rho \mathbf{g} , \qquad \mathbf{j} = \nabla \times \mathbf{B} ,$$

$$\mathbf{v} \cdot \nabla p + \gamma p \nabla \cdot \mathbf{v} = 0 , \qquad\qquad\qquad (5)$$

$$\nabla \times (\mathbf{v} \times \mathbf{B}) = 0 , \qquad \nabla \cdot \mathbf{B} = 0 .$$

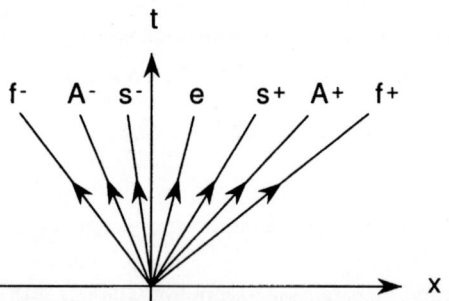

Figure 5. Space-time characteristics of the three MHD waves ($s^\pm$, $A^\pm$, $f^\pm$), travelling in forward and backward directions, and the entropy disturbances (e), which are just carried with the plasma flow.

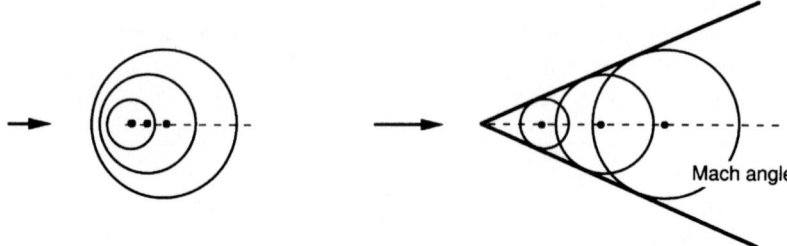

Figure 6. Sound in (a) subsonic and (b) supersonic gas flow about a point source.

For axi-symmetric geometries, like tokamaks and magnetic flux loops in the solar convection zone, or even complete accretion disks, these equations have solutions that basically correspond to nested surfaces of the magnetic field **B** (not of the current density **j** anymore) and of the plasma velocity **v**. Hence, the stronghold of magnetic confinement (related to the existence of magnetic surfaces) remains intact in the presence of arbitrary toroidal and poloidal plasma flows: *magnetic and flow surfaces coincide*! Before we start to indicate some blemishes on this beautiful edifice, let us first spell out the portent of this statement: Together with scale-invariance of the MHD equations, this implies that we can transfer the techniques and results on MHD spectroscopy, developed in laboratory tokamak research, directly to astrophysical problems like axi-symmetric winds, accretion flows, jets, etc.

Yet, we get stuck immediately if we try to do this. Invariably, when the plasma velocities are increased, the equilibrium solvers stop converging before relevant velocities are obtained. The hair in the soup comes from the velocity component in the symmetry-breaking direction, i.e. the poloidal direction. This is so because the equilibrium Equations (5) have a property, entering with the poloidal flow, that is completely lacking in their static counterparts (obtained from them in the limit $\mathbf{v} \to 0$). To appreciate it, we need to make a small detour in the topic of transonic flow.

A tacit assumption in the construction of the equilibria, including the ones with toroidal flows, has been that the governing Grad-Shafranov (nonlinear partial differential) equation is *elliptic*. The numerical techniques exploited need this property. In fact, all of the standard methods in use in MHD spectral analysis are based on the assumption that the equilibria are described by elliptic equations and the perturbations by hyperbolic ones. However, when the poloidal flow velocity increases beyond certain critical values, to be computed yet, the stationary equilibrium Equations (5) become hyperbolic (Zehrfeld and Green, 1972; Hameiri, 1983) and both the classical paradigm of a split in equilibrium and perturbations and the numerical techniques based on it break down. As a result, the standard equilibrium solvers, as used in tokamak computations, diverge and we need to rethink the problem completely.

Clearly, we have to go back to basics, in particular to the meaning of hyperbolicity. This concept is associated with the *characteristics* of the flow, which are the space-time manifolds along which perturbations propagate. For MHD, there are seven of such characteristics, as shown in Figure 5 for the case of one spatial dimension. Permitting two spatial dimensions, the temporal snapshots of the three MHD perturbations become the well-known figures of the Friedrichs group diagram. In two dimensions, these figures may exhibit an interesting new feature, depending on the magnitude of the background flow. This is illustrated in Figure 6 for the case of sound waves in ordinary fluids: When the flow velocity becomes supersonic, the spatial part of the characteristics forms envelopes where information accumulates and discontinuous solutions (shocks) are formed. Whereas in elliptic flows the solutions propagate everywhere in space, in hyperbolic flows these discontinuities separate space in regions where the solutions propagate and regions where they do not propagate. Unfortunately, although magnetic/flow surfaces exist in axi-symmetric MHD flows, the transitions from ellipticity to hyperbolicity occur somewhere, at a-priori unknown locations, on these surfaces and the elliptic solvers become useless.

The fundamental reason of the bankruptcy of the classical paradigm of equilibrium and perturbations is associated with the Lagrangian time derivative $D/Dt \equiv \partial/\partial t + \mathbf{v} \cdot \nabla$ in the MHD equations. Whereas, the Eulerian time derivative $\partial/\partial t$ produces the eigenfrequencies ω of the waves, the spatial derivative $\mathbf{v} \cdot \nabla$ not only produces the Doppler shifts of the perturbations but also the possibility of spatial discontinuities of the equilibria. [Note that this occurs through the poloidal, symmetry-breaking, part only since the toroidal derivative operator vanishes by assumption of axi-symmetry.] However, the two pieces of the Lagrangian time derivative really belong together so that *the waves and the stationary equilibria, with transitions from ellipticity to hyperbolicity, are no longer separate issues* (Goedbloed, 2002).

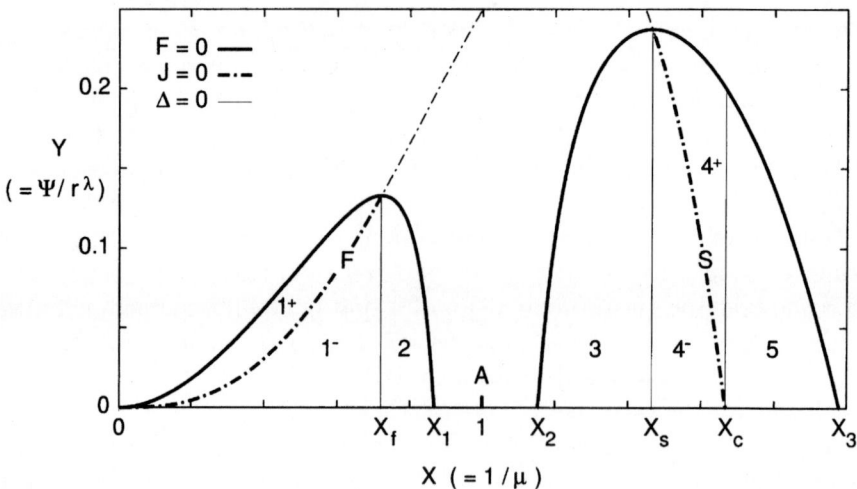

Figure 7. Four main flow regimes due to Alfvén gap (A) and fast (F) and slow (S) magnetoacoustic limiting lines.

7. Transonic Flow: Singularities

Since the transonic transitions present the basic problem, let us analyze some specific stationary equilibria in detail to see what is going on. For the present purpose, it is sufficient to consider 2D equilibria that are translation symmetric (Goedbloed and Lifschitz, 1997) so that the physical quantities are functions of the Cartesian x,y coordinates of the poloidal cross-section of the plasma. The stationary equilibrium states are then characterized by the poloidal magnetic flux $\psi(x, y)$ and by the square of the poloidal Alfvén Mach number, $M^2 \equiv \rho v_p^2 / B_p^2 = \mu(x, y)$. The flux ψ is determined by a partial differential equation (a generalization of the Grad–Shafranov equation) that is elliptic or hyperbolic depending on the value of μ, which is in turn determined by an algebraic equation (the Bernoulli equation). This pair of highly non-linear equations for ψ and μ admits solutions only for certain values of the parameters involved: The distinguishing feature of transonic flows is that *there are distinct flow regimes that cannot be connected by continuous flows when the speed is increased or decreased.*

A specific example is shown in Figure 7, obtained by imposing the following self-similarity in terms of the polar coordinates r, θ in the polidal plane:

$$M^2 \equiv \mu = [X(\theta)]^{-1}, \qquad \psi = r^\lambda Y(\theta). \tag{6}$$

This reduces the problem to its bare essentials, viz. the solution of a pair of autonomous differential equations for X and Y:

$$\frac{dX}{d\theta} = \pm \frac{H}{J}\sqrt{2F}, \qquad \frac{dY}{d\theta} = \pm\sqrt{2F}, \tag{7}$$

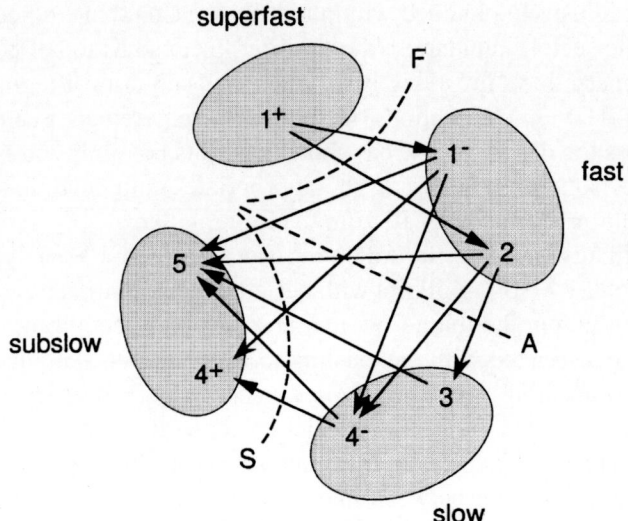

Figure 8. Connecting the four flow regimes: Fast, Alfvén, slow shocks.

where H, J, and F are explicit functions of X and Y. In this case, the different flow regimes show up as regions in the X-Y phase diagram that may be constructed without actually solving Equations (7). First of all, the condition $F(X, Y) = 0$ (the *Bernoulli boundary*) delineates two permissible flow regimes, viz. a *slow* ($X > 1$) and a *fast* ($X < 1$) one, where the poloidal field is real ($F > 0$). These two islands in phase space imply that there is no continuous path from static equilibria ($X = \infty$) to slow stationary equilibria, and also not from the slow to the fast equilibria since a gap at $X = 1$ (the Alfvén gap) interferes. Next, another algebraic condition, $\Delta(X) = 0$, separates the regions of *ellipticity* ($\Delta < 0$: no real characteristics) and *hyperbolicity* ($\Delta > 0$: two real characteristics). Finally, the most dramatic separation of flow regimes is due to the singularity $J(X, Y) = 0$, where the characteristics of the hyperbolic solutions exhibit *limiting line* behavior, i.e. both characteristics are 'reflected' there so that solutions can not propagate beyond the limiting line. Consequently, four *smooth* types of stationary 2D equilibrium solutions are obtained, viz. superfast (1^+), fast (1^-, 2), slow (3, 4^-), and subslow (4^+, 5) ones.

Of course, the explicit solutions of Equation (7) and the corresponding flow patterns have been investigated in detail (Goedbloed and Lifschitz, 1997). However, for our present discussion on the possibility of constructing spectral codes for transonic MHD flows, we just focus on one particular aspect of those solutions: their trajectories $dY/dX = J/H$ in phase space either cross or do not cross the limiting lines. In the latter case, smooth stationary flow solutions are obtained that have the requisite property of globally nested magnetic/flow surfaces. On the other hand, when trajectories cross the limiting lines, multiple solutions are obtained within a sector cutting through the magnetic/flow surfaces. More precisely, magnetic/flow surfaces are exclusively obtained within that sector. Could one reflect the solutions obtained at the sector boundaries (the limiting lines) so as to get periodic *discontinuous* stationary flows involving both super- and sub-critical regimes? Extensive study of the MHD jump conditions, including the requirement that entropy should increase across the discontinuity, has shown that this possibility must be excluded: Discontinuous solutions, satisfying the appropriate jump conditions, can only be found for solutions that stay away from the limiting lines.

We then finally come to the following state of affairs: Four types of smooth periodic stationary MHD equilibria with nested magnetic surfaces are obtained that strictly remain within the main flow regimes. For these equilibria, MHD spectral codes can be constructed with the existing tools. However, stationary *trans'sonic'* MHD flows, connecting two flow regimes, necessarily involve shock-type discontinuities, as illustrated in Figure 8. This picture shows that limiting lines and Alfvén gap are quite genuine obstacles in transonic stationary flows, but it also highlights the fascinating connection between linear waves and stationary states: In analogy to the three types of linear MHD waves, with their local singular asymptotics, in transonic flows also three types of MHD discontinuities appear that locally exhibit slow, Alfvén, and fast character at the singularity.

To sum up: For static or toroidally rotating tokamaks, the equilibria are complicated but essentially computable. When trans'sonic' poloidal flows are admitted, the determination of the stationary states becomes a fundamentally different and difficult problem because *discontinuities and singularities* appear manifesting that the waves and stationary states are entangled in a deep sense. An obvious way out is to drop the idea of a split in equilibrium and perturbations altogether and to employ a nonlinear time stepping code, e.g. the Versatile Advection Code (VAC), which we will discuss in Section 8. This should be considered as an aside though since we do not really wish to abandon the equilibrium–wave dichotomy because it has proved too useful. Therefore, in Section 9, we will return to it and show how to exploit the Frieman–Rotenberg formalism with the knowledge of the present section.

8. Large-Scale Nonlinear Computing

The development of a general set of state-of-the-art spectral codes for the analysis of MHD waves and instabilities for realistic laboratory experiments and astrophysical objects has been stimulated by our studies of resonant absorption in solar coronal flux tubes with inclusion of the geometric influence of line-tying (Halberstadt and Goedbloed, 1993–1995) and loop expansion (Beliën *et al.*, 1996–97). Visualization of coronal heating mechanisms proved to be instrumental for our transition to nonlinear MHD simulations of wave dissipation in flux tubes (Poedts *et al.*, 1996–97; Keppens *et al.*, 1997–98) and, finally, to simulations of SOHO observations (Beliën *et al.*, 1999). In the latter phase, operation of the VAC code was already in full swing.

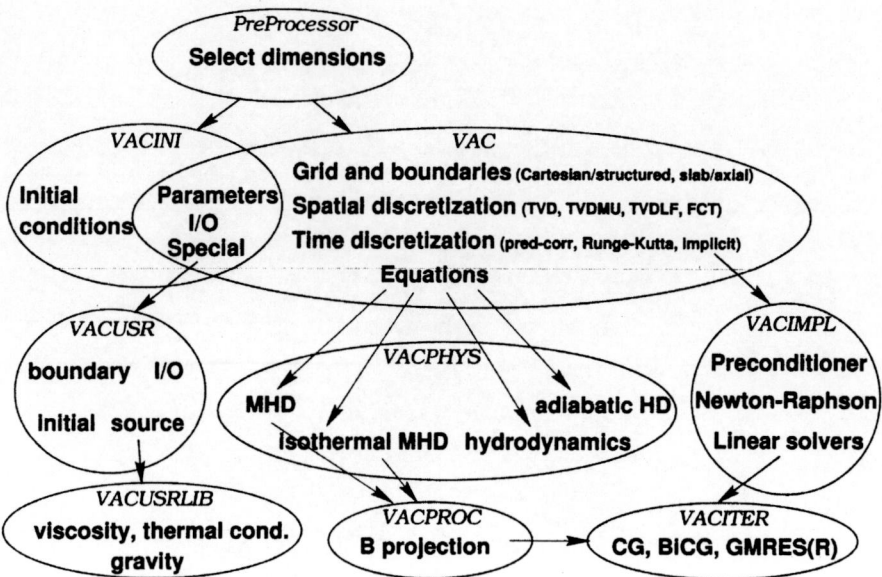

Figure 9. Structure of the Versatile Advection Code (VAC): A modular approach ensures the compatibility between the different code segments. As a result, several spatial and temporal (explicit, semi-implicit, and fully implicit) discretizations are applicable to all physics modules.

The Versatile Advection Code (Tóth, 1996) was developed as part of a Massively Parallel Computing project of NWO (Poedts, Keppens and Goedbloed, 1996–2000). It is a massively parallel MHD solver which is shock-capturing (through the use of conservative variables) and can bridge the huge time-scale disparities (from Alfvénic to dissipative) encountered in realistic astrophysical simulations by means of implicit time integration. Designed to permit inclusion of almost all present discretization methods, with a modular structure (Figure 9), it became an extremely versatile research instrument used by a rapidly increasing number of scientists. The code was steadily developed (Keppens and Tóth, 1999–2000), and applied to basic plasma dynamics like the Kelvin–Helmholtz instability and jets (Keppens *et al.*, 1999). Application to solar and stellar winds from axi-symmetric, rotating and gravitating, stars (Keppens and Goedbloed, 1999–2000) produced continuous acceleration from sub-slow flow at the surface to super-fast flow at large distances. Adding a 'dead' zone at the equator, anisotropy as observed by the Ulysses spacecraft was obtained (Figure 10). The recent extension with adaptive mesh refinement (AMR-VAC; Keppens *et al.*, 2002) is yet another step towards simulating realistic astrophysical plasma flows with small-scale structures.

In conclusion: With the new powerful tool VAC to compute the non-linear MHD evolution, we have a completely independent entry into the exciting field of transonic plasma dynamics.

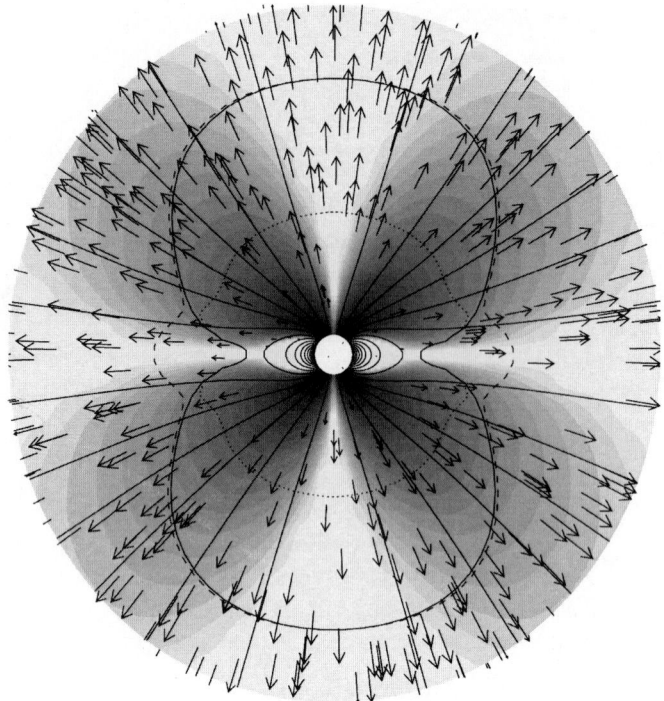

Figure 10. Axisymmetric magnetized wind with a 'wind' and 'dead' zone. Shown are the poloidal magnetic field lines and the poloidal flow field as vectors (parallel to the magnetic field, as they should). Also indicated are the slow (dotted), Alfvén (solid), and fast (dashed) critical surfaces. Shading indicates the toroidal field strength.

9. Waves in Astrophysical Objects Revisited

Returning now to our subject of spectral analysis of trans'sonic' astrophysical plasmas, where the intrinsic difficulty of lack of precise stationary equilibria in the hyperbolic regions appears to be near insurmountable, one might be inclined to settle for a cheap solution: Why not abandon the spectral approach altogether and exclusively exploit nonlinear MHD solvers like VAC? That would be an inferior solution indeed since it would amount to giving up the incredible precise and detailed information that spectral theory delivers on all 3D waves and instabilities *and* their dependence on the relevant physical parameters characterizing the stationary states. Clearly, the royal road is to keep both approaches operational, each in their respective domain of validity, and to try to approach the physical phenomena from the linear as well as from the nonlinear angle. For example, the prediction by a spectral code of exponential instability for a well-described equilibrium is already invaluable for the prescription of initial data for a nonlinear evolution code. However, there is more . . ..

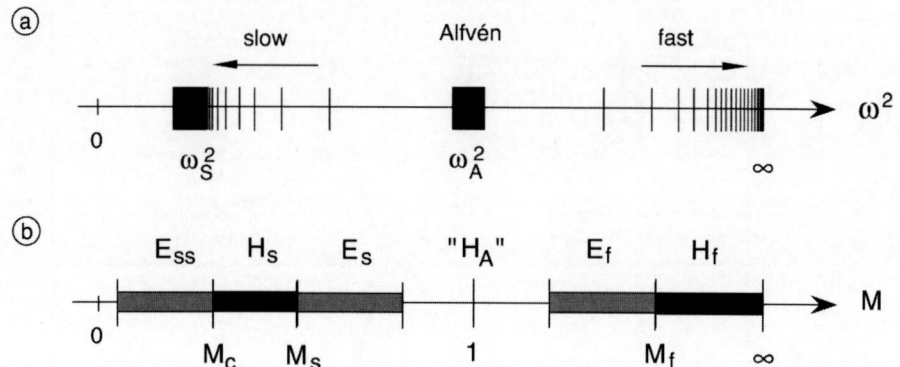

Figure 11. (a) Schematic spectrum of the three MHD waves for a *static* background equilibrium. For large wave numbers, the discrete eigenvalues accumulate at the continua $\{\omega_S^2\}$, $\{\omega_A^2\}$, and $\omega_F^2 \equiv \infty$; (b) Flow regimes characterized by the value of the poloidal Alfvén Mach Number $M \equiv v_p/v_{A,p}$ of a *stationary* equilibrium flow. The flow turns from elliptic to hyperbolic at the boundaries of the hatched regions H_s and H_f, whereas the Alfvén region 'H_A' has collapsed into the point $M \equiv 1$.

Consider again the phase space of our model trans'sonic' stationary states depicted in Figure 7: Clearly, to study the consequences of transition through the hyperbolic regions on the waves and instabilities one does not have to restrict the analysis to the sub-slow elliptic regime 5 (E_{ss}), since there are two more elliptic regimes, viz. the slow regime 3 (E_s), and the fast regime 2 (E_f). Hence, one may study the qualitative change of the spectra due to transition through the critical poloidal Alfvén Mach numbers by comparing the spectra *after* the transition through the slow or through the Alfvén critical value has been made. This may be done on the basis of the standard paradigm of a split in elliptic equilibrium and hyperbolic perturbations, and exploiting the numerical tools based on it. One essential complication must then be faced: The transition speeds M_c, M_s, and M_f depend on the local values of the physical variables, i.e. they are not known beforehand but are to be determined together with the solutions. Hence, staying in the elliptic flow regimes is a delicate numerical problem. This problem has been addressed and satisfactorily solved in the numerical equilibrium solver FINESSE (Beliën *et al.*, 2002). Hence, we can proceed now with the computation of waves and instabilities of trans'sonic' astrophysical plasmas with precise prescriptions of background flows.

Finally, we have argued in Section (5) that linear waves and nonlinear stationary states are not independent issues in trans'sonic' MHD flows. One may turn the coin and notice that this also implies that there is an incredibly beautiful connection between the two. As illustrated in Figure 11, somehow the asymptotic 'concentration' points of the wave spectra correspond to the hyperbolic regions of the equilibrium states, and their embedded singularities. Hence, studying the spectra by approaching the hyperbolic regimes while staying in the elliptic regimes,

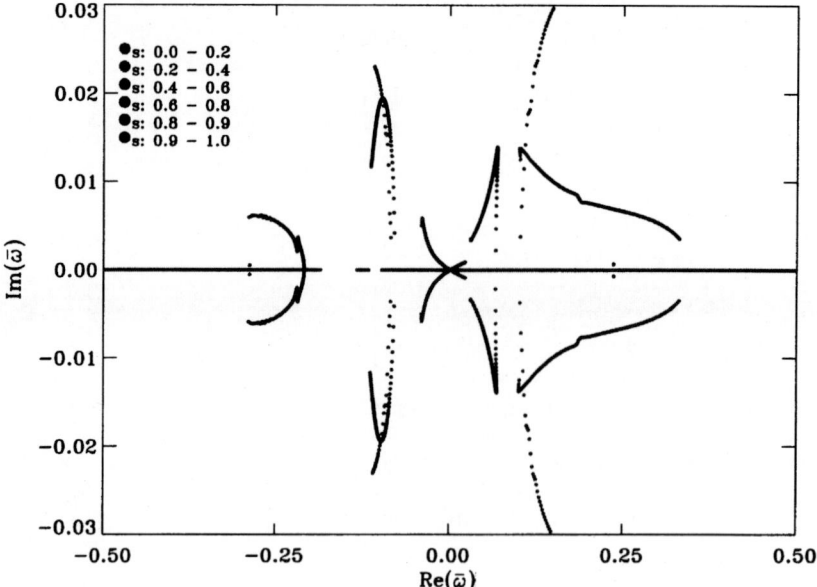

Figure 12. Instabilities in the 2nd elliptic flow regime E_S (slow and sub-Alfvénic: $M_S < M < 1$): Continuous spectrum of waves (real ω) and overstable modes (with an additional imaginary part of ω) of a thick accretion disk; the continuous distribution of the eigenvalue parameter ω is shown in the complex ω-plane with the radial location $s \equiv \sqrt{\psi}$ as a parameter.

undoubtedly will reveal important clues on the physical mechanisms of transonic flows. We will now give just one example to demonstrate this point.

The first spectral results for localized modes of a gravitating torus (a thick accretion disk or any closed flux loop) with both poloidal and toroidal magnetic fields and flows (Beliën *et al.*, 2001) are shown in Figure 12. [This is a corrected version of Figure 4 of Beliën *et al.*, 2001) and of Figure 9 of Goedbloed, 2002)).] The eigenvalues of the stable waves are located along the Re ω-axis, the curves in the complex ω-plane correspond to forward and backward propagating instabilities driven by the poloidal flow and gravity. The spectrum is quite characteristic for flows in the second elliptic flow regime and instability will generally occur when the value of the poloidal Alfvén Mach number for the flow has surpassed the critical value M_c. The instabilities are localized on magnetic/flow surfaces and occupy a large fraction of the outer part of the torus so that they may be considered as suitable candidates for anomalous dissipation by MHD turbulence, e.g. in accretion disks.

In conclusion: We have analyzed the waves and instabilities of tokamaks and toroidal astrophysical plasmas (like thick accretion disks or parts of solar magnetic loops) in the second elliptic flow regime (E_s, i.e. region 3 of Figure 7) and found that significant instabilities operate there that are absent in the first elliptic flow regime (E_{ss}, i.e. region 5). These instabilities should be ascribed to the transonic transition at $M = M_c$. Hence, there appears to be a strong correlation between the singularities and discontinuities that occur in the background nonlinear stationary states when the critical values of the poloidal Alfvén Mach number (lying in the hyperbolic flow regimes, which are as yet inaccessible for spectral studies) are surpassed and the instabilities that are found in the next elliptic flow regime.

The persistent development of the stationary equilibrium program FINESSE and the spectral code PHOENIX, and the accompanying in-depth analysis, have produced a new angle on the study of waves and instabilities in trans'sonic' plasma flows. Presently, the linear codes are operated in tandem with the nonlinear time-stepping code VAC to investigate both the linear and the nonlinear phases of the dynamics in the different flow regimes. They exhibit an abundance of new instabilities of interest for the different kinds of MHD turbulence operating in solar and astrophysical plasmas. Will be continued!

ACKNOWLEDGEMENTS

The authors wish to thank Sander Beliën, Bart van der Holst, and Sascha Lifschitz for many years of fruitful collaboration. This work was performed as part of the research program of the Euratom-FOM Association Agreement, with support from the Netherlands Science Organization (NWO) programs on Massively Parallel Computing and on Computational Science, and the EC Human Potential program under contract HPRN-CT-2000-00153 (PLATON). NCF is acknowledged for providing computer facilities.

References

Beliën, A. J. C., Botchev, M. A., Goedbloed, J. P., van der Holst, B. and Keppens, R.: 2002, *J. Comp. Phys.* **182**, 91.

Beliën, A. J. C., Goedbloed, J. P. and Van der Holst, B.: 2001, Proc. 28th Eur. Conf. on *Controlled Fusion and Plasma Physics*, Madeira, p. 1309.

Beliën, A. J. C., Martens, P. C. H. and Keppens, R.: 1999, in *Plasma Dynamics and Diagnostics in the Solar Transition Region and Corona*, 8th SOHO Workshop, ESA **SP-446**, 167–172.

Beliën, A. J. C., Poedts, S. and Goedbloed, J. P.: 1997, *Phys. Rev. Lett.* **76**, 567–570; in *Magnetodynamic Phenomena in the Solar Atmosphere–Prototypes of stellar magnetic activity*, IAU Coll. **153**, 423–424 (Kluwer Dordrecht, 1996); *Astron. Astrophys.* **322**, 995–1006 (1997); *Comp. Phys. Comm.* **106**, 21–38 (1997); in *The Corona and Solar Wind near Minimum Activity*, 5th SOHO Workshop, ESA **SP-404**, 193–197 (1997).

Bernstein, I. B., Frieman, E. A., Kruskal, M. D. and Kulsrud, R. M.: 1958, *Proc. Roy. Soc. London* **A244**, 17–40.

Frieman, E. and Rotenberg, M.: 1960, *Rev. Mod. Phys.* **32**, 898–902.

Goedbloed, J. P.: 2002, *Physica Scripta* **T98**, 43–47.

Goedbloed, J. P.: 1975, *Phys. Fluids* **18**, 1258–1268.

Goedbloed, J. P., Huysmans, G. T. A., Holties, H., Kerner, W. and Poedts, S.: 1993, *Plasma Phys. Contr. Fusion* **35**, B277–292.

Goedbloed, J. P. and Lifschitz, A.: 1997, *Phys. Plasmas* **4**, 3544.

Halberstadt, G. and Goedbloed, J. P.: 1995, *Astron. Plasmas* **301**, 577–592; **280**, 647–660 (1993); **286**, 265–301 (1994); **301**, 559–576 (1995).

Hameiri, E.: 1983, *Phys. Fluids* **26**, 230–237.

Huysmans, G. T. A., Goedbloed, J. P. and Kerner, W.: 1991, in *Comput. Phys.* **371** (World Scientific, 1991).

Keppens, R. and Goedbloed, J. P.: 2000, *Astron. Astrophys.* **343**, 251–260 (1999); *Space Sci. Rev.* **87**, 223–226 (1999); *Astrophys. J* **530**, 1036–1048.

Keppens, R. and Tóth, G.: 2000, in *Third International Conference for Vector and Parallel Processing*, Lecture Notes in Computer Science **1573**, 680–690 (Springer, 1999); Keppens, R., Tóth, G. and Goedbloed, J.P.: 2000, in *Pallallel Computing: Fundamentals and Applications*, ParCo99, 160–167, Imperial College Press, London.

Keppens, R., Casse, F. and Goedbloed, J. P.: 2002, *Astrophys. J.* **569**, L121–L126.

Keppens, R., Nool, M. and Goedbloed, J. P.: 2002, in *Pallallel Computational Fluid Dynamics – Practice and Theory*, 215–223 (Elsevier Science, 2002); Nool, M. and Keppens, R.: 2002, *Comp. Math. Appl. Math* **2**, 92–109.

Keppens, R., Tóth, G., Westermann, R. H. J. and Goedbloed, J. P.: 1999, *J. Plasma Phys.* **61**, 1–19 (1999); Keppens, R. and Tóth, G.: 1999, *Phys. of Plasmas* **6**, 1461–1469.

Keppens, R., Poedts, S., Meijer, P. M. and Goedbloed, J. P., in *High Performance Computing and Networking*, Lecture Notes in Computer Science **1225**, 190–199 (Springer, 1997); Keppens, R., Poedts, S. and Goedbloed, J. P., in *High Performance Computing and Networking*, Lecture Notes in Computer Science **1401**, 233–241 (Springer, 1998).

Kerner, W., Goedbloed, J. P., Huysmans, G. T. A., Poedts, S. and Schwarz, E.: 1998, *J. Comp. Phys.* **142**, 271–303.

MPP-CMFD team, ed. Poedts, S., Keppens, R. and Goedbloed, J. P.: 2000, 1st, 2nd, 3rd, Final Annual Report of the Massively Parallel Computing Cluster Project 95MPR04, Rijnhuizen Reports 96–231 (1996); 97–233 (1997); 98–234 (1998); 00–235 (2000).

Poedts, S. and Goedbloed, J. P., in *Magnetodynamic Phenomena in the Solar Atmosphere–Prototypes of Stellar Magnetic Activity*, IAU Coll. **153**, 425–426 (Kluwer Dordrecht, 1996); *Astron. Astrophys.* **321**, 935–944 (1997); Poedts, S., Tóth, G., Beliën, A. J. C. and Goedbloed, J.P., *Solar Physics* **172**, 45–52 (1997).

Sleijpen, G. L. G. and Van der Vorst, H. A.: 1996, *SIAM J. Matrix Anal. Appl.* **17**, 401.

Tóth, G.: 1996, *Astrophys. Lett. Comm.* **34**, 245.

Van der Holst, B., Beliën, A. J. C. and Goedbloed, J. P.: 2003, (in press).

Van der Holst, B., Beliën, A. J. C. and Goedbloed, J. P.: 2000, *Phys. Rev. Lett.* **84**, 2865; *Phys. Plasmas* **7**, 4208.

Van der Holst, B., Beliën, A. J. C., Goedbloed, J. P., Nool, M. and Van der Ploeg, A.: 1999, *Phys. Plasmas* **6**, 1554–1561.

Zehrfeld, H. P. and Green, B. J.: 1972, *Nuclear Fusion* **12**, 569–575.

HIGH-FREQUENCY RADIO SIGNATURES OF SOLAR ERUPTIVE FLARES

MARIAN KARLICKÝ

Astronomical Institute of the Academy of Sciences of the Czech Republic, CZ-25165 Ondřejov,
Czech Republic (e-mail: karlicky@asu.cas.cz)

Abstract. Several examples of the radio emission of eruptive solar flares with high-frequency slowly drifting structures and type II bursts are presented. Relationships of these radio bursts with eruptive phenomena such as soft X-ray plasmoid ejection and shock formation are shown. Possible underlying physical processes are discussed in the framework of the plasmoid ejection model of eruptive solar flares. On the other hand, it is shown that these radio bursts can be considered as radio signatures of eruptive solar flares and thus used for the prediction of heliospheric effects.

Key words: radio radiation, solar flares

1. Introduction

Eruptive (dynamic) solar flares are connected with the filament eruption and coronal mass ejection (Švestka *et al.*, 1992). On radio waves, in the frequency range below 300 MHz, these flares are usually associated with type II radio bursts indicating shocks driven by coronal mass ejection (Reiner and Kaiser, 1999). The frequency of type II bursts corresponds to the electron plasma frequency in the radio source and an upwards motion of the shock is expressed on radio spectrum by the slow negative frequency drift (~ -1 MHz s^{-1} at 200 MHz) corresponding to the vertical decrease of the atmospheric density. On the other hand, using specific models of the solar atmosphere shock velocities are determined from the frequency drift and used for the prediction of heliospheric disturbances (e.g. Dryer *et al.*, 1998). It is believed that these shocks are formed from rapid plasma motions evolving into MHD shocks. This idea was confirmed by various radio precursors of type II bursts. For example, in the 0.4–1.0 GHz frequency range a negatively drifting group of type U bursts (Karlický, 1992) and faint radio sources (Klassen *et al.*, 1999a) were found before the type II radio bursts.

Recently a new type of slowly drifting emission was recognized at frequencies above 1 GHz (Karlický and Odstrčil, 1994; Karlický, 1998; Hori, 1999): drifting pulsation structure (DPS). It is usually observed at the very beginning of eruptive solar flares (Karlický *et al.*, 2001). The radio observations of the October 5, 1992 flare reveal that the DPS was generated at times of a plasmoid ejection (Ohyama and Shibata, 1998; Kliem *et al.*, 2000). New positional measurements of the DPS

Space Science Reviews **107**: 81–88, 2003.

by the Nancay Radioheliograph (Khan *et al.*, 2002) confirmed this relationship. Based on the MHD numerical simulations, Kliem *et al.* (2000) suggested that every individual burst in the DPS is generated by superthermal electrons, accelerated at a maximum of the electric field in the quasi-periodic regime of the magnetic field reconnection. On the other hand, the global slow negative frequency drift of the DPS was explained by a plasmoid propagation upwards in the solar corona towards lower plasma densities. Furthermore, Hudson *et al.* (2001) identified a rapidly moving hard X-ray source, observed by the *Yohkoh*/HXT, associated with the moving microwave source and the plasmoid ejection seen in the *Yohkoh*/SXT images. The association with the high-frequency slowly drifting continuum was also reported. Therefore, it looks that not only the DPS but also the high-frequency drifting continua or other slowly drifting structures can indicate the plasmoid ejection.

In the present paper, first, examples of flares with slowly drifting structures are presented and their basic characteristics are summarized. The relationship of the drifting structure and the metric type II burst is elucidated in the case of the April 12, 2001 flare. Finally, the bursts under study are discussed using the model of solar eruptive flares (e.g. Yokoyama and Shibata, 2001).

2. Observations and Their Analysis

The August 18, 1998, $\sim$ 08:20 UT solar flare classified as X2.8/1B occurred at the east limb in NOAA AR8307 (N32E90) and was accompanied by an eruptive prominence, according to the Solar Geophysical Data event list. The hard X-ray emission as seen by the Hard X-ray Telescope (HXT) on board *Yohkoh* began as a gradual rise at $\sim$ 08:18 UT in all four energy bands of the instrument. An impulsive rise occurred at 08:19:30 UT in the M2 (33–53 keV) and H (53–93 keV) energy bands, followed by a series of pulses which lasted $\sim$ 2 min in total. The HXT images show a loop-shaped source in the M2 band (33–53 keV) with an indication of a loop-top source component; this loop-top source is indicated even in the H band image (53–93 keV). The *Yohkoh* Soft X-ray Telescope (SXT) took image series of the flare with different resolutions and filters, which reveal an ascending motion of blob structure (plasmoid) at times 08:19:54 to 08:21:32 UT. The blob had an average projected velocity along its upwards path $\sim$ 490 km s^{-1}.

At radio waves, in the 0.8–2.0 GHz frequency range, this flare commenced with a DPS at 08:18:30–08:20:46 UT (its most interesting part is shown in Figure 1). This time interval nearly coincides with the rise of the hard X-rays from onset to peak and includes the formation and initial acceleration of the ascending emission blob in the soft X-ray images. Remarkably, this DPS was well limited in frequency extent at both sides, most pulses were even amplified at the low- and high-frequency edges. Using the Fourier method the characteristic periods of the DPS at the frequency of 1.15 GHz were determined as: 24.0 s (the statistical probability of the period is 77.7%), and 8.8 s (72.5%) (Table I).

TABLE I

The basic characteristics of some drifting structures.

	Aug. 18, 1998	Apr. 12, 2001	Apr. 15, 2001
Start (UT)	8:18:30	10:17:20	13:37:27
Duration (s)	136	280	43
Global freq. drift (MHz s^{-1})	−4.4	−1.6	−4.7
Instantaneous bandwidth (MHz)	400	200–800	300
Period (s) (probability)	24.0 (77.7%)	75.0 (87.0%)	12.0 (89.3%)
	8.6 (72.5%)	25.0 (98.9%)	2.5 (80.7%)
		13.6 (85.4%)	1.5 (79.1%)
		9.4 (92.4%)	
Drift of pulses (MHz s^{-1})	Infinite	Infinite	−270
Accompanied bursts	dm-Continuum	Type II burst	−
Cross-Correl. Coeff.	0.3		0.6
Delay of hard X-rays (s)	−4		5–15

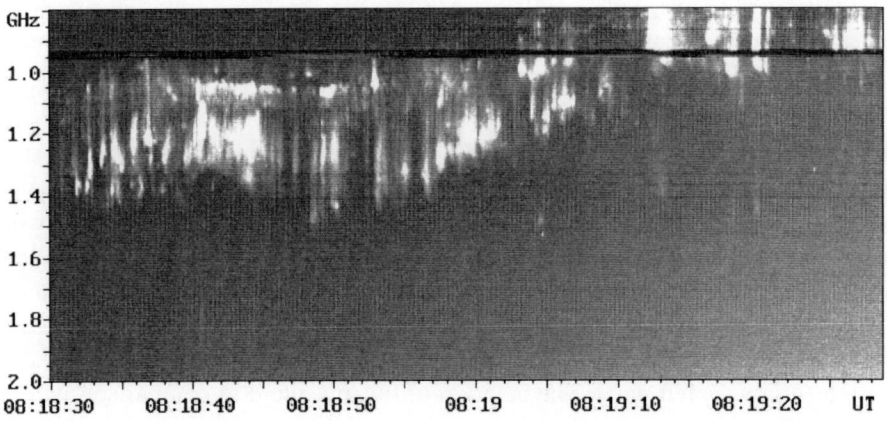

Figure 1. The 0.8–2 GHz radio spectrum observed by the Ondřejov radiospectrograph in August 18, 1998.

The April 15, 2001 flare belongs to the most intense flares in the present solar cycle. According to the GOES observations this flare started at 13:19 UT, reached maximum in soft X-rays at 13:50 UT, and ended at about 15:30 UT; its importance reached X14.4. Simultaneously, the Hα flare of the importance 2B was reported in the NOAA AR 9415 (at the position S20W85) at 13:36 UT, with maximum at 13:49 UT and ending at 15:35 UT. At the very beginning of the hard X-rays

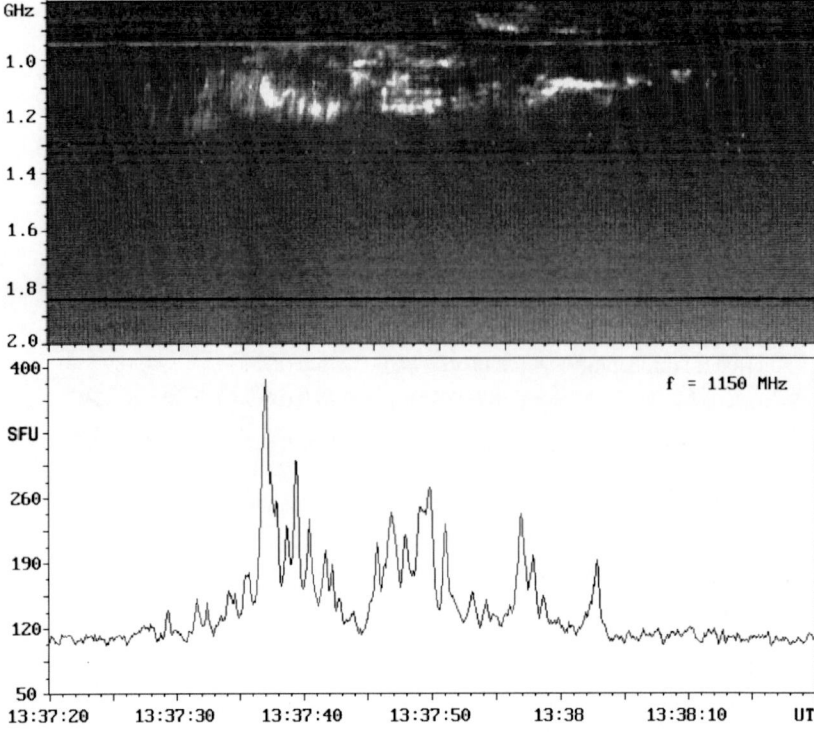

Figure 2. The 0.8–2.0 GHz radio spectrum showing the drifting structure at the very beginning of the April 15, 2001 flare (top) and the radio flux plot at the frequency of 1150 MHz (bottom) observed by the Ondřejov radiospectrograph.

(above 24 keV) of this flare, in the 0.8–1.3 GHz range the slowly drifting structure (DS) with the global negative frequency drift of -4.7 MHz s^{-1} was observed between 13:37:27 and 13:38:10 UT (Figure 2, Table I). Contrary to DPSs, in which pulses have infinite frequency drifts (see Kliem *et al.*, 2000, Karlický *et al.*, 2001), the individual bursts in this DS have the negative frequency drifts of about of -270 MHz s^{-1} (mainly at the beginning part of the DS). At the time of the DS the simultaneous TRACE observations (171 Å line, Fe IX, 0.9 MK) show the plasmoid ejection (Figure 3, left part). The position of this plasmoid in comparison with the *Yohkoh*/SXT bright loops is shown in Figure 3 (right part) by white contours. The speed of the plasmoid ejection in projection in the image plane was estimated to be 60 km s^{-1} in the upwards direction.

On April 12, 2001 a flare of the X2.0 importance (according to the GOES classification) was observed in the NOAA AR 9415 at 09:39 UT, with maximum at 10:28 UT, ending at 10:49 UT (NOAA Solar Events Report). The radio emission of this flare in the 40–4500 MHz range is shown in Figure 4. Here, not only type II burst bands can be seen in the 40–700 MHz range between 10:15 and 10:23 UT, but also the drifting pulsation-continuum structure in the frequency

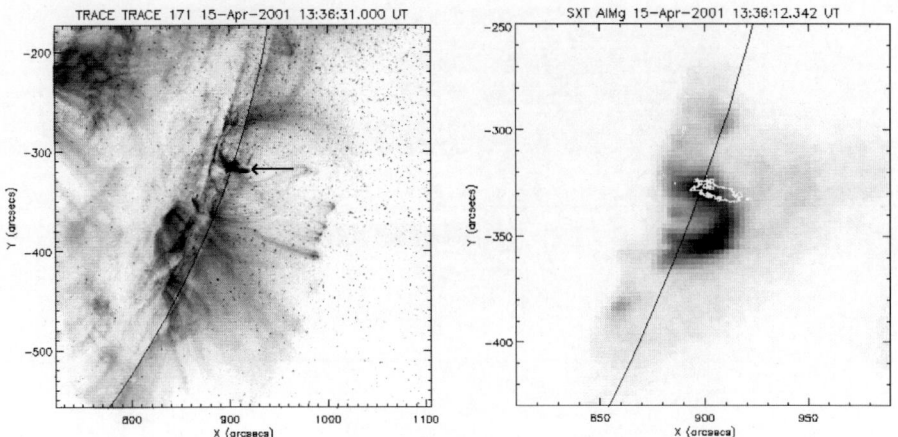

Figure 3. Left: TRACE 171 Å image of a plasmoid ejection (see the arrow) at the very beginning of the April 15, 2001 flare, at 13:36:41 UT, i.e. just before the DS observation. Right: *Yohkoh*/SXT image at 13:36:12 UT; the white contours show the position of the plasmoid observed by TRACE 171 Å at 13:38:36 UT.

range of 450–1500 MHz between 10:17:20 and 10:22:00 UT. The drift rate of the DS was -1.6 MHz s^{-1}. For basic characteristics of this DS, see Table I. In this flare the drifting structure is generated simultaneously with the type II radio burst, but in a different frequency range. Note also their similar frequency drift. At even higher frequencies these radio bursts are accompanied by fast drift bursts (electron beams?) at 10:18:40–10:21:00 UT in the 1.3–3.0 GHz range and by continuum in the 2.0–4.5 GHz range (low-frequency boundary of gyro-synchrotron emission).

For the drifting structures at the very beginning of flares (August 18, 1998 and April 15, 2001), where the radio emission is a relatively simple, their cross-correlations with the hard X-rays (35–57 keV) were made. While in the August 18, 1998 event no significant correlation was found, in the April 15, 2001 the correlation coefficient was 0.6 for the hard X-rays delayed of 5–15 s (Table I).

3. Discussion and Conclusions

Three typical examples of drifting structures (DSs) were presented. It looks that shorter DSs have higher absolute value of the frequency drifts, narrower bandwidths and shorter periods (Table I). It can be connected with the stability of the plasmoid.

In the model of solar eruptive flares with the plasmoid ejection (e.g. Yokoyama and Shibata, 2001), the magnetic field reconnection and plasma reconnection outflows form one self-consistent process, in which the current sheet is formed below the ejected plasmoid, the reconnection takes place between the slow mode shocks, and the plasmoid is simultaneously pushed upwards by the plasma reconnection

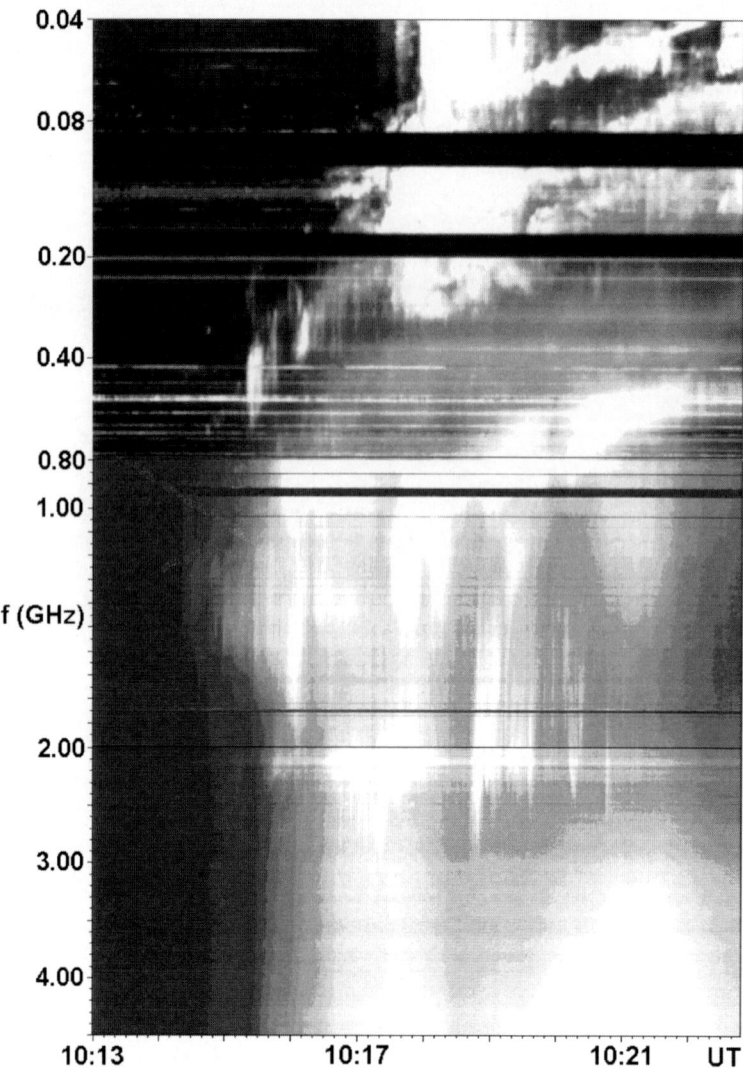

Figure 4. Potsdam radio spectrum in the 40–800 MHz band (courtesy Dr. A. Klassen) and Ondřejov radio spectrum in the 0.8–4.5 GHz band observed during the April 12, 2001 event. The drifting pulsation-continuum structure between 10:17:20 and 10:22:00 UT in the 0.45–1.5 GHz as well as the type II burst observed simultaneously in the metric frequency range are shown.

outflows. The bottom part of the plasmoid and the upper part of the lower-lying loops are obstacles for fast plasma reconnection outflows and thus there the fast mode (termination) shocks are formed. From the point of view of DSs and associated radio bursts the most important regions in the model are those where superthermal electrons are accelerated and trapped, i.e. the fast mode shocks, the MHD turbulence in the plasma reconnection outflows, the magnetically isolated plasmoid and the space limited by the slow and fast mode shocks. Further candidate for the radio emission is the fast mode shock which can be generated at the upper boundary of a rapidly ejected plasmoid. Accepting these ideas the electron density in the plasmoid is in the interval $2 \times 10^9 - 2 \times 10^{10}$ cm^{-3}. The cross-correlation of the DPS with hard X-rays indicate that the plasmoid is fully (August 18, 1998) or partially (April 15, 2001) magnetically closed. Very important observations were made during the April 12, 2001 flare where both the DS and the metric type II burst were observed simultaneously but in different frequency ranges. It indicates that the DS is not formed by large-scale travelling coronal shock front. Using all above mentioned ideas we propose that the DS of the April 12, 2001 flare was generated in the plasmoid and the type II burst in the shock above the plasmoid structure. This suggestion can be supported by a similar frequency drift of both the radio bursts (similar speed of a whole plasmoid structure). In the case of the DS at the very beginning of the X14.4 April 15, 2001 flare we found a clear association of this DS with the ejected plasmoid. The frequency drifts of individual bursts in this DS indicate density gradients inside the plasmoid.

We can see that not only the metric type II radio bursts, but also these new radio phenomena, especially the drifting structures at higher frequencies, indicate eruptive flare processes. Similarly as in the case of type II bursts we interpret their negative frequency drift as caused by a disturbance motion oriented upwards into higher solar atmospheric heights, i.e. towards lower plasma densities. Their frequency drifts express directly flare explosive motions and that is why these radio bursts are considered as radio signatures of the eruptive solar flares and can be used for the prediction of heliospheric effects.

Acknowledgements

The author thanks to Dr. A. Klassen for providing the Potsdam radio spectrum. This work was supported by the grant IAA3003202 of ASCR.

References

Dryer, M., Andrews M. D., Aurass, H. and 21 coauthors: 1998, 'The Solar Minimum Active Region 7978, its X2.6/1B Flare, CME, and Interplanetary Shock Propagation of 9 July 1996', *Solar Physics* **181**, 159–183.

Hori, K.: 1999, Study of Solar Decimetric Bursts with a Pair of Cutoff Frequencies, in *Proceedings of the Nobeyama Symposium, NRO Report*, **479**, 267–271.

Hudson, H. S., Kosugi, T., Nitta, N. and Shimojo, M.: 2001, 'Hard X-Radiation from a Fast Coronal Ejection', *Astrophys. J.* **561**, L211–L214.

Karlický, M.: 1992, Radio Emission of Eruptive Flares, in Z. Švestka, B.V. Jackson, and M.E. Machado (eds), Eruptive Solar Flares, Lecture Notes in Physics **399**, pp. 171–176.

Karlický, M.: 1998, 'Chromospheric Evaporation Shock and Reduced Optical Thickness Drifting in the 1–4.5 GHz Range', *Astron. Astrophysics* **338**, 1084–1088.

Karlický, M. and Odstrčil, D.: 1994, 'The Generation of MHD Shock Waves during the Impulsive Phase of the February 27, 1992 Flare', *Solar Physics* **155**, 171–184.

Karlický, M., Yan, Y., Fu, Q. *et al.*: 2001, 'Drifting Radio Bursts and Fine Structures in the 0.8–7.6 GHz Frequency Range observed in the NOAA 9077 AR (July 10–14, 2000) Solar Flares', *Astron. Astrophysics* **369**, 1104–1111.

Khan, J. I., Vilmer, N., Saint-Hilaire, P. and Benz, A. O.:2002, 'The Solar Coronal Origin of a Slowly Drifting Decimetric-Metric Pulsation Structure', *Astron. Astrophysics* **388**, 363–372.

Klassen, A., Aurass, H., Klein, K. L. *et al.*: 1999, 'Radio Evidence of Shock Wave Formation in the Solar Corona', *Astron. Astrophysics* **343**, 287–296.

Kliem, B., Karlický, M. and Benz, A. O.: 2000, Solar Flare Radio Pulsations as a Signature of Dynamic Magnetic Reconnection. *Astron. Astrophysics* **360**, 715–728.

Reiner, M.J. and Kaiser, M. L.: 1999, 'High-frequency type II Radio Emissions Associated with Shocks driven by Coronal Mass Ejection', *JGR* **104**, 16979–16992.

Ohyama, M. and Shibata, K.: 1998, 'X-ray Plasma Ejection Associated with an Impulsive Flare on 1992 October 5; Physical Conditions of X-ray Plasma Ejection', *Astrophys. J.* **499**, 934–944.

Švestka, Z., Jackson, B. V. and Machado, M. E.: 1992, Eruptive Solar Flares. *Lecture Notes in Physics* **399**. Springer-Verlag, Berlin, Heidelberg.

Yokoyama, T. and Shibata, K.: 2001, 'Magnetohydrodynamic Simulation of a Solar Flare with Chromospheric Evaporation Effect based on the Magnetic Reconnection Model', *Astrophys. J.* **549**, 1160–1174.

SOLAR MHD WAVES AND SOLAR NEUTRINOS

N. REGGIANI[1], M. M. GUZZO[2] and P. C. DE HOLANDA[3]

[1] *Faculdade de Matemática, Centro de Ciências Exatas Ambientais e de Tecnologias, Pontifícia Universidade Católica de Campinas, 13086-900 Campinas-SP, Brazil*
[2] *Instituto de Física Gleb Wataghin, Universidade Estadual de Campinas, UNICAMP, 13083-970 Campinas SP, Brazil*
[3] *The Abdus Salam International Centre for Theoretical Physics, I-34100 Trieste, Italy*

Abstract. We analyze here how solar neutrino experiments could detect time fluctuations of the solar neutrino flux due to magnetohydrodynamics (MHD) perturbations of the solar plasma. We state that if such time fluctuations are detected, this would provide a unique signature of the Resonant Spin-Flavor Precession (RSFP) mechanism as a solution to the Solar Neutrino Problem.

Key words: magnetic perturbations, neutrino, oscillation

1. Introduction

Assuming a non-vanishing magnetic moment of neutrinos, active electron neutrinos that are created in the Sun can interact with the solar magnetic field and be spin-flavor converted into sterile nonelectron neutrinos or into active nonelectron antineutrinos. This phenomenon is called Resonant Spin-Flavor Precession (RSFP). Particles resulting from RSFP interact with solar neutrino detectors significantly less than the original active electron neutrinos in such a way that this phenomenon can induce a depletion in the detectable solar neutrino flux (Cisneros, 1971; Volshin, Uysotskii and Okun, 1986; Lim and Marciano, 1988; Akhmedov, 1988; Balentekin, Hatchell and Loreti, 1990; Pulido, 1992; Akhmedov, Lanza and Petcov, 1993; Krastev, 1993).

If this interaction of the neutrinos with the solar magnetic field is the mechanism that explains the neutrino deficit on Earth (Lande *et al.*, 1998; Altmann *et al.*, 2000; Cattadori, 2001; Abdurashitov *et al.*, 2001; Fukuda *et al.*, 2001a, 2001b; Ahmad *et al.*, 2001), or, in other words, if the RSFP is the mechanism that solve the solar neutrino problem, solar magnetohydrodynamics (MHD) perturbations can lead to time fluctuations of the solar neutrino flux detected on Earth. This can be easily understood. The solar active electron neutrino survival probability based on the RSFP mechanism crucially depends on the values of four independent quantities. Two of them are related to the neutrino properties: its magnetic moment μ_ν and the squared mass difference of the physical eigenstates involved in the conversion mechanism divided by their energy $\Delta m^2/E$. The other two quantities are related to the physical environment in which neutrinos are inserted: the magnetic field profile $B(r)$ and the electron and neutron number density distribution $N(r)$ along the neutrino trajectory. MHD affects both the magnetic field profile as well as the matter density and therefore its effects will strongly influence the RSFP neutrino survival probability. Indeed we believe that such consequences can be thought as a test to this solution to the solar neutrino problem based on the RSFP mechanism (Guzzo, Reggiani and Colonia, 1997; Guzzo *et al.*, 2000; Reggiani *et al.*, 2000; Guzzo, de Holanda and Reggiani, 2002).

2. MHD Perturbations

The MHD perturbations were calculated deriving the MHD equations near the solar equator, the region relevant for solar neutrinos. Using cylindrical coordinates, considering also the effects of gravity, we obtained the Hain–Lüst equation with gravity (Priest, 1987; Goedbloed, 1971; Goedbloed and Sakanaka, 1974; Guzzo, Reggiani and Colonia, 1997):

$$\frac{\partial}{\partial r}\left(f(r)\frac{\partial}{\partial r}(r\xi_r)\right) + h(r)\xi_r = 0 \tag{1}$$

where

$$f(r) = \frac{\gamma p + B_o^2}{r}\frac{(w^2 - w_A^2)(w^2 - w_S^2)}{(w^2 - w_1^2)(w^2 - w_2^2)}, \tag{2}$$

$$h(r) = \rho_0 w^2 - k^2 B_0^2 + g\frac{\partial \rho_0}{\partial r}$$

$$-\frac{1}{D}g\rho_0^2(w^2\rho_0 - k^2 B_0^2)\left[gH + \frac{w^2}{r}\right] \tag{3}$$

$$-\frac{\partial}{\partial r}\left[\frac{1}{D}w^2\rho_0^2 g(w^2\rho_0 - k^2 B_0^2)\right]$$

and

$$w_A^2 = \frac{k^2 B_0^2}{\rho_0}, \quad w_S^2 = \frac{\gamma p}{\gamma p + B_0^2}\frac{k^2 B_0^2}{\rho_0}, \tag{4}$$

$$w_{1,2}^2 = \frac{H(\gamma p + B_0^2)}{2\rho_0}\left\{1 \pm \left[1 - 4\frac{\gamma p k^2 B_0^2}{(\gamma p + B_0^2)^2 H}\right]^{1/2}\right\}, \tag{5}$$

$$D = \rho_0^2 w^4 - H[\rho_0 w^2(\gamma p + B_0^2) - \gamma p k^2 B_0^2], \tag{6}$$

$$H = \frac{m^2}{r^2} + k^2 \tag{7}$$

where ξ_r is the radial component of the plasma displacement $\vec{\xi}$, g is the acceleration due to gravity, p is the pressure, $\gamma = C_p/C_v$ is the ratio of the specific heats, ρ_0 is the equilibrium matter density profile and B_0 is the magnetic equilibrium profile in the Sun. In this derivation we considered the equilibrium magnetic profile B_0 in the z direction for the cylinder coordinates we use.

For the equations above we considered for the solar matter density distribution, ρ_0, the standard solar model prediction, that is, approximately monotonically decreasing exponential functions in the radial direction from the center to the surface of the Sun (Bahcall and Ulrich, 1988; Bahcall, 1989; Bahcall and Pinsonneault, 1992; Bahcall, Basu and Pinsonneault, 2001). The density profile was used to calculate the acceleration of gravity. The pressure p is related with the density by the adiabatic equation of state $p \sim 5 \times 10^{14}\rho^\gamma$, which is obtained from the values of density and temperature of the solar standard model.

The Hain–Lüst equation shows singularities when $f(r)$ given in Equation (2) is equal to zero, that is, when $w^2 = w_A^2$ or $w^2 = w_S^2$, which regions in the w^2 space are called Alfvén and slow continua, respectively. In the interval $0 \leq r \leq 1$ the functions w_A^2 and w_S^2 take continuous values that define the ranges of the values of w^2 that correspond to improper eigenvalues, associated with localized modes. Eigenvalues of the Hain–Lüst equation must be searched, therefore, outside the regions where $w^2 = w_A^2$ or $w^2 = w_S^2$, and they define the global modes which are associated with magnetic and density waves along the whole radius of the Sun.

The global modes were obtained solving numerically the Hain-Lüst equation with gravity, imposing appropriate boundary conditions to $\vec{b}_1$ and ρ_1, the magnetic and density perturbations respectively, given by $\vec{b}_1 = \vec{\nabla} \times (\vec{\xi} \times \vec{B}_0)$ and $\rho_1 = \vec{\nabla} \cdot (\rho \vec{\xi})$ (Goedbloed, 1971; Goedbloed and Sakanaka, 1974).

The matter density fluctuations are very constrained by helioseismology observations. The largest density fluctuations ρ_1 inside the Sun are induced by temperature fluctuations δT due to convection of matter between layers with different local temperatures. According to an estimate of such an effect (Guzzo, Reggiani and Colonia, 1997), we assume density fluctuations ρ_1/ρ_0 smaller than 10%. The size of the amplitude b_1 is not very constrained by the solar hydrostatic equilibrium, since the magnetic pressure $B_0^2/8\pi$ is negligibly small when compared with the dominant gas pressure for the equilibrium profiles considered. Despite this fact, it cannot be arbitrarily large when we are solving the Hain–Lüst equation. This equation is obtained after linearization of the magnetohydrodynamics equations, which requires that the solution $\vec{\xi}$ must be very small, $|\vec{\xi}| << 1$, so that the nonlinear terms can be neglected. Moreover, we must have a clear distinction between the maximum and minimum magnetic field. In order to satisfy these criteria and have a significant effect, we choose the maximum value of the perturbation such that $|b_1|/|B_0| \sim 0.5$.

For the localized modes we adopted the following phenomenological assumption in our calculations: continuum modes introduce Gaussian-shaped magnetic fluctuations centered in r_s, with width δr, and amplitude given by a fraction of the equilibrium magnetic field in the position of the singularity, in such a way that the magnitude of the transverse component of the magnetic field, which is the relevant magnetic component for neutrino RSFP, will fluctuate in the following way:

$$|\vec{B}_{\perp}(r)| = |\vec{B}_0(r)| + b_0|\vec{B}_0(r_s)| \exp\left[-\left(\frac{r - r_s}{\delta r}\right)^2\right] \sin\left[w(r_s)t\right]. \tag{8}$$

where b_0 is the amplitude factor, a positive numerical value smaller than 1.

3. Solar Neutrino Evolution

If we consider a non-vanishing neutrino magnetic moment, the interaction of such neutrinos with this magnetic field will generate neutrino spin-flavor conversion which is given by the evolution Equations (1) in natural units, for ultra-relativistic neutrinos:

$$i\frac{d}{dr}\begin{pmatrix} \nu_L \\ \nu_R \end{pmatrix} = \begin{pmatrix} \frac{\sqrt{2}}{2}G_F N_{eff}(r) - \frac{\Delta m^2}{4E} & \mu_\nu|B_\perp(r)| \\ \mu_\nu|B_\perp(r)| & +\frac{\sqrt{2}}{2}G_F N_{eff}(r) + \frac{\Delta m^2}{4E} \end{pmatrix}\begin{pmatrix} \nu_L \\ \nu_R \end{pmatrix}, \tag{9}$$

where ν_L (ν_R) is the left- (right-) handed component of the neutrino field, Δm^2 is the squared mass difference of the corresponding physical fields, E is the neutrino energy, G_F is the Fermi constant, μ_ν is the neutrino magnetic moment and $|B_\perp(r)|$ is the transverse component of the perturbed magnetic field. Finally, we have $N_{eff} = N_e(r) - N_n(r)$ for Majorana neutrinos, where $N_e(r)$ ($N_n(r)$) is the electron (neutron) number density distribution, in which case the final right-handed states ν_R are active non-electron antineutrinos.

MHD magnetic and density fluctuations, b_1 and ρ_1, induced by global or localized modes, can alter the neutrino evolution since they can induce time variation of the transverse component of the magnetic field $|B_\perp(r)|$ as well as the matter density $N_e(r)$ appearing in the evolution equation above. Therefore, the MHD fluctuations can induce a time variation on the survival probability of the neutrinos, that can be detected in the experiments on Earth.

If we consider just the interaction of the neutrinos with a constant magnetic field, we will not be able to distinguish this effect from other neutrino conversion mechanism. And our goal is not just to predict the neutrino conversion but to test the mechanism responsible for it. Futhermore if the Ressonant Spin-Flavour Precession is the mechanism responsible for the neutrino conversion, the neutrino observations can be used as probes to the inner part of the Sun.

4. Localized Modes

The effect of the localized modes were estimated calculating the survival probability of an active solar neutrino to reach the solar surface after having interacted with the solar magnetic field perturbed by a localized MHD wave (Guzzo, Reggiani and Colonia, 1997; Guzzo et al., 2000). In this estimation, we considered for the equilibrium magnetic fields the following profiles:

$$B_0(r) = \begin{cases} 1 \times 10^6 \left(\frac{0.2}{r+0.2} \right)^2 \text{G} & \text{for } 0 < r \leq r_{convec} \\ B_C(r) & \text{for } r > r_{convec} , \end{cases} \tag{10}$$

where B_C is the magnetic field in the convective zone given by the following profiles:

$$B_C(r) = 4.88 \times 10^4 \left[1 - \left(\frac{r - 0.7}{0.3} \right)^n \right] \text{G} \quad \text{for } r > r_{convec} \tag{11}$$

or

$$B_C(r) = 4.88 \times 10^4 \left[1 + exp \left(\frac{r - 0.95}{0.01} \right)^{-1} \right] \text{G} \quad \text{for } r > r_{convec} \tag{12}$$

with $n = 2, 6$ and 8 and $r_{convec} = 0.7$. This profile was used by Akhmedov, Lanza and Petcov (1993) to show the consistency of the solar neutrino data with the RSFP phenomenon.

Note that $B_0^2 < \gamma p$, which is related to the fact that the magnetic pressure is negligibly small when compared to the gas pressure. Therefore, from Equations (4), $w_A^2 \sim w_S^2$ and for the equilibrium profiles considered above, the period of the fluctuation centered at the point of singularity varies from 1 to 10 days.

We note that the presence of localized magnetic fluctuations will modify the solar neutrino survival spectrum (Guzzo, Reggiani and Colonia, 1997; Guzzo et al., 2000). Fluctuations of the survival probability are maximal when r_s is close to a resonance region and are well localized in $\Delta m^2/4E$-space for $r_s \leq 0.7$. For $r_s = 0.9$ the position of the probability fluctuations in $\Delta m^2/4E$-space is less determined. Nevertheless localized magnetic waves near the solar surface lead to a picture for the survival probability $P(\nu_L \rightarrow \nu_L)$ which can be easily distinguishable from those ones generated by inner localized waves.

5. Global Modes

The global modes obtained with different magnetic fields used in the analysis of the effect of localized waves (Reggiani et al., 2000) are very similar. So, we present the effect of these modes on the neutrino RSFP phenomenon for just one of these magnetic fields, that we consider a good representative of the others:

$$B_0 = B_0(r) = \begin{cases} 1 \times 10^6 \left(\frac{0.2}{r+0.2}\right)^2 G & \text{for } 0 < r \leq r_{convec} \\ B_C(r) & \text{for } r > r_{convec}, \end{cases} \tag{13}$$

where B_C is the magnetic field in the convective zone given by the following profiles:

$$B_C(r) = 4.88 \times 10^4 \left[1 - \left(\frac{r - 0.7}{0.3}\right)^n \right] G \quad \text{for } r > r_{convec} \tag{14}$$

with $n = 6$ and $r_{convec} = 0.7$.

In order to illustrate the effect of the parametric resonance, we used other magnetic fields that have been used by different authors to solve the solar neutrino problem through the RSFP mechanism (Guzzo and Nunokawa,):

$$B_C(r) = \begin{cases} B_{initial} + \left[\frac{B_{max} - B_{initial}}{r_{max} - r_{convec}}\right](r - r_{convec}) & \text{for } r_{convec} < r < r_{max} \\ B_{max} + \left[\frac{B_{max} - B_{final}}{r_{max} - 1.0}\right](r - r_{max}) & \text{for } r > r_{max} \end{cases} \tag{15}$$

where $B_{initial} = 2.75 \times 10^5$ G, $B_{max} = 1.18 \times 10^6$ G, $B_{final} = 100$ G, $r_{convec} = 0.65$ and $r_{max} = 0.8$. Although the magnetic field in this configuration seems to be too strong to be present in the convective layer of the Sun, we extend our analysis for this configuration because it is very useful to illustrate the parametric resonance effect for different values of the perturbation oscillation length. It is also important to notice that the important quantity for the neutrino evolution is not the magnetic field itself, but the product $\mu_\nu |\vec{B}_\perp(r)|$, which we impose to be of the same magnitude in the Sun convective layer for all magnetic field configuration chose here.

We considered also a third field, constant all over r, given by (Minakata and Nunokawa, 1989; Balentekin, Hatchell and Loreti, 1990; Nunokawa and Minakata, 1993):

$$B_0 = 253 \text{ kG} \quad \text{for } 0 < r < 1.0 \tag{16}$$

We were interested in calculating the eigenfunctions of the Hain–Lüst equation out of the continua determined by the functions $w^2 = w_A^2$ and $w^2 = w_S^2$. The Hain–Lüst solutions calculated were found in the region of the MHD spectrum in which frequencies are smaller than the continuum frequencies: $w^2 < w_A^2 \approx w_S^2$, which period is of order of 10 days. The period of the solutions found above the continua are smaller than O (1 sec), very tiny, therefore, to be detected by present experiments.

In this case, we note that the range of the values of $\Delta m^2/4E$ for which the effect of the global magnetic perturbation is significant varies for each of the magnetic field profile considered. This is a direct consequence of the appearance of a parametric resonance (Ermilova, Tsarev and Chechin, 1986; Akhmedov, 1988; Krasten and Smirnov, 1989) in the evolution of the neutrino due to the MHD perturbations along its trajectory. To understand this effect we have to consider the neutrino oscillation length. When we have a neutrino oscillation length similar to the wavelength of the magnetohydrodynamic perturbations, a significant enhancement of the neutrino chirality conversion occurs. This is the parametric resonance which is clearly observed in the neutrino survival probability. If the perturbation wavelength is very different from the neutrino oscillation length than this effect will not be relevant.

6. Observing MHD Fluctuations in Solar Neutrino Detectors

According to these results and equilibrium profiles used, we conclude that neutrinos with energy of the order 1 MeV will be very sensitive to global MHD fluctuations if RSFP phenomenon is the solution of the solar neutrino anomaly (Guzzo, de Hollanda and Reggiani, 2002). Some operating solar neutrino detectors that are sensitive to such energy range, like Homestake (Lande *et al.*, 1998), Gallex/GNO (Hampel *et al.*, 1999; Altmann *et al.*, 2000; Cattaduri, 2001) and Sage (Abdurashitov *et al.*, 2001) can not detect these fluctuations because they do not operate in a real time basis, and such small fluctuations will be averaged out over the detection time. The Super-Kamiokande (Fukuda *et al.*, 2001) and SNO (Ahmad *et al.*, 2001) detectors operate in real-time basis but have a too high threshold in the neutrino energy to be sensitive to such fluctuations. So, it is necessary to consider (Guzzo, de Holanda and Reggiani, 2002) the detectors that operate in a real time basis and that have a low threshold in the neutrino energy, like Borexino (Malvezzi, 1998), Hellaz (Patzak, 1998) and Heron (Lanou, 1999).

The Borexino experiment (Malvezzi, 1998): this experiment will be able to measure the Berilium line neutrinos, in a real time basis. Since the Berilium neutrinos have a fixed energy ($E = 0.863$ MeV), it is quite easy to predict the time dependence of the neutrino signal in Borexino for a given Δm^2. Fixing the neutrino energy and taking the magnetic field normalization $f_{B_0} = 5$, within 99% C.L. no reasonable time fluctuation will be felt by this experiment. But although we can not use MHD perturbations to test the RSFP solution in Borexino, it has been recently discussed (Akhmedov and Pulido, 2002) how the low value of the expected rate of the Berilium line neutrinos on this experiment would be a clear indication of the RSFP mechanism.

The experiments Hellaz (Patzak, 1998) and Heron (Lanou, 1999): these experiments will utilize the elastic reaction, $\nu_{e,\mu,\tau} + e^- \rightarrow \nu_{e,\mu,\tau} + e^-$, for real-time detection in the energy region dominated by the pp and ^7Be neutrinos. Since the MHD fluctuations we found in (Reggiani *et al.*, 2000) appear to be affecting neutrinos with an energy range of the order of the pp-neutrinos energy, maybe Hellaz and/or Heron would be able to feel the time fluctuations on the neutrino signal generated by the MHD fluctuations. Since we expect something around ~ 7 pp-events/day on experiments like Hellaz or Heron we can see that, for one year (365 days, or ~ 2500 events) of data taking, in principle it is possible to distinguish such fluctuations in Hellaz experimental results.

Acknowledgements

The authors would like to thank FAPESP, CAPES and CNPq for several financial supports.

References

Abdurashitov, D. N. *et al.*: 2001, (SAGE Collaboration), *Nucl. Phys.* (Proc. Suppl.) **91**, 36; latest results from SAGE homepage: http://EWIServer.npl.washington.edu/SAGE/.
Ahmad, Q. R. *et al.* (SNO Collaboration): 2001, *Phys. Rev. Lett.* **87**, 071301.
Akhmedov, E.: 1988, *Yad. Fiz.* **47**, 475.
Akhmedov, E. Kh.: 1988, *Phys. Lett. B* **213**, 64.
Akhmedov, E. Kh., Lanza, A. and Petcov, S. T.: 1993, *Phys. Lett. B* **303**, 85.
Akhmedov, E. Kh., Lanza, A. and Petcov, S. T.: 1993, *Phys. Lett. B* **303**, 85.
Akhmedov, E. Kh. and Pulido, J.: hep-ph/0201089, *Phys. Lett. B* 2002/529 (2002), 193–198.
Altmann, M. *et al.*: 2000, (GNO Collaboration), *Phys. Lett. B* **490**, 16.
Bahcall, J. N.: 2001, *Neutrino Astrophysics*, Cambridge University Press, Cambridge, England.
Bahcall, J. N., Basu, S. and Pinsonneault, M. H.: 2001, *Astrophys. J.* **555**, 990.
Bahcall, J. N. and Pinsonneault, M. H.: 1992, *Rev. Mod. Phys.* **64**, 885.
Bahcall, J. .N. and Ulrich, R. K.: 1988, *Rev. Mod. Phys.* **60**, 297.
Balantekin, A. B., Hatchell, P. J. and Loreti, F.: 1990, *Phys. Rev. D* **41**, 3583.
Balantekin, A. B., Hatchell, P. J. and Loreti, F.: 1990, *Phys. Rev. D* **41**, 3583.
Cattadori, C.: 2001, on Behalf of GNO Collaboration, Talk Presented at TAUP2001, 8–12 September 2001, Laboratori Nazionali del Gran Sasso, Assergi, Italy.
Cisneros, A.: 1971, *Astron. Space Sci.* **10**, 87.
Ermilova, V. K., Tsarev, V. A. and Chechin, V. A.: 1986, *Kr. Soob, Fiz.*, Lebedev Inst. **5**, 26.
Fukuda, S. *et al.*, (SuperKamiokande Collaboration): 2001a, *Phys. Rev. Lett.* **86**, 5651.
Fukuda, S. *et al.*, (SuperKamiokande Collaboration): 2001b, *Phys. Rev. Lett.* **86** 5656.
Goedbloed, J. P.: 1974, *Physica* **53**, 412.
Goedbloed, J. P. and Sakanaka, P. H.: 1974, *Phys. Fluids* **17**, 908.
Guzzo, M. M. and Nunokawa, H.: 'Current Status of the Resonant Spin-flavor Solution to the Solar Neutrino Problem', HEP-PH/9810408.
Guzzo, M. M., de Holanda, P. C. and Reggiani, N.: 2002, 'MHD Fluctuations and Low Energy Solar Neutrinos', *Eur. Phys. J. C* (in press).
Guzzo, M. M., Reggiani, N., Colonia, J. H. and de Holanda, P. C.: 2000, *Braz. J. Phys.* **30**, 594.
Guzzo, M. M., Reggiani, N. and Colonia, J. H.: 1997, *Phys. Rev. D* **56**, 588.
Hampel, W. *et al.*, (GALLEX Collaboration): 1999, *Phys. Lett. B* **447**, 127.
Krastev, P. I.: 1993, *Phys. Lett. B* **303**, 75.
Krastev, P. I. and Smirnov, A. Yu.: 1989, *Phys. Lett. B* **226**, 341.
Lande, K. *et al.*: 1998, (Homestake Collaboration), *Astrophys. J.* **496**, 505.
Lanou, R. E.: 1999, *Nucl. Phys. Proc. Suppl.* **77**, 55–63; in M. Baldo Ceolin (ed.), Proceedings of the 8th International Workshop on Neutrino Telescopies, Venice, Italy, 1999, Vol. I, pp. 139.
Lim, C. S. and Marciano, W. J.: 1988, *Phys. Rev. D* **37**, 1368.
Malvezzi, S.: 1998, *Nucl. Phys. B* (Proc. Suppl.) **66**, 346.

Minakata, H. and Nunokawa, H.: 1989, *Phys. Rev. Lett.* **63**, 121.

Nunukawa, N. and Minakata, H.: 1993, *Phys. Lett. B* **314**, 371.

Patzak, T.: 1998, *Nucl.Phys. B* (Proc. Suppl.) **66**, 350.

Priest, E. R.: 1987, *Solar Magnetohydrodynamics*, D. Reidel Publishing Company, Dordrecht.

Pulido, J.: 1992, *Phys. Rep.* **211**, 167.

Reggiani, N., Guzzo, M. M., Colonia-Bartra, J. H. and de Holanda, P. C.: 2000, *Eur. Phys. J. C* **12**, 269.

Voloshin, M. B., Vysotskii, M. I. and Okun, L. B.: 1986, *Yad. Fiz.* **44**, 677, *Sov. J. Nucl. Phys.* **44**, 440.

VARIATIONS OF THE MAGNETIC FIELDS IN LARGE SOLAR FLARES

H. SCHUNKER and A. -C. DONEA

Department of Physics & Mathematical Physics,
The University of Adelaide, Adelaide, SA 5005, Australia

Abstract. We present preliminary results from high resolution observations obtained with the Michelson Doppler Imager (MDI) instrument on the SOHO of two large solar flares of 14 July 2000 and 24 November 2000. We show that rapid variations of the line-of-sight magnetic field occured on a time scale of a few minutes during the flare explosions. The reversibility/irreversibility of the magnetic field of both active regions is a very good tool for understanding how the magnetic energy is released in these flares. The observed sharp increase of the magnetic energy density at the time of maximum of the solar flare could involve an unknown component which deposited supplementary energy into the system.

Key words: magnetic fields, solar flares, sun

1. Introduction

Recent observations using SOHO-MDI magnetograms (Kosovichev and Zharkova, 2001) show clear signatures of magnetic energy release temporally coincident with large flares which suggests that the energy of solar flares is largely magnetic. In the corona, solar flares eject charged supersonic particles that hit the photosphere, causing a compression which in turn, may cause the Sun to vibrate and produce sunquakes (Kosovichev and Zharkova, 1998) and seismic signatures (Donea *et al.*, 1999). Therefore a significant fraction of the energy involved in the production of solar flares could be delivered back into the photosphere by fast particles.

The acoustic emission from solar flares cannot be easily detected and this may be due to the fact that even a weak, local magnetic field is sufficient to partially suppress the acoustic waves (Braun *et al.*, 1992). Since large solar flares are usually associated with complex active regions, it is expected that the acoustic power available for the production of a sunquake could be partially or totally engulfed by the large magnetic field strengths existing in active regions. Therefore, it is not clear whether the active regions responsible for large solar flares are the best candidates for detecting sunquakes, since the magnetic fields may act as a barrier and prevent the seismic waves from spreading.

This work is the first step in understanding what is the energy balance between the magnetic energy delivered into the flare and the energy possibly deposited back into the photosphere and responsible for acoustic signatures. The purpose of this

Space Science Reviews **107**: 99–102, 2003.
© 2003 *Kluwer Academic Publishers.*

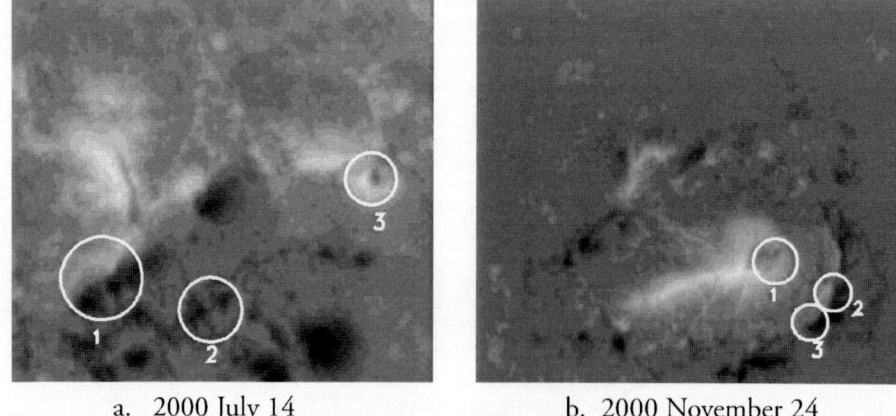

a. 2000 July 14   b. 2000 November 24

Figure 1. High resolution SOHO-MDI magnetograms obtained on 14 July 2000 (left) and 24 November 2000 (right). Circles indicate regions where reference masks are applied.

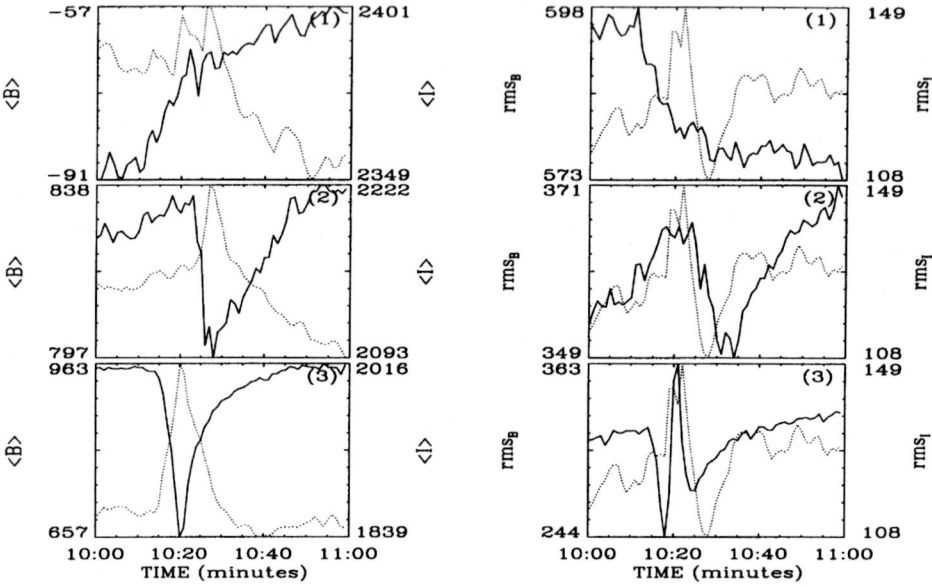

Figure 2. Solar flare of 14 July 2000. Left column: time series of $\langle B \rangle$ (solid curves) and $\langle I \rangle$ (dotted curves) in each of the three circular regions indicated by markers in Figure 1; Right column: times series of the rms_B and rms_I.

paper is to produce time series of the magnetic field and continuum intensity of two large active regions, in an effort to understand the sudden release of the magnetic energy in solar flares.

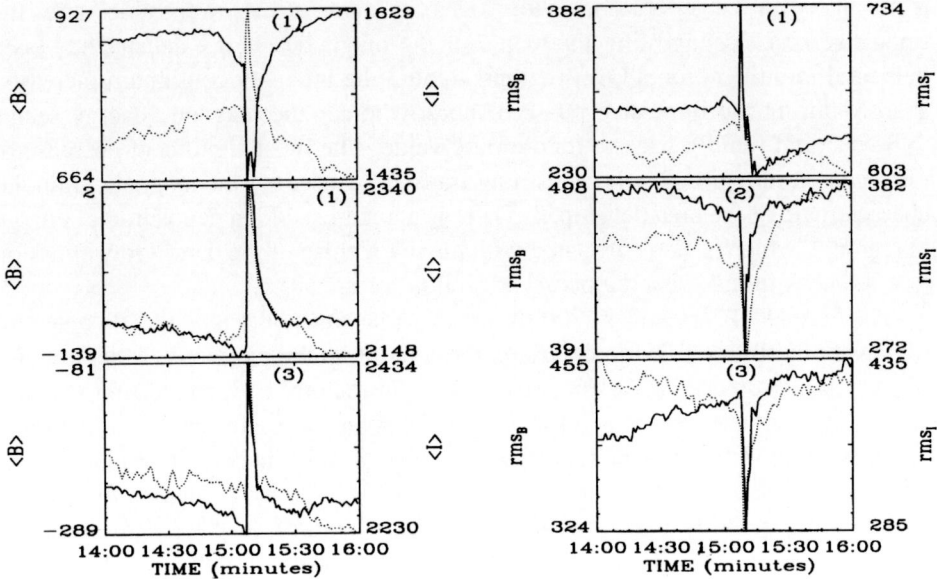

Figure 3. Solar flare of 24 November 2000: same as in Figure 2.

2. Temporal Character of Magnetic Fields During Flares

Here we analyse the observations of the 14 July 2000 (NOAA 9077) and 24 November 2000 (NOAA 9236) solar flares carried out by the MDI instrument on SOHO (Scherrer *et al.*, 1995) in the high-resolution mode with a pixel size of $0''.61(\approx 0.435$ Mm$)$. The data analysed here consist of time series of MDI line-of-sight magnetic field and continuum intensity lasting 1 hour for NOAA 9077 and 2 hours for NOAA 9236. We examine the temporal character of the areas indicated by the markers 1–3 (Figure 1) for magnetograms and intensity continuum diagrams of each flare.

The flare of 14 July 2000 started at about 10:10 UT and reached the peak soft X-ray flux (X5.7) at 10:24 UT. The first significant variations in the line-of-sight magnetic field are detected at around 10:10 UT. The flare of 24 November 2000 started at about 14:51 UT and reached the peak soft X-ray flux (X23) at 15:13 UT. The first significant variations in the line-of-sight magnetic field are detected at 14:20 UT, before the flare maximum. Figures 2 and 3 show the evolution during flare life time of the following quantities: the mean values of the line-of-sight magnetic field $\langle B \rangle$, the continuum intensity $\langle I \rangle$, the rms of the magnetic field rms$_B$ and the rms of continuum intensity rms$_I$, for regions 1–3 (as marked in Figure 1).

The line-of-sight magnetic field varies rapidly during the 14 July 2000 solar flare. At the impulse phase, $\langle B \rangle$ decreased by (34, 41, 306) G in regions 1–3, whereas rms$_B$ decreased by (25, 22, 70) G, respectively. Area 1 covers the neutral magnetic line where the flare is believed to occur and here the rms magnetic variations show an irreversible behaviour (rms_B remains at its lower values after the impulsive phase) suggesting that some of the magnetic energy density had been delivered into the corona. Over regions 2 and 3 the intensity continuum increased sharply during the impulsive phase of flare, whereas the magnetic energy seems to be restored rapidly back to its pre-flare value. The most significant decrease of $\langle B \rangle$ occured in region 3, where a strong transient caused by the magnetic unipolar characteristic of the area developed. There is a notable sudden deposition of energy in region 3, which is also correlated with the main pulse of the hard X-ray emission (see the peak in the rms_B temporary evolution for area 3).

The flare of 24 November 2000 shows the same signs of magnetic field reversibility as those observed in the first flare, throughout its three regions. At its impulse phase, rms_B decreased by $\sim (60, 107, 104)$ G in regions 1–3, suggesting that significant magnetic energy was released into the solar flare. We see that in region 1, the restoration of the magnetic energy to the pre-flare value is not completly done. The sharp increase of rms_B and rms_I in region 1 could be related to the sudden change of the polarity which occured at the maximum of the flare. Time series of $\langle B \rangle$ in regions 2 and 3 (Figure 3) reveal strong pulses at the moment of maximum of the flare.

In conclusion, the observed sharp peaks of the mean magnetic field $\langle B \rangle$ could be attributed to an increase of the magnetic shear at the location of the solar flare, contrary to what is expected if magnetic energy is released (Wang *et al.*, 1994). Also, entangled magnetic ropes may connect different locations within the same complex active region allowing for deposition of energy back into the photosphere. The mechanism that produces a surplus of energy at the moment of the solar flare (seen in the variation of rms_B) is still unclear, but surely this source could be relevant for the ignition of solar flare induced sunquakes.

Acknowledgement

We thank Dr. Charlie Lindsey for helpful discussions.

References

Kosovichev, A. G. and Zharkova, V. V.: 2001, *Astrohys. J.* **550**, L105.
Kosovichev, A. G. and Zharkova, V. V.: 1998, *Nature* **393**, 317.
Braun, D. *et al.*: 1992, *Astrophys. J.* **391**, 113.
Donea, A.-C., Braun, D. C. and Lindsey, C.: 1999, *Astrophys. J.* **513L**, 143.
Scherrer, P. H. *et al.*: 1995, *Solar Phys.* **162**, 129.
Wang, H. *et al.*: 1995, *Astrophys. J.* **424**, 436.

THE PROGRESS OF THE TAIWAN OSCILLATION NETWORK PROJECT

MING-TSUNG SUN[1], DEAN-YI CHOU[2] and THE TON TEAM*

[1]*Department of Mechanical Engineering, Chang-Gung University,
Tao-Yuan, 33333, Taiwan, R.O.C.*
[2]*Physics Department, Tsing Hua University, Hsinchu, 30043, Taiwan, R.O.C.*

Abstract. We describe the present status of the project of the Taiwan Oscillation Network (TON) and discuss a scientific result using the TON data. The TON is a ground-based network to measure solar intensity oscillations for the study of the solar interior. Four telescopes have been installed in appropriate longitudes around the world. The TON telescopes take K-line full-disk solar images of diameter 1000 pixels at a rate of one image per minute. The data has been collected since October of 1993. The TON high-spatial-resolution data are specially suitable for the study of local properties of the Sun. In 1997 we developed a new method, acoustic imaging, to construct the acoustic signals inside the Sun with the acoustic signals measured at the solar surface. From the constructed signals, we can form intensity map and phase-shift map of an active region at various depths. The direct link between these maps and the subsurface wave-speed perturbation suffers from the poor vertical resolution of acoustic imaging. Recently an inversion method has been developed to invert the measured phase travel time perturbation to estimate the distribution of wave-speed perturbation based on the ray approximation. This technique of acoustic imaging has been used to image the far-side of the Sun that could provides information on space weather prediction.

Key words: helioseismology, solar interior, solar magnetic fields

1. Taiwan Oscillation Network

The Taiwan Oscillation Network (TON) is a ground-based network measuring K-line intensity oscillation for the study of the internal structure of the Sun. The TON project has been funded by the National Research Council of ROC since 1991. The plan of the TON project is to install several telescopes at appropriate longitudes around the world to continuously measure the solar oscillations. So far, four telescopes have been installed. The first telescope was installed at the Teide Observatory, Canary Islands, Spain in 1993. The second and third telescopes were installed at the Huairou Solar Observing Station near Beijing and the Big Bear Solar Observatory, California, USA in 1994. The fourth telescope was installed in Tashkent, Uzbekistan in 1996.

*The TON Team includes: Antonio Jimenez (Instituto Astrofisica de Canarias, Spain); Guoxiang Ai and Honqi Zhang (Huairou Solar Observing Station, P.R.C.); Philip Goode and William Marquette (Big Bear Solar Observatory, U.S.A.); Shuhrat Ehgamberdiev and Oleg Ladenkov (Ulugh Beg Astronomical Institute, Uzbekistan).

Space Science Reviews **107**: 103–106, 2003.
© 2003 *Kluwer Academic Publishers.*

The TON is designed to obtain informations on high-degree solar p-mode oscillations, along with intermediate-degree modes. A discussion of the TON project and its instrument is given by Chou et al. (1995). Here we give a brief description. The TON telescope system uses a 3.5-inch Maksutov-type telescope. The annual average diameter of the Sun is 1000 pixels. A K-line filter, centered at 3934Å, of FWHM $= 10$ Å and a prefilter of FWHM $= 100$ Å are placed near the focal plane. The measured amplitude of intensity oscillation is about 2.5%. A 16-bit water-cooled CCD is used to take images. The image size is 1080 by 1080 pixels. The exposure time is 800–1500 ms. The TON full-disk images have a spatial sampling window of 1.8 arcseconds per pixel, and they can provide information of modes up to $l \approx 1000$. The TON high-resolution data are specially suitable for the studies of local properties of the Sun. Here we briefly discuss a new method, acoustic imaging, first developed by the TON group for the study of the local property of the solar interior, such as active regions (Chang et al., 1997; Chou et al., 1999). We also discuss the inversion problem of acoustic imaging (Sun and Chou, 2002).

2. Acoustic Imaging

The most commonly used analysis in helioseismology is the mode decomposition. A time series of intensity or Doppler images is decomposed into eigenmodes to find the dispersion relation of solar p-mode waves to study the solar interior. In acoustic imaging, the acoustic signals measured at the surface are coherently added, based on the time-distance relation (Duvall et al., 1993), a relation between the travel time and horizontal travel distance of wavepacktes, to construct the signal at a target point and a target time (Chou, 2000)

$$\Psi_{\mathrm{out,in}}(t) = \sum_{\tau=\tau_1}^{\tau_2} W(\tau, \theta) \cdot \bar{\Psi}(\theta, t \pm \tau) \,, \tag{1}$$

where $\Psi_{\mathrm{out}}(t)$ and $\Psi_{\mathrm{in}}(t)$ are the constructed signals at the target point at time t, $\bar{\Psi}(\theta, t \pm \tau)$ is the azimuthal-averaged signal measured at the angular distance θ from the target point at time $t \pm \tau$, where τ and θ satisfy the time-distance relation, and $W(\tau, \theta)$ is the weighting function. The positive and negative sign corresponding to Ψ_{out} and Ψ_{in} are constructed with the waves propagating outward from and inward toward the target point respectively. For a target point on the surface, one can use the measured time-distance relation. For a target point located below the surface, one has to use the time-distance relation computed with a standard solar model for sound speed and the ray approximation (Chang et al., 1997).

The constructed signals contain information on intensity and phase. The intensity at the target point can be computed by summing $|\Psi_{\mathrm{out,in}}(t)|^2$ over time. The intensity is lower in magnetic regions than that in the sourrounding quiet Sun due to the absorption of p-mode waves in magnetic regions (Chang et al., 1997).

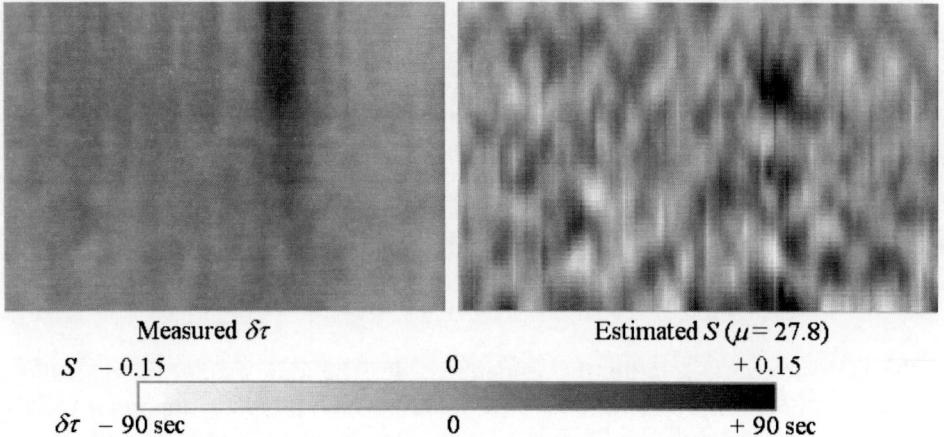

Figure 1. Measured phase travel time perturbation $\delta\tau$ (left) and estimated relative sound-speed perturbation S for a smooth factor $\mu = 27.8$ (right) in a vertical plane cutting through the large sunspot (rearranged from Figure 10 in Sun and Chou, 2002). The horizontal size is 169 Mm and the vertical size 56 Mm.

The phase of constructed signals can be studied by the cross-correlation function between $\Psi_{\text{out}}(t)$ and $\Psi_{\text{in}}(t)$ (Chen *et al.*, 1998). The phase of the cross-correlation function, denoted as τ, relates to the phase travel time of the wave packet. The changes in τ relative to the quiet Sun is called the phase travel time perturbation $\delta\tau$ (or phase shift). The acoustic imaging technique has been used to image the phase-shift map of active regions at the far-side of the Sun (Lindsey and Braun, 2000). Imaging of the far-side of the Sun could help space weather predictions because it provides information on large active regions before they reach the near-side.

3. The Inversion Problem of Acoustic Imaging

Although the goal of acoustic imaging is to construct the signal at the target point, the constructed signal contains both the signal at the target point, and the signals along the ray path (Chou *et al.*, 1999). The contribution of nonlocal effects depends on the density of the ray distribution in the ray approximation. The phase travel time perturbation at $\vec{r}$, $\delta\tau(\vec{r})$, can be expressed as (Chou and Sun, 2000)

$$\delta\tau(\vec{r}) = \int K(\vec{r}, \vec{r}')S(\vec{r}')\,d^3r' \tag{2}$$

where $S \equiv \delta c/c$ is the relative change in wave speed. The kernel $K(\vec{r}, \vec{r}')$ is proportional to the density of the ray distribution. The kernel can be computed with a standard solar model based on the ray theory (Chou and Sun, 2000). We have used the regularized least-square fit to invert the measured $\delta\tau$ to estimate the distribution of S in Equation (2). The technique is discribed in detail in Sun and Chou (2002). The inversion method has been applied to the $\delta\tau$ of active region NOAA 7981

(1–3 August 1996) measured with the TON data. The measured $\delta \tau$ and estimated S are shown in Figure 1. The estimated S peaks at a depth of about 14 Mm with a value of about 0.16. Notice that the estimated sound-speed perturbation is small near the surface. This phenomenon is consistent with the inversion result from the p-mode absorption coefficients in sunspots (Chen and Chou, 1997), and the inversion result from time-distance analysis (see Figure 8 in Kosovichev, Duvall and Scherrer, 2000).

4. Conclusion

The TON project is entering the second decade. The long span of the data allows studying of solar-cycle variations of p-modes, which provides important inform-ation on the properties and origin of the solar cycle, in addition to short-term phenomena. Acoustic imaging is a promising method to probe the subsurface mag-netic fields at a depth of 0–30 Mm. It can be used to study various tpyes of active regions and the evolution of active regions, such as newly emerging regions and decaying old regions. Because of the nonlocal effects in acoustic imaging, the inversion is necessary to improve its spatial resolution. An inversion based on the ray theory has been developed. For further improvement, an inversion based on the wave theory is necessary.

Acknowledgements

The authors MTS and DYC were supported by NSC of ROC under grants NSC-90-2112-M-182-001 and NSC-90-2112-M-007-036, respectively.

References

Chang, H. -K., Chou, D. -Y., LaBonte, B. and the TON Team: 1997, *Nature* **389**, 825.

Chen, H. -R., Chou, D. -Y. and the TON Team: 1997, *Astrophys. J.* **490**, 452.

Chen, H. -R., Chou, D. -Y., Chang, H. -K., Sun, M. -T., Yeh, S. -J., LaBonte, B. and the TON Team: 1998, *Astrophys. J.* **501**, L139.

Chou, D. -Y., Sun, M. -T., Huang, T. -Y. *et al.*: 1995, *Solar Phys.* **160**, 237.

Chou, D. -Y., Chang, H. -K., Sun, M. -T., LaBonte, B., Chen, H. -R., Yeh, S. -J. and the TON Team: 1999, *Astrophys. J.* **514**, 979.

Chou, D. -Y. and Duvall, T. L., Jr.: 2000, *Astrophys. J.* **533**, 568.

Chou, D. -Y. and Sun, M. -T.: 2000, in A. Wilson (ed.), *SOHO10/GONG2000: Helio- and Astero-seismology at the Dawn of the Millennium*, ESA SP-464, ESA, Noordwijk, p. 157.

Duvall, T. L., Jr., Jefferies, S. M., Harvey, J. W., and Pomerantz, M. A.: 1993, *Nature* **362**, 430.

Kosovichev, A. G., Duvall, T. L., Jr. and Scherrer, P. H.: 2000, *Solar Phys.* **192**, 159.

Lindsey, C. and Braun, D. C.: 2000, *Science* **287**, 1799.

Sun, M. -T. and Chou, D. -Y.: 2001, in A. Wilson (ed.), *SOHO10/GONG2000: Helio- and Astero-seismology at the Dawn of the Millennium*, ESA SP-464, ESA, Noordwijk, p. 251.

Sun, M. -T. and Chou, D. -Y.: 2002, *Solar Phys.* **209**, 5.

MAGNETIC RECONNECTION PHENOMENA IN INTERPLANETARY SPACE

FENGSI WEI, QIANG HU, XUESHANG FENG and QUANLIN FAN

Laboratory for Space Weather, Center for Space Science and Applied Research, Chinese Academy of Sciences, P.O. Box 8701, Beijing 100080, China

Abstract. Interplanetary magnetic reconnection(IMR) phenomena are explored based on the observational data with various time resolutions from Helios, IMP-8, ISEE3, Wind, etc. We discover that the observational evidence of the magnetic reconnection may be found in the various solar wind structures, such as at the boundary of magnetic cloud, near the current sheet, and small-scale turbulence structures, etc. We have developed a third order accuracy upwind compact difference scheme to numerically study the magnetic reconnection phenomena with high-magnetic Reynolds number ($R_M = 2000\text{--}10000$) in interplanetary space. The simulated results show that the magnetic reconnection process could occur under the typical interplanetary conditions. These obtained magnetic reconnection processes own basic characteristics of the high R_M reconnection in interplanetary space, including multiple X-line reconnection, vortex velocity structures, filament current systems, splitting, collapse of plasma bulk, merging and evolving of magnetic islands, and lifetime in the range from minutes to hours, etc. These results could be helpful for further understanding the interplanetary basic physical processes.

Key words: interplanetary magnetic reconnection, observational evidence, numerical simulation

1. Introduction

The magnetic reconnection is an important physical mechanism in explaining many physical processes occurring in the solar physics and the magnetospheric physics (Hones, Jr., 1984). Many good suggestions were proposed to describe and identify the existence of magnetically closed structures or flux rope in interplanetary space (Gosling *et al.*, 2001). Recently, the magnetic hole in the solar wind raised scientists's interest (Zurbuchen *et al.*, 2001; Farrugia *et al.*, 2001). However, direct observational evidence and further knowledge about magnetic reconnection in interplanetary space is still insufficient.

Recently, we have examined in detail the IMR on basis of magnetic and plasma data from the various spacecrafts and have made numerical tests under the typical interplanetary conditions. Their partial results are briefly presented in this paper.

Space Science Reviews **107**: 107–110, 2003.
© 2003 *Kluwer Academic Publishers*.

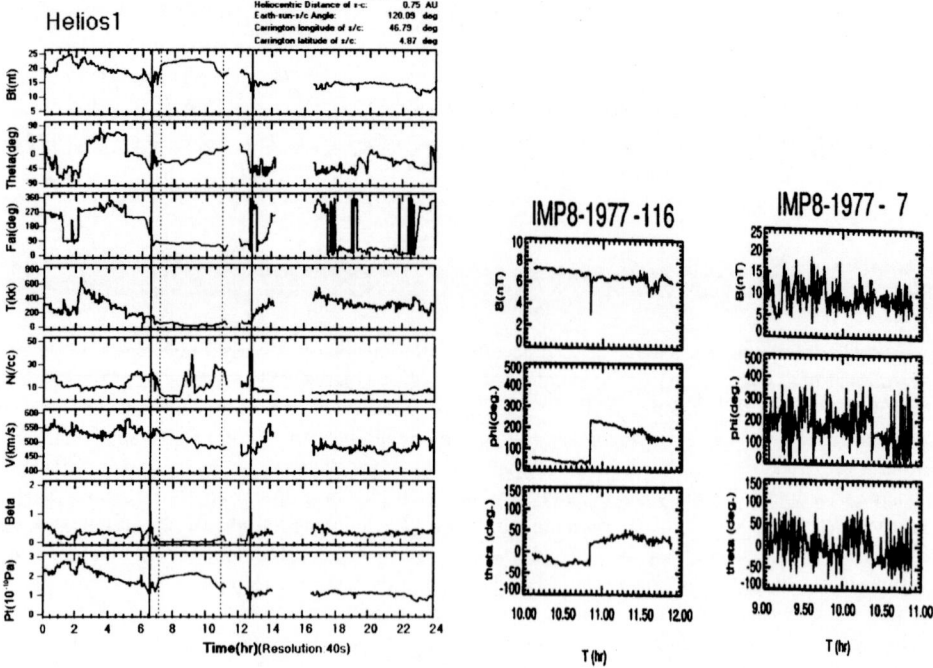

Figure 1. A magnetic cloud observed by Helios 1 during June 1979 at 0.75 AU (left); a typical reconnection example near the current sheet (middle); and another turbulent magnetic reconnection example in the solar wind (right).

2. Observational Examples

We know that the variations of the field direction and strength versus time are fundamental significance in identifying a magnetic reconnection phenomena possibly occurring in interplanetary space, although other plasma parameters, such as temperature T, density N, β parameter and so on, are also necessary.

Figure 1 (left) shows a magnetic cloud, with measurement resolution of 40 seconds, observed by Helios 1 during June 1979 at 0.75 AU, where the two vertical solid lines indicate the positions of the two magnetic reconnection regions possibly occurring at the boundary layers of the cloud. The basic feature of magnetic reconnection is obvious both in the magnetic field signatures including strength drop, large changes in Φ and θ angles and in the plasma parameters, such as higher temperature, higher density and higher β relatively to that of the cloud body. Figure 1 (Middle) is a typical reconnection example occurring near a current sheet, which was observed by IMP8 in the 116 day, 1977. The abrupt drop in **B** and the correspondent $\sim 180°$ change in Φ and $\sim 50°$ change in θ are clearly displayed in the data; Figure 1 (right) shows another turbulent magnetic reconnection example possibly occurring in the turbulent solar wind. There exist good correspondence among the field strength drop, $\sim 180°$ change in Φ and large change in θ for a lot of

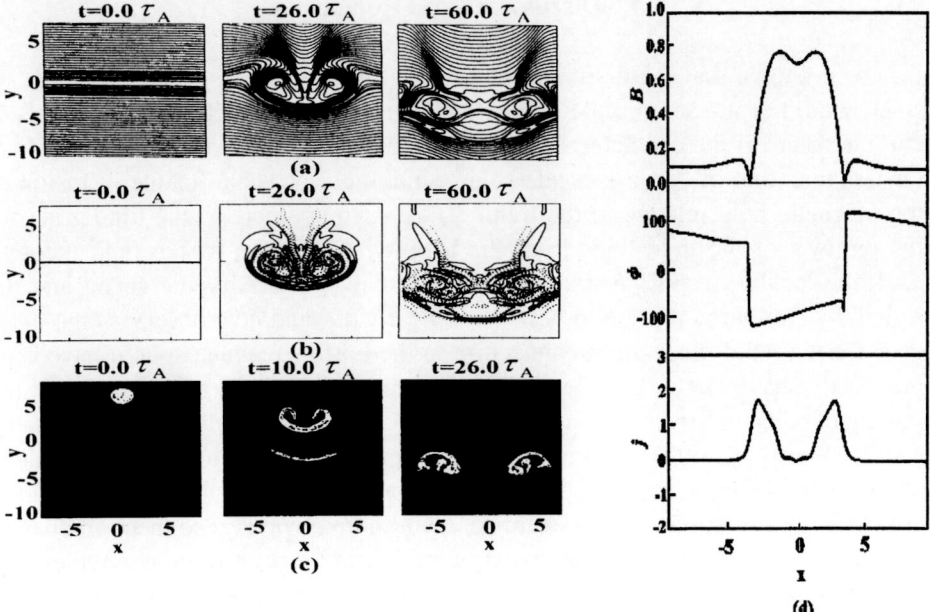

Figure 2. Some preliminary simulation results for magnetic reconnection under the typical interplanetary conditions. (a), (b), (c) and (d) represent the evolutions of field line configuration, filament current system, density, and the sampling for **B**, Φ and $\vec{j}$ at $y = -2.5$, $t = 22.0\tau_A$, respectively.

fluctuations, and this is also a manifestation of the turbulent reconnection. Partial examples here have shown that the magnetic reconnection phenomena could be seen in the various interplanetary structures.

3. Preliminary Simulation

The third order accurate upwind compact difference scheme (Ma and Fu, 1991) has been applied for the numerical study of the magnetic reconnection process near the interplanetary current sheet, under the framework of the two-dimensional compressible magnetohydrodynamics (MHD) and the typical interplanetary conditions, $\mathbf{B}_0 = 8.33$ nT, $N_0 = 5$ proton/cm^3, $T_0 = 1 \times 10^5$ K, $\gamma = 5/3$, $\beta = 2.0$. Results here confirm that the driven reconnection near the current sheet can occur within minutes–hours when $R_M = 2000-10000$. Some basic properties, such as the multiple X–line reconnections, vortical velocity structures, filament current systems, splitting and collapse of the plasma bulk are seen obviously in Figure 2(a)–(c). Figure 2(d) gives the profiles of magnetic field strength **B**, azimuthal angle Φ and current density $\vec{j}$ recorded along $y = -2.5$, $t = 22.0\tau_A$, which is basically consistent with the observations (see Figure 1 (left)).

4. Discussions and Conclusion

It is well known that the field and the plasma flow are 'frozen in' the large scale solar wind. For the small scale interaction region with magnetic field reversals or current sheets, if the characteristic length scale (L) and speed $(\mathbf{V})$ investigated are far less than that of the large scale solar wind, the time scale for the diffusion of the magnetic field relative to the fluid, L_2/η, is comparable to the time scale for the removal of the field lines from the locality, L/V_A, the frozen field theorem could be locally invalid (Axford, 1984), where V_A, η is Alfvenic speed and the resistivity; furthermore, due to new theoretical work and new observational evidence for the rapid magnetic reconnection in the Earth's magnetosphere have been proposed (Shay *et al.*, 1999; Deng and Matsumoto, 2001), the requirement for the scaling of collisionless magnetic reconnection could be deduced; and the simulated results (Wei *et al.*, 2001) also indicate that the evolution lifetime of the magnetic reconnection is in the range of minute ∼hour order of magnitude for $R_M = 2000–10000$. So, those observational examples with more obvious reconnection characteristics would be the manifestations of interplanetary reconnection phenomena.Of course, further identification will be necessary, especially from the view point of the microscopic mechanism.

Acknowledgements

NNSFC's support (Grant No.49990450, 49925412) and MBRPC (Grant No. G200078405) for this work is appreciated.

References

Axford, W. I.: 1984, in E. W. Hones, Jr. (ed.), *Magnetic Reconnection in Space and Laboratory Plas-mas*, Geophysical Monograph, **30**, 1.
Deng, X. H. and Matsumoto, H.: 2001, *Nature* **410(29)**, 557.
Farrugia, C. T., Vasquez, B., Richardson, I. G. *et al.*: 2001, *Adv. Space Res.* **28(5)**, 759.
Gosling J., Pizzo, V. and Bame, S.: 2001, *J. Geophys. Res.* **78**, 1973.
Hones, Jr. E. W.: 1984, in E. W. Hones Jr.(ed.), Geophysical Monograph **30**, 1.
Ma, Y. and Fu, D. S.: 1991, *Chin. J. Compu. Phys.* **8**, 194.
Shay, M. A. *et al.*: 1999, *Geophys. Res. Lett.* **26**, 2163.
Wei, Fengsi, Qiang Hu and Xueshang Feng: 2001, *Chin. Sci. Bull.* **46(2)**, 111.
Zurbuchen, T. H. *et al.*: 2001, *J. Geophys. Res.* **106(A8)**, 16001.

RECONSTRUCTED 3-D MAGNETIC FIELD STRUCTURE AND HARD X-RAY TWO RIBBONS FOR 2000 BASTILLE-DAY EVENT

YIHUA YAN[1] and GUANGLI HUANG[2]

[1] *National Astronomical Observatories, Chinese Academy of Sciences, Beijing 100012, China*
[2] *Purple Mountain Observatory, Chinese Academy of Sciences, Nanjing 210008, China*

Abstract. The Bastille-day event in 2000 produced energetic 3B/X5.6 flare with a halo CME, which had great geo-effects consequently. This event has been studied extensively and it is considered that it follows the two-ribbon flare model. The flare/CME event was triggered by an erupting filament and TRACE observations showed formation of giant arcade structures during the flare process. Hard X-ray (HXR) two ribbons revealed for the first time in this flare event (Masuda *et al.*, 2001). The reconstruction of 3-D coronal magnetic fields revealed a magnetic flux rope structure, for the first time, from extrapolation of observed photospheric vector magnetogram data and the flux rope structure was co-spatial with portion of the filament and a UV bright lane (Yan *et al.*, 2001a, 2001b). Here we review some recent work related to the flux rope structure and the HXR two ribbons by comparing their locations and the flux temporal profiles during the flare process so as to understand the energy release and particle accelerations. It is proposed that the rope instability may have triggered the flare event, and reconnection may occur during this process. The drifting pulsation structure in the decimetric frequency range is considered to manifest the rope ejection, or the initial phase of the coronal mass ejection. The HXR two ribbons were distributed along the flux rope and the rope foot points coincide with HXR sources. The energy dissipation from IPS observations occurred within about 100 $R_{\bigodot}$ is consistent with the estimate for the flux rope system.

Key words: sun: corona, sun: flares, sun: magnetic fields

1. Introduction

The X5.7/3B flare at position N22W07 in NOAA 9077 active region at 10:24 UT on 14 July 2000 (so called Bastille-Day event) was well-observed with ground- and space-based instruments in hard and soft X-rays, extreme ultraviolet wavelengths (EUV), Hα, radio and magnetic fields, etc. Thus it provides perhaps an unique example to examine the mechanism of triggering, energy release, and related dynamic effects in the geo-effective flare/CME process (c.f., *Solar Physics* topic issue, **204**).

According to the observed features for the Bastille-Day flare, its energy release mechanism is generally considered to follow the Kopp and Pneuman (1976) magnetic reconnection model for two-ribbon flares. A filament eruption due to magnetic instability is supposed to take place above the magnetic neutral line, accompanied by a coronal mass ejection at the beginning of the flare process. As the filament goes up, a current sheet forms during the magnetic reconnection process

Space Science Reviews **107**: 111–118, 2003.
© 2003 *Kluwer Academic Publishers.*

and acts as source of particle acceleration. These nonthermal particles precipitate down to the two flare ribbons in parallel with the neutral line and the precipitation sites of nonthermal particles along the two flare ribbons are for the first time mapped as ribbon-like features in hard X-rays (HXR) during this flare (Masuda *et al.*, 2001). This filament is co-spatial with an inferred magnetic flux rope which is not formed by magnetic reconnection at the time of eruption but has existed long before the eruption as a part of the global structure (Yan *et al.*, 2001a, 2001b).

Kosovichev and Zharkova (2001) analyzed high-resolution observations from the Michelson Doppler Imager (*MDI*) instrument on the *SOHO* spacecraft and found rapid variations of the magnetic field in the lower solar atmosphere during the flare. Some of these variations were irreversible, occurred in the vicinity of magnetic neutral lines, and likely were related to magnetic energy release in the flare. The HXR two ribbons were located along the footpoints of the TRACE arcades that across the neutral line, and the ribbons evolved from west to east (Masuda *et al.*, 2001). The co-alignment of the HXR sources with TRACE emissions in east region are studied in Fletcher and Hudson (2001). It is interesting to note that the above region is the place where one foot point of the flux rope is located (Deng *et al.*, 2001). Karlický *et al.* (2001) studied the radio spectra of the four most intense flares observed in NOAA 9077 region including the Bastille-Day flare, and found that the drifting pulsation structures were typical signatures of these flares, which are interpreted as the upward plasmoid motion (or the flux rope eruption) by (Kliem *et al.* (2000)). The radio fine structures observed in 1–7.6 GHz range with associations to the flux rope structure were discussed in Wang *et al.* (2001).

Therefore, in this review, we examine the role of the inferred magnetic flux rope in the Bastille-day event. In Section 2 we describe the 3-D magnetic field structures in the NOAA 9077 active region. Then in Section 3 we discuss the associations with the HXR two ribbons in that region and draw our conclusions in Section 4.

2. The Coronal Magnetic Field Structure

It is well-accepted that the magnetic field plays a central role in the solar activity (Parker, 2001). Using the reconstruction method of Yan and Sakurai (1997, 2000), Yan *et al.* (2001a) have reconstructed the general force-free magnetic field of the region before the X5.7/3B flare, and revealed for the first time the presence of a magnetic rope from the extrapolation of the 3D magnetic field structure. This magnetic rope was located in a space above the magnetic neutral line of the filament, where the features as described by Zhang *et al.* (2001) and Kosovichev and Zharkova (2001) occurred. Overlying the rope are multi-layer magnetic arcades with different orientations. These arcades are in agreement with TRACE observations. Such a magnetic flux rope structure provides a favorable model for the interpretation of the energetic flare processes as revealed by Hα, EUV, and radio observations. In particular, the intermittent co-spatial brightening of the rope in UV

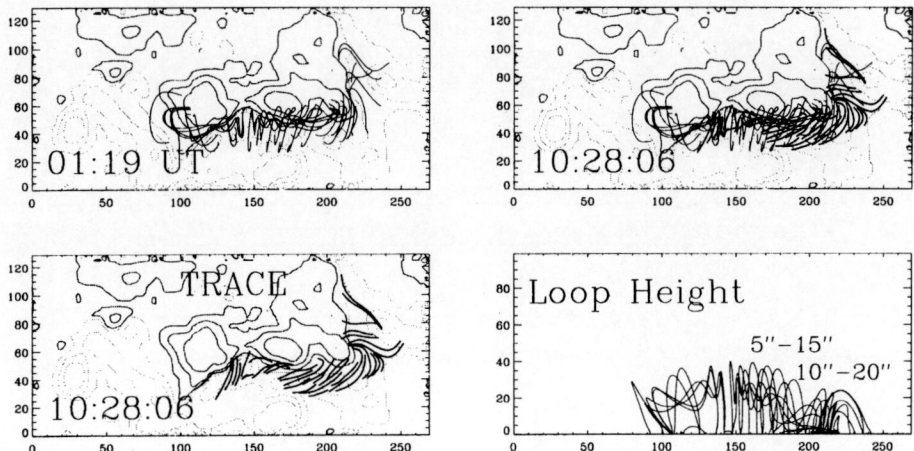

Figure 1. Up-left: Selected field lines in coincidence with *TRACE* EUV flare loops (*lower-left*) at 10:28:06. The light dotted (dahsed) contours represent S (N) polarity of the longitudinal magnetic field with 100, 480 and 1200 G. The thin solid lines are projected field lines and the thick solid lines are EUV loops. *Up-right*: overlay of the projected field lines and the flare loops. *Low-right*: front view of field lines showing loop heights (Yan *et al.*, 2001b).

1600 Å image leading to the onset of the flare suggests that the rope instability may have triggered the flare event, and the drifting pulsation structure in the decimetric frequency range is considered to manifest the rope ejection, or the initial phase of the coronal mass ejection (Yan *et al.*, 2001a). Aschwanden and Alexander (2001) analyzed the evolution of the thermal flare plasma during the 3B/X5.7 flare event by using spacecraft data from *Yohkoh*, *GOES*, and *TRACE*. They demonstrate that the spatial structure of this two-ribbon flare in *TRACE* 171 Å consists of a wound arcade with some 100 flare loops which light up in a sequential manner from highly-sheared low-lying to less-sheared higher-lying bipolar loops (Aschwanden and Alexander, 2001).

By comparing the calculated field lines and the EUV flare loops that roughly have similar orientations and locations, Yan *et al.* (2001b) found that the spatial structure of this two-ribbon flare in *TRACE* 171 Å indeed light up in a sequential manner from highly-sheared low-lying to less-sheared higher-lying bipolar loops, as shown in Figure 1 (Yan *et al.*, 2001b). Since the magnetic field lines are obtained from pre-flare state which is highly nonpotential, they cannot deduce loop heights for potential-like post-flare loops. According to UV/EUV observations the topological structure in NOAA 9077 was in similar arcade forms after the 3B/X5.7 flare until 01 UT on 15 July although there was other minor flare occurred during this period. It is noted that the height evolution from 10:28:06 UT to 10:37:34 UT just involves the height of the magnetic flux rope, suggesting that the flux rope was erupting during this period, which is in agreement with the radio observations of drifting pulsation structure at 10:27–10:35 UT (Yan *et al.*, 2001a; Karlický *et al.*, 2001). Overlay of the flux rope and HXR ribbons during 10 min in flare maximum

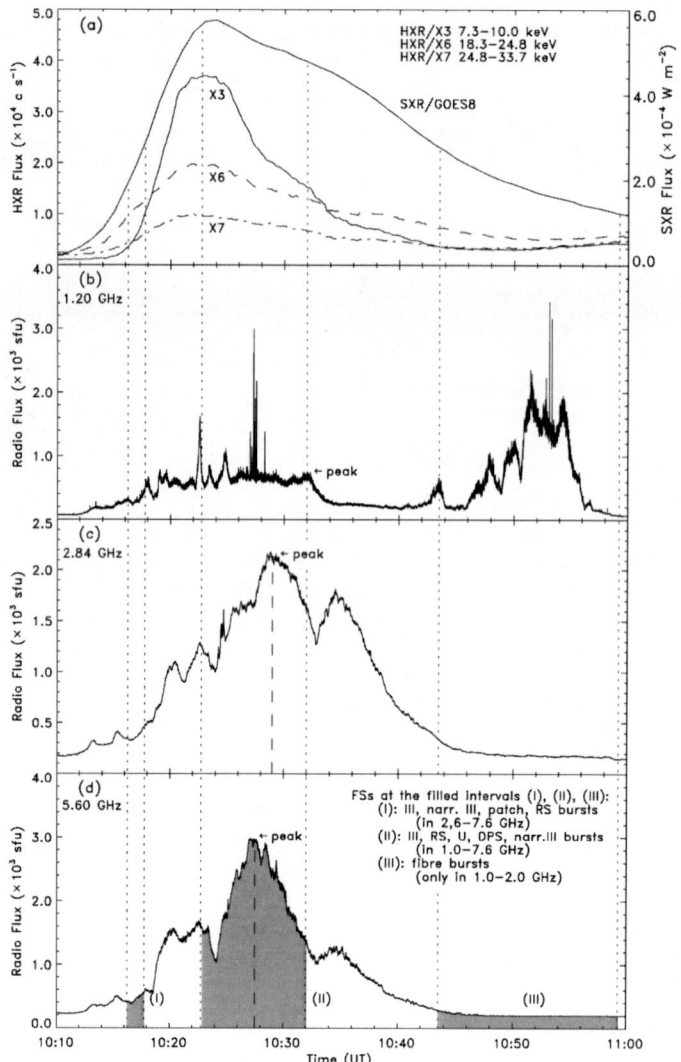

Figure 2. Time profiles of the hard X-ray (HXR), soft X-ray (SXR) and the radio emission at several frequencies during 10:10–11:00 UT. (a) HXRs by Chinese FY-2 satellite with a time resolution of 8.2s and SXRs from GOES-8; (b) 1.20 GHz; (c) 2.84 GHz; and (d) 5.60 GHz. The shaded ranges in (d) indicate three intervals when many radio fine structures occurred (Wang *et al.*, 2001).

shows that the HXR ribbons were distributed along the flux rope and the rope foot points coincide with HXR sources (Huang, 2002).

3. Associations with Hard X-ray Two-Ribbons

The Solar Radio Broadband Fast Dynamic Spectrometers at Huairou, Beijing (1–2, 2.6–3.8, and 5.2–7.6 GHz frequency range) with high temporal (100, 8, and 5 ms) and spectral (20, 10, and 20 MHz) resolutions (Fu *et al.*, 1995) recorded very rich and interesting radio bursts and radio spectral fine structures in the active region NOAA 9077. In the 1.0–7.6 GHz range: type III burst, narrowband type III burst, type RS burst, type U burst, drifting pulsation structures, fiber and Patch were recorded (Wang *et al.*, 2001). They all occurred in 3 distinct regimes: interval I during the impulsive phase in 10:16–10:18 UT; interval II around the flare maximum in 10:22–10:31 and interval III in the decay phase in 10:43–10:59 UT, as shown in Figure 2. Since there was a *Yohkoh* gap during the rising phase of the flare, the HXR time profile from Chinese *FY-2* satellite (Lin *et al.*, 2000) is demonstrated.

Kosovichev and Zharkova (2001) analyzed high-resolution observations from *SOHO/MDI* and found good correlation between the rapid variations of the magnetic field in the lower solar atmosphere and the double-peak HXR flux temporal profiles during the flare maximum. Some of these variations were irreversible, occurred in the vicinity of magnetic neutral lines, and likely were related to magnetic energy release in the flare.

The radio fine structures are considered to associate with the flux rope eruption as described in Wang *et al.* (2001). They found that: (a) during the impulsive phase (interval I), the microwave fine structures mainly in higher frequency ranges (low altitude) occurred, and SXR flux increased quickly; (b) just after the flare maximum in SXR (interval II), rich radio fine structures occurred in all spectral bands, which may correspond to the ejection of flux rope (Yan *et al.*, 2001a); (c) in the later phase of the flare (interval III), fine structures at high altitude (1–2 GHz band) were observed. The observed drifting pulsation structure is interpreted as the upward plasmoid motion (or the flux rope eruption) by Kliem *et al.* (2000).

Manoharan *et al.* (2001) employed LASCO, metric Radioheliograph and interplanetary scintillation (IPS) observations to study the expansion of the CME, formation of the halo in the low corona, and its speed history in the interplanetary medium. It suggests a process involving conversion of stored magnetic energy into kinetic energy, which would also assist in the propagation of the CME disturbance into the interplanetary medium. It is more likely that the necessary energy to drive the eruption and open up the field lines is provided by the twisted magnetic flux rope. They concluded that the energy dissipation occurred within about 100 $R_\odot$ is consistent with the estimate given by Yan *et al.* (2001a) for the flux rope system.

The spatial distribution of HXR sources is very important for understanding the energy release and particle acceleration processes (Brown, 1975). Emslie (1981, 1983) predicated the spatial differences between the thermal and non-thermal models in HXR emission. For non-thermal model, the HXR emission should appear firstly at footpoints of the low density loop, or the whole dense loop. For thermal case, the HXR emission should appear firstly a loop-top source and two foot point sources, as confirmed by Yohkoh observations (Masuda *et al.*, 1994).

For the Bastille-day event, the HXR observation revealed for the first time the two-ribbon structure (Masuda *et al.*, 2001). The HXR two ribbons were located along the footpoints of the TRACE arcades that across the neutral line, and the ribbons evolved from west side to east side (Masuda *et al.*, 2001). The co-alignment of the HXR sources with TRACE emissions at east side are studied in Fletcher and Hudson (2001), and it is suggested that emission in the EUV ribbons is caused by electron bombardment of the lower atmosphere, supporting the hypothesis that flare ribbons map out the chromospheric footpoints of magnetic field lines newly linked by reconnection. The region is the place where one foot point of the flux rope is located (Deng *et al.*, 2001). Fletcher and Hudson (2001) have found that the weaker-field footpoint was not the brighter one in hard X-rays. We know from microwave/HXR comparisons that mirroring and trapping do occur, so this discrepancy suggests the complication in coronal dynamics. It is suggested that there exists a more complicated coronal restructuring than the standard model requires. Aschwanden and Alexander (2001) have analyzed the plasma cooling process for the flare event. Huang (2002) recently compared the reconstructed flux rope with the HXR two ribbons and it is found that there are several HXR sub-sources located in the footpoints of the flux rope. This comparison is very helpful to understand the thermal and non-thermal radiation processes in this event. For example, there is a sub-source located in the observed arcades, which would be wrongly considered as a loop-top thermal source according to the general thermal/non-thermal models above mentioned. However, it is actually a non-thermal source and located at the foot point of the inferred flux rope, in agreement with the spectral analysis of the source. Huang (2002) also estimated the low cut-off energy by a common cross point of the power-law distribution lines at different time of the event. The low cut-off energy is 20–40 Kev, which may be due to the enhancement of the thermal component. It is found that a quasi-periodic oscillation of the low cut-off energy is well consistent with the time profile, which may provide an evidence for the periodic acceleration process.

4. Conclusion

In summary, the 3D magnetic field structure of the active region NOAA 9077 is characterized by the presence of a magnetic rope which is located above the magnetic neutral line in space of the filament. Overlying the rope are multi-layer magnetic arcades with different orientations.

For the Bastille-Day flare, its energy release mechanism is considered to follow the Kopp and Pneuman (1976) magnetic reconnection model for two-ribbon flares. A filament eruption due to magnetic instability is supposed to take place above the magnetic neutral line, accompanied by a coronal mass ejection at the beginning of the flare process. As the filament goes up, a current sheet forms during the magnetic reconnection process and acts as source of particle acceleration. These nonthermal particles precipitate down to the two flare ribbons in parallel with the neutral line and the precipitation sites of nonthermal particles along the two flare ribbons are for the first time mapped as ribbon-like features in hard X-rays during this flare (Masuda *et al.*, 2001). This filament is co-spatial with an inferred magnetic flux rope which is not formed by magnetic reconnection at the time of eruption but has existed long before the eruption as a part of the global structure (Yan *et al.*, 2001a). However, the eruption of the flux rope may cause reconnection of overlaying magnetic arcades as well as nearby interacting magnetic structures.

The flux rope structure plays a central role in the mechanism of CME formations. For the Bastille-day event, comparison of the inferred magnetic flux rope and overlying arcades show that the spatial structure of this two-ribbon flare in *TRACE* 171 Å indeed light up in a sequential manner from highly-sheared low-lying to less-sheared higher-lying bipolar loops. It suggests that the flux rope was erupted and reconnection may occur during the flare process, which is in agreement with the radio observations of drifting pulsation structure at 10:27–10:35 UT.

Acknowledgements

This work is supported by the Ministry of Science and Technology of China (G2000078403), NNSF of China (10225313, 19973008) and Chinese Academy of Sciences.

References

Aschwanden, M. J. and Alexander, D.: 2001, *Solar Phys.* **204**, 93.
Brown, J. C. and McClymont, A. N.: 1975, *Solar Phys.* **41**, 135.
Deng, Y., Wang, J., Yan, Y. and Zhang, J.: 2001, *Solar Phys.* **204**, 9.
Emslie, A. G.: 1981, *Astrophys. J.* **245**, 711.
Emslie, A. G.: 1983, *Solar Phys.* **86**, 133.
Fletcher, L. and Hudson, H.: 2001, *Solar Phys.* **204**, 71.
Fu, Q. J., Qin Z. H., Ji, H. R. and Pei, L. B.: 1995, *Solar Phys.* **160**, 97.
Huang, G. L.: 2002, *Solar Phys.* submitted.
Karlický, M., Yan, Y., Fu, Q., Wang, S., Jiricka, K., Mészárosová, H. and Liu, Y.: 2001, *Astron. Astrophys.* **369**, 1104.
Kliem, B., Karlický, M. and Benz, A. O.: 2000, *Astron. Astrophys.* **360**, 715.
Kopp, R. A. and Pneuman, G.: 1976, *Solar Phys.* **50**, 85.
Kosovichev, A. G. and Zharkova, V. V.: 2001, *Astrophys. J.* **550**, L105.
Lin, H. A., Zhu, G. W. and Wang, S. J.: 2000, *Chinese J. Space Sci.* **20**, 251.
Manoharan, P. K., Tokumaru, M., Pick, M., Subramanian, P., Ipavich, F. M., Schenk, K., Kaiser, M. L., Lepping, R. P. and Vourlidas, A.: 2001, *Astrophys. J.* **559**, 1180.
Masuda, S., Kosugi, S., Hara, H., Tsuneta, S. and Ogawara, Y.: 1994, *Nature*, **371**, 495.
Masuda, S., Kosugi, S. and Hudson, H.: 2001, *Solar Phys.* **204**, 57.

Parker, E. N.: 2001, *Chinese J. Astron. Astrophys.* **1**, 99.

Wang, S. J., Yan, Y., Zhao, R. Z., Fu, Q., Tan, C. M., Xu, L., Wang, S. and Lin, H.: 2001, *Solar Phys.* **204**, 155.

Yan, Y., Deng, Y., Karlický, M., Fu, Q., Wang, S. and Liu, Y.: 2001a, *Astrophys. J.* **551**, L115.

Yan, Y., Aschwanden, M. J., Wang, S. and Deng, Y.: 2001b, *Solar Phys.* **204**, 15.

Yan, Y. and Sakurai, T.: 1997, *Solar Phys.* **174**, 65.

Yan, Y. and Sakurai, T.: 2000, *Solar Phys.* **195**, 89.

Zhang, J., Wang, J., Deng, Y. and Wu, D.: 2001, *Astrophys. J.* **548**, L99.

ON THE APPLICATION OF THE BOUNDARY ELEMENT METHOD IN CORONAL MAGNETIC FIELD RECONSTRUCTION

YIHUA YAN

National Astronomical Observatories, Chinese Academy of Sciences, Beijing 100012, China

Abstract. Solar magnetic field is believed to play a central role in solar activities and flares, filament eruptions as well as CMEs are due to the magnetic field re-organization and the interaction between the plasma and the field. At present the reliable magnetic field measurements are still confined to a few lower levels like in photosphere and chromosphere. Although IR technique may be applied to observe the coronal field but the technique is not well-established yet. Radio techniques may be applied to diagnose the coronal field but assumptions on radiation mechanisms and propagations are needed. Therefore extrapolation from photospheric data upwards is still the primary method to reconstruction coronal field. Potential field has minimum energy content and a force-free field can provide the required excess energy for energy release like flares, etc. Linear models have undesirable properties and it is expected to consider non-constant-alpha force-free field model. As the recent result indicates that the plasma beta is sandwich-ed distributed above the solar surface (Gary, 2001), care must be taken in modeling the coronal field correctly. As the reconstruction of solar coronal magnetic fields is an open boundary problem, it is desired to apply some technique that can incorporate this property. The boundary element method is a well-established numerical techniques that has been applied to many fields including open-space problems. It has also been applied to solar magnetic field problems for potential, linear force-free field and non-constant-alpha force-free field problems. It may also be extended to consider the non-force-free field problem. Here we introduce the procedure of the boundary element method and show its applications in reconstruction of solar magnetic field problems.

Key words: sun: corona, sun: flares, sun: magnetic fields

1. Introduction

Solar activities such as flares and coronal mass ejection (CME), etc., are believed due to sudden energy release and re-organization of coronal magnetic fields. Magnetic field plays a central role in these solar activities (Priest and Forbes, 2000; Parker, 2001). At present reliable magnetic field measurements are still confined to a few lower levels such as the photosphere and the chromosphere (Ai, 1987; Zhang, 1994; Deng *et al.*, 1997). Near infrared technique may be applied to observe the coronal magnetic field (e.g., Lin *et al.*, 2000) but it is not well-established yet for routine observations. Radio techniques can be applied to diagnose the coronal field with assumptions on radiation mechanisms and propagations (Fu, 1999). Coronal observations in EUV/UV and/or SXR wavelengths provide information on coronal magnetic structures. The loop or thread-like structures shown in EUV/UV and/or

Space Science Reviews **107**: 119–138, 2003.
© 2003 *Kluwer Academic Publishers.*

SXR are believed to resemble the coronal magnetic field (Klimchuk, 2000). However, what we observe are plasmas, not magnetic field itself. Therefore, extrapolation from photospheric data upwards is still the primary tool for understanding the nature of coronal magnetic fields. The potential and linear (or constant-α) force-free models are well-known examples and they have provided some instructive information in realistic active regions (Sakurai, 1989; Gary, 1990). The potential model has a minimum energy content and a force-free field can provide the required free energy (in excess of potential energy) for energy releases like flares and CMEs, etc. In the past 4 decades, almost all methods reconstructing non-potential magnetic field from boundary data assume the coronal field to be force-free (Sakurai, 1989; Gary, 1990; McClymont et al., 1997; Yan, 1998; Wang, 1999). Nevertheless, there are some other attempts for non-force-free field models (Low, 1992; Neukirch, 1995; Gary, 2001).

Observation indicates that the magnetic fields are not force-free on the photosphere but gradually become so above ~ 600 km high in the solar atmosphere (Metcalf et al., 1995). However, Moon et al. (2002) using a different method show that the photospheric magnetic fields are not so far from force-free as conventionally regarded. Linear force-free models have undesirable properties, e.g., non-uniqueness of the solution and infinite energy content in open space above the sun if the asymptotic condition at infinity is not incorporated into formulation properly (but see Aly (1992) and Yan (1995b) for finite energy linear force-free fields where decay of the field faster than distance squared is employed at infinity).

The methods available at present for non-constant-α force-free fields can perhaps be classified into 4 kinds: a direct numerical discretization of the force-free field equations (Wu et al., 1990), or in terms of the vector potential (Amari et al., 1997) by the finite difference method (FDM); a numerical treatment of the variational problem of the force-free field (Sakurai, 1981), or the finite element method (FEM); a numerical treatment of the boundary integral equation of the force-free field (Yan and Sakurai, 1997, 2000), or the boundary element method (BEM); and the quasi-physical evolution from MHD equations to the force-free state (Mikic and McClymont, 1994; Roumeliotis, 1996). Sakurai (1989) and Gary (1990) gave excellent reviews mainly on the potential and constant-α force-free fields. McClymont et al. (1997) discussed practical methods for reconstructing non-constant-α force-free fields from observed photospheric data. Yan (1998) compared the numerical implementation of the practical methods that can reconstruct non-constant-α force-free fields from boundary data. Wang (1999) recently reviewed the measurements and analyses of solar vector magnetic fields, with the aim to bridge the observations and theories.

Since the BEM was first employed in solving solar magnetic field problem (Yan et al., 1991), it has found many applications in solar activity studies. In particular, when the boundary integral equation of the non-constant-α force-free field problem is derived (Yan and Sakurai, 1997, 2000), various solar magnetic fields above solar active regions in the corona are reconstructed by the BEM numerically as a

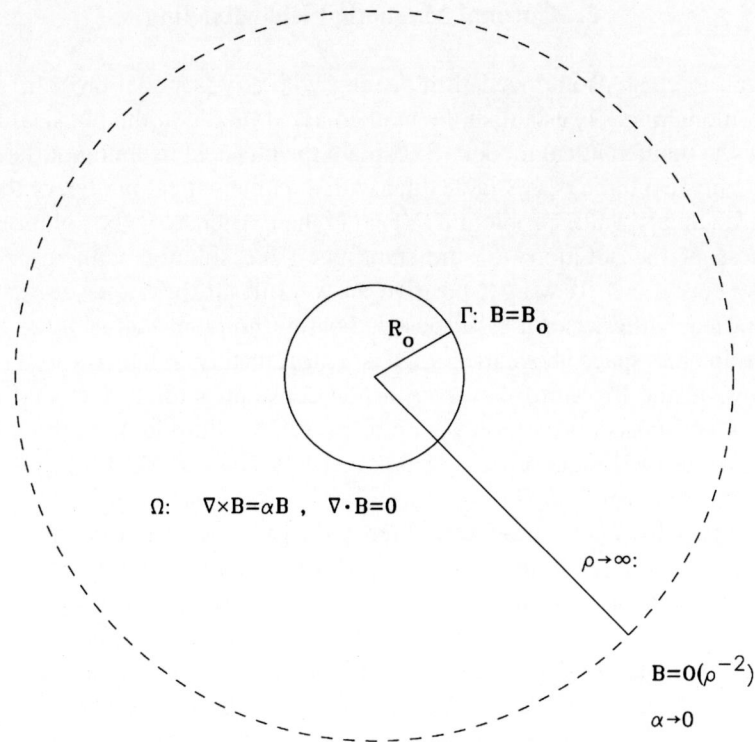

Figure 1. The exterior boundary value problem of the general force-free magnetic field problem (Yan and Sakurai, 2000).

good approximation to the non-constant-α force-free condition. In order to obtain a better understanding of the solar magnetic fields and to promote the application of the BEM for solar activity and space weather studies, here we present results mainly obtained by the BEM of the non-constant-α force-free field problem for flare/CME events. Although the BEM can also be applied to deal with Low (1992) non-force-free field model (Yan, 2002), we will not consider this situation in the present paper. For the sake of clarity and consistency we will introduce the model and provide the derivation of the formulations. The paper is organized as follows. In the next section, we first introduce the boundary value problems for the 3-d non-linear force-free fields. Then in Section 3 the boundary integral equation is derived and in Section 4 the BEM is introduced to solve the integral equation numerically. The results by the BEM are demonstrated in Section 5. Finally, we summarize our conclusions in Section 6.

2. Coronal Magnetic Field Modeling

In general, three steps are needed in dealing with any physical probelm by mathematical modelling: (1) establish the mathematical model of the physical problem; (2) solve the mathematical model; (3) explain the physical meaning of the solution.

There are also three issues in dealing with a mathematical model, or the corresponding boundary value problem (BVP): (1) the existence of the solution; (2) the uniqueness of the solution; (3) the stability of the solution with respect to the boundary conditions. If we get positive answers to all these questions, the BVP is well-posed. Otherwise, it is ill-posed. For the non-constant-α force-free field problems in open space above the Sun, it is, unfortunately, still an ill-posed problem up to now. In the literature, however, a non-constant-α force-free field model is often said well-posed, or not, only with respect to the third item, without knowing anything about the first two items (Sakurai, 1989; Gary, 1990; McClymont et al., 1997; Yan, 1998; Wang, 1999).

The current-free, constant-α force-free and non-constant-α force-free fields are the three different equilibium states reduced from the magnetohydrostatic equilibria for a low β plasma, such as the solar corona (Sakurai, 1989; Priest, 1994; Parker, 2001). The potential, or current-free field model has a minimum energy content and it cannot provide the required free energy for energy releases like flares and CMEs, etc. Therefore, as mentioned above, almost all methods in the past forty years reconstructing non-potential magnetic field from photospheric magnetograms assume the coronal field to be force-free (Sakurai, 1989; Gary, 1990; McClymont et al., 1997; Yan, 1998; Wang, 1999). In addition, there are some other attempts for non-force-free field models recently (Low, 1992; Neukirch, 1995; Gary, 2001).

As mentioned in the preceding section, the magnetic fields are not force-free at the bottom of the photosphere but gradually become so above a few hundrad kilometers high in the solar atmosphere (Metcalf et al., 1995). However, Moon et al. (2002) using a different method show that the photospheric magnetic fields are not so far from force-free as conventionally regarded. Linear force-free models have undesirable properties though they have provided some instructive information in realistic geometry as mentioned above. In this paper we confine our discussion on the practical method for non-linear force-free field equations.

In modeling the solar coronal magnetic field above active regions, it is generally assumed, instead of $\Gamma = \{(x, y, z) | |x\mathbf{i} + y\mathbf{j} + z\mathbf{k}| = R_\circ\}$ and $\Omega = \{(x, y, z) | |x\mathbf{i} + y\mathbf{j} + z\mathbf{k}| > R_\circ\}$ for a spheric geometry as shown in Figure 1, that $\Gamma = \{(x, y, z) | z = 0\}$ plane corresponds to the photosphere and semi-space: $\Omega = \{(x, y, z) | z > 0\}$ to the solar atmosphere. In either case, under the force-free condition, the magnetic field yields the following equations,

$$\nabla \times \mathbf{B} = \alpha \mathbf{B} , \quad \nabla \cdot \mathbf{B} = 0 \quad \text{in} \quad \Omega, \tag{1}$$

where α is a function of spatial location and is determined consistently from the boundary conditions since it is constant along field lines. If α is constant everywhere, it is called a linear, or constant-α, force-free field; otherwise it is called a non-constant-α, or nonlinear force-free field.

The boundary condition is

$$\mathbf{B} = \mathbf{B}_o \quad \text{on} \quad \Gamma. \tag{2}$$

$\mathbf{B}_o$ here denotes known boundary values which can be supplied from vector magnetograph measurements.

Since there are three independent variables involved in the general force-free field equations (1), specifying both the vector field on the boundary and the connectivity of field lines will over-determine the boundary value problem. Therefore, different methods have different choices of the boundary condition in order to avoid this over-specification. In general, the normal component of the magnetic field, B_n, is always prescribed over the boundary. In addition, Sakurai (1981) employed normal current component, J_n, (or curl of the tangential components of the field) over one polarity whereas Mikic and McClymont (1994) specified J_n over the whole boundary. Wu *et al.* (1990) and Roumeliotis (1996) used all three components of the field. Yan and Sakurai (1997, 2000) also use the vector field as boundary conditions and assume that the boundary values are prescribed consistently with the force-free field in space. Physically the magnetic field should tend to zero identically at infinity and no current is there. Mathematically the boundary value problem (1,2) has a well-known ill-posed feature as mentioned in many papers (see, e.g., Sakurai, 1989; McClymont *et al.*, 1997, etc.). Therefore an asymptotic condition at infinity must be added in order to avoid this ill-posed property (Amari *et al.*, 1997). Yan and Sakurai (1997, 2000) have proposed to use an asymptotic condition which ensures a finite energy content in open space above the Sun,

$$\mathbf{B} = o(\rho^{-2}), \quad \text{when} \quad \rho \to \infty. \tag{3}$$

The point is how to formulate the boundary value problem (1, 2, 3) because the ill-poseness will appear if the asymptotic condition is not incorporated into the formulation properly. To quote Priest (1994): the force-free field equation 'looks disarmingly simple, but very little has been done so far in understanding the nature of its solution in general.' Our motiviation is to provide a method that can take into account the finite energy content in open space and solvable from boundary data $\mathbf{B}_o$ directly without numerical difficulties. In this paper we present the results obtained by the BEM for solar flare and/or CME events.

In the next section we will introduce the boundary integral equation representation of the non-constant-α force-free field in solar corona (Yan and Sakurai, 2000).

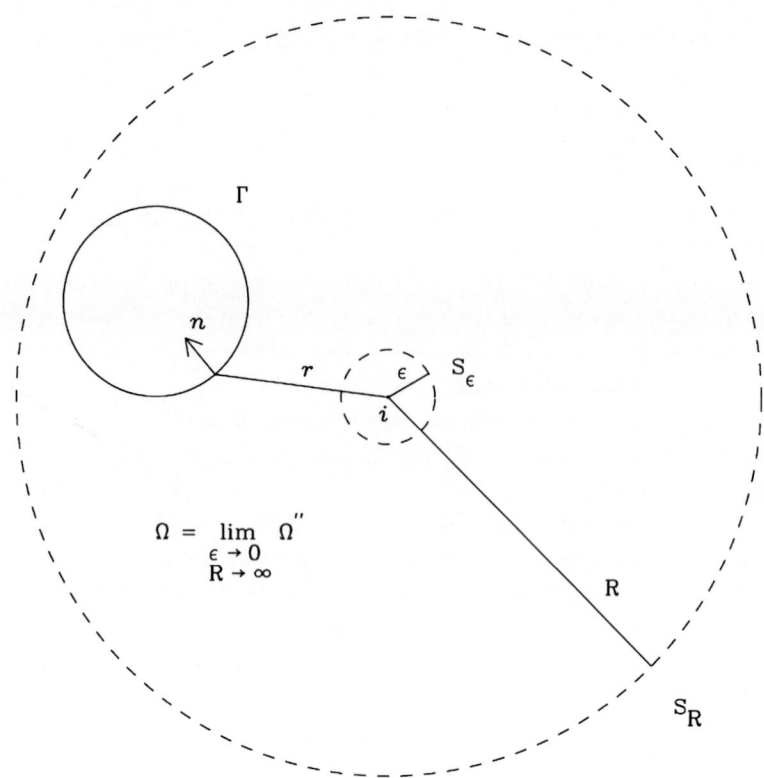

Figure 2. The geometry for application of Green's second identity in the exterior of the Sun (Yan and Sakurai, 2000).

3. Boundary Integral Equation

Let us choose a sufficiently large radius R, as shown in Figure 2, forming a sphere $S_R = \{\mathbf{r} | \, |\mathbf{r} - \mathbf{r}_i| = R\}$ to bound a finite volume V'. The volume V' is then bounded internally by S and externally by the sphere S_R. We also consider a reference function Y chosen as (Yan, 1995a)

$$Y = \frac{\cos(\lambda \rho)}{4\pi \rho}, \tag{4}$$

where $\rho = |\mathbf{r} - \mathbf{r}_i|$ is the distance between the field point $\mathbf{r}$ and the source point $\mathbf{r}_i$, both in V, and λ is a parameter to be found by the following procedure. The reference function is the fundamental solution of the Helmholtz equation,

$$\nabla^2 Y + \lambda^2 Y = \delta(\mathbf{r} - \mathbf{r}_i), \tag{5}$$

where δ is the Dirac function.

The function Y, however, fails to satisfy the necessary condition of continuity at $\mathbf{r}_i$ where $\rho = 0$. To exclude the singularity, a small sphere of radius ϵ, $S_\epsilon =$

$\{\mathbf{r}|\ |\mathbf{r} - \mathbf{r}_i| = \epsilon\}$, is circumscribed about $\mathbf{r}_i$. The volume is then bounded internally by S and externally by S_R, excluding the small volume V_ϵ bounded by S_ϵ. Within $V'' = V' - V_\epsilon$ both $\mathbf{B}$ and Y now satisfy the requirements of Green's second identity and furthermore $\nabla^2 Y + \lambda^2 Y = 0$. Thus we have

$$\int_{V''} (Y \nabla^2 \mathbf{B} - \mathbf{B} \nabla^2 Y)\, dV$$

$$= \int_{S+S_R+S_\epsilon} \left(Y \frac{\partial \mathbf{B}}{\partial n} - \mathbf{B} \frac{\partial Y}{\partial n} \right) dS. \tag{6}$$

The surface integral over the small sphere S_ϵ can be written as

$$\lim_{\epsilon \to 0} \int_{S_\epsilon} \left(Y \frac{\partial \mathbf{B}}{\partial n} - \mathbf{B} \frac{\partial Y}{\partial n} \right) dS = -\mathbf{B_i}, \tag{7}$$

where $\mathbf{B}_i$ is the value of $\mathbf{B}$ at $\mathbf{r}_i$. On the other hand, over the sufficiently big sphere S_R we obtain

$$\frac{\partial Y}{\partial n} = \frac{dY}{dr} = -\frac{\lambda R \sin(\lambda R) + \cos(\lambda R)}{4 \pi R^2}. \tag{8}$$

Since R is constant, the contribution of the sphere, denoted by U, is

$$U = \int_{S_R} \left(Y \frac{\partial \mathbf{B}}{\partial n} - \mathbf{B} \frac{\partial Y}{\partial n} \right) dS$$

$$= \int_{S_R} \left[\frac{\cos(\lambda R)}{4 \pi R} \frac{\partial \mathbf{B}}{\partial n} + \frac{\lambda R \sin(\lambda R) + \cos(\lambda R)}{4 \pi R^2} \mathbf{B} \right] dS$$

$$= \frac{\cos(\lambda R)}{4 \pi R^2} \int_{S_R} \left(R \frac{\partial \mathbf{B}}{\partial n} + \mathbf{B} \right) dS$$

$$+ \frac{\lambda \sin(\lambda R)}{4 \pi R^2} \int_{S_R} R \mathbf{B}\, dS. \tag{9}$$

For U to vanish, it is sufficient that the above two integrals vanish. Therefore we have,

$$R \frac{\partial \mathbf{B}}{\partial n} + \mathbf{B} = O(R^{-1}) \qquad \text{and} \qquad R \mathbf{B} = O(R^{-1}). \tag{10}$$

In the limit $R \to \infty$, the second condition indicates that the field must have finite energy content, and the first condition can be deduced from the second one.

By substituting (1) and (3), the volume integral of (6) reduces to

$$\int_{V''} (Y\nabla^2\mathbf{B} - \mathbf{B}\nabla^2 Y)dV$$

$$= \int_{V''} \{Y[\nabla(\nabla \cdot \mathbf{B}) - \nabla \times (\alpha\mathbf{B}) + \lambda^2\mathbf{B}]$$

$$-\mathbf{B}(\nabla^2 Y + \lambda^2 Y)\} \, dV \tag{11}$$

$$= \int_{V''} Y[\lambda^2\mathbf{B} - \alpha^2\mathbf{B} - \nabla\alpha \times \mathbf{B}] \, dV$$

$$= \mathbf{F}(\lambda).$$

Here we have denoted the volume integral as $\mathbf{F}(\lambda)$, which is a bounded function of α and $\nabla\alpha$, and is finite due to (3) when $R \to \infty$.

By taking the limit of $\epsilon \to 0$ and $R \to \infty$, the field value at any point $\mathbf{r}_i$ is given as

$$\mathbf{B}_i + \mathbf{F}(\lambda) = \int_S \left(Y\frac{\partial\mathbf{B}}{\partial n} - \frac{\partial Y}{\partial n}\mathbf{B} \right) dS. \tag{12}$$

In order that the field vanishes at infinity, we must have

$$\mathbf{F}(\lambda) = \int_V Y[\lambda^2\mathbf{B} - \alpha^2\mathbf{B} - \nabla\alpha \times \mathbf{B}] \, dV = 0. \tag{13}$$

Thus the parameter λ, which is an implicit function of $\mathbf{r}_i$, will be determined. Actually (13) is a vector equation, and for each component we introduce different λ and Y function.

If λ is chosen so that $\mathbf{F}$ vanishes, the boundary integral equation is expressed as (Yan and Sakurai, 2000)

$$c_i\mathbf{B}_i = \int_S \left(Y\frac{\partial\mathbf{B}}{\partial n} - \frac{\partial Y}{\partial n}\mathbf{B} \right) dS, \tag{14}$$

where c_i is a constant depending upon the location of the point $\mathbf{r}_i$. If $\mathbf{r}_i$ is in V, $c_i = 1$. If $\mathbf{r}_i$ is on S, only a semi-sphere rather than a full sphere is needed to exclude the singularity of (4), and $c_i = 1/2$.

This equation indicates that if the values of $\mathbf{B}$ and $\partial\mathbf{B}/\partial n$ over S are known, the field value $\mathbf{B}$ at any point in V is determined by the integration of products of the proposed fundamental solution Y and the values of $\mathbf{B}$ and $\partial\mathbf{B}/\partial n$ over S. Although the value of λ has to be selected to satisfy the constraint (13), it is not practical to seek for the exact values of λ. Rather, we arbitrarily set the value of λ, for example, to the average value of observed α, and still we assume that the constraint (13) holds.

The quantity $\partial\mathbf{B}/\partial n$ is not supplied from boundary data. When (14) is applied to the point $\mathbf{r}_i$ on S, we obtain a set of linear equations for $\partial\mathbf{B}/\partial n$. By solving this equation, we obtain a consistent relation between the boundary values and their

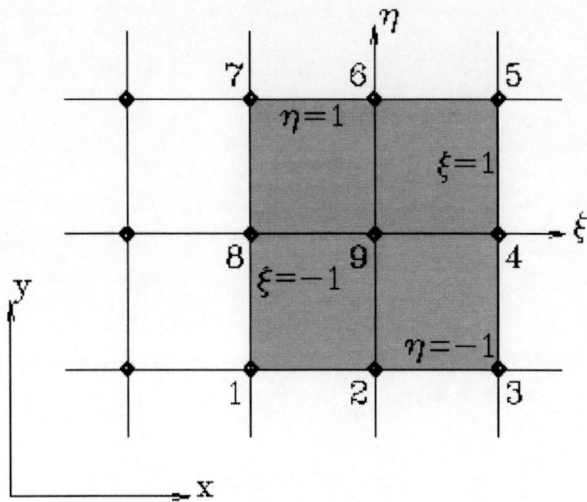

Figure 3. Description of a 9-point bi-quadratic element on the boundary with tangential directions (ξ, η) forming a local intrinsic coordinates in the unit area $\{(\xi, \eta)| - 1 \leq \xi \leq 1, -1 \leq \eta \leq 1\}$. The local nodes are numbered as $k = 1, 2, \ldots, 9$ as indicated by '·' (Yan and Sakuvai, 2000).

normal derivatives. This is a standard procedure in the boundary element method (BEM, see Brebbia *et al.*, 1984; Yan *et al.*, 1993).

4. Numerical Implementation-BEM

4.1. DISCRETE SYSTEM OF BOUNDARY INTEGRAL EQUAION

Although the integral representation (14) is established in a spheric geometry, there is no any *a priori* assumption on the shape of the boundary in the above derivations.

By the boundary element method to solve Equation (14) numerically, we first subdivide Γ into a number of elements and the approximation functions are assumed over each element. In our case the element dimensions should be adequate to the observational resolution. Probably the most common way is to use iso-parametric elements (usually bi-quadratic shape functions) to represent both the geometry and functions on each element (Brebbia *et al.*, 1984). A local intrinsic coordinate system (ξ, η) is defined for each element, as shown in Figure 3, where the normal direction of the boundary outward from the volume concerned is defined by the right-hand convention. Both the boundary location, and **B** and $\partial \mathbf{B}/\partial n$ functions over each element in the intrinsic system (ξ, η) are represented by the sums of products of nodal values and shape functions. Thus

$$
\begin{cases}
x(\xi, \eta) = \sum_{k=1}^{9} N_k(\xi, \eta) x_k^e \\
y(\xi, \eta) = \sum_{k=1}^{9} N_k(\xi, \eta) y_k^e \\
z(\xi, \eta) = \sum_{k=1}^{9} N_k(\xi, \eta) z_k^e
\end{cases}
\tag{15}
$$

and

$$
\mathbf{B}(\xi, \eta) = \sum_{k=1}^{9} N_k(\xi, \eta) \mathbf{B}_k^e \ , \quad \frac{\partial \mathbf{B}}{\partial n} = \sum_{k=1}^{9} N_k(\xi, \eta) \left(\frac{\partial \mathbf{B}}{\partial n} \right)_k^e
\tag{16}
$$

where x_k^e, y_k^e, and z_k^e, are nodal global coordinates, $\mathbf{B}_k^e$ and $(\frac{\partial \mathbf{B}}{\partial n})_k^e$ are nodal values of $\mathbf{B}$ and $\partial \mathbf{B}/\partial n$ respectively, superscript e denotes the element, and $N_k(\xi, \eta)$ are the bi-quadratic shape functions defined as follows,

$$
\begin{cases}
N_k(\xi, \eta) = \dfrac{1}{4} \xi \eta (\xi + \xi_k)(\eta + \eta_k) \ , & k = 1, 3, 5, 7, \\[2mm]
N_k(\xi, \eta) = \dfrac{1}{2} \eta (1 - \xi^2)(\eta + \eta_k) \ , & k = 2, 6, \\[2mm]
N_k(\xi, \eta) = \dfrac{1}{2} \xi (\xi + \xi_k)(1 - \eta^2) \ , & k = 4, 8, \\[2mm]
N_k(\xi, \eta) = (1 - \xi^2)(1 - \eta^2) \ , & k = 9,
\end{cases}
\tag{17}
$$

in which (ξ_k, η_k) are nodal intrinsic coordinates as shown in Figure 3. Introducing these representations into Equation (14), and transforming from the surface to the intrinsic coordinates (ξ, η), we obtain the following equation for a boundary node, i,

$$
c_i \begin{Bmatrix} B_{xi} \\ B_{yi} \\ B_{zi} \end{Bmatrix} + \sum_{e=1}^{M} \sum_{k=1}^{9} [\int_{-1}^{+1} \int_{-1}^{+1} \begin{bmatrix} \frac{\partial Y_x}{\partial n} & 0 & 0 \\ 0 & \frac{\partial Y_y}{\partial n} & 0 \\ 0 & 0 & \frac{\partial Y_z}{\partial n} \end{bmatrix} N_k J d\xi d\eta] \begin{Bmatrix} B_{xk}^e \\ B_{yk}^e \\ B_{zk}^e \end{Bmatrix}
$$
$$
= \sum_{e=1}^{M} \sum_{k=1}^{9} [\int_{-1}^{+1} \int_{-1}^{+1} \begin{bmatrix} Y_x & 0 & 0 \\ 0 & Y_y & 0 \\ 0 & 0 & Y_z \end{bmatrix} N_k J d\xi d\eta] \begin{Bmatrix} (\frac{\partial B_x}{\partial n})_k^e \\ (\frac{\partial B_y}{\partial n})_k^e \\ (\frac{\partial B_z}{\partial n})_k^e \end{Bmatrix} ,
\tag{18}
$$

where M is the number of total boundary elements, $N_k = N_k(\xi, \eta)$, and $J = J(\xi, \eta)$ denotes the Jacobian. In the above equation (18), the reference functions Y_x, Y_y and Y_z contain the parameters λ_x, λ_y and λ_z corresponding to the present point, i, which need to be determined. We have derived in Section 3 that they satisfy Equation (13) which needs information in the whole space. Since it is a complicated nonlinear equation it is not obvious that (13) has a solution λ_i at each point i in Ω. But in the following we propose an iterative procedure so that λ_i can always be defined during the iteration and the goal is to find an approximation solution of the field.

4.2. MATRIX EQUATION AND NUMERICAL SOLUTION

Now, we at first assume an initial value of the parameters λ_x, λ_y and λ_z so that the integrand in Equation (18) can be calculated for the first approximation of the field, **B**.

Constructing Equation (18) for each boundary node by changing the location of the point i over the boundary with the corresponding parameters λ_x, λ_y and λ_z initially assumed, we generate a set of simultaneous equations. This set of equations can be written in a matrix form, with the result of each integration in (18) forming a coefficient of the following matrices,

$$
\begin{bmatrix} \mathbf{H_x} & \mathbf{0} & \mathbf{0} \\ \mathbf{0} & \mathbf{H_y} & \mathbf{0} \\ \mathbf{0} & \mathbf{0} & \mathbf{H_z} \end{bmatrix} \begin{Bmatrix} \hat{\mathbf{B}}_x \\ \hat{\mathbf{B}}_y \\ \hat{\mathbf{B}}_z \end{Bmatrix} = \begin{bmatrix} \mathbf{G_x} & \mathbf{0} & \mathbf{0} \\ \mathbf{0} & \mathbf{G_y} & \mathbf{0} \\ \mathbf{0} & \mathbf{0} & \mathbf{G_z} \end{bmatrix} \begin{Bmatrix} \hat{\mathbf{Q}}_x \\ \hat{\mathbf{Q}}_y \\ \hat{\mathbf{Q}}_z \end{Bmatrix},
\tag{19}
$$

where we have used Q to denote $\partial B / \partial n$, and $\{\hat{\mathbf{B}}\} = \{\hat{\mathbf{B}}_x, \hat{\mathbf{B}}_y, \hat{\mathbf{B}}_z\}^T$ and $\{\hat{\mathbf{Q}}\} = \{\hat{\mathbf{Q}}_x, \hat{\mathbf{Q}}_y, \hat{\mathbf{Q}}_z\}^T$ are vectors of nodal values. In the present case as the boundary is plane, e.g., $z = 0$ surface, the above Equation (19) actually reduces to,

$$
\frac{1}{2} \begin{bmatrix} \mathbf{I} & \mathbf{0} & \mathbf{0} \\ \mathbf{0} & \mathbf{I} & \mathbf{0} \\ \mathbf{0} & \mathbf{0} & \mathbf{I} \end{bmatrix} \begin{Bmatrix} \hat{\mathbf{B}}_x \\ \hat{\mathbf{B}}_y \\ \hat{\mathbf{B}}_z \end{Bmatrix} = \begin{bmatrix} \mathbf{G_x} & \mathbf{0} & \mathbf{0} \\ \mathbf{0} & \mathbf{G_y} & \mathbf{0} \\ \mathbf{0} & \mathbf{0} & \mathbf{G_z} \end{bmatrix} \begin{Bmatrix} \hat{\mathbf{Q}}_x \\ \hat{\mathbf{Q}}_y \\ \hat{\mathbf{Q}}_z \end{Bmatrix},
\tag{20}
$$

where $\mathbf{I}$ is the identity matrix, and the Jacobian $J(\xi, \eta)$ in (18) reduces to merely a constant: one quarter of the element area (Yan *et al.*, 1993). At each node **B** is known (as given by the boundary condition). Solution of the above Equation (20) permits unknown values to be determined properly.

Therefore we have obtained a first approximation for the boundary solution, and the field in space can be calculated from (14), or its discrete form (18), with $c_i = 1$ as a first approximation as well. Then we may modify the parameters λ_x, λ_y and λ_z in terms of the three components of the field **B** and the initial value of λ's as follows,

$$
\begin{cases}
\lambda_x^2 = \displaystyle\int_\Omega Y \left(\alpha^2 B_x + \frac{B_x B_y}{B_z} \frac{\partial \alpha}{\partial x} + \frac{B_y^2 + B_z^2}{B_z} \frac{\partial \alpha}{\partial y} \right) d\Omega \Big/ \int_\Omega Y B_x d\Omega \\[4mm]
\lambda_y^2 = \displaystyle\int_\Omega Y \left(\alpha^2 B_y - \frac{B_x^2 + B_z^2}{B_z} \frac{\partial \alpha}{\partial x} - \frac{B_x B_y}{B_z} \frac{\partial \alpha}{\partial y} \right) d\Omega \Big/ \int_\Omega Y B_y d\Omega \ , \\[4mm]
\lambda_z^2 = \displaystyle\int_\Omega Y \left(\alpha^2 B_z + B_y \frac{\partial \alpha}{\partial x} - B_x \frac{\partial \alpha}{\partial y} \right) d\Omega \Big/ \int_\Omega Y B_z d\Omega
\end{cases}
\tag{21}
$$

where α is obtained from equations (1, 2) as follows,

$$
\alpha = \left(\frac{\partial B_y}{\partial x} - \frac{\partial B_x}{\partial y} \right) / B_z,
\tag{22}
$$

and we have substituted (1) into (21) to eliminate $\partial \alpha / \partial z$ so that the divergence free condition in (1) holds.

With the updated parameters λ_x, λ_y and λ_z from (21), the coefficient matrices in (20) can be re-generated and the solution of (20) gives a new result of nodal unknowns' vector $\{\hat{Q}\}$. Therefore an iterative procedure is readily available and the approximate solution $\{\hat{Q}\}$ of boundary unknowns is found if the difference between two consecutive solutions is small.

Note that the volume integrals in (21) do not increase the scale (or dimension) of the matrix Equation (20) but contribute additional computations in the iterative procedure. These volume integrals can be evaluated by the finite element technique (Sakurai, 1981). In this case, we may choose a sufficiently large box with a 3-d grid above the problem region. Only the quantities at the grid points are calculated and the approximation functions can be assumed over each sub-volume, or finite element. Therefore, the volume integrals can be obtained by summing up all contributions from each sub-volume. However, the above procedure will take will take longer computer time and need large computer facilities.

Yan and Sakurai (1997, 2000) proposed to arbitrarily set the value of λ, for example, to the average value of the boundary α distribution, and still we assume that the constraint (13) holds. In this case (13) does not hold necessarily but merely contribute an error term in (14) to the field **B**. The discrepancy in this assumption is checked later when the solution is evaluated. Applications indicate that with properly-chosen λ values this error can be neglectable small. This strategy has been applied to a number of practical problems and very good agreements with observations have been obtained as demonstrated in the next section. Next we provide an error analysis and give an explanation for the understanding of numerical results.

4.3. APPROXIMATION TO SULUTION AND ERROR ANALYSIS

In the Cartesion coordinates we will see that on the plane boundary Γ one should have (another term in (14) vanishes on the plane $z = 0$):

$$\frac{1}{2}\mathbf{B}_{ex} = \int_{\Gamma} Y_{ex}\mathbf{Q}_{ex}d\Gamma, \tag{23}$$

where $\mathbf{Q} = \partial\mathbf{B}/\partial n$, and the subscript 'ex' indicates the exact solution. We also use subscript 'num' to indicates the numerical solution. When we arbitrarily set a value for λ in Y_{num} the volume integral in (13) will not vanish, but this merely contribute an error term to (14) on the boundary,

$$\frac{1}{2}\mathbf{B}_{ex} + \Delta\mathbf{B} = \int_{\Gamma} Y_{num}\mathbf{Q}_{ex}d\Gamma. \tag{24}$$

By numerical solution, we solve the boundary normal gradient $\mathbf{Q}_{num}$ from,

$$\frac{1}{2}\mathbf{B}_{ex} = \int_{\Gamma} Y_{num}\mathbf{Q}_{num}d\Gamma. \tag{25}$$

By (24)–(23) and (24)–(25), we get,

$$\int_{\Gamma} Y_{num}\Delta\mathbf{Q}d\Gamma = \int_{\Gamma} \Delta Y\mathbf{Q}_{ex}d\Gamma, \tag{26}$$

where $\Delta\mathbf{Q} = \mathbf{Q}_{ex} - \mathbf{Q}_{num}$ and $\Delta Y = Y_{num} - Y_{ex}$. The above Equation (26) indicates that the error in the boundary normal gradient $\Delta\mathbf{Q}$ is introduced due to the difference ΔY in Green's function. This error will further transfer to the field computation in space through (14) as follows,

$$\Delta\mathbf{B}(x, y, z) = \int_{\Gamma}\left[\Delta Y\mathbf{Q}_{ex} - \frac{\partial\Delta Y}{\partial n}\mathbf{B}_{ex}\right]d\Gamma - \int_{\Gamma} Y_{num}\Delta\mathbf{Q}d\Gamma, \quad z > 0. \tag{27}$$

The above equation indicates that the errors come from two sources. The first integral shows that even the boundary field and the normal gradient are true, the error is introduced due to the difference in Green's function Y. The second integral shows that the error is transferred from the approximation of the boundary normal gradient function. However, it should be pointed out that both Y_{num} and ΔY are bounded functions declining with distance away from the current position and the error from (27) will not diverge for sure. The practical applications show that this error is in general small. Therefore when we choose a parameter λ to solve the BEM Equation (18), we actually obtain an approximation of the original non-constant-α force-free field solution.

The BEM was first applied to solar magnetic field problem in Yan *et al.* (1991) for linear force-free fields. A detailed procedure of the BEM but for the potential and linear problems can be found in a tutorial paper in an edited book (Yan *et al.*, 1993).

5. Applications

The good coincidence between the extrapolated field lines and observations have
been obtained in previous studies by using the BEM for nonlinear force-free field
problems (e.g., Yan and Sakurai, 1997; Liu *et al.*, 1998a, 1998b; Zhang *et al.*,
2000; Wang *et al.*, 2000; Yan *et al.*, 2001a–c). Quantitative analyses of errors were
also demonstrated in Yan and Sakurai (2000) and Wang *et al.* (2001), showing
that this extrapolation method is very effective. As mentioned before, the aim of
this paper is to demonstrate results obtained by the BEM for flare/CME events
so as to promote its applications in solar activity and space weather studies for a
better understanding of the solar magnetic fields. In the following we will show the
reconstructed coronal magnetic fields for both active regions and the global field.

5.1. CORONAL MAGNETIC FIELDS ABOVE ACTIVE REGIONS

The active region NOAA 8100 was the first one in the new 23rd solar cycle that
produced GOES X-ray X-class flares during its pass on the Sun in November 1997.
A 2B/X2.1 flare occurred on 4 November 1997 in this active region with the maximum GOES SXR flux at 05:58 UT (beginning time: 05:52 UT and ending time:
06:02 UT). This event had a 690 s.f.u. burst at 10 cm and Type II/IV bursts. Type III
microwave bursts were also observed. A halo CME was observed following this
event. The flare was preceded by an M1.3/1N at 01:35 UT and an M4.1/1F at
02:42 UT in the region. There were many CMEs produced during the disk passage
of this active region. NOAA 8100 produced many C-class flares as well. Yan *et al.*
(2001c) reconstructed the 3-D finite energy non-linear force-free magnetic field
structures before and after a 2B/X2 flare at 05:58 UT in this region. The magnetic
field lines were extrapolated in close coincidence with the *Yohkoh* soft X-ray (SXR)
loops accordingly. It is found that the active region is composed of an emerging flux
loop, a complex loop system with differential magnetic field shear, and the large
scale, or open field lines. The similar magnetic connectivity have been obtained
for both instants but apparent changes of the twisting situations of the calculated
magnetic field lines can be observed that align with the corresponding SXR coronal
loops properly (Yan *et al.*, 2001c). Yan *et al.* (2001c) concluded that this flare was
triggered by the interaction of an emerging flux loop and a large loop system with
differential magnetic field shear, as well as large scale, or open field lines. The onset
of the flare was at the common footpoints of several interacting magnetic loops and
confined near the footpoints of the emerging flux loop. The sheared configuration
remained even after the energetic flare, as demonstrated by calculated values of
the twist for the loop system, which means that the active region was relaxed to
a lower energy state but not completely to the minimum energy state (two days
later another X-class flare occurred in this region) (Yan *et al.*, 2001c). This result
is helpful for understanding the magnetic topology of the region that is associated
with the CME generations (Green *et al.*, 2001, 2002).

Recently, Liu *et al.* (2002) has applied the BEM for the non-linear force-free field problem to study the formation of the sigmoid-structure in the solar corona in NOAA 8100 on 3–4 November 1997. They found that the sigmoidal structure appeared after the occurrences of a series of flares accompanied by new magnetic flux emergence. This implies that reconnection may play a role in formation of this sigmoid structure. Liu *et al.* (2002) calculated the self-helicity (twist) and mutual helicity of the active region before and after the formation of the sigmoidal structure and found that the mutual helicity decreased. The twist of the sigmoidal structure was higher than the twist of the emerging magnetic flux and exceeded the critical twist for kink instability (Hood and Priest, 1981). This result suggests that the reconnection increased the twist of magnetic flux tubes by converting mutual helicity to self-helicity when magnetic helicity is conserved, supporting the previous theoretical predictions (Berger, 1999; Priest, 1999). It was the first time to actually estimate the converse of mutual helicity and self-helicity from real observation.

The X5.7/3B flare at position N22W07 in NOAA 9077 active region at 10:24 UT on 14 July 2000 (so called Bastille-Day event) was well-observed with ground- and space-based instruments in hard and soft X-rays, extreme ultraviolet wavelengths (EUV), Hα, radio and magnetic fields, etc. (e.g., Karlický *et al.*, 2001; Deng *et al.*, 2001; Masuda *et al.*, 2001; Aschwanden and Alexander, 2001; Wang *et al.*, 2001; Maia *et al.*, 2001). Thus it provides a good chance to examine the mechanism of triggering, energy release, and related dynamic effects in the flare process.

For the 2000 Bastille-Day flare, Yan *et al.* (2001a, 2001b) have for the first time reconstructed a flux rope structure from the observed vector magnetogram data of the active region NOAA 9077. The inferred magnetic rope is located above the magnetic neutral line in space of the filament. Overlying the rope are multi-layer magnetic arcades with different orientations, as shown in Figure 4. The flux rope structure is helpful to understand the hard X-ray two ribbons in this event (Masuda *et al.*, 2001), i.e., the thermal and non-thermal processes for this flare (Aschwanden and Alexander, 2001; Masuda *et al.*, 2001; Fletcher and Hudson, 2001; Yan and Huang, 2002).

5.2. GLOBAL MAGNETIC FIELDS IN CORONA

NOAA Active Region 8210 was a remarkable flare-productive region that produced several intense flares and was associated with energetic CMEs during its disk passage from later April to early May in 1998 (Warmuth *et al.*, 2000; Thompson *et al.*, 2000). On 1998 May 2 the great X1.1/3B flare occurred at 13:42 UT in AR 8210 near disk center, which was followed by a halo coronal mass ejection (CME) at 15:03 UT observed by SOHO/LASCO. Manifestations of the radio sources over the solar disk by the Nançay Radioheliograph suggested that the CME development may be involved in the large-scale coronal field restructuring (Pohjolainen *et al.*, 2001).

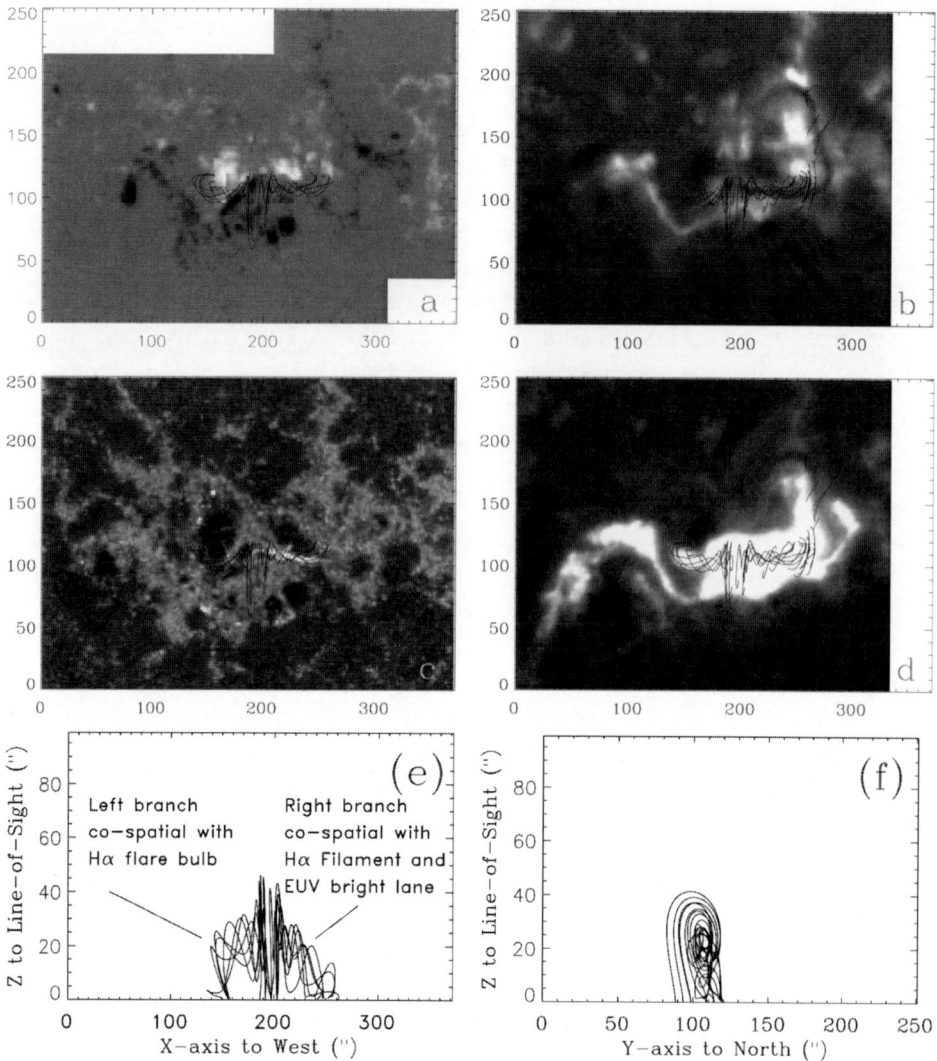

Figure 4. The magnetic rope overlaid with the magnetogram (a), EUV image (c), and Hα images (b, d); and the front (e) and side (f) views of the rope. Some low-lying field lines (2∼33″ high) forming arcades across the filament are also displayed over Hα images (b, d) (Yan *et al.*, 2001b, reproduced by permission of the AAS).

Using the BEM on a global potential model, the large-scale coronal field structure was reconstructed (Wang *et al.*, 2002) from a composite boundary by SOHO/MDI and Kitt Peak magnetograms. The extrapolated large field lines well model a transequatorial interconnecting loop seen in the soft X-ray (SXR) between AR 8210 and AR 8214, which disappeared after the CME. The SOHO/EIT observed the widely extending dimmings, which noticeably deviate from the SXR transequatorial interconnecting loop in position. Wang *et al.* (2002) found that the

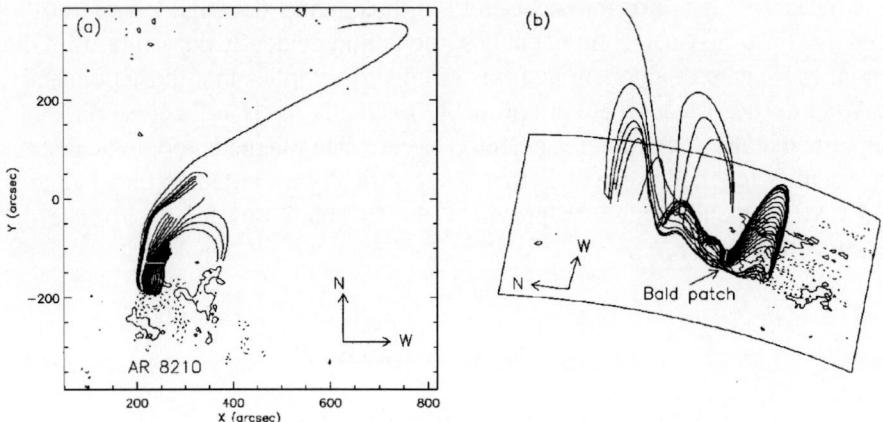

Figure 5. (a) Projection from the top of the computed bald patch field lines that form the separatrix. The origin of the Cartesian coordinates is set at the disk center. (b) Perspective view of the same field lines. A factor times 10 radial stretching is applied in order to get a better view of these very flat field lines (Wang *et al.*, 2002, reproduced by permission of the AAS).

major dimmings are magnetically linked to the flaring active region but some dimmings are not. From the spatial relationships of these features it is suggested that the CME may be led by a global restructuring of multipolar magnetic systems due to flare disturbances. Wang *et al.* (2002) also estimated the mass, magnetic energy, and flux of the ejected material from the dimming regions which are comparable to the output of large CMEs, derived from the limb events. At the CME source region, Huairou vector magnetograms show that a strong shear was rapidly developed in a newly emerging flux region (EFR) near the main spot before the flare (Wang *et al.*, 2002). Magnetic field reconstruction by the BEM revealed the presence of a 'bald patch' (defined as the locations where the magnetic field is tangent to the photosphere) at the edge of the EFR (Wang *et al.*, 2002). The preflare features such as EUV loop brightenings and SXR jets appearing at the bald patch suggest a slow reconnection between the transequatorial interconnecting loop system and a preexisting overlying field above the sheared EFR flux system.

Figure 5 shows the bald-patch obtained from the potential field reconstruction for May–2–1998 event by the BEM, which helps to understand the flare/CME processes for the event (Wang *et al.*, 2002).

More recently, Liu and Zhao (2001) have studied the global non-constant-α force-free field for CR1968 and compared that with the potential field. Evident difference is found. In Liu and Zhao (2001) there were 6 areas where the observed vector field replaces the calculated potential field. The force-free field appears sheared in 4, flares and CMEs associated, areas. The potential magnetic field shows additional connectivity in other 2 areas, while such connectivities do not show up in force-free field configuration.

Checked with vector magnetic field in those areas (Liu and Zhao, 2001), it is seen that the transverse component has orientation evidently departure from the potential transverse component and this orientation implies that those connectivities may not exist; instead, there are probably magnetic interfaces between them. This suggests that the vector field is essential to calculate magnetic topological structure and identify magnetic separatrix that are specifically important to understand flares and CMEs as such locations provide suitable conditions for occurrence of fast reconnection.

6. Conclusions

In summary, several points can be listed below:

(1) For the sake of clarity and consistency, the boundary integral equation representation for non-constant-α force-free field with finite energy content in open space above the sun is derived in detail. The numerical implementation of the boundary integral equation, or the so-called BEM is also introduced in detail to show how to establish the systematic matrix equation and incorporate discrete boundary conditions into formulation so as to solve the problem numerically. An error analysis is provided in order to understand the numerical approximations.

(2) Some applications of the BEM in the reconstruction of coronal magnetic fields from photospheric magnetograph data are demonstrated. They mostly employed the assumption that the field is force-free in the corona under non-constant-α condition. These applications show how the reconstructed coronal fields are associated with solar flare and/or CME activities.

– Using the BEM, it was able to actually estimate the converse of mutual helicity and self-helicity from real observation for the first time.

– The flux rope structure plays a central role in the mechanism of CME formations. For the famous 2000 Bastille-Day flare/CME event, a magnetic rope was, for the first time, reconstructed from the observed vector magnetogram data under the finite energy non-constant-α force-free field assumption by the BEM.

– The global non-constant-α force-free field was able to be reconstructed from a synoptical vector magnetogram on photosphere recently, and the comparison with the potential field indicates that the vector field is essential to calculate magnetic topological structure and identify magnetic separatrix that are especially important to understand flares and CMEs.

– As a general numerical technique, the BEM can also deal with global potential field in the corona. For the 2 May 1998 flare/CME event with transequatorial interconnecting loop structure, a 'bald-patch' was reconstruction by the BEM which helps to understand the flare/CME process for this event.

The succesful applications indicate that the BEM is a prospective and practical tool in the coronal field modeling for solar activity and space weather studies.

(3) It is demonstrated from applications that the vector magnetograms are essential in coronal magnetic field reconstruction. However, the intrinsic 180-degree uncertainty in transverse field measurements can, at present, only be resolved theoretically with different assumptions. This will put further uncertainties for any method to extrapolate the coronal field upwards. Therefore the stereo observations of vector magnetograms are essential to settle the 180-degree ambiguity in transverse fields, and more realistic models are needed.

Acknowledgements

The work was supported by National Natural Science Foundation of China (10225313, 19973008), Chinese Academy of Sciences and the Ministry of Science and Technology of China (G2000078403).

References

Ai, G.: 1987, *Publ. Beijing Astron. Obs.* **9**, 27.
Aly, J. J.: 1992, *Sol. Phys.* **138**, 133.
Amari, T., Aly, J. J., Luciani, J. F., Boulmezaoud, T. and Mikic, Z.: 1997, *Sol. Phys.* **174**, 129.
Aschwanden, M. J. and Alexander, D.: 2001, *Sol. Phys.* **204**, 93.
Berger, M. A.: 1999, in M. R. Brown, R. C. Canfield, and A. A. Pevtsov (eds.), *Magnetic Helicity in Space and Laboratory Plasma*, Geophysical Monograph 111, p. 1.
Brebbia, C. A., Telles, J. C. F. and Wrobel, L. C.: 1984, *Boundary Element Techniques*, Springer-Verlag, Berlin.
Deng, Y., Ai, G., Wang J., Song, G., Zhang, B. and Ye, X.: 1997, *Sol. Phys.* **173**, 207.
Deng, Y., Wang, J., Yan, Y. and Zhang, J.: 2001, *Sol. Phys.* **204**, 9.
Fletcher, L. and Hudson, H.: 2001, *Sol. Phys.* **204**, 71.
Fu, Q. J.: 1999, *Progr. Astron.* **15**, 199.
Gary, G. A.: 1990, *Mem. Soc. Ast. It.* **61**, 457.
Gary, G. A.: 2001, *Sol. Phys.* **203**, 71.
Green, L. M., Harra, L. K., Matthews, S. A. and Culhane, J. L.: 2001, *Sol. Phys.* **200**, 189.
Green, L. M., López Fuentes, M. C., Mandrini, C. H., Démoulin, P., Van Driel-Gesztelyi, L. and Culhane, J. L.: 2002, *Sol. Phys.* **208**, 43.
Hood, A. W. and Priest, E. R.: 1981, *Geophys. Astrophys. Fluid Dynamics* **17**, 297.
Karlický, M., Yan, Y., Fu, Q., Wang, S., Jiricka, K., Mészárosová, H. and Liu, Y.: 2001, *Astron. Astrophys.* **369**, 1104.
Klimchuk, J. A.: 2000, *Sol. Phys.* **193**, 53.
Lin, H. S., Penn, M. J. and Tomczyk, S.: 2000, *Astrophys. J.* **541**, 83.
Liu, Y., Akioka, M., Yan, Y. and Ai, G.: 1998a, *Sol. Phys.* **177**, 395.
Liu, Y., Akioka, M., Yan, Y. and Sato, J.: 1998b, *Sol. Phys.* **180**, 377.
Liu, Y. and Zhao, X.: 2002, American Geophysical Union, Fall Meeting 2001, #SH12B-0756 (http://sun.stanford.edu/~yliu/muri/nlfff.html).
Liu, Y., Zhao, X., Hoeksema, J. T., Scherrer, P. H., Wang, J. and Yan, Y.: 2002, *Sol. Phys.* **206**, 333.
Low, B. C.: 1992, *Astrophys. J.* **399**, 300.
Maia, D., Pick, M., Hawkins III, S. E., Formichev, V. V. and Jiricka, K.: 2001, *Sol. Phys.* **204**, 199.
Masuda, S., Kosugi, S. and Hudson, H.: 2001, *Sol. Phys.* **204**, 57.

McClymont, A. N., Jiao, L. and Mikic, Z.: 1997, *Sol. Phys.* **174**, 191.

Metcalf, T. R., Jiao, L., McClymont, A. N., Canfield, R. C. and Uitenbroek, H.: 1995, *Astrophys. J.* **439**, 474.

Mikic, Z. and McClymont, A. N.: 1994, in K. S. Balasubramaniam and G. Simon (eds), *Solar Active Region Evolution: Comparing Models with Observations*, A.S.P., San Francisco, p. 225.

Moon, Y.-J., Choe, G. S., Yun, H. S., Park, Y. D. and Mickey, D. L.: 2002, *Astrophys. J.* **568**, 422.

Neukirch, T.: 1995, *Astron. Astrophys.* **274**, 319.

Parker, E. N.: 2001, *Chinese J. Astron. Astrophys.* **1**, 99.

Pohjolainen, S., Maia, D., Pick, M., Vilmer, N., Khan, J. I., Otruba, W., Warmuth, A., Benz, A., Alissandrakis, C. and Thompson, B. J.: 2001, *Astrophys. J.* **556**, 421.

Priest, E. R.: 1994, in A. O. Benz and T. J. -L. Courvoisier (eds), *Plasma Astrophysics*, Springer-Verlag, Berlin, p. 1.

Priest, E. R.: 1999, in M. R. Brown, R. C. Canfield and A. A. Pevtsov (eds), *Magnetic Helicity in Space and laboratory Plasma*, Geophysical Monograph 111, p. 141.

Priest, E. R. and Forbes, T.: 2000, *Magnetic Reconnetion-MHD Theory and Applications*, Cambridge University Press, Cambridge, 359.

Roumeliotis, G.: 1996, *Astrophys. J.* **473**, 1095.

Sakurai, T.: 1981, *Sol. Phys.* **69**, 343.

Sakurai, T.: 1989, *Space Sci. Rev.* **51**, 11.

Thompson, B. J., Cliver, E. W., Nitta, N., Delannée, C. and Delaboudinière, J. -P.: 2000, *Geophys. Res. Lett.* **27**, 1431.

Wang, H. N., Yan, Y., Sakurai, T. and Zhang M.: 2000, *Solar Phys.* **197**, 263.

Wang, H. N., Yan, Y. and Sakurai, T.: 2001, *Sol. Phys.* **201**, 323.

Wang, J. X.: 1999, *Fund. Cosmic Phys.* **20**, 251.

Wang, S. J., Yan, Y., Zhao, R. Z., Fu, Q., Tan, C. M., Xu, L., Wang, S. and Lin, H.: 2001, *Sol. Phys.* **204**, 155.

Wang, T. J., Yan, Y. H., Wang, J. L., Kurokawa, H. and Shibata, K.: 2002, *Astropohys. J.* **572**, 580.

Warmuth, A., Hanslmeier, A., Messerotti, M., Cacciani, A., Moretti, P. F. and Otruba, W.: 2000, *Sol. Phys.* **194**, 10.

Wu, S. T., Sun, M. T., Chang, H. M., Hagyard, M. J. and Gary, G. A.: 1990, *Astrophys. J.* **362**, 698.

Yan, Y.: 1995a, *Sol. Phys.* **159**, 97.

Yan, Y.: 1995b, *Lett. Math. Phys.* **34**, 365.

Yan, Y.: 1998, *Astrophysics Reports-Publ. Beijing Astron. Obs.* **32**, 69.

Yan, Y.: 2002, in The 10th European Meeting on Solar Physics, Prague, Czech Republic, 9–15 September 2002 (ESA SP506, 2002), 401.

Yan, Y., Aschwanden, M. J., Wang, S. and Deng, Y.: 2001a, *Sol. Phys.* **204**, 15.

Yan, Y., Deng, Y., Karlický, M., Fu, Q., Wang, S. and Liu, Y.: 2001b, *Astrophys. J.* **551**, L115.

Yan, Y. and Huang, G.: 2002, *Space Sci. Rev.*, this issue.

Yan, Y., Liu, Y., Akioka, M. and Wei, F.: 2001c, *Sol. Phys.* **201**, 337.

Yan, Y. and Sakurai, T.: 1997, *Sol. Phys.* **174**, 65.

Yan, Y. and Sakurai, T.: 2000, *Sol. Phys.* **195**, 89.

Yan, Y., Yu, Q. and Kang, F.: 1991, *Sol. Phys.* **136**, 195.

Yan, Y., Yu, Q. and Shi, H.: 1993, in J. H. Kane, G. Maier, N. Tosaka and S. N. Atluri (eds), *Advances in Boundary Element Techniques*, Springer-Verlag, Berlon, p. 447.

Zhang, C., Wang, H. N., Wang, J. and Yan, Y.: 2000, *Sol. Phys.* **195**, 135.

Zhang, H. Q.: 1994, *Sol. Phys.* **154**, 207.

II: MAGNETOSPHERE/BOW SHOCK

ENERGY FLUX IN THE EARTH'S MAGNETOSPHERE: STORM – SUBSTORM RELATIONSHIP

IGOR I. ALEXEEV

Scobeltsyn Institute of Nuclear Physics, Moscow State University, Moscow, 119992, Russia
(e-mail: Alexeev@dc1.sinp.msu.ru)

Abstract. Three ways of the energy transfer in the Earth's magnetosphere are studied. The solar wind MHD generator is an unique energy source for all magnetospheric processes. Field–aligned currents directly transport the energy and momentum of the solar wind plasma to the Earth's ionosphere. The magnetospheric lobe and plasma sheet convection generated by the solar wind is another magnetospheric energy source. Plasma sheet particles and cold ionospheric polar wind ions are accelerated by convection electric field. After energetic particle precipitation into the upper atmosphere the solar wind energy is transferred into the ionosphere and atmosphere. This way of the energy transfer can include the tail lobe magnetic field energy storage connected with the increase of the tail current during the southward IMF. After that the magnetospheric substorm occurs. The model calculations of the magnetospheric energy give possibility to determine the ground state of the magnetosphere, and to calculate relative contributions of the tail current, ring current and field–aligned currents to the magnetospheric energy. The magnetospheric substorms and storms manifest that the permanent solar wind energy transfer ways are not enough for the covering of the solar wind energy input into the magnetosphere. Nonlinear explosive processes are necessary for the energy transmission into the ionosphere and atmosphere. For understanding a relation between substorm and storm it is necessary to take into account that they are the concurrent energy transferring ways.

Key words: magnetospheric energy, solar wind–magnetosphere interactions

1. Introduction

Magnetospheric disturbances are very complex geophysical phenomena. Magnetospheric substorms and storms are accompanied by many features which differ one disturbance from another. It is necessary to find a compromise between the driven and unloading processes, and to take into account nonlinear dynamical models (see for example Vassiliadis *et al.*, 1995; Baker *et al.*, 1997b). The energy balance is very important for the understanding of the magnetospheric dynamics.

Main conclusion of the review by Weiss *et al.* (1992) is that the total energy dissipated as Joule heat is about twice the ring current injection term, about five the energy expended in the auroral electron precipitation as well as the energy expending in the refilling and heating the plasma sheet, and about ten the energy lost due to the plasmoid ejection. The models and issues associated with the magnetotail energy storage during substorms were presented by Baker *et al.* (1997a). It is shown that the tail lobe energy changing can provide an adequate power for the intense

Space Science Reviews **107:** 141–148, 2003.
© 2003 *Kluwer Academic Publishers.*

substorm event. This conclusion is supported by the well observed case study for May 3, 1986 (Baker *et al.*, 1997a). As pointed out by Alexeev *et al.* (1996), the tail current contribution to D_{st} index has the same value as the ring current field strength in the course of the strong magnetic storm. This conclusion is supported by a storm weakening during substorm expansion phase (tail current disruption) which was discovered by Iyemory and Rao (1996). Siscoe and Petschek (1997) explained the Iyemory and Rao (1996) results by combining the virial theorem and a principle of energy partitioning between the energy storage elements.

In this paper the magnetospheric energy balance for the ground state of the magnetosphere as well as in the course of substorm and storm events is reexamined. Magnetospheric magnetic field energy is calculated by using paraboloid model of the magnetosphere (Alexeev *et al.*, 1996).

2. Energy Flows into the Magnetosphere

At first, it is evident that the solar wind MHD generator is an unique energy source for all magnetospheric processes. The different energy flows into the magneto-sphere are distinguished by the different transforming processes and by different energy storage reservoirs. However, a real dissipation (in the sense of the en-tropy increase) may happens only in the collision ionosphere. The ionospheric and plasma sheet ion accelerations in the course of magnetospheric plasma convection, the auroral electron and ion accelerations, and the amplification of the ring current by the injection of plasma sheet particles – all processes which are realized in colli-sionless magnetospheric plasma – only transform energy between different plasma populations and/or between plasma and magnetic field. In the present paper a plas-moid ejection, energetic particle escaping, waves emission into the magnetosheath and another processes which return energy to the solar wind are not taken into account, but they do provide a contributions (small) to the magnetospheric energy balance.

The ionospheric Joule heating, auroral particle precipitation, ring current particle loss by the charge exchange with the exosphere neutrals, and their precipitation to the atmosphere after wave–particles interactions are the main contributors to the energy losses. Before losses in the atmosphere the energy can be transformed and accumulated as the tail magnetic field energy, as the plasma convection energy, as the ring current particle energy.

The Region 1 field-aligned currents directly transport energy and momentum of the solar wind plasma to the Earth's ionosphere and upper atmosphere. It is the first mostly simple way for the energy input to the Earth's ionosphere and upper atmosphere. This way does not include any reservoir for energy storage. Time scale for the energy transform is the Alfvenic time – lesser than about ten minutes. Typical power for this process is equal to the product of the polar cap potential drop (100 kV) on the total strength of the Region 1 field-aligned currents

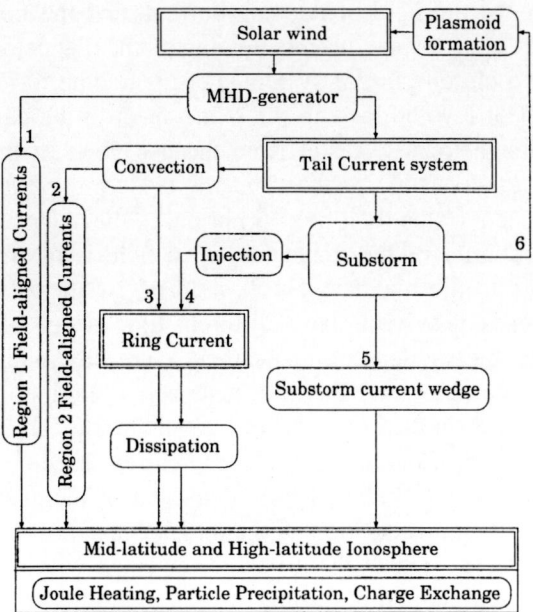

Figure 1. Solar wind energy transfer to magnetosphere and relation between field–aligned currents, substorm, and storm.

(1 MA) and is of the order of 10^{11} W. For extremely disturbed conditions it may be ten times bigger (see Figure 1).

The magnetospheric plasma sheet and tail lobe plasma convection generated by the solar wind is the reservoir for energy storage. The convection is accompanied by the plasma sheet and cold ionospheric polar wind ions acceleration.

The enhancement of the tail current accompanied by the tail lobe magnetic field increase is the main energy storage reservoir. This enhancement is caused by the southward IMF turning and increasing of the solar wind dynamo efficiency near the magnetopause and in the boundary layer.

The explosive release of the accumulated magnetic energy by substorms is the second channel for the energy transfer from the solar wind to the Earth's ionosphere and atmosphere. As a result of the above mentioned processes, the ring current injection occurs during the disturbed intervals. After recovery phase of the magnetic storm, the energy is released in the form of a ring current particle charge exchange and precipitation to the atmosphere. It is the third energy transfer way.

3. Energy of the Magnetosphere and its Ground State

Let us calculate the magnetic energy which is stored in the magnetosphere. The magnetospheric field is a sum of the geodipole field, the field of the screening currents created by the solar wind plasma on the magnetopause, the field of the magnetotail currents, the field of the ring current, and the field of the Region 1 field–aligned currents. Closure currents flowing on the magnetopause and the cross tail current form a closed current system. Let us assume the latter to be located on the paraboloid of revolution with the plane sheet in the equatorial plane. The currents on the magnetopause result from the net effect of the dipole screening current system closed on this surface and the closure current of the tail current sheet. The component of the total field $\vec{B}$ normal to the magnetopause is zero.

The main peculiarity of the magnetospheric field is its time variability. Different contributors to the magnetospheric field or, by the other words, different magnetospheric current systems have different time scales of their time variability. For example, the sudden commencement connected with the change of the Chapman–Ferraro currents have the time scale about ten minutes (Alfvenic time). In the course of substorm the inner part of the tail current can change dramatically during one hour. The time evolution of the ring current occupies several days (several tens of hours). A good method for calculation of the energy transfer during these processes is the representation of the magnetospheric magnetic field, $\vec{B}_m$, as a sum of several terms each of them being connected with some current system (magnetopause current, tail current, ring current, field-aligned currents, and etc.):

$$\vec{B}_m = \vec{B}_d + \vec{B}_{sd} + \vec{B}_t + \vec{B}_r + \vec{B}_{sr} + \vec{B}_{fac}. \tag{1}$$

Here $\vec{B}_d$ is the dipole magnetic field; $\vec{B}_{sd}$ is the field of current on the magnetopause screening the dipole field; $\vec{B}_t$ is the field of the magnetospheric tail current system (cross tail currents and closure magnetopause currents); $\vec{B}_r$ is the field of the ring current; $\vec{B}_{sr}$ is the field of current on the magnetopause screening the ring current field; $\vec{B}_{fac}$ is the field of the field–aligned currents. The closure currents of the tail currents are placed in the magnetopause, that is the same surface where the Chapman–Ferraro currents are located. Because one can observe only both currents simultaneously, a model distinguishing these currents could be made by arbitrary way. In the paraboloid model each current system conserves the condition $B_n = 0$ at the magnetopause.

In general, the energy stored in the magnetospheric volume V_M may be written as

$$\int_{V_M} \left(\frac{\vec{B}_m^2}{2\mu} + p_\perp + \frac{p_\parallel}{2} + \frac{\rho v^2}{2} \right) dV. \tag{2}$$

This is a general formulation which does not include only a wave energy contribution assuming that it is small. Below it will be proposed that the magnetospheric plasma flow is subalfvenic, and we may omit $\rho v^2/2$ in integral (2). The plasma pressure term represents the plasma sheet plasmas and the ring current energetic particles.

Now the energy of the magnetic field of magnetospheric current systems is examined. The energy of the interaction of the current systems between each others as well as their own current system energies will be taken into account. Let us calculate the magnetic energy which is stored in the magnetosphere. To calculate the magnetospheric field energy, the Green's formula will be used (see Maguire and Carovillano, 1996 or Carovillano and Siscoe, 1973). The magnetospheric magnetic field is defined by Equation 1. At first, let $\vec{B}_t + \vec{B}_r + \vec{B}_{sr} + \vec{B}_{fac} = 0$ (cross tail current, ring current, and field–aligned currents magnetic fields were omitted), and the integral of the magnetic field energy density over the entire magnetospheric volume V_M may be written as:

$$\varepsilon_1 = \frac{1}{2\mu} \int_{V_M} (\vec{B}_d + \vec{B}_{cfd})^2 dV. \tag{3}$$

The $\vec{B}_d$ and $\vec{B}_{cfd}$ may be written as gradients of the scalar potential inside the magnetosphere. After Maguire and Carovillano (1996) the integral (3) can be re-written as integral taken over the magnetopause and the Earth's surface. After some algebraic transformations, the magnetospheric energy may be written as a function of the strength of the magnetic field $\vec{B}_{cfd}$ of the screening dipole currents created by the solar wind plasma on the magnetopause. This value calculated at the Earth's center, B_1, defines the integral (3):

$$\varepsilon_1 = 3\varepsilon_d \left(1 - \frac{B_1}{2B_0} \right). \tag{4}$$

For the usual solar wind condition the paraboloid model gives $B_1 = 20$ nT and the difference between ε_1 and $3\varepsilon_D$ (the dipole magnetic field energy equals $2.4 \cdot 10^{18}$ J) gives ε_m – the magnetospheric magnetic field energy. Numerical value of ε_m for the paraboloid model is

$$\varepsilon_m = \delta\varepsilon_D = \varepsilon_1 - 3\varepsilon_D = -0.001\varepsilon_D = -8 \cdot 10^{14} \text{ J}. \tag{5}$$

If the tail magnetic field adds to the integrand of integral (3) then the magnetic field energy may be written as

$$\varepsilon_1 = \frac{1}{2\mu} \int_{V_M} \left(\vec{B}_d + \vec{B}_{cfd} + \vec{B}_t \right)^2 dV. \tag{6}$$

This integral gives us a possibility to calculate change of the magnetospheric energy caused by the tail currents as $\varepsilon_m = \varepsilon_1 - 3\varepsilon_D = \delta\varepsilon_D + \delta\varepsilon_T$, here

$$\delta\varepsilon_T = \frac{F_{pc}I}{2} - \frac{2B_{TE}}{B_1}\delta\varepsilon_D. \tag{7}$$

The first term defines the own tail current energy as the product of the tail lobe magnetic flux F_{pc} on the tail current I. The second term defines the energy of the interaction between the tail current and the dipole. The energy of the interaction between the tail current and the dipole shielding currents is equal to zero, because the magnetic flux of the tail current system field across the magnetopause equals to zero. The dipole shielding currents are placed on magnetopause.

Magnetospheric tail lobe length is infinity. It indicates that the integral (6) goes to infinity. This divergence of the magnetospheric energy can be corrected by taking into account the solar wind plasma thermal energy. Interaction between the solar wind plasma flow and the geomagnetic dipole creates a cavity in the solar wind flow (the magnetosphere). For calculation of the magnetospheric energy ε_m we must extract from ε_1 (Equation (6)) not only the dipole energy ($3\varepsilon_D$), but also the solar wind plasma thermal energy inside the magnetosphere. For determination of the magnetospheric energy we must compare the energy of the sum of the undisturbed dipole and the undisturbed solar wind plasma and the energy of the sum of the solar wind flow past the magnetosphere and the magnetospheric energy. In the distant tail from the pressure balance it follows that the solar wind thermal plasma energy density is equal to the tail lobe magnetic field energy density. We limited the tail lobe length by 60 R_E. Besides this distance we propose that the difference between the magnetic energy density and the solar wind thermal plasma energy density is zero and $\delta\varepsilon_m = 0$.

Tail current magnetic field at the Earth's center $\vec{B}_{TE}$ is directed opposite to $\vec{B}_{cfd}$ and second term in Equetion (7) has a positive sign contrary to $\delta\varepsilon_D$. We can define the ground state of the magnetosphere as the state which corresponds to the magnetospheric magnetic field energy equal to zero. The ground state of the magnetosphere corresponds to $\varepsilon_m = 0$ and $B_T \neq 0$ ($\delta\varepsilon_T = -\delta\varepsilon_D$). Numerical estimations give for the ground state of the magnetosphere $F_{pc} = 521$ MWb, $I = 5$ MA, and $B_{TE} = 4$ nT.

The adding of the plasma sheet energy to integrand of the integral (6) does not change the result of calculation. If $p + B^2/2\mu$ =constant then for $V_M = V_B + V_p$ =constant the sum of the both terms $\dfrac{1}{2\mu} \displaystyle\int_{V_B} \vec{B}_T^2 dV + \int_{V_p} p\,dV$ is equal to the integral (6).

The calculation of the ring current magnetic field energy can be made by the similar way as in previous description. However, the ring current magnetic field energy is positive as well as the ring current particles kinetic energy K_{rc}. Both terms are connected by the Dessler–Parker relationship. For this reason the ring current energy contribution is every time positive:

$$\delta\varepsilon_{rc} = \delta\varepsilon_{Brc} + K_{rc} > 0. \tag{8}$$

In the course of any storm event one can calculate the ring current energy contribution using a paraboloid magnetospheric model. For the great storm of 6–11 February 1986 with the minimum of the hourly averaged $D_{st} = -300$ nT, based on AMPTE data Hamilton et al. (1988) estimated the ring current ions energy to be about $K_{rc} = 8 \cdot 10^{15}$ J. It corresponds to $D_{st} = -200$ nT. In this case $K_{rc} = -10\delta\varepsilon_D$ if we do not take into account the change of the magnetospheric sizes during the disturbance. Unfortunately, in the course of this storm we have no permanent solar wind data set. We can estimate the solar wind dynamic pressure from the data set for another similar storms (Alexeev and Feldstein, 2001). In this case Chapman–Ferraro magnetopause current give the magnetic field at the Earth's center $B_1 = 100$ nT and we receive $\delta\varepsilon_D = -0.5K_{rc}$.

Similar as in the case of the ring current, adding of the field–aligned current amplifies the magnetospheric magnetic field energy by:

$$\delta\varepsilon_{fac} = \int_{V_M} \frac{B_{fac}^2}{2\mu} dV > 0. \qquad (9)$$

For calculation of the field–aligned magnetic field energy, one can use the Equation:

$$\delta\varepsilon_{fac} = \frac{F_{fac}I_{fac}}{2}. \qquad (10)$$

Here F_{fac} is the magnetic flux across the Region 1 field-aligned current loop which includes the ionospheric and magnetopause closure currents. The total Region 1 field–aligned current strength is I_{fac}. For numerical estimations of F_{fac} and I_{fac} we use the model by Alexeev *et al.* (2001):

$$F_{fac} = 140 \text{ MWb} \quad I_{fac} = 5 \text{ MA}, \quad \delta\varepsilon_{fac} = 4.5 \cdot 10^{14} \text{J}. \qquad (11)$$

This value $\delta\varepsilon_{fac}$ is close to $\delta\varepsilon_D$ and it must be taken into account for the energy balance calculations during the substorm and storm events.

4. Conclusions

The magnetospheric magnetic field energy is calculated. Studying contributions to the total magnetospheric energy budget of the magnetopause current, the tail current, the ring current, and the field–aligned currents we determine the ground state of the magnetosphere. This state depends on the solar wind pressure and the interplanetary magnetic field. At the magnetospheric ground sate the tail lobe magnetic flux is of the order of 500 MWb for quiet solar wind conditions. In this case at the Earth's center the sum of the magnetopause current magnetic field and the tail current magnetic field is equal to zero. The key role of the tail current system at the energy balance during substorm and storm is demonstrated by the model calculations. The pressure balance in the distant tail gives the relationship between the lobe magnetic field and the plasma sheet pressure. It allows us to include the plasma sheet plasma in the energy budget calculations. The divergence of the magnetospheric tail energy has been corrected by taking into account the solar wind plasma thermal energy which forced out by tail lobe magnetic field. The Region 1 field–aligned currents, the tail current, and the ring current represent three different ways for the transfer of the solar wind energy into the magnetosphere-ionosphere. At different phases of the magnetic storm, relation between the tail and ring current contributions to the magnetospheric energy budget can be different. During a main phase of the magnetic storm the tail current contribution can be the same as the ring current one.

Acknowledgements

This research was supported by Russian Foundation for Basic Research Grants 01-05-65003, 00-15-96623, and 02-05-74643. The author is grateful to the Organizing Committee of the WSEF2002 chaired by A. C.-I. Chian for the financial support.

References

Alexeev, I. I. *et al.*: 1996, 'Magnetic Storms and Magnetotail Currents', *J. Geophys. Res.* **101**, 7737 –7748.

Alexeev, I. I. *et al.*: 2000, 'A Model of Region 1 Field–aligned Currents dependent on Ionospheric Conductivity and Solar Wind Parameters', *J. Geophys. Res.* **103**, 21, 119–21, 130.

Alexeev, I. I. and Y. I. Feldstein: 2001, 'Modeling of Geomagnetic Field during Magnetic Storms and Comparison with Observations', *J. Atmos. Sol. Terr. Phys.* **63**, 331–340.

Baker, D. N. *et al.*: 1997a, 'A Quantitative Assessment of Energy Storage and Release in the Earth's Magnetotail', *J. Geophys. Res.* **102**, 7169–7178.

Baker, D. N. *et al.*: 1997b, 'Reexamination of Driven and Unloading Aspects of Magnetospheric Substorms', *J. Geophys. Res.* **102**, 7159–7168.

Carovillano, R. L. and G. L. Siscoe: 1973, 'Energy and Momentum Theorems in Magnetospheric Processes', *Rev. Geophys. Space Phys.* **11**, 289–353.

Hamilon, D. C. *et al.*: 1988, 'Ring Current Development during the Great Magnetic Storm of February 1986', *J. Geophys. Res.* **93**, 14, 343–14, 356.

Iyemory, T. and D. R. K. Rao: 1996, 'Decay of the Dst Field of Geomagnetic Disturbance after Substorm Onset and its Implication to Storm – substorm relation', *Ann. Geophysicae* **14**, 608–618.

Maguire, J. J. and R. L. Carovillano: 1966, 'Energy Principles for the Confinement of a Magnetic Field, *J. Geophys. Res.* **71**, 5533–5539.

Siscoe, G. L. and H. E. Petschek: 1997, 'On Storm Weakening during Substorm Expansion Phase', *Ann. Geophysicae* **15**, 211–216.

Vassiliadis, D. *et al.*: 1995, 'A Description of Solar Wind-magnetosphere Coupling based on Nonlinear Filters', *J. Geophys. Res.* **100**, 3495–3506.

Weiss, L. A. *et al.*: 1992, 'Energy Dissipation in Substorms', in *Substorms 1*, ESA SP-335, pp. 309 –317.

RECENT DEVELOPMENTS IN MAGNETOSPHERIC DIAGNOSTICS USING ULF WAVES

B. J. FRASER

Cooperative Research Centre for Satellite Systems, School of Mathematical and Physical Sciences, University of Newcastle, Callaghan, NSW 2308, Australia

Abstract. One of the most ubiquitous indicators of the state and topology of the magnetosphere are ultra-low frequency (ULF) waves. These may be continuously and inexpensively monitored from the ground using networks of magnetometers. The most robust measurable quantity provided by magnetometer networks is signal phase and this paper emphasizes the usefulness of this parameter in a variety of ULF wave diagnostic processes ranging from equatorial to high latitudes.

Key words: magnetosphere, plasma, waves

1. Introduction

The key to better understanding space weather is the development of realistic models of particle populations and their dynamics in the magnetosphere. Ultra low frequency (ULF) waves play a vital role as they transport energy from the solar wind through the magnetosphere. It is thought that ULF waves may play a direct role in providing a mechanism to accelerate particles to MeV 'killer electron' energies (Rostoker *et al.*, 1998; O'Brien *et al.*, 2001). In a passive role ULF waves in the Pc3-5 (1–100 mHz) band are generally observed as resonances and therefore can be important indicators of magnetosphere geometry and boundary region dynamics. By studying the characteristics of ULF waves and using modern data processing techniques we can observe the properties and dynamics of the plasma regions of the magnetosphere through which they propagate. Our emphasis in analysis has been on the measurement of signal phase, which is the most reliable and stable measured parameter at ULF frequencies (Baransky *et al.*, 1985, 1989; Waters *et al.*, 1991).

2. The Plasma Environment

The cold magnetised plasma in the Earth's magnetosphere supports two distinct modes of ULF wave propagation, the Alfven mode and the fast mode. The Alfven mode is field aligned and establishes a field-line resonance (FLR) by forming standing waves between conjugate ionospheres, while the fast magnetisonic or

Space Science Reviews **107**: 149–156, 2003.
© 2003 *Kluwer Academic Publishers*.

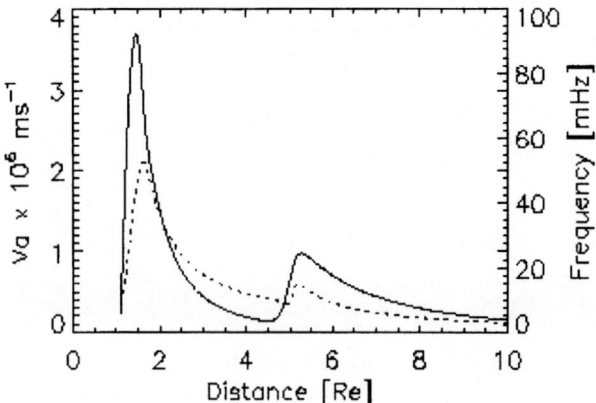

Figure 1. The Alfven velocity (solid line) and FLR frequency (dotted line) variations with radial distance in the ionosphere-plasmasphere system (Waters *et al.*, 2000).

compressional wave mode may propagate isotropically across the field. The propagation speed of both types of waves is determined by the density of the cold plasma (ρ) and the flux density of the Earth's magnetic field (**B**) and is given by the Alfven speed $V_a = B/(\mu\rho)^{1/2}$. The cold ion plasma density is predominately the proton ion population but should include heavy ions such as helium (He^+) and oxygen (O^+) as minor but variable constituents. Oxygen (O^+) in the ionosphere contributes at very low latitudes. The regimes under which we may study ULF diagnostics of the magnetosphere-ionosphere system are illustrated by the Alfven velocity-latitude profile shown in Figure 1.

The maximum in both V_a and the FLR frequency at 5 R_e is due to the presence of a steep plasmapause gradient. Recently, a peak in V_a and the FLR frequency has been observed at very low latitudes (Poulter *et al.*, 1988; Menk *et al.*, 2000; Waters *et al.*, 2000). With a decrease in field line length towards the equator more of the field line becomes embedded in the conjugate ionospheres and this leads to significant mass loading near the feet of the fieldlines by ionospheric ions, mainly oxygen and a corresponding decrease in V_a and FLR frequency at latitudes ≤ 1.5 R_e. This contrasts with the higher latitudes where the mass loading has greatest effect on V_a and the FLR frequency at the top of the field line in the equatorial region.

The regions of the magnetosphere controlled by the radial distribution of V_a conveniently provide four areas of interest; namely the equatorial region (L < 1.5), the low and middle latitude region (1.5 $\leq$ L $\leq$ 5), the plasmapause (L $\sim$ 5) where V_a maximises, and the high latitude plasma trough region from L $\sim$ 5 out to the last closed field line near the magnetopause. This paper will briefly describe the different ULF wave diagnostic techniques developed by the University of Newcastle Space Physics Group to study these four regions.

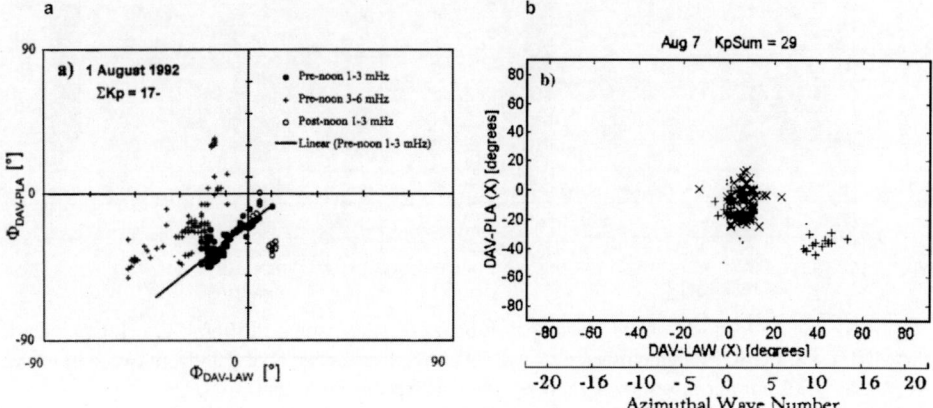

Figure 2. Phase difference scatter plots. (a) Crossphase measurements for Davis–Law are plotted against concurrent crossphase measurements for Davis–Plateau. The regression coefficient for 1 August 1992 is 0.7. (b) Corresponding data for 7 August.

3. High Latitude Cusp Observations

Field line topology and magnetospheric convection is dramatically different under conditions of positive and negative B_Z IMF (Cowley *et al.*, 1991). The cusp and the open/closed boundary have generally been defined in terms of particle measurements (e.g. Newell *et al.*, 1997). In this context it is of interest to consider a wave determined boundary between open and closed field lines (OCB) and a so-called 'wave cusp' (Potemra *et al.*, 1992).

The concept that FLRs do not occur on open field lines provides a very simple method to define the OCB as the latitude at which of the last closed field line where FLR are observed. The diurnal FLR signature, or the absence of it can provide a useful indicator of when magnetometers are located under open or closed field lines (Waters *et al.*, 1996; Lanzerotti *et al.*, 1999).

Ables *et al.* (1998), using quantitative cross-phase difference techniques, illustrated how a combination of the geomagnetic north-south (N–S) and east-west (E–W) field components may be used to determine open and closed field line conditions.

In Figure 2 the polarized power cross-phase difference between two N-S stations spaced 110 km apart is plotted against the cross-phase difference between two E-W stations a similar distance apart, over the 1–6 mHz Pc5 band. The criterion of phase closure around the triangular array ensured that only waves propagating across the triangle were included. The N-S phase difference in Figure 2a is due to the FLR phase signature (e.g. Southwood, 1974) and the E-W phase difference changes sign around local noon and is due to azimuthal wave propagation. In the prenoon sector the second harmonic in the 3–6 mHz band can also be seen.

In the phase scatterplot for 7 August the points are clustered around zero phase with a slightly negative N-S phase difference. Since geomagnetic activity was high

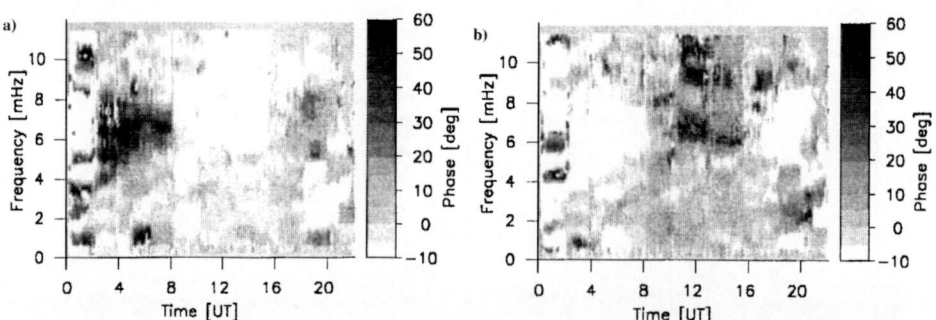

Figure 3. Precise location of the plasmapause using FLR crossphase difference techniques on the SAMNET and IMAGE magnetometer arrays. (a) Crossphase analysis of data from two stations near L = 4.75. (b) A similar plot showing a reversal in crossphase between 08–16 UT.

on this day ($\Sigma\ K_p = 29$) we consider that the array was under open field lines and the FLR and propagation phase structure were absent.

It can be seen from these phase observations that studying the phase of FLR using magnetometers provides a relatively inexpensive means of studying the OCB.

4. Plasmapause Observations

FLR theory shows that a change in the sign of slope of the V_a profile is associated with the sudden plasma density change at the plasmapause (Figure 1). However, equally important is the change in the sign of the crossphase spectrum at the plasmapause. Waters (2000) recognised this phase reversal as a very sensitive technique to monitor the location of the plasmapause using FLR crossphase spectral observations.

Figure 3a, between 02–20 UT shows a FLR frequency at 4–7 mHz, typical near the plasmapause. Here phases are essentially positive with negative phase difference plotted in white. This shows the phase difference changes sign between 08 and 18 hr UT. In contrast, Figure 3b which plots only negative phase is a mirror image of Figure 3a. These two panels illustrate a phase difference reversal at 08 and 18 hr UT which indicate plasmapause crossings. It can be concluded that the plasmapause was observed at L $\sim$ 4.75 at 08 and 18 hr UT.

5. Equatorial and Low Latitude Observations

From a modelling perspective, although the geomagnetic field at low latitudes may be considered dipolar, the propagation of waves through the ionosphere must consider a tilted magnetic field and mixed downgoing wave modes. The extreme situation here is at the equator where the geomagnetic field is horizontal and FLR should not be observed.

The equatorial region is generally considered an ideal location to observe fast compressional mode waves. Phase properties of waves at the equator show a distinct ionospheric influence. With the passage of dawn polarisation properties and interstation phase (Waters *et al.*, 2001) may change dramatically.

As seen in Figure 1, off the equator at very low latitudes, the heavy ion ionosphere mass loading lowers the Alfven velocity and the FLR eigenfrequency so that a frequency maximum occurs around L = 1.4–1.6 (Hattingh and Sutcliffe, 1987; Menk *et al.*, 2000). The extent to which the position of the peak is dependent on ionospheric and magnetospheric conditions is unknown. However, the diurnal variation in resonant frequency over the dawn-to-dusk interval follows the expected trend in O^+ density along the low latitude flux tubes. These FLR are low Q resonances varing from 1.3–2.5 with a normalised damping factor from 0.15–0.3 to 0.3–0.4 at the lowest latitude (Menk *et al.*, 2000). The absence of FLR below L = 1.3 may be associated with increased damping of waves in the ionosphere.

6. Plasma Density Measurements

A latitudinal chain of ground based magnetometers is capable of monitoring the radial distribution of the mass density in the magnetosphere from the last closed field line through the plasmatrough and the plasmapause, into the low latitude plasmasphere. As the chain rotates with local time a map of plasma mass density covering the dayside from dawn to dusk is generated. Typical latitudinal pair spacing of 75–200 km has been found appropriate for the cross-phase and amplitude ratio and difference techniques (eg. Baransky *et al.*, 1985; Waters *et al.*, 1991).

The plasma mass density determination is a four stage process (Waters, 2000). Firstly the FLR frequency distribution with latitude is obtained using the cross-phase method often supplemented by the power and amplitude difference or ratio methods (Baransky *et al.*, 1985). A dipole field is used for the plasmasphere-plasmapause but the Tsyganenko T89 or T96 models must be used in the plasmatrough. Thirdly a suitable functional form for the variation of the plasma mass density along the magnetic field line must be chosen. This is generally an R^{-3} or R^{-4} radial dependence and there is very little difference in the results obtained by these two functions. Fourthly, the FLR shear Alfven wave equation must be solved for the plasma mass density (Singer *et al.*, 1981).

An example of the plasma mass density on 3 October 1990, plotted as a function of time and radial distance is shown in Figure 4a (Waters *et al.*, 1996). At distances greater than about $7R_e$ over 11–21 UT densities are typically less than 10 H^+ cm^{-3} (assuming all ions are H^+). These are characteristic of the plasmatrough with the plasmapause showing a reasonably steep gradient. The dense region seen between 22–01 UT with densities of 20–60 H^+ cm^{-3} is typical of the plasmasphere. Local noon is 18 UT and the evening pre-midnight plasmapause bulge is seen over this time. It is apparent from this example that the FLR plasma density measurements

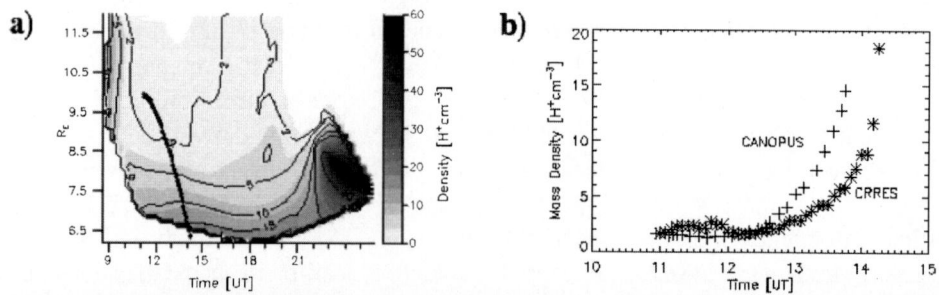

Figure 4. (a) Plasma mass densities for 3 October, 1990. (b) Comparison between the densities from FLR crossphase difference and CRRES at the locations shown in (a).

are a powerful tool for monitoring plasmasphere, plasmapause and plasmatrough topology. More examples can be seen on http./plasma.newcastle.edu.au/spwg/ on the Plasma Density Page.

7. Satellite-ground Comparisons

ULF wave plasma density observations cannot be considered reliable until checked by intercalibration against in-situ plasma density measurements. On 3 October, 1990, plasma mass densities from the CANOPUS FLR data coincided with a day-side over-pass of the CRRES spacecraft (Loto'aniu *et al.*, 1999). This is illustraterd in Figure 4a by the track commencing at 6 R_e and 14 UT. The spacecraft density data were obtained from the Iowa Plasma Wave Experiment (PWE)(Anderson *et al.*, 1992) which measures the upper hybrid resonance line, providing an indirect measurement of the electron density. Charge neutrality and a hydrogen plasma is assumed so the estimated electron density is equivalent to the proton density. Figure 4b shows excellent agreement between the two sets of data, especially over 11–13 UT (Loto'anui *et al.*, 1999).

This suggests that the ion component of the plasma is essentially composed of protons. The observed higher CANOPUS densities after $\sim$ 13 UT could be due to heavy ion mass loading by He^+ or O^+ ions. This example illustrates how a comparison between electron density and FLR ion mass measurements may provide an indication of heavy ion mass loading in the magnetosphere. Electron density data could be provided by VLF whistler or satellite measurements.

8. Conclusions

This paper has briefly highlighted the potential of using inexpensive groundbased magnetometer latitudinal arrays to monitor the plasmasphere-magnetosphere mass loaded ion density from dawn to dusk. There is also the capability of studying the local time and latitudinal dynamics of the plasmapause motion. It is expected that international collaboration over the next year or two will see the establishment of dedicated and strategically located magnetometer arrays in both hemispheres in order to improve the spatial and temporal resolution of mass density measurements.

Acknowledgements

C. L. Waters, F. W. Menk, S. T. Ables and P. Loto'aniu are thanked for helpful discussions. This research was carried out with financial support from the Commonwealth of Australia through the Cooperative Research Centres Program.

References

Ables, S. T., Fraser, B. J., Waters, C. L., Neudegg, D. A. and Morris, R. J.: 1998, 'Monitoring Cusp/Cleft Topology Using Pc5 ULF Waves', *Geophys. Res. Lett.* **25**, 1507–1510.

Anderson, R. R., Gurnett, D. A. and Odem, D. L.: 1992, 'CRRES Plasma Wave Experiment', *J. Spacecraft and Rockets* **29**, 570.

Baransky, L. N., Borovkov, J. E., Gokhberg, M. B., Krylov, S. M. and Troitskaya, V. A.: 1985, 'High Resolution Method of Direct Measurement of the Magnetic Field Lines' Eigenfrequencies', *Planet. Space Sci.* **33**, 1369.

Baransky, L. N., Belokris, S. P., Borovkov, Y. E., Gokhberg, M. B., Federov, E. N. and Green, C. A.: 1989, 'Restoration of the Meridional Structure of Geomagnetic Pulsation Fields from Gradient Measurements', *Planet. Space Sci.* **37**, 859.

Cowley, S. W. H., Morelli, J. P. and Lockwood, M.: 1991, 'Dependence of Convective Flows and Particle Precipitation in the Dayside High Latitude Ionosphere on the X and Y Components of the Interplanetary Field', *J. Geophys. Res.* **99**, 17323–17342.

Hattingh, S. K. F. and Sutcliffe, P. R.: 1987, 'Pc3 Pulsation Eigen Period Determination at Low Latitudes', *J. Geophys. Res.* **92**, 12,433.

Lanzerotti, L. J., Shino, A., Fukunishi, H. and Maclennan, C. G.: 1999, 'Long Period Hydromagnetic Waves at Very High Geomagnetic Latitudes', *J. Geophys. Res.* **104**, 28,423–28,435.

Loto'anui, T. M., Waters, C. L. and Fraser, B. J.: 1999, 'Plasma Mass Density in the Plasmatrough: Comparison Using ULF Waves aqnd CRRES', *Geophys. Res. Lett.* **26**, 3277–3280.

Menk, F. W., Waters, C. L. and Fraser, B. J.: 2000, 'Field Line Resonances and Waveguide Modes at Low Latitudes; 1. Observations', *J. Geophys. Res.* **105**, 7747–7761.

Newell, P. T., Xu, D., Meng, C.-I. and Kivelson, M. G.: 1997, 'Dynamical Polar Cusp: A Unifying Approach', *J. Geophys. Res.* **102**, 127–140.

O'Brien, T. P., McPherron, R. L., Sornette, A. Reeves, G. D., Freidel, R. and Singer, H. J.: 2001, 'Which Magnetic Storms Produce Relativistic Electrons at Geosynchronous Orbit', *J. Geophys. Res.* **106**, 15533–153344.

Potemra, T. A., Erlandson, R. E., Zanetti, L. J., Arnoldy, R. L., Woch, J. and Friis-Christensen, E.: 1992, 'The Dynamic Cusp', *J. Geophys. Res.* **97**, 2835–2844.

Poulter, E. M., Allan, W. and Bailey, G. J.: 1988, 'ULF Pulsation Eigenperiods Within the Plasmasphere', *Planet. Space Sci.* **36**, 185.

Rostoker, G., Skone, S. and Baker, D. N.: 1998, 'The Origin of Relativistic Electrons in the Magnetosphere Associated with Some Geomagnetic Storms', *Geophys. Res. Lett.* **25**, 3701–3704.

Singer, H. J., Southwood, D. J., Walker, R. J. and Kivelson, M. G.: 1981, 'Alfven Wave Resonances in a Realistic Magnetospheric Magnetic Field Geometry', *J. Geophys. Res.* **86**, 4589.

Southwood, D. J.: 1974, 'Some Features of Field Line Resonance in the Magnetosphere', *Planet. Space Sci.* **22**, 483.

Waters, C. L., Menk, F. W. and Fraser, B. J.: 1991, 'The Resonance Structure of Low Latitude Field Line Resonances', *Geophys. Res. Lett.* **18**, 2293–2296.

Waters, C. L., Samson, J. C. and Donovan, E. F.: 1996, 'Variation of Plasmatrough Density Derived from Magnetospheric Field Line Resonances', *J. Geophys. Res.* **101**, 24737.

Waters, C. L., Harrold, B. G., Menk, F. W., Samson, J. C. and Fraser, B. J.: 2000, 'Field Line Resonances and Waveguide Modes at Low Latitudes; 2. A model', *J. Geophys. Res.* **105**, 7763–7774.

Waters, C. L.: 2000, 'ULF Resonance Structure in the Magnetosphere', *Adv. Space Res.* **25**, 1541.

Waters, C. L., Sciffer, M. D., Fraser, B. J., Brand, K., Foulkes, K., Menk, F. W., Saka, O. and Yumoto, K.: 2001, 'The Phase Structure of Very Low Latitude ULF Waves Across Dawn', *J. Geophys. Res.* **106**, 15,599.

NEW INSIGHTS ON GEOMAGNETIC STORMS FROM MODEL SIMULATIONS USING MULTI-SPACECRAFT DATA

VANIA K. JORDANOVA

Space Science Center, University of New Hampshire, Durham, NH 03824, U.S.A.
(e-mail: vania.jordanova@unh.edu)

Abstract. The forecast of the terrestrial ring current as a major contributor to the stormtime *Dst* index and a predictor of geomagnetic storms is of central interest to 'space weather' programs. We thus discuss the dynamical coupling of the solar wind to the Earth's magnetosphere during several geomagnetic storms using our ring current-atmosphere interactions model and coordinated space-borne data sets. Our model calculates the temporal and spatial evolution of H^+, O^+, and He^+ ion distribution functions considering time-dependent inflow from the magnetotail, adiabatic drifts, and outflow from the dayside magnetopause. Losses due to charge exchange, Coulomb collisions, and scattering by EMIC waves are included as well. As initial and boundary conditions we use comple-mentary data sets from spacecraft located at key regions in the inner magnetosphere, Polar and the geosynchronous LANL satellites. We present recent model simulations of the stormtime ring current energization due to the enhanced large-scale convection electric field, which show the transition from an asymmetric to a symmetric ring current during the storm and challenge the standard theories of (a) substorm-driven, and (b) symmetric ring current. Near minimum *Dst* there is a factor of ~ 10 variation in the intensity of the dominant ring current ion specie with magnetic local time, its energy density reaching maximum in the premidnight to postmidnight region. We find that the O^+ content of the ring current increases after interplanetary shocks and reaches largest values near *Dst* minimum; $\sim 60\%$ of the total ring current energy was carried by O^+ during the main phase of the 15 July 2000 storm. The effects of magnetospheric convection and losses due to collisions and wave-particle interactions on the global ring current energy balance are calculated during different storm phases and intercompared.

Key words: magnetic storms, magnetosphere-ionosphere coupling, ring current, solar wind-magne-tosphere coupling, space weather

Abbreviations: EMIC – electromagnetic ion cyclotron; IMF – interplanetary magnetic field; LANL – Los Alamos National Laboratory

1. Introduction

A major goal of space science research and NASA's 'space weather' program is understanding the causes of large geomagnetic storms well enough to be able to predict them. The ring current is an essential element of all geomagnetic storms and a major contributor to the stormtime *Dst* index. A large fraction of solar wind energy is extracted by the magnetosphere through reconnection at the low-latitude

Space Science Reviews **107**: 157–165, 2003.
© 2003 *Kluwer Academic Publishers.*

dayside magnetopause (Dungey, 1961) and is stored in the stormtime ring current. The temporal evolution of the ring current and its relation to various interplanetary conditions are thus of central interest in all studies of geomagnetic disturbances.

The terrestrial ring current consists mostly of energetic (1–300 keV) H^+ and O^+ ions (e.g., Daglis, 1997). Minor but important components of the ring current include He^+, He^{++}, and N^+ ions, and electrons. The strength of the ring current increases as the storm develops and its variation is mainly controlled by magnetospheric convection. Radial diffusion affects mostly the local time variations of higher energy (> 160 keV) particles (Chen *et al.*, 1994). Ring current intensification during the main phase of the storm causes a sharp decrease of the horizontal component of the terrestrial magnetic field at Earth, namely of the *Dst* index. During the recovery phase of the storm the ring current decays due to various loss mechanisms – adiabatic drifts through the dayside magnetopause, charge exchange, Coulomb collisions, wave-particle interactions, and collisions at low altitude with the dense atmosphere. The energy transferred from the solar wind into the ring current is thus released to the plasmasphere and upper atmosphere, and the magnetic field at Earth returns to normal conditions. Recent numerical modeling, theory, and observations indicate that the ring current is a highly dynamic region and to understand its complexity, global scale studies are needed. New advances in ring current modeling using multiple satellite observations, related to ring current formation and the relative importance of ring current losses caused by drift, collisions, and plasma waves scattering to ring current decay are presented in this paper.

2. Ring Current Model

We investigate the development of the ring current ion population during several geomagnetic storms using our global physics-based model (Jordanova *et al.*, 1996, 2001a). The model solves numerically the bounce-averaged kinetic equation for the distribution function F of charged particles:

$$\frac{\partial F}{\partial t} + \frac{1}{R_o^2}\frac{\partial}{\partial R_o}\left(R_o^2\langle\frac{dR_o}{dt}\rangle F\right) + \frac{\partial}{\partial\varphi}\left(\langle\frac{d\varphi}{dt}\rangle F\right) + \frac{1}{\sqrt{E}}\frac{\partial}{\partial E}\left(\sqrt{E}\langle\frac{dE}{dt}\rangle F\right)$$
$$+ \frac{1}{h(\mu_o)\mu_o}\frac{\partial}{\partial\mu_o}\left(h(\mu_o)\mu_o\langle\frac{d\mu_o}{dt}\rangle F\right) = \langle\left(\frac{\partial F}{\partial t}\right)_{loss}\rangle. \tag{1}$$

Here R_o is the radial distance in the equatorial plane from $2\,R_E$ to $6.5\,R_E$ and φ is the magnetic local time (MLT) in degrees with $0°$ at midnight. We consider H^+, O^+, and He^+ particles with kinetic energy E from 100 eV to 400 keV, and equatorial pitch angle α_o from $0°$ to $90°$, where $\mu_o = \cos(\alpha_o)$. The function $h(\mu_o) = S_B/2R_o$, where S_B is the half-bounce path length.

The variations of F due to transport are described by the left-hand side of Equation (1). Adiabatic drifts in time-dependent magnetospheric electric fields and a three-dimensional dipole model of the Earth's magnetic field are considered. We use a Volland–Stern (Volland, 1973; Stern, 1975) type semi-empirical potential model which consists of a convection potential

$$U_{conv} = \frac{U(t)}{2} \left(\frac{L}{L^*}\right)^{\gamma} \sin(\varphi - \varphi_o) \tag{2}$$

and a corotation potential $U_{cor} = -C/R_o$, where $C = 1.44 \times 10^{-2} R_E^2$ V m^{-1} and $L^* = 8.5$. For the time-dependent potential $U(t)$ we use either the Kp-dependent formulation of Maynard and Chen (1975) or the polar cap potential drop calculated with the IMF-dependent model of Weimer (1996). The magnetospheric inflow on the nightside is modeled after geosynchronous LANL data. Losses through the dayside magnetopause are taken into account in the model, allowing free outflow of charged particles from the dayside boundaries.

All major loss processes of ring current ions are included in the right-hand side of Equation (1); charge exchange with geocoronal hydrogen, Coulomb collisions with thermal plasma, wave-particle interactions, and absorption of ring current particles at low altitude in the atmosphere. To calculate losses due to charge exchange our model is coupled with a hydrogen geocoronal model of Rairden, Frank, and Craven (1986). The plasmaspheric densities used to calculate losses due to Coulomb collisions are obtained from the time-dependent plasmasphere model of Rasmussen, Guiter and Thomas (1993). The atmospheric loss cone is from $0°$ to α_{oc}, where α_{oc} corresponds to a 200 km mirror altitude in the atmosphere. The growth rate of EMIC waves is self-consistently calculated and multicomponent plasma diffusion coefficients are used to include the effect of wave-particle interactions.

3. Results

In this paper we study the global response of the ring current to the interplanetary triggers observed during several geomagnetic storms with different peak intensity and duration. The passage at Earth of the 10–11 January 1997 magnetic cloud induced a storm of moderate geomagnetic activity with $Dst \sim -80$ nT. Interplanetary plasma and field measurements were made by the SWE and MFI instruments on the Wind spacecraft and are shown in detail by Farrugia et al. (1998). In Figure 1 we plot the north-south B_z component of the IMF, the solar wind dynamic pressure, and the Dst index. An interplanetary shock arrived at Wind at hour ~ 25, driven by a magnetic cloud which extends from hour ~ 29 until hour ~ 51. The IMF reaches minimum $B_z \sim -15$ nT at hour ~ 31.5. The solar wind dynamic pressure is highly variable reaching high (~ 50 nPa) values near the end of the cloud due to the high density prominence material observed at Wind (Burlaga et al., 1998). The 18-station measured Dst (dashed-dotted line) and corrected $Dst*$ (solid line) are shown in Figure 1c. Corrections for magnetopause currents (Burton, McPherron,

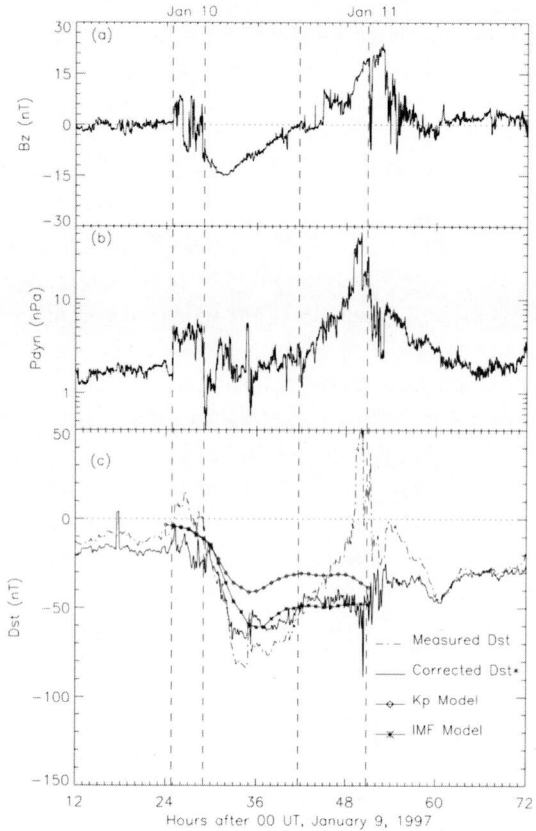

Figure 1. (a) IMF B_z data from Wind; (b) solar wind dynamic pressure; and (c) *Dst* index on 9–11 January 1997. The *vertical guidelines* indicate from left to right the interplanetary shock, the cloud front boundary, the B_z negative to positive transition inside the cloud, and the cloud rear boundary.

and Russell, 1975) and currents induced in the solid Earth are included. The storm has a rapid main phase, which starts on cloud arrival and lasts for the first ~ 4 hours of cloud passage; moderately strong and constant activity persists until the rear of the cloud passes Earth, followed by a slow storm recovery.

We simulated ring current evolution during the 10–11 January 1997 storm period using our physics-based model and initial conditions inferred from measurements from the HYDRA, TIMAS, and CAMMICE instruments on Polar spacecraft (Jordanova *et al.*, 1999). We calculated the ring current contribution to the *Dst* index using the relation derived by Dessler and Parker (1959) and generalized by Sckopke (1966). A comparison of the model results with the corrected, Dst^*, values is shown in Figure 1c. To illustrate the effect of the large-scale magnetospheric electric field on ring current formation, we compare ring current development during this storm using two formulations of the cross-tail electric field: (a) Kp-dependent (diamond line) and (b) IMF-dependent (starred line). Although our model does not resolve the small-scale Dst^* fluctuations, the ring current perturbation calculated with the IMF-dependent model follows the large-scale temporal variations of the observed Dst^* index reasonably well. The Weimer (1996) model predicted a polar cap potential drop with a steep rise starting after cloud arrival and reaching maximum values of ~ 150 kV, while the Kp-dependent Volland–Stern model (Volland, 1973; Stern, 1975; Maynard and Chen, 1975) predicted a monotonic increase of the polar cap potential drop up to ~ 120 kV. The ring current injection rate was thus larger when Weimer model was used and resulted in stronger ring current buildup and better agreement with observations.

We should note that besides the convection electric field as a mechanism for energizing particles, another important factor contributing to ring current formation is the plasma sheet ion population as a direct source for ring current ions. During magnetic storms, plasma sheet particles are transported earthward, energized, and become trapped, thus increasing the preexisting ring current population. Stormtime plasma sheet density enhancements are therefore important in understanding ring current buildup (e.g., Chen *et al.*, 1994; Jordanova *et al.*, 1998; Kozyra *et al.*, 1998). Jordanova *et al.* (1998) modeled ring current development during the October 1995 magnetic storm comparing simulations from a test run using prestorm plasma sheet density with simulations considering stormtime enhancements in the plasma sheet. They demonstrated that the increase in the convection electric field alone was not sufficient to reproduce the stormtime Dst index; the strength of the ring current doubled when the stormtime enhancement of plasma sheet density was considered and good agreement with Dst index was achieved.

In addition to calculating the ring current distribution functions and Dst, we can produce global images of the ring current energy and number densities during all phases of the storms. This is illustrated in Figure 2 showing theoretical results from our model for the 14–19 May 1997 and the 13–18 July 2000 geomagnetic storms. The May 1997 storm was caused partly by $B_z < 0$ fields in the sheath region behind an interplanetary shock and partly by the magnetic cloud driving the shock (Jordanova *et al.*, 2001a). Peak values of $Dst* \sim -85$ nT were measured ~ 2 hours after the front boundary of the cloud with its strong negative $B_z = -25$ nT reached Earth. On the other hand, the period of intense geomagnetic activity during 13–18 July 2000 was caused by three interplanetary coronal mass ejecta each driving interplanetary shocks, the last ejection containing a negative to positive B_z excursion from about -55 nT to 25 nT (Jordanova *et al.*, 2001b). The corrected $Dst*$ reflects the 3 depositions of solar wind energy at the shocks. There are minor to moderate $Dst*$ depressions during the intervals following the first two shocks, and a very large decrease to values of about -220 nT after the third shock.

The globally averaged O^+ energy density percentage obtained from our model during the two storms is shown in the top panel of Figure 2 with a solid line. The O^+ density percentage at geosynchronous orbit from the statistical study of Young, Balsiger, and Geiss (1982) is shown with a dash-dot line. We find that O^+ content increases after each shock and reaches the largest values near minimum

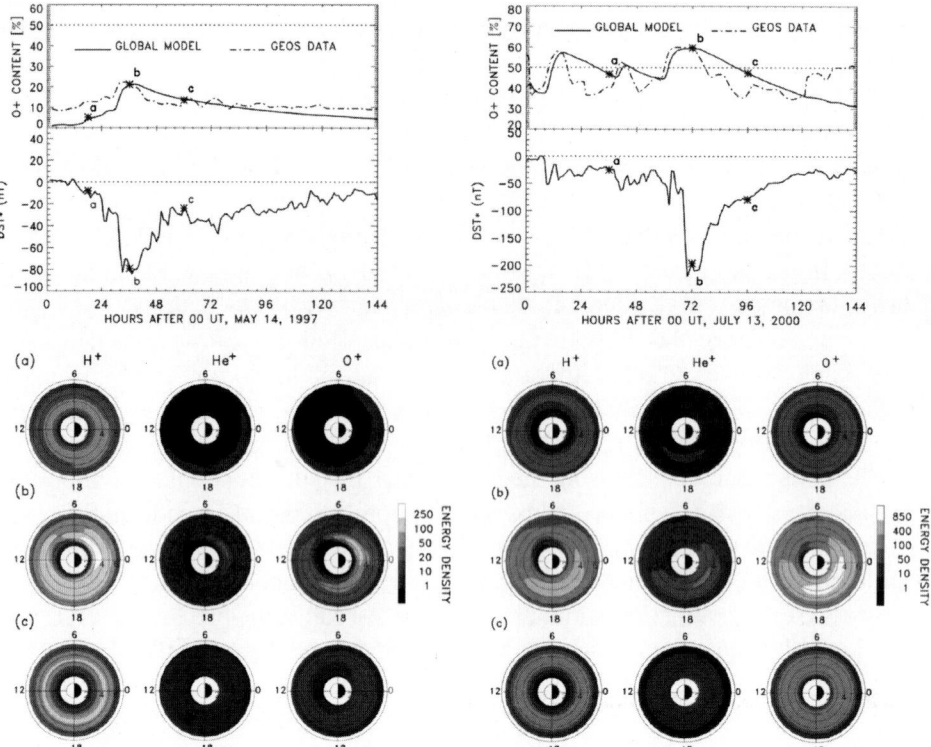

Figure 2. Ring current parameters during the May 1997 (*left*) and July 2000 (*right*) geomagnetic storms. From *top* to *bottom* are shown the content of ring current O^+, the corrected $Dst*$ index, and H^+, He^+, and O^+ ring current ion energy density (keV/cm^3) at selected hours indicated with (a), (b), and (c) on the $Dst*$ plot.

Dst. During the May 1997 storm it remained below 20% and most of the energy was in the proton population, H^+ being thus the dominant ring current ion. In contrast, during the July 2000 solarmax storm period O^+ content did not drop below 45% during the passage of the three shock-ejecta system reaching $\sim 60\%$ near Dst minimum of the great storm. The dominance of O^+ ring current ion was found as well in CRRES observations during a storm of comparable strength which occurred on 24 March 1991 (Daglis, 1997).

The energy densities of the major ring current ion species H^+, He^+, and O^+ are shown as dial plots in Figure 2 as a function of radial distance in the equatorial plane and MLT during the initial (a), main (b), and recovery (c) phases of the storms. Pronounced ring current asymmetry is evident during the main phase of the storms (b) for all species. During these times most of ring current particles are on open drift paths and ring current energy is concentrated in the dusk-to-postmidnight sector, at L values from 3 to 5. As noted above, H^+ (O^+) is the dominant specie during the May 1997 (July 2000) storm and there is a factor of ~ 10 variation in its intensity with magnetic local time, its energy density reaching maximum near

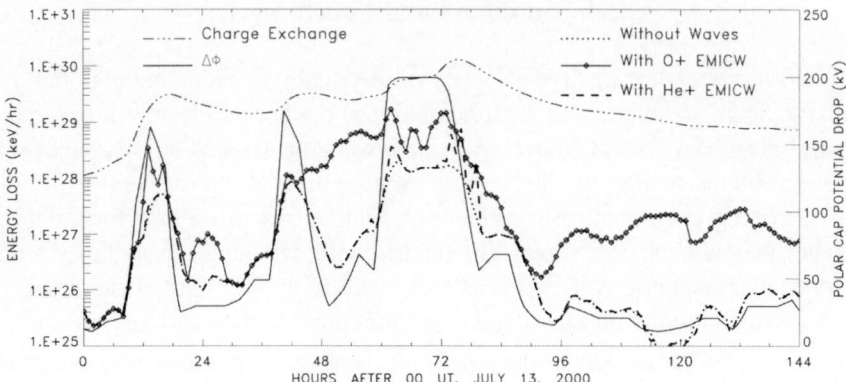

Figure 3. The polar cap potential drop for the July 2000 storm, on which are superposed H^+ ring current energy losses due to charge exchange and ion precipitation.

postmidnight (premidnight). Such ring current asymmetry is not present during less disturbed times shown in Figures 2a and 2c. During the recovery phase of the storms the convection electric field decreases and causes particles to move from open to closed drift paths and to become trapped. Particles that are not trapped are lost as they drift through the dayside magnetopause. The trapped population evolves into a symmetric ring current.

Finally, we discuss the relative contribution to ring current decay of various loss mechanisms included in our model, charge exchange, Coulomb collisions, and ion precipitation. Jordanova *et al.* (2001b) calculated the wave growth of EMIC waves considering, in addition to previous studies, wave excitation not only in the frequency range between the He^+ and O^+ gyrofrequencies (He^+ band), but also at frequencies below the O^+ gyrofrequency (O^+ band). The curves in Figure 3 show the ring current energy losses due to charge exchange (dashed-dotted line), ion precipitation at low altitude when no wave-particle interactions are considered (dotted line), ion precipitation when scattering by O^+ band EMIC waves is included (diamond line), and ion precipitation when scattering by He^+ band EMIC waves is included (dashed line), computed as the 13–18 July 2000 storm progresses. Energy losses due to Coulomb collisions are about two orders of magnitude smaller than charge exchange losses and follow their variations in time (Jordanova *et al.*, 1998), and are thus not shown. The ion precipitation at low altitudes increases during periods of strong magnetospheric convection identified with enhancements in the polar cap potential drop (solid line). Intense EMIC waves are generated near *Dst* minima, the wave gain of O^+ band exceeding the magnitude and extending over larger MLT sector than the He^+ band (Jordanova *et al.*, 2001b). The proton precipitation losses increase by more than an order of magnitude when wave-particle interactions are considered. The energy losses due to charge exchange are, however, predominant.

4. Summary and Conclusions

Ring current dynamics involve a variety of physical processes that couple the solar wind, the magnetosphere, and the ionosphere, and whose understanding is a crucial component of Sun-Earth Connection 'space weather' studies. In this paper we discussed problems related to ring current development during several geomagnetic storms. Giving examples from numerical simulations using two formulations of the inner magnetospheric convection electric field, we demonstrated the important role of magnetospheric convection in ring current buildup; a realistic electric field model is needed to obtain quantitative agreement with observations. We found that an asymmetric ring current forms during the main and early recovery phases of all geomagnetic storms; this asymmetry is seen in all major ring current ion species, H^+, O^+, and He^+. The ring current becomes symmetric during the late recovery phase. We showed that the distribution of ring current energy among the three ion species varies with storm intensity and solar cycle. The dominant ion during the May 1997 storm was H^+, while during the July 2000 storm the ring current was dominated by the O^+ population. Finally, we compared proton ring current energy losses due to charge exchange, Coulomb collisions, and ion precipitation. We found that although scattering by EMIC waves enhanced H^+ precipitation loss by more than an order of magnitude so it exceeded Coulomb collisions loss near Dst minima, the charge exchange loss was predominant during all storm phases. The effect of wave-particle interactions on the heavy ion components will be investigated in future studies, since as discussed above O^+ is the dominant ring current ion specie during great storms and wave-induced heavy ion precipitation may have an important impact on their rapid first-phase recovery.

Acknowledgements

This work is supported in part by NASA under grants NAG5-7804 and NAG5-12006, and NSF under grant ATM-0101095.

References

Burlaga, L. F. *et al.*: 1998, *J. Geophys. Res.* **103**, 277.
Burton, R. K., McPherron, R. L. and Russell, C. T.: 1975, *J. Geophys. Res.* **80**, 4204.
Chen, M. W., Lyons, L. R. and Schulz, M.: 1994, *J. Geophys. Res.* **99**, 5745.
Daglis, I. A.: 1997, in B. T. Tsurutani, W. D. Gonzalez, Y. Kamide, J. K. Arballo (eds.), *Magnetic Storms*, American Geophysical Union, Washington, p. 107.
Dessler, A. J. and Parker, E. N.: 1959, *J. Geophys. Res.* **64**, 2239.
Dungey, J. W.: 1961, *Phys. Res. Lett.* **6**, 47.
Farrugia, C. J. *et al.*: 1998, *J. Geophys. Res.* **103**, 17261.
Jordanova, V. K., Kistler, L. M., Kozyra, J. U., Khazanov, G. V. and Nagy, A. F.: 1996, *J. Geophys. Res.* **101**, 111.

Jordanova, V. K. *et al.*: 1998, *J. Geophys. Res.* **103**, 79.

Jordanova, V. K., Torbert, R. B., Thorne, R. M., Collin, H. L., Roeder, J. L. and Foster, J. C.: 1999, *J. Geophys. Res.* **104**, 24895.

Jordanova, V. K., Farrugia, C. J., Thorne, R. M., Khazanov, G. V., Reeves, G. D. and Thomsen, M. F.: 2001a, *J. Geophys. Res.* **106**, 7.

Jordanova, V. K., Thorne, R. M., Farrugia, C. J., Dotan, Y., Fennell, J. F., Thomsen, M. F., Reeves, G. D. and McComas, D. J.: 2001b, *Solar Phys.* **204**, 361.

Kozyra, J. U., Jordanova, V. K., Borovsky, J. E., Thomsen, M. F., Knipp, D. J., Evans, D. S., McComas, D. J. and Cayton, T. E.: 1998, *J. Geophys. Res.* **103**, 26285.

Maynard, N. C. and Chen, A. J.: 1975, *J. Geophys. Res.* **80**, 1009.

Rairden, R. L., Frank, L. A. and Craven, J. D.: 1986, *J. Geophys. Res.* **91**, 13613.

Rasmussen, C. E., Guiter, S. M. and Thomas, S. G.: 1993, *Planetary Space Sci.* **41**, 35.

Sckopke, N.: 1966, *J. Geophys. Res.* **71**, 3125.

Stern, D. P.: 1975, *J. Geophys. Res.* **80**, 595.

Volland, H.: 1973, *J. Geophys. Res.* **78**, 171.

Weimer, D. R.: 1996, *Geophys. Res. Lett.* **23**, 2549.

Young, D. T., Balsiger, H. and Geiss, J.: 1982, *J. Geophys. Res.* **87**, 9077.

A NEW PARADIGM FOR 3D COLLISIONLESS MAGNETIC RECONNECTION

GIOVANNI LAPENTA

Plasma Theory Group, Los Alamos National Laboratory, Los Alamos, NM 87545, U.S.A.

Abstract. A new paradigm is suggested for 3D magnetic reconnection where the interaction of reconnection processes with current aligned instabilities plays an important role. According to the new paradigm, the initial equilibrium is rendered unstable by current aligned instabilities (lower-hybrid drift instability first, drift-kink instability later) and the non-uniform development of kinking modes leads to a compression of magnetic field lines in certain locations and a rarefaction in others. The areas where the flow is compressional are subjected to a driven reconnection process. In the present paper we illustrate this series of events with a selection of simulation results.

Key words: magnetotail, reconnection

1. Introduction

We consider a new paradigm for 3D magnetic reconnection where the interaction of reconnection processes with current aligned instabilities plays an important role. The new paradigm has been first suggested based on MHD models where the presence of velocity shear induces a Kelvin–Helmholtz instability (KHI) that drives reconnection by locally compressing field lines (Brackbill, 1993; Brackbill and Knoll, 2001; Knoll and Brackbill, 2002; Lapenta and Knoll, 2002). The new paradigm is now emerging also in kinetic models where the lower-hybrid drift instability (LHDI) and kink modes (KM) (Lapenta and Brackbill, 2002; Daughton, 2002) drive field lines together and promote the onset of reconnection (Lapenta and Brackbill, 2000; Lapenta et al., 2002).

According to the new paradigm, reconnection in 3D is eminently a naturally driven process. The driving force is determined by the instabilities developing in the current aligned direction. For the typical magnetotail configuration with the current in the dawn dusk direction (y) and the field mostly in the tailward direction (x) and the gradients mostly in the north-south direction (z), the tearing instability develops in the (x, z) plane. The collisionless tearing instability has long been considered the best hope to explain reconnection onset. However, accurate studies have not yet resolved the issue of the instability of the tearing mode in actual realistic magnetotail configurations. Most of the results seem to conclude that the tearing mode is stable in realistic configurations and that reconnection onset has other causes (Quest et al., 1996).

Space Science Reviews **107:** 167–174, 2003.
© 2003 *Kluwer Academic Publishers.*

The new paradigm for 3D reconnection cited above deals with the issue of reconnection onset. It is a new explanation of how reconnection can start, of what mechanism breaks the frozen-in condition in the first place. In a spatially varying magnetic field configuration (such as the Earth magnetotail (Lapenta and Brackbill, 2002) or a corona arcade Lapenta and Knoll, 2002)) or in presence of spatially varying flow shear (such as at different latitudes along the magnetopause (Brackbill and Knoll, 2001; Knoll and Brackbill, 2002)) the non-uniform development of kinking modes leads to compression of magnetic field lines in certain locations and rarefaction in others. The areas where the flow is compressional are subjected to a driven reconnection process. In the new paradigm, two main physics processes must be considered.

First, the kink modes, including both the KHI and the drift kink instability (DKI) (Lapenta and Brackbill, 1997; Daughton, 1999), can drive field lines together, causing a localized compression that drives field lines to reconnect (Brackbill and Knoll, 2001).

Second, new oblique modes are excited and contribute to the process of reconnection (Lapenta and Brackbill, 2000; Wiegelmann and Buchner, 2000).

This new paradigm is distinct but complementary to the other recent remarkable progress in understanding fast reconnection through the role played by the Hall physics (Biskamp, 2000). The Hall fast reconnection that has attracted so much recent attention is relevant to the fully developed reconnection process in the non-linear phase. The progress in that area has brought about understanding of the fast reconnection rates observed naturally. But still leaves us at a loss in trying to understand reconnection onset. It is the new paradigm for the role of current aligned instabilities that gives us the tools to understand onset of reconnection.

2. Physical System

The Earth's magnetotail is described as usual with a modified Harris equilibrium. The equilibrium is described by the following particle distribution function for species s

$$f_{0s} = n(z) \left(\frac{m_s}{2\pi k T_s} \right)^{3/2} \cdot \exp\left[-\frac{m_s}{2k T_s} \left(v_x^2 + \left(v_y - u_s \right)^2 + v_z^2 \right) \right] \tag{1}$$

where u_s is the drift velocity and $v_{th,s} = \sqrt{k T_s/m_s}$ is the thermal velocity of species s. The ratio of the ion drift velocity to ion thermal velocity is related to the current sheet thickness L: $u_i/v_{th,i} = 2\rho_i/L$ where ρ_i is the ion gyroradius. The Alfven time is defined as $\tau_A = L/v_A$ where v_A is the Alfven speed.

The usual magnetotail reference frame is used with x-axis in the tail direction, y-axis in the dawn-dusk direction and z-axis along the Earth's polar axis (neglecting the tilt in the Earth's magnetic axis). The initial current $\mathbf{J}_0$ is aligned with the

y axis. Gradients of the current and magnetic field are aligned with the z axis. The initial magnetic field $\mathbf{B}_0$ has a dominant component aligned with the x axis:

$$B_x(x, z) = -B_o v(x) \tanh\left(\frac{z v(x)}{L}\right) \qquad (2)$$

and a vertical component:

$$B_z(x, z) = B_o v'(x) \left(z \tanh\left(\frac{z v(x)}{L}\right) - \frac{L}{v(x)}\right) \qquad (3)$$

The variation of the field along the tail is determined by the function $v(x)$. In the present work, we consider an empirical profile chosen to best fit the actual magnetotail (Birn and Schindler, 1983):

$$v(x) = \left(1 + \frac{b_n x}{\gamma L}\right)^{-\gamma} \qquad (4)$$

where $\gamma = 0.6$ and $b_n = 0.05$. The equilibrium charge density is:

$$n(x, z) = n_s \frac{v^2(x)}{\cosh^2(v(x)z/L)} \qquad (5)$$

To avoid any confusion with velocity-gradient driven modes, no background plasma is added.

The most general perturbation of the initial profile described above can be expressed as:

$$\psi(x, y, z, t) = \tilde{\psi}(z) exp(ik_x x + ik_y y - i\omega t) \qquad (6)$$

where ψ is the generic field component and $\mathbf{k} = (k_x, k_y)$ is the direction of propagation in the equatorial plane. Many different modes have been identified in the magnetotail. The most relevant for the present discussion are the tearing modes (Biskamp, 2000) with $k_x \neq 0$ and $k_y = 0$, the LHDI and the kink modes (Lapenta and Brackbill, 1997; Lapenta and Brackbill, 2002; Daughton, 1999) with $k_x = 0$ and $k_y \neq 0$ and the oblique modes (Lapenta and Brackbill, 2000) that propagate at a non zero angle with both the x and the y axis: $k_x \neq 0$ and $k_y \neq 0$.

The system described above is simulated using the Vlasov–Maxwell model, solved using the CELESTE 3D implicit particle in cell code. A detailed description of the implicit moment method used in CELESTE 3D can be found in the review paper (Brackbill and Dorslund, 1985) and the details of the implementation can be found in (Ricci et al., 2002).

For comparison, we will also consider simulations conducted with the 3D MHD code FLIP–MHD (Brackbill, 1991).

3. Velocity Shear

The initial equilibrium considered above is unstable to a number of current aligned
instabilities, propagating along y. Two are of particular importance for understanding reconnection onset: the lower hybrid drift instability (LHDI) and kinking
modes (KM).

Recent simulation work (Lapenta and Brackbill, 2002; Daughton, 2002) has
shown that the early dynamic of current sheet is dominated by the LHDI. It has
been observed (Lapenta and Brackbill, 2002; Daughton, 2002) that the nonlinear
evolution of the LHDI changes the initial density and current profile and modifies
the initial flat velocity profile by creating a velocity shear. Figure 1 shows the
velocity profile after saturation of the LHDI. The creation of a robust velocity shear
is evident. It should also be noted that the LHDI growth rate and saturation level is
directly proportional to the $\sqrt{T_i/T_e}$ just like the velocity shear observed in Figure 1,
an indication that the velocity shear is caused by the LHDI.

It is interesting to note that the process of creation of velocity shear by the
LHDI is reminiscent of the creation of zonal flows (Drake *et al.*, 1992) observed in
tokamaks in relation with the L-H transition (Diamond *et al.*, 2000).

Once a velocity shear is created in the short time scales of the LHDI a much
slower fluid instability arises, the Kelvin–Helmholtz instability (KHI). The process
can be modeled very simply using MHD. Figure 2a shows a fully kinetic simulation where the velocity shear is created initially by the LHDI and is destabilized
later by the KHI. At the end of the simulation the distinctive kinking caused by
the KHI is observed. Figure 2b shows a simple MHD simulation with the same
initial equilibrium used for the kinetic simulation but with the addition of an initial
velocity shear equal to the velocity shear formed naturally by the LHDI in the
kinetic simulation. Clearly, the comparison of Figure 2a and Figure 2b proves that
the evolution following the creation of shear is purely a fluid instability. However,
note that in the kinetic simulation the velocity shear arises naturally while in the
MHD simulation the shear is artificially introduced as an initial condition.

4. Reconnection Onset

The evolution of the KHI and the kinking of the initial current sheet has an important consequence on reconnection onset. The presence of kinking causes the
compression of the field lines in some regions and the rarefaction in others. Compression of field lines can drive reconnection. Such mechanism has already been
observed in MHD simulations of the evolution of the KHI and tearing instability (Brackbill, 1993) for application to the Earth magnetopause (Brackbill and
Knoll, 2001; Knoll and Brackbill, 2002) and to the solar corona (Lapenta and
Knoll, 2002). The present paper is the first instance where the same mechanism
is observed in the magnetotail.

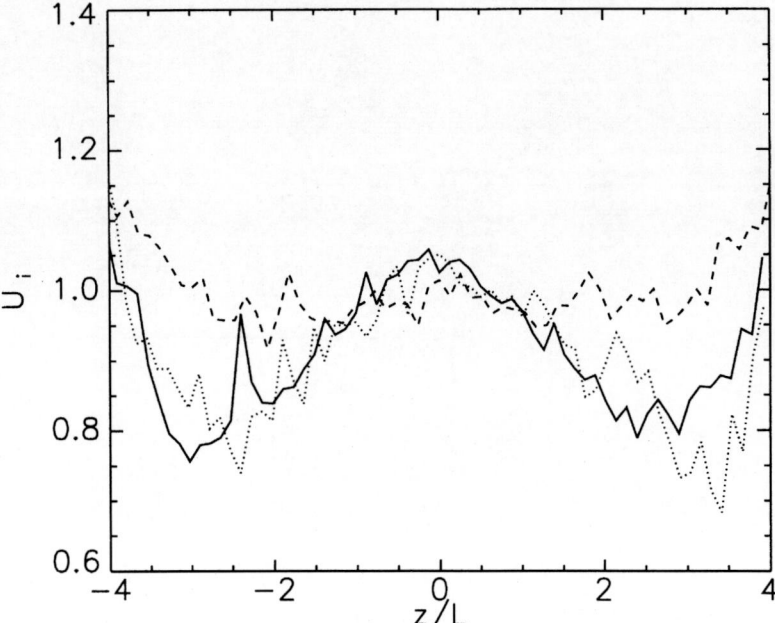

Figure 1. Velocity shear produced by the LHDI in a kinetic simulation. The velocity profile is shown at time $t/\tau_A = 12$ for a system with $u_i/v_i = 1$, $m_i/m_e = 180$. Three different temperature ratios are considered: $T_i/T_e = 10$ (solid); $T_i/T_e = 4$ (dotted); $T_i/T_e = 2$ (dashed). Details in Lapenta and Brackbill (2002).

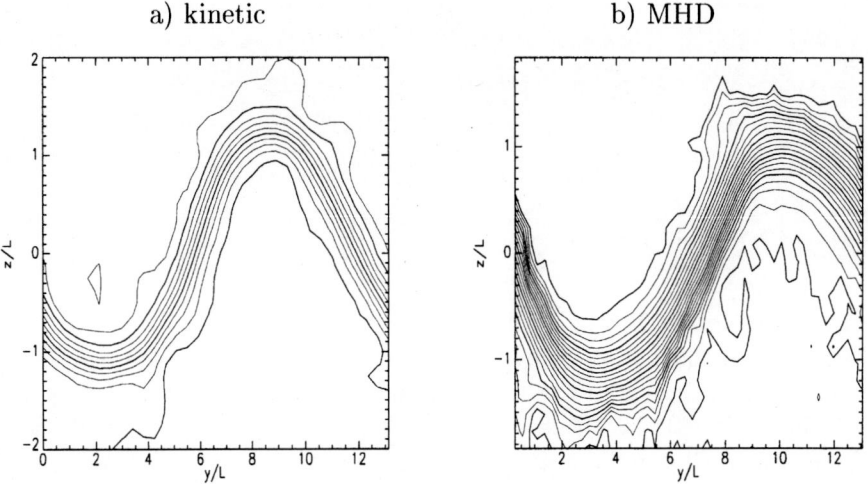

Figure 2. Contour plot of $B_x(y, z)$ at the end of two simulation ($t/\tau_A = 110$): a) a fully kinetic simulation with no initial shear; b) a MHD simulation with an initial shear equal to the shear observed in the kinetic simulation after saturation of the LHDI. The case considered has $m_i/m_e = 180$, $u_i/v_{th,i} = 1$ and $T_i/T_e = 2$. Details in Lapenta and Brackbill (2002).

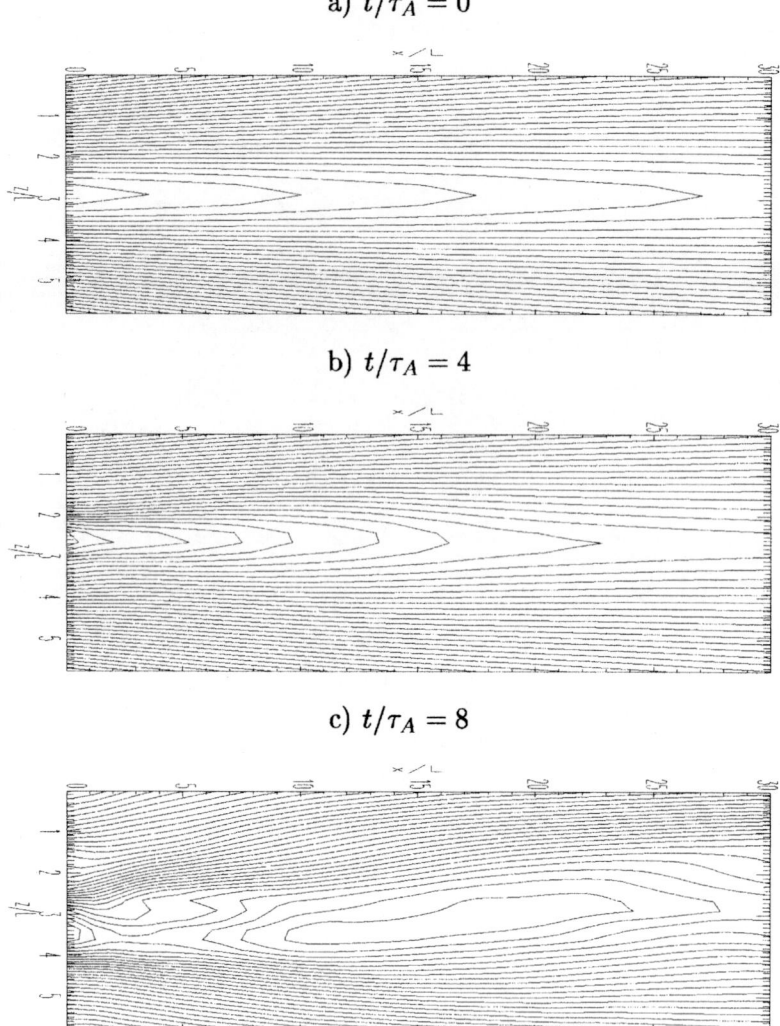

Figure 3. Cross section of the flux surfaces for a 3D MHD simulation at $y = L_y/2$ and three different times: initial ($t/\tau_A = 0$), $t/\tau_A = 4$, $t/\tau_A = 8$.

In the references just cited, the mechanism was proposed in presence of externally driven shears (driven by the solar wind in the case of the magnetopause (Brackbill and Knoll, 2001) or driven by the photosphere in the case of the corona (Lapenta and Knoll, 2002)). In the magnetotail we suggest that, instead, the shear is created naturally by the LHDI without requiring any external action. Once the current sheet is thin enough for the LHDI to cause the velocity shear the chain of events sets in place and causes the onset of reconnection. The only required external action is the transfer of flux from the dayside to cause the thinning of the magnetotail.

An illustration of how the KHI can drive reconnection onset is presented in Figure 3. A typical magnetotail (Figure 3a) described by Equation (4) is rendered unstable to the KHI by the velocity shear induced by the LHDI. The KHI moves the current sheet up and down, compressing field lines in some regions and rarefying in others. A region of field line compression is evident in the northern side of the near Earth tail in Figure 3b. Eventually, field line compression drives reconnection (Figure 3c). Note that a 2D simulation of the same initial configuration cannot capture the presence of the KHI and the mechanism just proposed would not be observed. Indeed, in a 2D simulation of the equilibrium shown in Figure 3a, the tearing mode is stable due to the electron compressibility (Quest *et al.*, 1996) and reconnection never sets in. Only 3D simulations can capture the chain of events proposed in the new paradigm for naturally driven reconnection onset.

5. Conclusions

A new scenario for magnetotail reconnection has been presented. The scenario is based on a sequence of three events.

First, the LHDI grows and saturates changing the initial equilibrium. A consequence of this change is the creation of a velocity shear that renders the equilibrium unstable to the KHI.

Second, the KHI grows and kinks the current sheet causing regions of compression of field lines.

Third, local compression drives field lines together and causes the onset of reconnection. Once reconnection is started by the driving force of the KHI it can progress via the action of the Hall term mediated fast reconnection process of recent discovery (Biskamp, 2000).

Acknowledgements

The author is grateful to J.U. Brackbill, L. Chacon, W.S. Daughton and D.A. Knoll for the useful discussions on the physics of reconnection. The supercomputers used in this investigation were provided by funding from JPL Institutional Computing and Information Services and the NASA Offices of Space Science and Earth Science. Research supported by the United States Department of Energy, under Contract No. W-7405-ENG-36 and by NASA, under the 'Sun Earth Connection Theory Program'.

References

Birn, J., Schindler, K.: 1983, *J. Geophys. Res.* **88**, 6969.

Biskamp, D.: 2000, *Magnetic Reconnection in Plasmas*, Cambridge University Press, Cambridge.

Brackbill, J. U., Forslund, D. W.: 1985, 'Simulation of Low-Frequency, Electromagnetic Phenomena in Plasmas', in J. U. Brackbill and B. I. Cohen (eds.), *Multiple Time Scales*, Academic Press, Orlando, pp. 271–310.

Brackbill, J. U.: 1991, *J. Computat. Phys.* **96**, 163.

Brackbill, J. U.: 1993, *J. Computat. Phys.* **108**, 38.

Brackbill, J. U. and Knoll, D. A.: 2001, *Phys. Rev. Lett.* **86**, 2329.

Daughton, W. S.: 1999, *Phys. Plasmas* **6**, 1329–1343.

Daughton, W. S.: 2002, *Phys. Plasmas* **9**, 3668.

Diamond P. H., Rosenbluth, M. N., Sanchez, E., Hidalgo, C., VanMilligen, B., Estrada, T., Branas, B., Hirsch, M., Hartfuss, H. J. and Carreras, B. A.: 2000, *Phys. Rev. Lett.* **84**, 4842.

Drake, J. F., Finn, J. M., Guzdar, P. Shapiro, F. and Shevchenko, V.: 1992, *Phys. Fluids B* **4**, 488.

Knoll, D. A. and Brackbill, J. U.: 2002, *Phys. Plasmas* **9**, 3775.

Lapenta, G. and Brackbill, J. U.: 1997, *J. Geophys. Res.* **102**, 27099.

Lapenta, G. and Brackbill, J. U.: 2000, *Nonlinear Processes Geophys.* **7**, 151.

Lapenta, G. and Brackbill, J. U.: 2002, *Physics Plasmas* **9**, 1544.

Lapenta, G. and Knoll, D. A.: 2003, *Solar Phys.*, to appear.

Lapenta, G. Brackbill, J. U. and Daughton, W. S.: 2003, *Physics Plasmas* **10**, 1577.

Quest, K. B., Karimabadi, H. and Brittnacher, M.: 1996, *J. Geophys. Res.* **101**, 179.

Ricci, P., Lapenta, G. and Brackbill, J. U.: 2002, *J. Computat. Phys.* **183**, 117.

Wiegelman, T. and Buchner, J.: 2000, *Nonlinear Processes Geophys.* **7**, 141.

INNER MAGNETOSPHERIC MODELING WITH THE RICE CONVECTION MODEL

FRANK TOFFOLETTO, STANISLAV SAZYKIN, ROBERT SPIRO and
RICHARD WOLF

Department of Physics and Astronomy, Rice University, Houston, TX 77005, U.S.A.

Abstract. The Rice Convection Model (RCM) is an established physical model of the inner and middle magnetosphere that includes coupling to the ionosphere. It uses a many-fluid formalism to describe adiabatically drifting isotropic particle distributions in a self-consistently computed electric field and specified magnetic field. We review a long-standing effort at Rice University in magnetospheric modeling with the Rice Convection Model. After briefly describing the basic assumptions and equations that make up the core of the RCM, we present a sampling of recent results using the model. We conclude with a brief description of ongoing and future improvements to the RCM.

Key words: magnetosphere-inner, numerical modeling, plasma convection, ring current

1. Introduction

Earth's inner magnetosphere is home to an interesting variety of particle populations and plasma processes. While radiation-belt particles (> 1 MeV energy) have been elegantly described by the adiabatic theory developed in the earliest days of the space age e.g., (Northrop, 1963), this picture can be disrupted by severe solar-wind disturbances (Li *et al.*, 1993). The ring current, consisting mainly of ions and electrons in the 10–200 KeV energy range, carries a large fraction of the total particle energy of the magnetosphere and enough current to substantially affect the overall magnetic configuration. Coexisting in the same region of space as the radiation belt and ring current is the plasmasphere, consisting mainly of particles with energies < 1 eV. While the plasmasphere does not directly affect the magnetospheric magnetic field configuration, it still contains most of the mass of the magnetosphere. The plasmasphere exhibits a wide range of interesting plasma phenomena which affect wave propagation, particle loss, and heating processes in the ring current and radiation belt populations.

Because many space-based assets are located in the inner magnetosphere and the underlying ionosphere, understanding the physical processes that control this region of space has important space weather implications. For example, the performance of geosynchronous communications spacecraft can be impaired by the effects of surface charging and the resultant arcing in solar panels. MeV outerbelt 'killer electrons' constitute another important space-weather hazard; they are

Space Science Reviews **107**: 175–196, 2003.
© 2003 *Kluwer Academic Publishers.*

known to have seriously damaged several spacecraft, and they could also be hazardous to astronauts on the long space-walks planned for the International Space Station program. Energetic ions that can pierce the skin of spacecraft pose an additional radiation hazard to astronauts. The prompt penetration of magnetospherically-associated electric fields to low ionospheric latitudes can have important implications for the onset and development of equatorial spread-F instability and related radio scintillation.

The different particle populations of the inner magnetosphere cannot be treated as a single fluid because of the large range of energies present. Even within the ring-current population the differential drift of particles with different energies can be critical for the electrodynamics. The Rice Convection Model (RCM) was specifically designed to treat this unique and complicated system (Wolf, 1970; Jaggi and Wolf, 1973; Harel et al., 1981a. Erickson et al., 1991; Sazykin et al., 2002). The RCM represents the plasma population in terms of multiple fluids, typically about 100. RCM equations and numerical methods have been chosen for accurate treatment of the inner magnetosphere, including the flow of electric currents along magnetic field lines to and from the conducting ionosphere. The RCM computes these currents and associated electric fields self-consistently, assuming perfectly conducting field lines and employing pre-computed time-dependent magnetic field information and associated induction electric fields. Recent efforts to self-consistently compute the magnetic field in force-balance with RCM-computed pressure distributions are described in (Toffoletto et al., 2001) and will be discussed later in this article.

A description of the basic equations and formulation of the RCM was given by (Harel et al., 1981a). The RCM has been compared extensively with spacecraft and ground observations over a period of many years (Harel et al., 1981b; Spiro et al., 1981; Chen et al., 1982; Wolf et al., 1982; Spiro et al., 1988; Fejer et al., 1990). In the early 1980's, the model was used to simulate several specific, well-studied magnetospheric events. Some observational data were used as model inputs, while other observations were used to test predictions (Chen et al., 1982; Harel et al., 1981b; Karty et al., 1982; Spiro et al., 1981; Spiro and Wolf, 1984; Wolf et al., 1982). While model predictions agreed with observations in most major aspects, there have been significant disagreements on various details. In the mid-80's, the code was completely rewritten, and several of the numerical procedures were upgraded. Magnetospheric particle loss was added to the code, and a new algorithm was developed for computing ionospheric conductance from model-computed electron precipitation (Erickson et al., 1991; Wolf et al., 1991). The program was also applied to the problem of the generation of region-1 sense Birkeland currents on closed middle-magnetosphere field lines (Yang et al., 1994a). Sazykin (2000) rewrote the code in Fortran-95 and introduced a new electric field solver based on a GMRES algorithm. A version of the RCM has been also used to model the rotationally-dominated magnetosphere of Jupiter (Yang et al., 1994b; Pontius et al., 1998).

2. Model Description

A well-developed theory of adiabatic particle motion exists for the inner and middle magnetosphere (Northrop, 1963; Roederer, 1970; Wolf, 1983). It is applicable to the quasi-dipolar part of the magnetosphere (<10 R_E), but it can also be applied to the middle plasma sheet out to ~ 30 R_E. Here, the magnetospheric particle populations that carry most of the pressure, energy, and current, are non-relativistic.

The RCM assumes an isotropic particle distribution, which allows the description of the motion of the plasma to be represented as the motion of a whole flux tube. By averaging over an isotropic distribution and over a flux tube one gets the equation for bounced-averaged motion of a particle in an electric field and magnetic field that includes also the motion due to gradient and curvature drift (Wolf, 1983)

$$\vec{v}_k(\lambda_k, \vec{x}, t) = \frac{[\vec{E} - \frac{1}{q_k}\nabla W(\lambda_k, \vec{x}, t)] \times \vec{B}(\vec{x}, t)}{B(\vec{x}, t)^2} \tag{1}$$

where $W(\lambda_k, \vec{x}, t)$ is the particle kinetic energy, q_k is the charge and λ_k is an energy invariant defined as

$$|\lambda_k| = W(\lambda_k, \vec{x}, t) V^{\frac{2}{3}} \tag{2}$$

which is conserved along a drift path. The subscript k describes a given charge, mass and energy invariant λ_k species where negatively charged particles, such as electrons, are labeled with a negative value for the invariant λ_k and V is the flux tube volume

$$V \equiv \int_{sh}^{nh} \frac{ds}{B(\vec{x}, t)} \tag{3}$$

where nh and sh refer to the northern and southern hemisphere endpoints of the fieldline. It can be shown that the quantity $\eta_k(\vec{x}, t)$, defined as the number of particles per unit magnetic flux, follows the conservation law (Wolf, 1983)

$$(\frac{\partial}{\partial t} + \vec{v}_k(\lambda_k, \vec{x}, t) \cdot \nabla)\eta_k = S(\eta_k) - L(\eta_k) \tag{4}$$

where $S(\eta_k)$ and $L(\eta_k)$ represent sources and sinks respectively. In the case of no sources and sinks η_k is conserved along a drift path. The flux tube content η_k is related to thermodynamic pressure P via the relation

$$PV^{5/3} = \frac{2}{3} \sum_k \eta_k |\lambda_k| \tag{5}$$

The flux tube content η_k can be related to the plasma distribution function $f_k(\lambda)$ by the relation

$$\eta_k = \frac{4\pi\sqrt{2}}{m_k^{3/2}} \int_{\lambda_{min}}^{\lambda_{max}} \sqrt{\lambda} f_k(\lambda) d\lambda \tag{6}$$

where $(\lambda_{max} - \lambda_{min})$ is the width of the invariant energy channel associated with species k.

There are several components contributing to the electric field in Equation (1). Equation (1) can be re-expressed as the sum of a potential component and a component due to inductive effects

$$\vec{E} = -\nabla\Phi - \vec{v}_{inductive} \times \vec{B} \tag{7}$$

where the inductive component results from changes in the magnetic field. In the RCM, inductive effects are included implicitly through time-dependent magnetic field mappings.

The potential field Φ of Equation (7) has contributions from several sources

$$\Phi = \Phi_i + \Phi_{corotate} + \Phi_{\|} \tag{8}$$

The effect of the driving solar wind electric field is included in the Φ_i term. Low-order models of this electric field predate the RCM. In its simplest form this component has been modeled as a constant dawn-dusk electric field

$$\Phi_i = -E_0 y \tag{9}$$

where y is in GSM coordinates. A commonly used variation of this model has been the Stern–Volland electric field model (Volland, 1973; Stern, 1975) that includes a parameterization of the shielding of the inner magnetospheric electric field

$$\Phi_i = -A_0 y r \tag{10}$$

where A_0 is a constant and r is the radial distance from the Earth. Neither Equations (9) nor (10) capture the true behavior of the magnetosphere, especially when the coupling of the magnetosphere to the ionosphere is considered. Physically, this coupling is achieved through field-aligned currents computed by the RCM.

The field-aligned currents are computed using the so-called Vasyliunas equation (Vasyliunas, 1970). The derivation of this equation can be obtained quite elegantly from MHD (Heinemann and Pontius, 1990) or from the distribution function (Birmingham, 1992). MHD force balance ($\vec{j} \times \vec{B} = \nabla P$) implies that

$$\vec{j}_\perp = \frac{\vec{B} \times \nabla P}{B^2} \tag{11}$$

By using the current conservation relation ($\nabla \cdot \vec{j} = 0$) and integrating over a flux tube one arrives at the Vasyliunas equation

$$\frac{j_{\|nh} - j_{\|sh}}{B_i} = \frac{\hat{b}}{B} \cdot \nabla V \times \nabla p \tag{12}$$

which relates field-aligned currents in the ionosphere to pressure gradients in the magnetosphere, where B_i is the magnetic field at the southern- and northern-ionospheric footprints of the field line (assumed the same). The derivation makes use of the fact that the right hand side of (12) can be evaluated anywhere along the field line. In RCM variables, (12) can be re-expressed as

$$\frac{j_{\|nh} - j_{\|sh}}{B_i} = \frac{\hat{b}}{B} \cdot \sum_k \nabla \eta_k(\vec{x}, t) \times \nabla W(\lambda_k, \vec{x}, t) \tag{13}$$

A version of Equation (13) for gyrotropic distribution is described in (Fok *et al.*, 2001) in which case $W(\lambda_k, \vec{x}, t)$ is replaced with its gyrotropic counterpart $W(\mu_k, J_k, \vec{x}, t)$ where μ_k and J_k are the first and second adiabatic invariant respectively.

In the RCM, the ionospheric electric potential Φ_i in the rotating frame is computed as a solution to the ionospheric current conservation equation. Current conservation, in the 2D thin-shell approximation, can be expressed as (Wolf, 1983)

$$\nabla_i \cdot \left[\overleftrightarrow{\Sigma} \cdot (\nabla_i \Phi_i) \right] = -(j_{\|nh} - j_{\|sh}) sin(I) \tag{14}$$

where $\overleftrightarrow{\Sigma}$ is the field-line integrated conductivity tensor due to both hemispheres, I is the dip angle of the magnetic field in the ionosphere and $j_{\|nh} - j_{\|sh}$ is the ionospheric field aligned current density, determined from Equation (12). Subscript i refers to ionospheric-computed quantities. The high-latitude boundary condition is a Dirichlet boundary condition where the solar wind potential is specified as a function of local time. The low-latitude boundary is more complex and incorporates the effects of the equatorial electrojet. At the magnetic equator, where the magnetic field of the Earth is close to horizontal, east-west electric fields associated with the dynamo action of the neutral winds create a strong polarization vertical electric field, which in turn causes a strong east-west current to flow, mostly on the day side [e.g., (Kelley, 1989)]. The physical condition on the equatorward boundary is (Sazykin, 2000)

$$J_{\theta eb} = \frac{1}{R_I} \frac{\partial I_{EEJ}}{\partial \varphi} \tag{15}$$

where $J_{\theta eb}$ is the current per unit length flowing through the equatorial boundary, I_{EEJ} is the current carried by the equatorial electrojet (EEJ), θ and φ are spherical coordinates and R_I is the radial location of the ionosphere.

The transformation from Φ_i to Φ (Equation (8)) is through the effect of the corotation which transforms the calculation to a coordinate frame that does not rotate with the Earth. It takes the form

$$\Phi_{corotate} = -\frac{\omega_E B_M R_E^3}{r} \approx -(89,500 volts)\frac{R_E}{r} \tag{16}$$

where B_M is the magnitude of the dipole moment and ω_E is the rotation rate of the Earth. For completeness, $\Phi_{\|}$ in Equation (8) represents the potential due to field-aligned potential drops which are not presently included in the RCM.

Table I compares the equations of ideal MHD with corresponding equations used in the RCM. In place of the MHD mass-conservation relation, the RCM has a separate particle-conservation law for each species k. The RCM's drift laws

TABLE I

Comparison of equations of ideal MHD with those used in the RCM

Ideal MHD	RCM
$\frac{\partial \rho}{\partial t} + \nabla \cdot (\rho v) = 0$	$(\frac{\partial}{\partial t} + \vec{v}_k(\lambda_k, \vec{x}, t) \cdot \nabla)\eta_k = S(\eta_k) - L(\eta_k)$
$(\frac{\partial}{\partial t} + \vec{v} \cdot \nabla)(\rho \vec{v}) = \vec{j} \times \vec{B} - \nabla P$	$\vec{j}_k \times \vec{B} = \nabla P_k$
$(\frac{\partial}{\partial t} + \vec{v} \cdot \nabla)(P\rho^{-5/3}) = 0$	$P = \frac{2}{3} \sum_k \eta_k |\lambda_k| V^{-5/3}, \lambda_k = constant$
$\nabla \cdot \vec{B} = 0$	Part of the magnetic field model.
$\nabla \times \vec{B} = \mu_0 \vec{j}$	Included in magnetic field, but $\vec{j} \neq \sum_k \vec{j}_k$.
$\nabla \times \vec{E} = -\frac{\partial \vec{B}}{\partial t}$	Included implicitly in mapping.
$\vec{E} + \vec{v} \times \vec{B} = 0$	$\vec{E} \cdot \vec{B} = 0$ and $\vec{E}_\perp + \vec{v}_k \times \vec{B} = \frac{\nabla W(\lambda_k, \vec{x}, t)}{q_k}$

guarantee that the current density carried by each species k, crossed with the magnetic field, balances the gradient of the partial pressure for species k, under the assumption that the inertial terms in the momentum equation are negligible. In the RCM, the adiabatic-compression condition takes the form of assuming that the energy invariant λ_k is conserved. Magnetic field models used as input to the classical RCM always assume $\nabla \cdot \vec{B} = 0$. While those input models also generally satisfy a condition $\nabla \times \vec{B} = \mu_0 \vec{j}$, the current used in the magnetic field model is, in general, not constrained to be equal to the sum of the RCM-computed partial currents $\vec{j}_k$; however, some RCM runs are now being performed using a magnetic field that is consistent with RCM-computed currents (Section 5). As discussed previously Faradays law is included in the RCM implicitly, through the time-dependent mapping between ionosphere and magnetosphere. Like MHD, the RCM generally assumes that there is no electric field component along the magnetic field; however, the RCM drift equation includes the effect of gradient-curvature drifts, which is equivalent to writing a separate Ohm's law for each species k and including an extra gradient/curvature term.

3. RCM Algorithms

Figure 1 outlines the essential logical structure of the RCM. This chart elaborates the basic scheme first proposed by (Vasyliunas, 1970). The differences between the scheme shown in Figure 1 and that given by Vasyliunas are the additional boxes representing various model inputs and outputs as well as modifications to include such effects as self-consistently computed ionospheric conductance. The core of the calculation, in the center of the figure, displays the basic algorithmic time loop. The RCM steps in time (typically with steps of 1–5 sec.), iteratively solving Equations (4) and (12), with time-dependent inputs. Equation (4) is used to advance the particles, using the velocity computed from Equation (1). Equation (1)

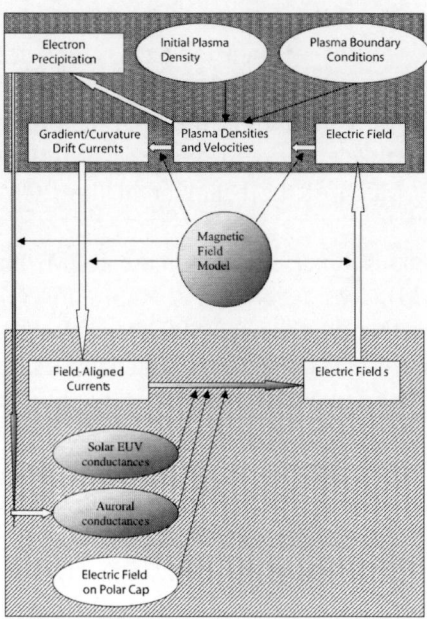

Figure 1. Flowchart that outlines the essential logical structure of the RCM. The upper part of the figure represent magnetospheric quantities and the lower part ionospheric. Rectangles are computed quantities, and ovals are model inputs. Thick white lines represent computations while black thin lines are for model inputs. Adapted from Sazykin (2000).

includes the effect of gradient and curvature drift as well as the ionospheric electric field. Gradient and curvature drifts depend primarily on the particle energy and the magnetic field, both of which are included as model inputs. The ionospheric electric field is determined by solving the current conservation equation for the ionosphere (Equation (14)), including field-aligned currents calculated from Vasyliunas' equation (Equation 12)). The magnetospheric electric field is calculated by mapping the ionospheric electric field along field lines, assuming no field-aligned potential drop and adding a corotation field to convert to a frame that does not rotate with the Earth. The poleward-boundary electric potential is usually specified as an input, with the total potential drop being an input controlled by the solar wind magnetic field and plasma parameters. Ion losses in (4) are due to charge-exchange only. Electron loss is assumed to be by precipitation into the loss cone (see Wolf *et al.*, 1991)). Solar-EUV-generated conductances are estimated from the IRI-90 empirical ionosphere; electric fields due to atmospheric dynamo are usually not calculated. Electron precipitation is typically assumed to be 30% of the strong-pitch-angle-scattering limit. The average energy and flux of precipitating electrons are computed from the distribution of plasma sheet electrons as in Wolf *et al.* (1991), and auroral conductances are then estimated according to Robinson *et al.* (1987). Field-aligned potential drops are usually neglected.

In a basic time-step the following occurs: particles are moved using the computed electric field as well as gradient and curvature drift, the new distribution of particles is then used to compute field-aligned currents from the Vasyliunas Equation (12), which in turn are used as the right-hand-side of Equation (14). This system has inherent feedback and is the essence of the logical loop in the RCM. This feedback imparts much of the behavior of the RCM that will be discussed later.

A stripped-down, operational, cousin of the RCM, the Magnetospheric Specification Model (MSM), was developed at Rice University and delivered to the US Air Force, Air Weather Service in 1990. As a design goal, the MSM had to meet the following criteria: (1) be responsive to changes in geophysical conditions on time scales of 15–30 minutes; (2) contain as much of the essential physics and cover as much of the physical system as possible; (3) make maximum use of the extensive near-real-time data stream available to the Air Force; (4) run faster than real time on computers available at the time. The MSM was designed primarily for computing particle fluxes near geostationary orbit in the 1–100 KeV range. The simplest version of the MSM can run using only Kp as input. It can also be run using a more complete suite of data including IMF, and polar cap electric field data, solar wind density and velocity, high-latitude ion drift parameter data, and the auroral boundary index. The Magnetospheric Specification and Forecast Model (MSFM), is similar to the MSM but was designed and tested for greater accuracy over a wider range of geocentric distances. The MSFM provides limited, short-term forecast capabilities using the neural network Kp prediction model by Costello (1997) to make predictions up to 1 hour ahead. Components of the MSM, include the magnetic field model which is the Hilmer–Voigt magnetic field Model (Hilmer and Voigt, 1995), electric field model that is based on combination of RCM results and a Heppner–Maynard model (Heppner and Maynard, 1987), initial condition, boundary particle fluxes and loss rates. While the RCM computes the electric field distribution self-consistently with the plasma distribution, the MSM/MSFM relies on data-driven empirical E-field models as well as parameterized rules for the penetration of the electric field to low-latitudes based on experience with RCM results.

4. Inner Magnetosphere Electric Fields

4.1. INNER MAGNETOSPHERIC SHIELDING

In order to set the stage for the detailed results presented in the following subsections, we first illustrate the most basic electrodynamic phenomenon of the inner magnetosphere – the shielding of the innermost magnetosphere from the cross-tail electric field. Pressure gradients associated with the inner edge of the plasmasheet produce region-2 currents (generally, downward into the ionosphere on the dusk

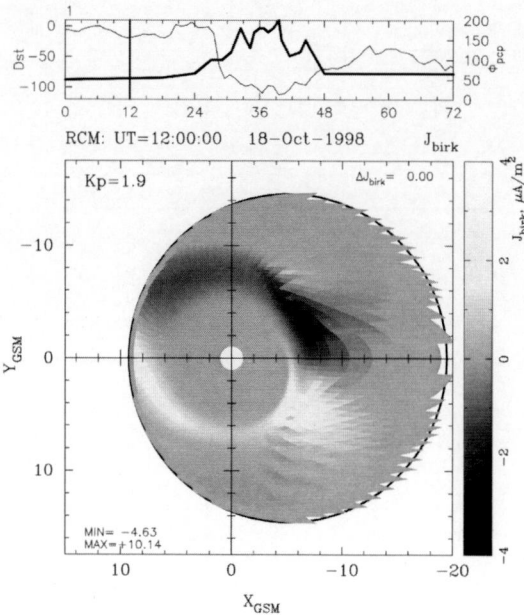

Figure 2. Plot of field-aligned currents into the ionosphere (both hemispheres together) mapped to the equatorial plane. The outer dark line is the RCM modeling boundary. The top panel shows some of the RCM inputs such as the cross polar cap potential (dark line) and *Dst*. Dotted contours represent currents up from the ionosphere.

side and upward on the dawn side). RCM results depicting the distribution of these currents are shown in Figure 2. These pressure gradients produce an electric field that is opposite to the general dawn-to-dusk convection electric field. While further away from the Earth, the electric field is dominated by the cross-tail convection electric field responsible for driving sunward flow in the tail (Figure 3), earthward of the region of Birkeland currents, the two fields, to a large degree, cancel each other, leaving the inner magnetospheric electric field much weaker than the outer region field. In Figure 3, this can be seen as a much lower density of the equipotentials close to the Earth.

This equilibrium picture is disrupted if the electric field along the outer boundary (determined, to a large extent, by the interaction of the solar wind plasma with the magnetosphere in the boundary regions) increases, resulting in the inner magnetosphere seeing the convection electric field (Figure 4). After the plasma re-adjusts itself, the region-2 currents increase in strength and come closer to the Earth, re-establishing shielding with a time constant of 10–20 minutes. Figure 5 illustrates time decay of the two perpendicular components of the electric field at 45° invariant latitude and local times of 00 (thin curves) and 09 magnetic local time. It is common in this type of modeling to assume, for simplicity, that the magnetic field does not change with time. However, in reality the magnetic field lines stretch,

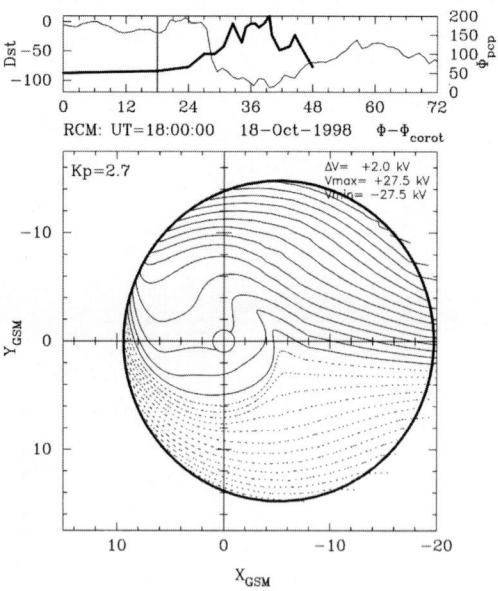

Figure 3. RCM-computed equipotentials as plotted in the equatorial plane, not including the effects of corotation. The top panel is the same as in Figure 2.

making the shielding time constants longer compared to the illustrative case of a constant magnetic field (dashed curves).

If the potential difference across the polar boundary of the RCM suddenly decreases, region-2 currents are suddenly too strong, creating the characteristic pattern of overshielding electric fields (Figure 6). As the shielding currents readjust and weaken, again with a 10–20 min time constant, the overshielding electric field decays (Figure 7). Overshielding penetration electric fields are thought to play an important role in setting off the generation of a host of plasma instabilities; at low ionospheric latitudes, these produce the phenomenon commonly referred to as equatorial spread-F. In addition, overshielding electric fields, predicted with the RCM and parameterized in the MSM, have been shown to play an important role in determining the shape of the plasmapause (Goldstein *et al.*, 2002).

4.2. SUBAURORAL ION DRIFT

The typical pattern of the electrostatic potential illustrates how particle pressure gradients at the inner edge of the plasma sheet modify sunward flow in the central plasma sheet, rotating equipotentials along the edge of the plasma sheet by almost 90° counterclockwise. This picture, typical of steady-state quiet times, becomes more complicated during times of increased magnetic activity. It has been discovered, first from *in-situ* satellite electric field measurements, that sometimes there is a narrow ($\sim1°$ wide in invariant latitude) region extended in magnetic local time, on the equatorward edge of the diffuse aurora where the electric field

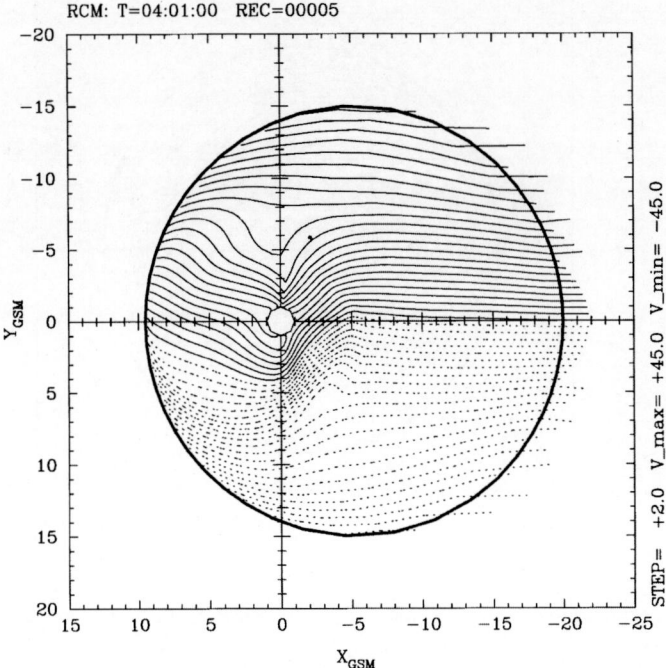

Figure 4. Equatorial plot of RCM computed equipotentials shortly after an increase in the solar wind electric field.

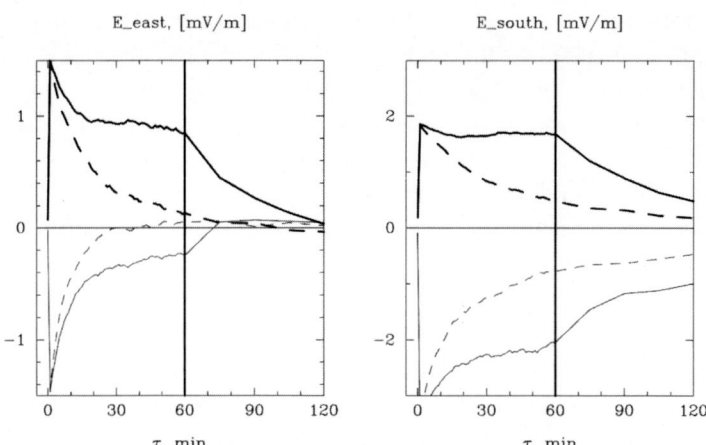

Figure 5. This figure illustrates time decay of the two perpendicular components of the electric field at 45° invariant latitude and local times of 00 (thin curves) and 09 magnetic local time (thick curves). The dashed curves illustrate the case of using a constant magnetic field.

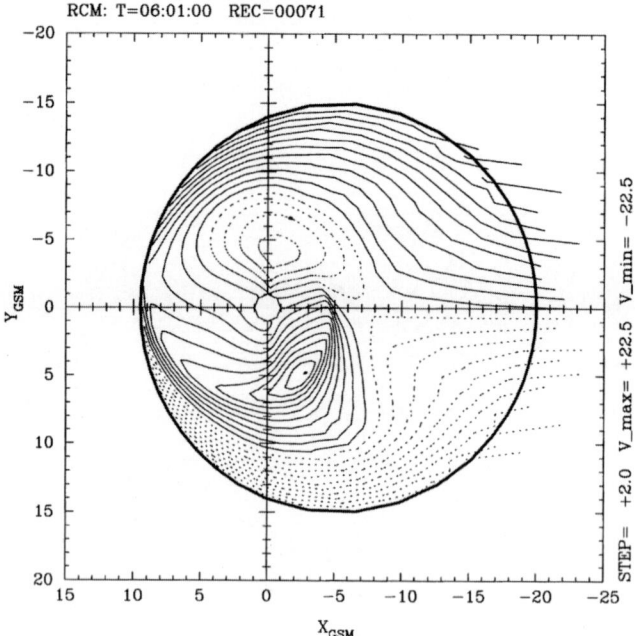

Figure 6. This equatorial plot of equipotentials illustrates the characteristic pattern of overshielding electric fields in the RCM.

is strong (sometimes exceeding 100 mV/m) and directed poleward, resulting in rapid westward plasma drift, counter to the general flow in that region (Galperin *et al.*, 1973, 1974; Spiro *et al.*, 1978). These strong flow regions in the ionosphere have been termed Polarization Jets (Galperin *et al.*, 1974) or SubAuroral Ion Drift (SAID) events (Spiro *et al.*, 1978). Figure 8 shows such an RCM-predicted narrow jet of fast flow mapped to the magnetic equatorial plane during the main phase of a magnetic storm. The region of fast westward flow is where equipotentials are concentrated with greater density, extending from dusk to about 03 MLT. The SAID in Figure 8 was formed in response to the total potential drop across the modeling region increasing from 50 kV (Figure 3) to 100 kV.

Recently, this phenomenon, as well as other large ion drifts observed with ground-based incoherent-radar measurements (Foster and Vo, 2002), have been termed SubAuroral Polarization Streams, or SAPS. These regions of large poleward electric field, located on the equatorward edge of the auroral zone and sometimes extending well into the subauroral ionosphere, can persist for as short as 10 minutes or as long as hours, and are frequently observed by ground-based, rocket, and spacecraft probes. The RCM has in the past successfully predicted SAID events (Harel *et al.*, 1981a). A physical mechanism for the generation of SAID events has been proposed by (Southwood and Wolf, 1978), and the RCM-predicted picture of them appears consistent with that mechanism. Here, we explain it using the

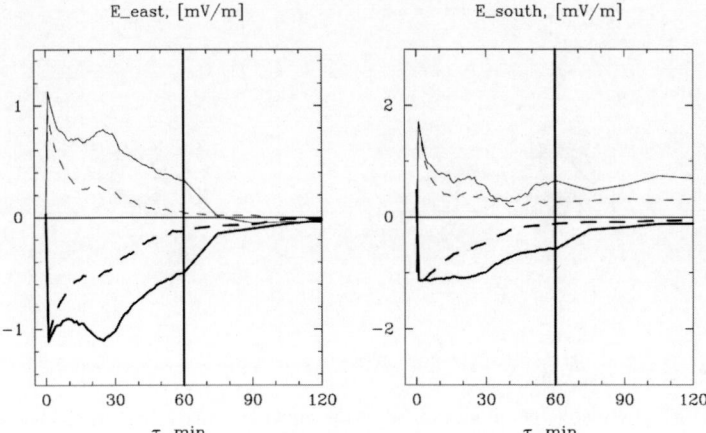

Figure 7. As in Figure 5 showing the 10–20 min time constant decay of the overshielding electric field.

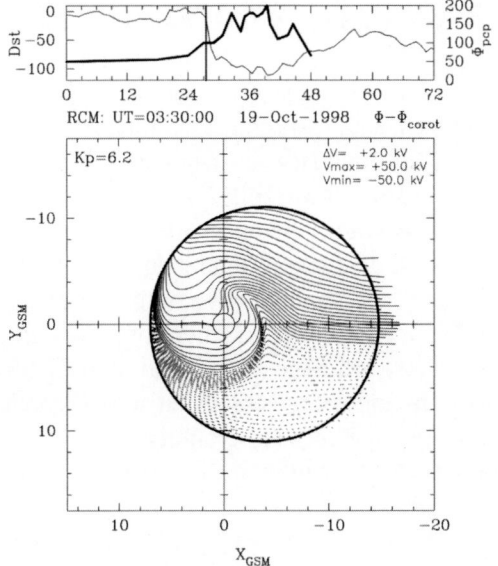

Figure 8. Equatorial equipotential diagram showing an RCM-predicted narrow jet of fast flow at $L = 3$–4, mainly in the dusk-midnight sector.

SAID example of Figure 8. (The RCM-computed SAID event persisted for many hours throughout the storm.) The charge and energy-dependent drift trajectories of magnetospheric particles provide the basic mechanism for the generation of SAIDs along with the fact that positive ions are responsible for most of the field-aligned currents while electrons are the principal source of enhanced ionization in the auroral ionosphere. Figure 9 shows contour plots of the effective potential

$$\Phi_{EFF} = \Phi_i + \lambda_s V^{-\frac{2}{3}} \tag{17}$$

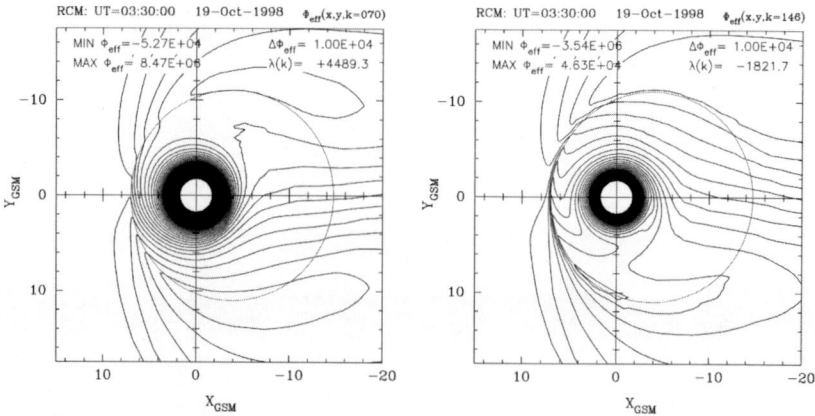

Figure 9. (a) RCM effective potential plotted in the equatorial plane for ions with energy invariant $\lambda_S = 4489$. (b) Effective potential plotted in the equatorial plane for electrons with energy invariant $\lambda_S = 1822$.

for a component of the ion distribution function (a) and for an electron component (b). According to (4), contours of constant effective potential, represent instantaneous trajectories for particles with energy invariant λ_k (note that all figures of the potential, are shown in the frame that rotates with the Earth). Positive ions, injected along the night side of the boundary, drift toward the Earth and turn westward, while electrons turn eastward. As a result, ions, which tend to be a factor of ~ 7.8 hotter, come closer to the Earth on the dusk side, while the electrons are closer on the dawn side. Electrons are lost (right-hand side of (4) through precipitation, enhancing the conductance along the inner edge of the electron plasma sheet. It is the ions who provide most of the pressure, and therefore, field-aligned currents. On the dusk side, then, the equatorward edge of the field-aligned currents is closer to the Earth than the equatorward edge of the enhanced auroral conductance (identified in simulations with the diffuse auroral). Horizontal current closure across this narrow region, which, in the absence of sunlight, has very low conductance (ionospheric trough), requires a large electric field.

This phenomenon is further illustrated in Figure 10, where we show latitudinal profiles of field-aligned currents, perpendicular southward electric field, and Pedersen conductance (currents and conductances are sums for Northern and Southern hemispheres), for the configuration of Figure 8. The left panel is for MLT = 21 and cross-sections the region of SAID; the right panel, for MLT = 05, is for a region without SAID. The secondary peak in the electric field profile on the left panel (equatorward of the main electric field convection) is the characteristic signature of SAID observed with polar-orbiting satellites. It can be clearly seen how the field-aligned current, extending into the region of very low (<2 S) conductance, generates this strong 70 mV/m electric field structure. On the dawn side (right panel), the enhanced conductance region is closer to the Earth. Field-aligned cur-

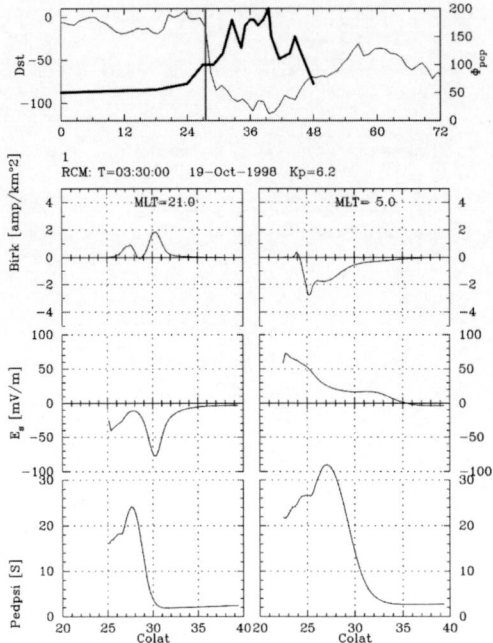

Figure 10. This plot shows latitudinal profiles of field-aligned currents, perpendicular southward electric field, and Pedersen conductance (currents and conductances are sums for Northern and Southern hemispheres), for the configuration of Figure 8. See text for details.

rents map into the high-conducting ionosphere for all latitudes at this local time, and no SAPS or SAIDs is observed.

4.3. ASYMMETRY OF THE RING CURRENT PRESSURE DISTRIBUTION

One capability that the RCM provides is to compute and follow the distribution of the total particle pressure in the ring current region. The traditional, prevailing view is that during magnetic storms, when fresh particles are being injected into the ring current, the pressure peaks around dusk. This picture is based on assumed electric fields used in ring current calculations. In the last couple of years, global images of the ring current flux distribution have become available, obtained by the energetic neutral atom (ENA) imagers on board the IMAGE spacecraft. These global images indicate, in fact, that the flux distribution maximizes close to midnight, and sometimes well past midnight, close to dawn.

Since the RCM self-consistently computes particle motion and electric fields controlling particle flow, we are able to follow the dynamics of the ring current pressure. For the magnetic storm of October 19, 1998 already discussed above, Figures 11–13 show two-dimensional distributions of the total pressure from Equation (5) in the innermost part of the modeling region. The vertical line on the upper panel shows at what time the plot was made in relation to the time series of the *Dst*

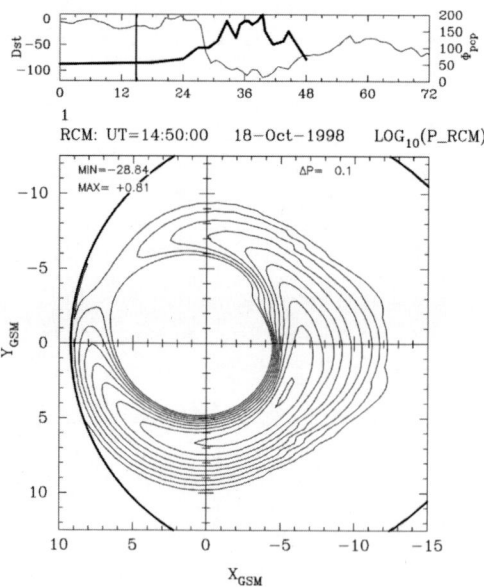

Figure 11. RCM computed logarithm of pressure plots of the magnetic storm of October 19, 1998 at 14:50 UT in the innermost part of the modeling region, magnetospheric equatorial plane. The vertical line on the upper panel shows where the plot was made in relation to the time series of the Dst index, which indicates development of the ring current as measured by its magnetic field signature at low latitudes on the ground.

index, which is a measurement of the ring current strength. Shown are contours of the logarithm of the pressure p.

In Figure 11, corresponding to a pre-storm, quiet-time steady-state condition, the pressure distribution indicates the lack of substantial ring current. Later, when the storm-time ring current is injected during the main phase of the storm (Figure 12), the pressure distribution is asymmetric, with the peak being close to, but before, midnight. The contours also indicate a strong dusk-dawn asymmetry in the ring current, in agreement with multiple observations. Figure 13 represents the ring current well into the recovery phase of the storm. The pressure distribution symmetrizes due to: (1) lack of fresh injection, (2) trapping of already injected particles on closed trajectories as the convection strength decreases (thick curve on the upper panels), and (3) charge-exchange loss with a characteristic lifetime of many hours. The pressure peak is located at midnight.

5. Coupling to Force Balanced Magnetic Field Models

The computation of realistic magnetospheric equilibria has long been a goal in magnetospheric physics, a goal that has only recently been achieved. Cheng (1995) developed an elegant and sophisticated numerical scheme for computing 3D magnetic field configurations that satisfy the equilibrium condition ($\vec{j} \times \vec{B} = \nabla p$) for

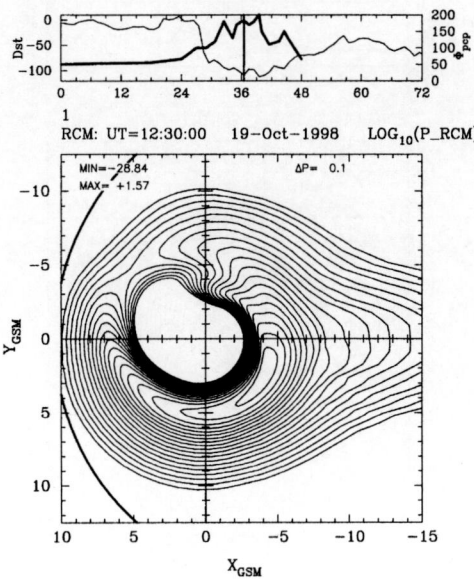

Figure 12. Same as Figure 11 for 12:30 UT.

an assumed distribution of pressure among the different field lines. This approach, however, does not couple naturally to the RCM. Hesse and Birn (1993) used a magnetofriction (MF) method (Chodura, 1981) to relax the empirical Tsyganenko (Tsyganenko, 1989) field model for $x \leqslant 5R_E$ into a force-balanced equilibrium. Toffoletto *et al.* (1996) adopted the MF approach for computing equilibria. The basic equations are MHD, described in detail by (Hesse and Birn, 1992), but the momentum equation takes the form:

$$\frac{\partial \rho \vec{v}}{\partial t} = \vec{j} \times \vec{B} - \nabla P - \nabla \cdot [\rho \vec{v} \vec{v}] - \alpha \rho \vec{v} + \nu \nabla^2 \rho \vec{v} \tag{18}$$

where the density (ρ) is not the actual density as computed by the RCM, but rather an artificial parameter that is set to keep the fast-mode speed constant and the code stable near the Earth. The last two terms in Equation (18) represent the friction (α) that is employed to remove kinetic energy from the system and a viscous term (ν) that parameterizes the flux limiting routine in the time integration algorithm. Since the desired solution does not depend on any of these terms, they represent a set of convenient parameters to allow faster calculations of the equilibrium. (See Hesse and Birn, 1993; Toffoletto *et al.*, 1996, 2001; Lemon *et al.* 2002, for more details.)

The magnetic and plasma pressure distribution from a converged equilibrium code is used to specify the initial condition for the RCM. The pressure and magnetic field computed from the equilibrium code is used to compute η_k using (4) and an assumed distribution function. For simplicity, the results presented here used a single species (s) of fixed energy invariant. The RCM and MF code are coupled by the following algorithm:

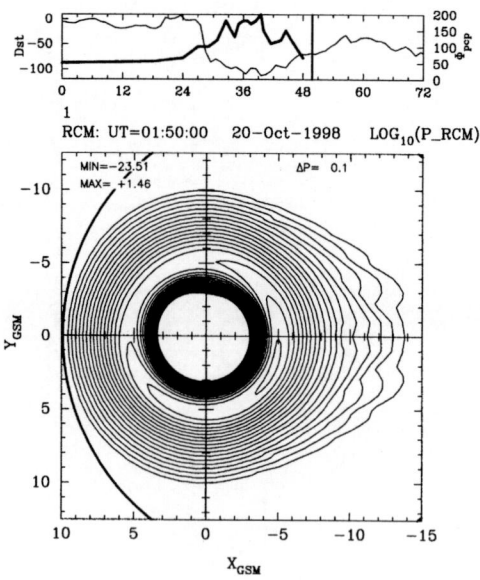

Figure 13. Same as Figure 11 for October 20, 1998 01:50 UT.

1. Compute an equilibrated magnetic field and corresponding pressure distribution starting from an empirical model, in this case a (Tsyganenko, 1989) magnetic field model and an initial pressure distribution taken from an empirical summary of nightside equatorial observations (Spence *et al.*, 1989), extended uniformly along field lines (as required by the model assumptions) and assumed (arbitrarily) to extend uniformly in local time.
2. Compute flux tube volumes and values of $PV^{5/3}$ to set up the RCM's energy channels from the results of step 1.
3. Run the RCM for a specified interval (typically 5 minutes of magnetosphere time) using an assumed electric potential distribution around the polar cap with a maximum potential of 100 kV).
4. Use the RCM results to override the pressure at friction-code grid points within the RCM modeling region. Pressures are computed from the RCM using Equation 5.
5. Recompute a new equilibrated configuration.
6. Compute new flux-tube volumes and poleward boundary of the RCM.
7. Repeat from 3. Where new field lines have been added to the RCM specify $PV^{5/3}$ from either the MF code or some pre-specified algorithm.

The first results of these calculations showed that a B_z minimum formed in the inner magnetosphere at around x $= -12$ R_E (Toffoletto *et al.*, 1996). However, in those runs, stability requirements limited the far-tail boundary of the calculation to x $= -20$ R_E downstream. Recent technical improvements in the equilibrium code have permitted the boundary to be moved out to x $= -50$ R_E downstream (Toffoletto *et al.*, 2001). This extended region allows higher content flux tubes

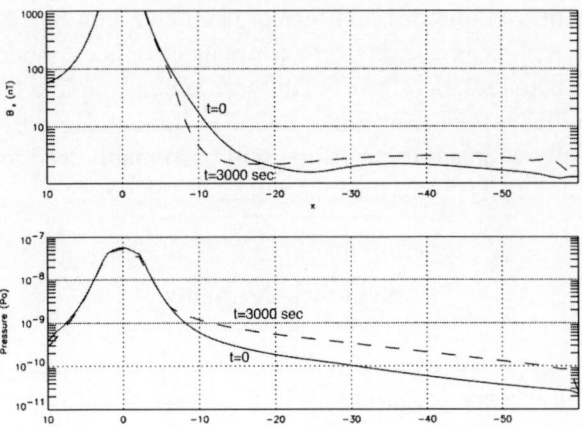

Figure 14. Magnetic field B_z and pressure plots along the x-axis from a growth-phase simulation, from Toffoletto *et al.* (2002).

(larger $pv^{5/3}$) to be carried into the inner magnetosphere resulting in a more dramatic effect on the inner region. (Earlier runs using the smaller modeling region (Toffoletto *et al.*, 1996) required a substantially longer time for the formation of the B_z minimum.) Here we present some results from a simulation using the coupled RCM/MF code investigating a substorm growth phase. Figure 14 is a composite plot at $t = 0$ seconds and $t = 3000$ seconds showing the change in the magnetic field along the $x-$axis, where a deep B_z minimum has formed at around $x \simeq -12\ R_E$. The magnetic field stretching in the inner tail results from the system trying to maintain pressure balance, given the $pv^{5/3}$ distribution dictated by the RCM. The code has been run for longer intervals with the result that the tail field strengthens and the magnetic field in the inner plasma sheet gets more and more stressed until finally the numerical method breaks. This happens when the current sheet gets too thin for the MF-code grid, and the grid spacing in the RCM gets too wide. Nature presumably solves the problem by creating a substorm and thereby violates the adiabatic-drift laws. These results are consistent with 2D calculations (Erickson, 1992; Hau, 1991; Pritchett and Coroniti, 1996), where the earthward convection of plasma resulted in stretching of tail field lines in the inner region of the tail, while field lines that map to the distant plasma sheet become increasingly dipolar.

6. Summary

This paper has been an attempt to summarize the current state of inner magnetospheric modeling using the RCM, as well as describing the current state of work. We have presented and discussed some recent results of the Rice Convection Model including the effects of magnetospheric shielding, subauroral ion drifts, the physics of ring current asymmetries and results from coupling to a Magneto-Friction equilibrium solver. It is an exciting time for magnetospheric modeling of the inner magnetosphere, with current efforts being focussed on coupling the RCM to global MHD codes. Initial results from these coupling studies are only now starting to appear. When fully coupled, these efforts will presumably lead to much improved understanding of the physics of the inner magnetosphere.

Acknowledgements

This work was supported by NSF GEM grant ATM-9900983 and NASA Sun-Earth-Connection Theory program NAG5-8136. We thank the referee for careful reading of the manuscript.

References

Birmingham, T. J.: 1992, 'Birkeland Currents in an anisotropic, magnetostatic plasma', *J. Geophys. Res.* **97**, 3907–3917.

Chen, C. -K., Wolf, R. A., Harel, M. and Karty, J. L.: 1982, 'Theoretical Magnetograms Based on Quantitative Simulation of a Magnetospheric Substorm', *J. Geophys. Res.* **87**, 6137.

Cheng, C. Z.: 1995, 'Three-dimensional Magnetospheric Equilibrium with Isotropic Pressure', *Geophys. Res. Lett.* **22**, 2401–2404.

Chodura, R. and Schlüter, A.: 1981, 'A 3D Code for MHD Equilibrium and Stability', *J. Comp. Phys.* **41**, 68–88.

Costello, K. A.: 1997, *Moving the Rice MSFM Into a Real-time Forecast Mode Using Solar Wind Driven Forecast Modules*, PhD Thesis, Rice University.

Erickson, G. M., Spiro, R. W. and Wolf, R. A.: 1991, 'The Physics of the Harang Discontinuity', *J. Geophys. Res.* **96**, 1633–1645.

Erickson, G. M.: 1992, 'A Quasi-static Magnetospheric Convection Model in Two Dimensions', *J. Geophys. Res.* **97**, 6505–6522.

Fejer, B. G., Spiro, R. W., Wolf, R. A. and Foster, J. C.: 1990, 'Latitudinal Variation of Perturbation Electric Fields During Magnetically Disturbed Periods: 1986 SUNDIAL Observations and Model Results', *Ann. Geophys.* **8**, 441–454.

Fok, M. -C., Wolf, R. A., Spiro, R. W. and Moore, T. E.: 2001, Comprehensive Computational Model of Earth's Ring Current', *J. Geophys. Res.* **106**, No. A5, 8417–8424.

Foster, J. C. and Vo, H. B.: 2002, 'Average Characteristics and Activity Dependence of the Subauroral Polarization Stream', *J. Geophys. Res.*, in press.

Galperin, Y. I., Ponomarev, V. N. and Zosimova, A. G.: 1973, 'Direct Measurements of Ion Drift Velocity in the Upper Atmosphere During a Magnetic Storm', *Cosmic Res.* **11**, 273–283.

Galperin, Y. I., Ponomarev, V. N. and Zosimova, A. G.: 1974, 'Plasma Convection in Polar Ionosphere', *Ann. Geophys.* **4A**, 301.

Goldstein, J., Sandel, B. R., Forrester, W. T. and Reiff, P. H.: 2002, 'IMF-driven Plasmasphere Erosion of 10 July 2000', *Geophys. Res. Lett.* (submitted Oct.).

Harel, M., Wolf, R. A., Reiff, P. H., Spiro, R. W., Burke, W. J., Rich, F. J. and Smiddy, M.: 1981a, 'Quantitative Simulation of a Magnetospheric Substorm 1, Model Logic and Overview', *J. Geophys. Res.* **86**, 2217–2241.

Harel, M., Wolf, R. A., Spiro, R. W., Reiff, P. H., Chen, C. -K., Burke, W. J., Rich, F. J. and Smiddy, M.: 1981b, 'Quantitative Simulation of a Magnetospheric Substorm 2, Comparison With Observations', *J. Geophys. Res.* **86**, 2242–2260.

Hau, L. -N.: 1991, 'Effects of Steady State Adiabatic Convection on the Configuration of the Near-Earth Plasma Sheet, 2', *J. Geophys. Res.* **96**, 5591–5596.

Heinemann, M. and Pontius, D. H., Jr.: 1990, 'Representations of Currents and Magnetic Fields in Isotropic, Magnetohydrostatic Plasma', *J. Geophys. Res.* **95**, 251–257.

Heppner, J. P. and Maynard, N. C.: 1987, 'Empirical High-latitude Electric Field Models', *J. Geophys. Res.* **92**, 4467–4489.

Hesse, M. and Birn, J.: 1992, 'Three-dimensional MHD Modeling of Magnetotail Dynamics for Different Polytropic Indices', *J. Geophys. Res.* **97**, 3965.

Hesse, M. and Birn, J.: 1993, 'Three-dimensional Magnetotail Equilibria by Numerical Relaxation Techniques', *J. Geophys. Res.* **98**, 3973–3982.

Hilmer, R. V. and Voigt, G. -H.: 1995, 'A Magnetospheric Magnetic Field Model with Flexible Current Systems Driven by Independent Physical Parameters', *J. Geophys. Res.* **100**, 5613–5626.

Jaggi, R. K. and Wolf, R. A.: 'Self-consistent Calculation of the Motion of a Sheet of Ions in the Magnetosphere', *J. Geophys. Res.* **78**, 2842.

Karty, J. L., Chen, C. -K., Wolf, R. A., Harel, M. and Spiro, R. W.: 1982, 'Modeling of High-latitude Currents in a Substorm', *J. Geophys. Res.* **87**, 777.

Kelley, M. C.: 1989, *The Earth's Ionosphere*, Academic Press, San Diego.

Lemon, C., Tololetto, F., Hesse, M. and Birn, J.: 2002, 'Computing Global Magnetospheric Force Equilibria using MHD', *J. Geophys. Res.* (submitted).

Li, X., Roth, I. Temerin, M., Wygant, J. R., Hudson, M. K. and Blake, J. B.: 1993, 'Simulation of the Prompt Energization and Transport of Radiation Belt Particles During the March 24, 1991 SSC', *Geophys. Res. Lett.* **20**, 2423–2426.

Lui, A. T. Y., Meng, C. -I. and Ismail, S.: 1982, 'Large Amplitude Undulations on the Equatorward Boundary of the Diffuse Aurora', *J. Geophys. Res.* **87**, 2385.

Northrop, T. G.: 1963, *The Adiabatic Motion of Charged Particles*, J. Wiley and Sons, New York.

Pontius, D. H., Jr., Wolf, R. A., Hill, T. W., Yang, Y. S. and Smyth, W. H.: 1996, 'Velocity Shear Impoundment of the Io Plasma Torus', *J. Geophys. Res.* **103**, 23,551.

Pritchett, P. L. and Coroniti, F. V.: 1998, Formation of Thin Current Sheets During Convection', *J. Geophys. Res.* **100**, 19935–19946.

Robinson, R. M., Vondrak, R. R., Miller, K., Dabbs, T. and Hardy, D.: 1987, 'On Calculating Ionospheric Conductances from the Flux and Energy of Precipitating Electrons', *J. Geophys. Res.* **92**, 2565–2569.

Roederer, J. G.: 1970, *Dynamics of Geomagnetically Trapped Radiation*, Springer-Verlag, Berlin.

Sazykin, S.: 2000, *Theoretical Studies of Penetration of Magnetospheric Electric Fields to the Ionosphere*, PhD Dissertation, University of Utah.

Sazykin, S., Wolf, R. A., Spiro, R. W., Gombosi, T. I., De Zeeuw, D. L. and Thomsen, M. F.: 2002, 'Interchange Instability in the Inner Magnetosphere Associated with Geosynchronous Particle Ux Decreases', *Geophys. Res. Lett.* **10**, 1029/2001GL014416.

Stiles, G. S., Hones, E. W., Bame, S. J. and Asbridge, J. R.: 1978, 'Plasma Sheet Pressure Anisotropies', *J. Geophys. Res.* **83**, 3166–3172.

Southwood, D. J. and Wolf, R. A.: 1978, 'IAn Assessment of the Role of Precipitation in Magnetospheric Convection', *J. Geophys. Res.* **83**, 5227.

Spence, H. E., Kivelson, M. G., Walker, R. J. and McComas, D. J.: 1989, 'Magnetospheric Plasma Pressures in the Midnight Meridian: Observations from 2.5 to 35 R_E', *J. Geophys. Res.* **94**, 5264–5272.

Spiro, R. W., Heelis, R. A. and Hanson, W. B.: 1978, 'Ion Convection and the Formation of the Midlatitude F Region Ionization Trough', *J. Geophys. Res.* **83**, 4255.

Spiro, R. W., Harel, M., Wolf, R. A. and Reiff, P. H.: 1981, 'Quantitative Simulation of a Magnetospheric Substorm, 3, Plasmaspheric Electric Fields and Evolution of the Plasmapause', *J. Geophys. Res.* **86**, 2261–2272.

Spiro, R. W., Wolf, R. A. and Fejer, B. G.: 1988, 'Penetration of High-latitude-electric-field Effects to Low Latitudes During SUNDIAL 1984', *Ann. Geophys.* **6**, 39–50.

Spiro, R. W. and Wolf, R. A.: 1984, 'Electrodynamics of Convection in the Inner Magnetosphere', in T. A. Potemra (ed.), *Magnetospheric Currents*, American Geophysical Union, Washington, D.C., 247.

Stern, D. P.: 1970, 'Euler Potentials', *Am. J. Phys.* **38**, 494–501.

Stern, D. P.: 1975, 'The Motion of a Proton in the Equatorial Magnetosphere', *J. Geophys. Res.* **80**, 595–599.

Toffoletto, F. R., Spiro, R. W., Wolf, R. A., Hesse, M. and Birn, J.: 1996, 'Self-consistent Modeling of Inner Magnetospheric Convection', in E. J. Rolfe nd B. Kaldeich (eds.), *Third International Conference on Substorms (ICS-3)*, ESA Publications Division, Noordwijk, The Netherlands, pp. 223–230.

Toffoletto, F. R., Birn, J., Hesse, M., Spiro, R. W. and Wolf, R. A.: 2001, 'Modeling Inner Magnetospheric Electrodynamics', in P. Song, H. Singer and G. Siscoe (eds.), *Space Weather*, Geophysical Monograph Series, 125, p. 265.

Tsyganenko, N. A.: 1989, 'A Magnetospheric Magnetic Field Model with a Warped Tail Current Sheet', *Planet. Space Sci.* **37**, 5–20.

Vasyliunas, V. M.: 1970, 'Mathematical Models of Magnetospheric Convection and its Coupling to the Ionosphere', in B. m. McCormac (ed.), *Particles and Fields in the Magnetosphere*, D. Reidel, Hingham, MA, Hingham, MA, pp. 60–71.

Volland, H. A.: 1973, 'A Semiempirical Model of Large-scale Magnetospheric Electric Fields', *J. Geophys. Res.* **78**, 171–180.

Wolf, R. A.: 1970, 'Effects of Ionospheric Conductivity on Convective Flows of Plasma in the Magnetosphere', *J. Geophys. Res.* **75**, 4677–4698.

Wolf, R. A., Harel, M., Spiro, R. W., Voigt, G. -H., Reiff, P. H. and Chen, C. K.: 1982, 'Computer Simulation of Inner Magnetospheric Dynamics for the Magnetic Storm of July 29, 1977', *J. Geophys. Res.* **87**, 5949–5962.

Wolf, R. A.: 1983, 'The Quasi-static (Slow-flow) Region of the Magnetosphere', in R. L. Carovillano and J. M. Forbes (eds.), *Solar Terrestrial Physics*, Series, D. Reidel, Hingham, MA, pp. 303–368.

Wolf, R. A., Spiro, R. W. and Rich, F. J.: 1991, 'Extension of the Rice Convection Model into the high-latitude ionosphere, *J. Atm. Terrest. Phys.* **53**, 817–829.

Yang, Y. S., Spiro, R. W. and Wolf, R. A.: 1994, 'Generation of Region-1 Current by Magnetospheric Pressure Gradients', *J. Geophys. Res.* **99**, 223–234.

Yang, Y. S., Wolf, R. A., Spiro, R. W., Hill, T. W. and Dessler, A. J.: 1994, 'Numerical Simulation of Torus-driven Plasma Transport in the Jovian Magnetosphere', *J. Geophys. Res.* **99**, 8755–8770.

SOME OBSERVATIONAL EVIDENCE OF ALFVEN SURFACE WAVES INDUCED MAGNETIC RECONNECTION

C. UBEROI

Department of Mathematics, Indian Institute of Science, Bangalore 560 012, India

Abstract. The surface wave induced magnetic reconnection (SWIMR) model based on Alfven Resonance theory will be discussed briefly both for collisional and collisionless plasmas. It is shown that the spatial scales and time delays associated with Flux Transfer Events and Pulsed Ionospheric Flows, as observed by satellites and SuperDARN radars and the magnetic bubbles, observed at the high latitude boundary of the magnetopause, can be explained by the SWIMR model.

Key words: magnetic reconnection, surface waves

1. Introduction

Magnetic reconnection and Alfven wave resonances are key processes in various astrophysical and geophysical phenomena involving tapping of energy resources in order to produce large dissipative events. In case of magentic reconnection energy is tapped from mangetic field and in Alfven resonant phenomena it is tapped from an external source or Alfven surface waves along the tangential discontinuities in the system. The concept of surface wave induced magnetic reconnection (SWIMR) unifies these two key processes.

From the observations made by the Earth orbiting satellites it is now evident that both the magnetic reconnection processes and the propagation of surface waves are a common occuring plasma phenomena along the magnetopause, the Earth's magnetospheric boundary. In this review therefore, we shall first give the basic concept of the SWIMR model and then take three examples of magnetic reconnection processes recently observed by various space missions argueing that these can be explained by evoking the concept of surface wave induced reconnection.

2. Basic Concept of the SWIMR Model

The SWIMR model is based on the Alfven wave resonant theory near the neutral point (Uberoi, 1994). In this paper it is seen that though the mathematical structure of the equation governing the dynamics of the MHD waves in non-uniform media remains the same at zero and non-zero singular points, the role played by surface waves in the resonant absorption mechanism at these two points is different.

Space Science Reviews **107:** 197–206, 2003.
© 2003 *Kluwer Academic Publishers.*

The theories of resonant absorption of Alfven waves consider surface waves propagating along a sharp discontitunity separating two infinitely extended plasma regions. The structural discontinuities are not taken into account. Near a neutral point the structure of the discontinuity becomes important as waves are now of long wavelengths. In this case wave propagation is to be considered for a plasma layer, which can support two types of surface modes of oscillations: symmetric and asymmetric. It is argued that at the neutral point, the long wavelength symmetric modes of the plasma layer couple to the low-frequency end of the Alfven continuum. When the resistivity is switched on, the long wavelength couples with the tearing mode of the layer. This mode is unstable for certain wavelengths greater than a critical value and begins to grow until magnetic islands are formed. An estimate of the linear dimensions of the islands that are formed was discussed by Uberoi $et\ al.$ (1996), and is given as $\lambda = 2\pi a/0.64$, where a is the thickness of the plasma layer.

The important time scales found in this model while studying the time evolution of the MHD resistive system are:

$$t_h = \tau_A^{2/3}\tau_R^{1/3} \quad \text{or} \quad t_h = \tau_A S^{1/3}, \tag{1}$$

where $\tau_R = \frac{4\pi a^2}{\eta}$, $\tau_A = a/V_A$, and $S = \tau_R/\tau_A$, is the Lundquist number. Here $2a$ is the width of the plasma layer around the zero-singular point, η is the finite conductivity and V_A is the Alven speed. The resistivity effects begin to play role for $t \geq t_h$.

The other importnat scale is the reconnection time:

$$t_r = \tau_A^{2/5}\tau_R^{3/5} \quad \text{or} \quad t_r = \tau_A S^{3/5} \tag{2}$$

The width of the resistive layer scales as

$$\Delta x \propto S^{-1/3}$$

It is of interest to point out here that the concept of the SWIMR model is consistent with the forced magnetic reconnection induced by perturbing the boundary of a simple Taylor model, a simple slab equilibrium of an incompressible plasma with a resonant surface inside, investigated by Hahm and Kulsrud (1985). They studied the time evolution of the MHD system near the neutral point for linear magnetic field variation. In fact the SWIMR model is a general formulation of Taylor's model as was discussed in detail by Uberoi and Zweibel (1999).

The SWIMR model can be extended to include the finite Larmor radius and electron inertial effects using the generalised Ohm's law in the MHD equations. It is now well understood that the dispersion relation for kinetic Alfven wave, both when Hall-current effect and/or electron inertial effects are important, can be obtained from the Hall-MHD model in the limit of low-β and $k_\perp \gg k_\parallel$ (Uberoi, 1995). Taking, compressibility into account it can be easily seen that finite ρ_s and λ_e in an inhomogeneous sheared media couple the compressional surface waves and the kinetic Alfven wave (KAW). Here $\lambda_e = \omega_{pe}/c$ and ρ_s is the ion-sound Larmor radius with $\beta = V_S/V_A$. The compressional surface waves, however, show strong depence on β and $(k_{\parallel}/k_{\perp})$ (Uberoi, 1989). This also holds good for the surface waves propagating in the layered system. Following the same arguments as in the case of collisional plasma we find that near the neutral layer, the symmetric compressional surface modes for the plasma layer can resonantly excite the collisionless tearing mode.

Following the work on the collisionless tearing mode instability (Ding *et al.*, 1992), the reconnection time-scale t_{rc} can be written as

$$t_{rc} = \left(\frac{\pi}{4}\right)^{1/2} \tau_A \frac{1}{\beta}(\lambda_e a)^2 (\rho_s a)^{1/2}.$$

3. Flux Transfer Events

The multi-spacecraft ISEE mission (Song *et al.*, 1988) has given detailed observations of surface waves with periods of about 2 min or more along the magnetopause. It was also seen that magnetopause is more oscillatory for southward interplanetary magnetic field (IMF) than for northward IMF. However evidence of magnetic reconnection at the magnetopause came with the ISEE observations (Russell *et al.*, 1997) of magnetic flux tubes of cross section $1 R_E$ near Earth's magnetosphere which were isolated from each other. It was identified that the observed B_n polarity reverses with the passing of a localized reconnected flux tube. These were called flux transfer events (FTEs), as these events seem to be responsible for flux or momentum transfer into the magnetosphere during the solar-wind geomagnetic field interaction. FTEs are characterized by prominent isolated bipolar (+ve then −ve or vice versa) perturbations in the component normal to the magnetopause. They occur intermittently every 5 to 15 minutes. A recent survey confirms that onset of FTE's is controlled by southward component of IMF. It is inferred from observations (Figure 1) that flux transferring reconnection does not occur unless the magnetic field at the magnetopause has been southward for the order of 7 min or more. This time delay is real and not due to spacecraft not being in the right location at the right time to get the observations correct. From observations it is noted additionally that there is no apparent solar wind or other interplanetary driver for the observed initial delay of the first FTE (Russell *et al.*, 1997).

There are different mathematical models for FTE's. Tearing instability is an integral part of all the models. However, these required either perpendicular velocity flow or $V > V_c$, the critical velocity for K-H instability (see Uberoi *et al.*, 1996). The SWIMR concept does not require high plasma flow velocities along the magnetopause. The surface waves, even if they arise from the K-H instability can exist for the case of $V < V_c$. Further, this model requires only low frequency surface waves. As pointed above, observations of magnetopause surface waves with

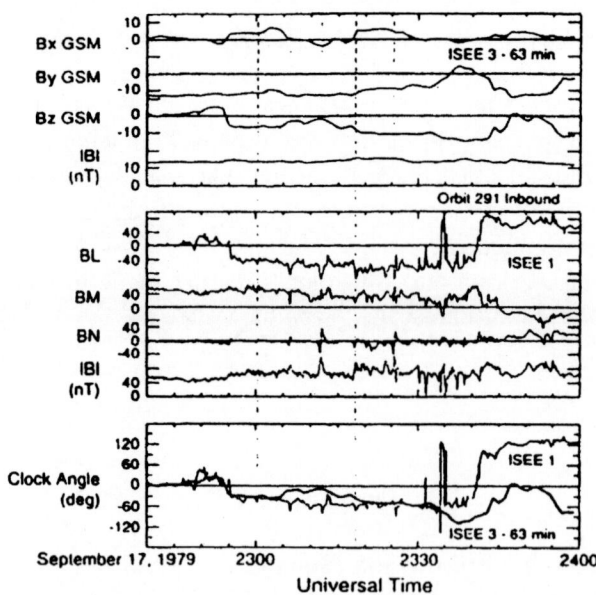

Figure 1. Flux transfer events detected by ISEE-I in the magnetosheath under steady IMF conditions. The bottom panel shows the clock angle measured by ISEE-1 and 3 that was used to determine the relative time delay of the observations (Russell *et al.*, 1997).

periods $\geq$ 2 min have been reported. Theoretical curves (Uberoi *et al.*, 1996) drawn from the dispersion of symmetric surface waves for a layered medium show that wave periods of order of 50–500s (about 1 to 8 min) can exist along the magnetopause for values of 'a' ranging from 100 to 300 kms. Here 'a' is the half thickness of the neutral layer which separates the two plasma regimes, the magnetosheath and the magnetosphere. The value of Alfven speed being 300 km s^{-1}. The width of the islands formed by reconnection ranges from 900 to 2700 km for this range of a. This also agrees well with the observational values for the smallest scale FTE's.

To explain the observations of delay time between the southward turning of the interplanetary magnetic field and the onset of a flux transfer event at the magnetopause Uberoi *et al.* (1999) plot the values of t_h as a function of 'a', for several values of the resistivity in the reconnection region and for V_A ranging from 200 to 400 km s^{-1} at each resistivity value. It is evident that ther are a range of reasonable values for the layer thickness, resistivity and V_A that will produce initial delays in reconnection of the order of 5 to 10 min consistent with the observations (Figure 2). Larger values of V_A require lower values of the resistivity in order to achieve similar values of the onset delay time t_h. Other than the B_z condition, the onset of reconnection does not depend upon any external interplanetary driver (Mach number, dynamic pressure) in accordance with the conclusions drawn from the observations. The time scale t_h is an intrinsic time scale for the SWIMR process.

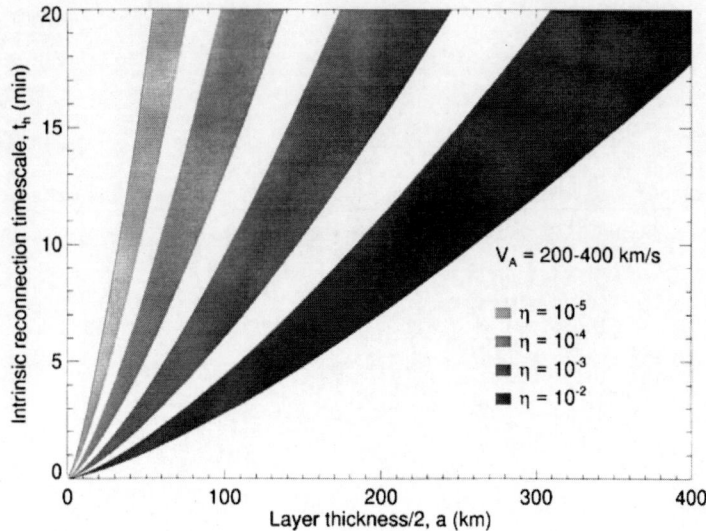

Figure 2. Variation of t_h with 'a' for different value of η and V_A. The value of η increases to the right V_A increases from 200 at the top of each band to 400 at the lower edge (Uberoi *et al.*, 1999).

4. Pulsed Ionospheric Flows

Recently Prikryl *et al.* (2002) reported observations of pulsed ionospheric flows (PIF's) in the cusp footprint by the SuperDARN radars with periods between a few minutes and several tens of minutes (Figure 3). PIFs are believed to be ionospheric signatures of FTEs. The quasiperiodic PIFs are correlated with Alfvenic fluctuations observed in the upstream solar wind. From this they conclude that on these occasions the FTEs were driven by Alfven waves coupling to the dayside magnetosphere. Case studies for two events are presented in which the dawn-dusk component of the Alfven wave electric field modulates the reconnection rate as evidenced by the radar observations of the ionospheric cusp flows. The arrival of the IMF southward turning at the magnetopause is determined from multi-point solar wind magnetic field and/or plasma measurements assuming plane phase fronts in solar wind. The cross-correlation lag between the solar wind data and ground magnetograms that were obtained near the cusp footprint exceeded the estimated spacecraft-to-magnetopause propagation time by up to several minutes thus accounting for exceeding the Alfven propagation time between the magnetopause and ionosphere. In addition for the case of short-period (2–13 min) PIFs, the onset times of the flow transients appear to be further delayed by at most a few more minutes after the IMF southward turning arrived at the magnetopause. For the case of long period (30–40 min) PIFs, the observed additional delays were 10–20 min. They interpret the excess delay in terms of an intrinsic time scale for reconnection, as in the case of FTE's, which can be explained by the SWIMR model. The compressional fluctuations in solar wind and those generated in the magnetosheath

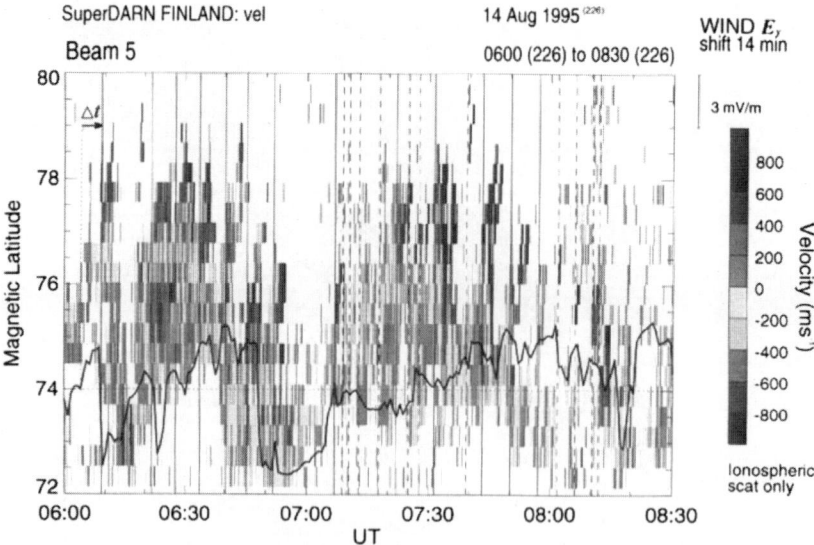

Figure 3. Finland radar high resolution line-of-sight velocity data. The color-coded velocity (beam 5) shows quasi-periodic poleward progressing PIFs. The dawn-dusk electric field E_Y is superposed shifted by 14 min. The solid/broken vertical lines show the onset times of major/minor PIFs and the horizontal arrow indicates the inferred delay (Δt) of the reconnection onset after the IMF southward turning at the magnetopause (Prikryl *et al.*, 2002).

through the interaction between the solar wind Alfven waves and the bow shock were the source of magnetopause surface waves inducing reconnection.

For two events the PIF frequencies are (0.4–0.6 mHz) for event 1 and (1.3–8.3 mHz) for event 2. Assuming that this is the range of frequencies (periods) of surface waves inducing reconnection at the magnetopause the Figure 4 plots t_h against frequencies. The observed PIF delay time decreases with frequency similar to the decrease of intrinsic reconnection time scale as calculated from the SWIMR model.

5. Magnetic Bubbles

Recenlty a detailed analysis of the observations of magnetic holes or magnetic bubbles made by the POLAR satellite crossing the magnetopause current layer and entering the magnetosheath region is given by Stasiewicz *et al.* (2001). Figure 5 shows the depression of the magnetic field in the bubble region as seen by the reversal of B_z. The minimum field measured during this event was 1.4 nT which corresponds to 98% depression of the ambient field of $\sim$100 nT. Using particle moment measurements they ruled out the mirror instability condition to be the cause of these depressions. It was seen that the parallel ion and electron pressure

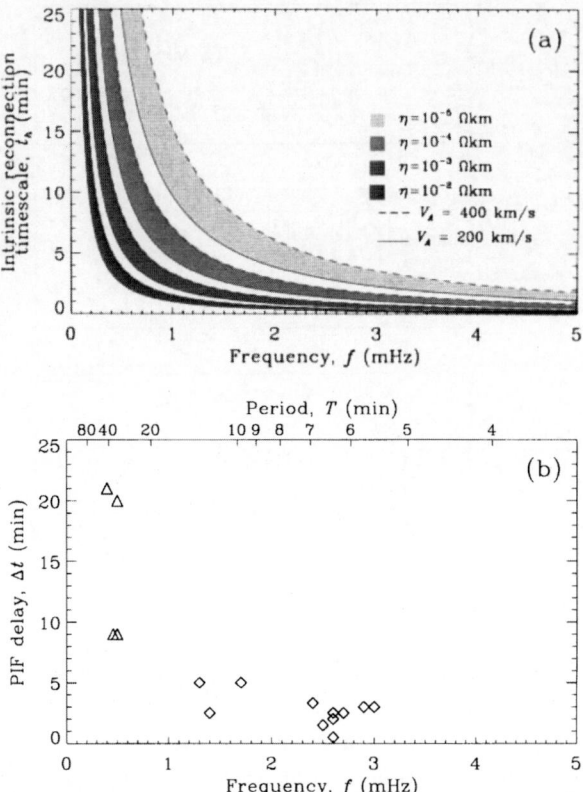

Figure 4. (**a**) The intrinsic timescale for the onset of surface-wave-induced magnetic reconnection (Uberoi *et al.*, 1999) as a function of wave frequency with $ka = 0.001$, and for different values of magnetic resistivity η and Alfven velocity V_A. η increases from the top band to the lower band. (**b**) Observed delay of the PIF onset (ionospheric signature of reconnection onset) after the IMF southward turning at the magnetopause for several PIF events. The Alfven propagation time between the magnetopause and ionosphere has been subtracted (Prikryl *et al.*, 2002).

exceeds the perpendicular components and, therefore, mirror instability condition does not arise.

The nature of the observed magnetic bubbles is revealed with a higher resolution plot of two orthogonal components of the magnetic and electric fields. High degree of correlation between the B_z and E_y components is characteristic for Alfven waves. For example for amplitudes $\delta E \approx 30$ mV m^{-1} and $\delta B \approx 150$ nT, the ratio $\delta E / \delta B \simeq 200$ km s^{-1}, which is close to the value for Alfven velocity inside the bubble layer. Further analysis of the wave modes in the frequency range 0–30 Hz at the magnetopause or the bubble layer it was seen that the broad band waves in this frequency range represent most likely spatial turbulence of kinetic Alfven waves, Doppler shifted to higher frequencies (in the satellite frame) by convective plasma flows. In addition KAW fluctuation concentrated around the bubble layer were observed. The results of the numerical simulation were interpreted that the

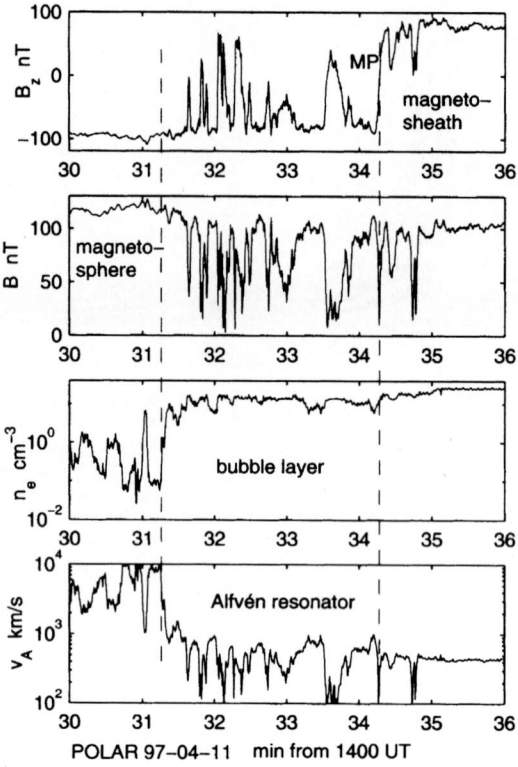

Figure 5. Details of a bubble layer: magnetic field B_z and B, the electron density n_e, and the Alfven velocity V_A (Stasiewicz *et al.*, 2001).

bubbles are produced by a tearing mode reconnection process (Figure 6) and the KAW fluctuation are related to the Hall instability created by macroscopic pressure and magnetic field gradients.

We like to argue here that observations of magnetic bubbles and the KAW fluctuations can be related to surface waves induced reconnection. The high latitude magnetopause boundary is known to support surface waves. The numerical simulation carried to understand the bubbles use the Hall-MHD model equations. This set of equations for sheared magnetic fields show that the magnetsonic surface waves can induce collisionless tearing instability. The non-linear analysis of the SWIMR model for collisionless plasmas can explain the formation of the bubbles.

For the observed fluctuations concentrated on the bubble boundaries we like to draw attention to the work by Nagasaki and Itoh (1991) on the decay process of magnetic islands by forced reconnection using the Taylor's model. Their results show that the evolution time in the decay process depends not only on S, as in the case of growth process but also on the island width. The time evolution of the current density shows oscillations unlike in the growth process. From the similarity between the SWIMR model and Taylor's model we can say that when the width of

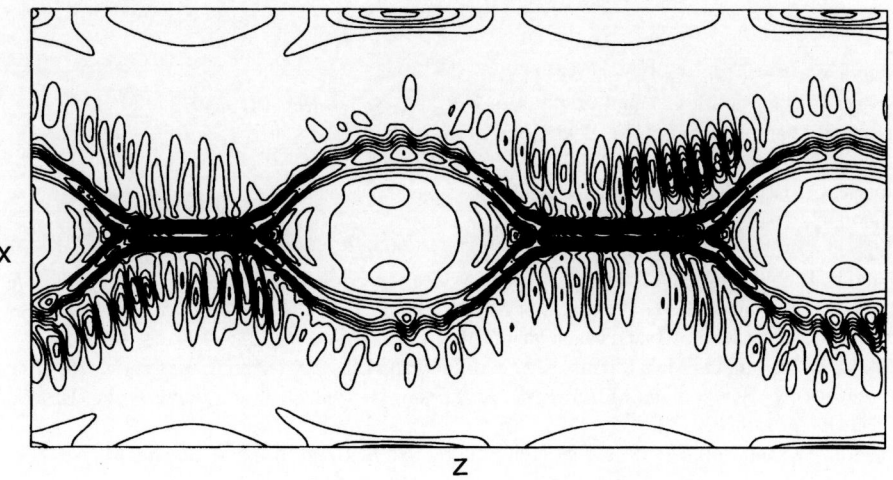

Figure 6. Current density contours taken at time 100 showing the region of most intense current which bounds the magnetic bubble/island region (Stasiewicz *et al.*, 2001).

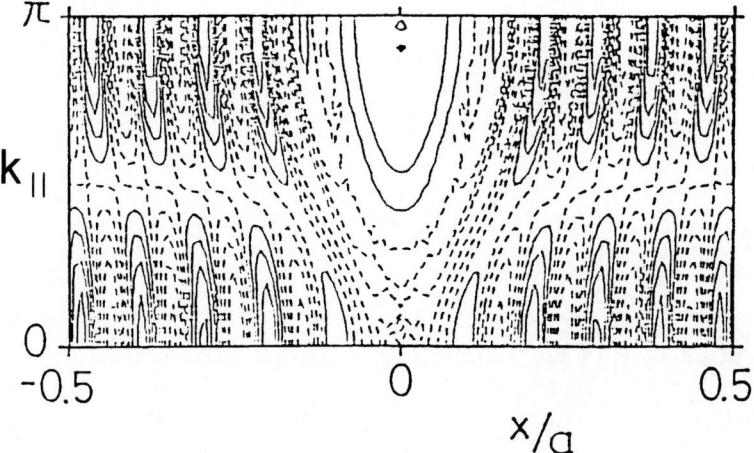

Figure 7. Time history of the current density contours at one oscillation period (Nagasaki and Itoh, 1991).

the island is wide compared to the current layer, the surface waves interacts with the islands. The numerical simulation results for the time history of the current density contours at one oscillation period are shown in Figure 7. Comparing these with the Figure 6, it appears that the bubbles were observed during the decay process of the magnetic islands formed by reconnection process.

References

Uberoi, C.: 1994, *Plasma Physics* **52**, 215.

Uberoi, C., Lanzerotti, L. J. and Wolfe, A.: 1996, *J. Geophys. Res.* **101**, 24979.

Hahm, T. S. and Kulsrud, R. M.: 1985, *Phys. Fluids* **28**, 2412.

Uberoi, C. and Zweibel, E. G.: 1999, *J. Plasma Phys.* **62**, 345.

Uberoi, C.: 1995, *Physica Scripta* **T60**, 20.

Uberoi, C.: 1989, *J. Geophys. Res.* **94**, 6941.

Ding, D. Q., Lee, L. C. and Kennel, C. F.: 1992 *J. Geophys. Res.* **97**, 8257.

Song, P., Elphic, R. C. and Russell, C. T.: 1988, *Geophys. Res. Lett.* **15**, 744.

Russel, C. T., Kawno, Le H., Petninec, S. M. and Zhang, T. I.: 1997, *Adv. Space Res.* **19**, 1913.

Uberoi, C., Lanzerotti, L.J. and Maclennan, C.G.: 1999, *J. Geophys. Res.* **104**, 25, 153.

Prikryl, P., Provan, G., McWilliams, K.A. and Yeoman, T.K.: 2002, *Ann. Geophysicae* **20**, 161.

Stasiewicz, K., Seyler, C.F., Mozer, F.S., Gustafsson, G., Pickett, J. and Popielawska, B.: 2001, *J. Geophys. Res.* **106**, 29, 503.

Nagasaki, K. and Itoh, K.: 1991, Research Report, Nat. Inst. for Fusion Sc., Japan, May 1991.

SOME ASPECTS OF THE LOW LATITUDE GEOMAGNETIC RESPONSE UNDER DIFFERENT SOLAR WIND CONDITIONS*

U. VILLANTE, P. FRANCIA, M. VELLANTE and P. DI GIUSEPPE

Dipartimento di Fisica, Università and Area di Ricerca in Astrogeofisica, L'Aquila, Italy.

Abstract. We review some aspects of low latitudes (L≤2) geomagnetic field variations associated with magnetospheric pulsations as well as with continuous and impulsive variations of the solar wind (SW) pressure.

Key words: impulsive variations, magnetospheric pulsations

1. Introduction

Measurements of the geomagnetic field variations in a wide frequency band represent a useful tool for monitoring several dynamical processes which originate from the interaction between the SW flow and the magnetosphere. At low latitudes (L ≤ 2) they are particularly important in that they also allow to monitor the magnetospheric (and ionospheric) conditions in a region of space where spacecraft observations, in general, are not available. In the present paper we review some experimental aspects of the low latitude geomagnetic field variations (approximately between sec. and several min.) under different SW conditions. Since other papers in this meeting are addressed to the storm and substorm aspects, this review mostly looks at 'closed' magnetospheric conditions, i.e. northward interplanetary magnetic field (IMF) orientation. Most results summarized hereafter have been obtained at L'Aquila (AQU, Italy, corr. geom. lat. 36.2°, L = 1.55). For previous reviews of low latitude pulsations the reader is referred to Yumoto (1986), Menk *et al.* (1994), Takahashi (1998), Villante and Vellante (1997) and for the aspects of the geomagnetic response to impulsive SW variations to Nishida (1978), and Araki (1994).

2. Mid-frequency Pulsations (15–100 mHz)

A familiar manifestation of the interaction between the SW and the magnetosphere is represented by the occurrence of geomagnetic pulsations. Typically, at low latitudes, they are a Pc3 (10–45 s) daytime phenomenon characterized by two

*This research activity is supported by Ministero dell'Istruzione, dell'Università e della Ricerca.

dominant oscillation modes which can be respectively interpreted in terms of a magnetospheric manifestation of an external driving source ('upstream' waves), and of a signal amplification ('resonant' waves) at the frequency of the fundamental oscillation of the local field line (Villante and Vellante, 1997). By long term analysis (1985–1994), we identified a clear solar cycle variation of both dominant frequencies: in particular, the decreasing frequency of the resonant mode (f_r, Figure 1a) with the increasing solar cycle is consistent with the expected general increase of the plasmaspheric density along the field line. On the other hand, many aspects of non resonant pulsations appear to be regulated by the SW parameters. In this sense, the relationship between frequency (f_d, Figure 1a) and IMF strength, the correspondence between power and SW velocity (Figure 1b), the maximum occurrence when the IMF is close to the radial direction (Figure 1c), are basic arguments in support of the idea that these pulsations are mostly related to ion cyclotron waves generated upstream of the bow shock that penetrate deep into the magnetosphere (Figure 1d). Two mechanisms for the entry of the ULF wave energy into the magnetosphere have been proposed. One mechanism suggests a direct transmission from the subsolar bow shock (Russell et al., 1983), while the other mechanism suggests an indirect process involving particle and current modulation at the near-cusp ionosphere (Engebretson et al., 1991). In this sense a comparative study of Pc3 occurrences at different latitudes and under different IMF conditions could be helpful for evaluating the relative importance of these two mechanisms.

The polarization characteristics provide useful indications on the propagation aspects in that for a wave mode propagating westward (eastward), the sense of polarization in the horizontal plane would be LH (RH, Figure 2a). The expected reversal of the polarization pattern around noon was observed at low latitudes by Lanzerotti et al. (1981), Saka and Kim (1985), Ansari and Fraser (1986). Recently, we conducted a statistical analysis over 15 years and found four polarization reversals (Figure 2b), at $\sim 0LT$, $\sim 4LT$, $\sim 11LT$, $\sim 21LT$. It suggests the presence of two dominant pulsation sources, one located 1–2 hours before noon and the other around midnight. Obviously, the daytime source can be mostly related with the penetration of upstream waves on the morning side of the magnetosphere (Figure 1d). To make more clear this aspect, we investigated the effects of the IMF orientation and found (Figure 2c) that for a radial IMF orientation the polarization reversal occurs closer to noon, as expected for a more symmetric wave penetration. The nighttime reversals are likely related to the less frequent occurrence of transient events generated in the magnetotail during substorm onset and propagating Earthward. These irregular events (Pi2, 40–150 s), in general, show maximum amplitudes at ground auroral latitudes, as expected for a source related to a high latitude current system close to the region of substorm enhanced electrojets. At low latitudes Pi2 were investigated in particular by Li et al. (1998) who interpreted their characteristics in terms of cavity/waveguide modes resonances, while higher latitudes Pi2 are most likely interpreted in terms of field line mode resonances.

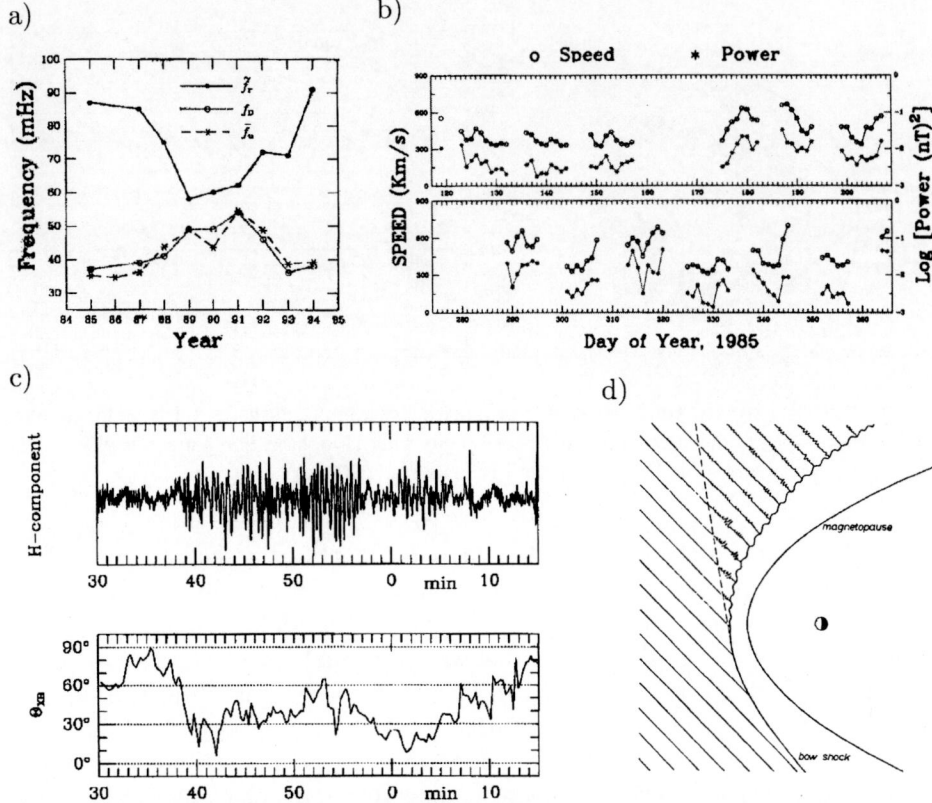

Figure 1. a) The solar cycle variation of the frequency of the 'resonant' mode (f_r), of the 'upstream' mode (f_d) and its predicted value (f_u, which is proportional to the IMF strength). (Vellante *et al.*, 1996). b) A comparison between pulsations power and SW speed (Yedidia *et al.*, 1991). c) A comparison between pulsation activity and cone angle (Vellante, 1993). d) Under nominal conditions, upstream waves are generated on the morning side by protons reflected off the bow shock along the spiral IMF lines. Obviously, for a radial IMF orientation there is a wider foreshock region symmetric around the subsolar point.

3. Low Frequency Variations (0.5–10 mHz)

The physical characteristics of Pc5 pulsations (150–600 s) at L = 1.1–2.3 were investigated by Ziesolleck and Chamalaun (1993) who found that the frequency of selected signals was independent of latitude and LT, while amplitude significantly decreased with decreasing latitude. They considered the experimental observations consistent with signatures of global compressional modes or large scale cavity resonances. An interesting aspect of low latitude observations is represented by the occurrence of power enhancements at discrete frequencies (for example, ∼1.1, 1.7, 2.3, 2.8 and 3.7 mHz, Figure 3a) which appear on horizontal components in the statistical analysis of the daytime power spectra (Francia and Villante, 1997; Villante *et al.*, 2001; see Takahashi, 1997, for a comprehensive review). These

a)

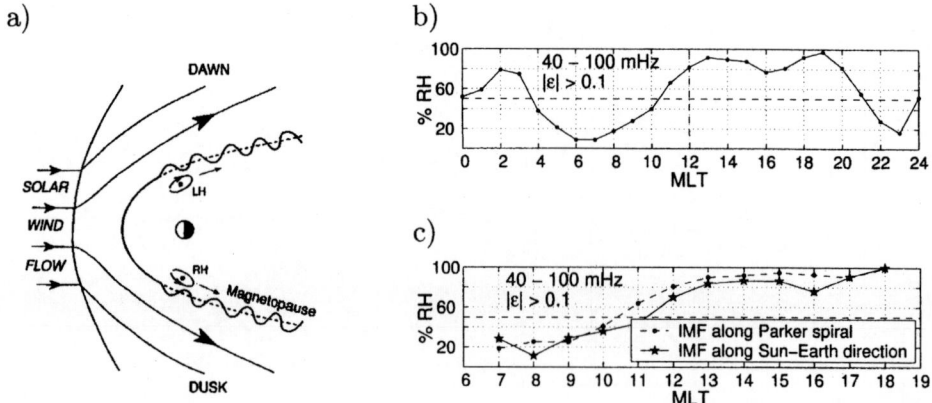

b)

c)

Figure 2. a) The expected polarization for a symmetric generation (left-handed, LH, in the morning, right-handed, RH, in the afternoon, Hughes, 1994). b) Diurnal polarization pattern at AQU. c) The same under radial and spiral IMF conditions (Vellante *et al.*, 2002b)

discrete frequencies are approximately the same identified in the F-region drift velocities and in the geomagnetic field components at auroral latitudes (Ruohoniemi *et al.*, 1991; Samson *et al.*, 1991, 1992; Walker *et al.*, 1992; Ziesolleck and McDiarmid, 1994, Mathie *et al.*, 1999). Moreover, in several cases, the same oscillation mode was simultaneously detected in the magnetosphere as well as at low and Antarctic latitudes (Figure 3d). According to theoretical models (Radoski, 1974; Kivelson and Southwood, 1985, 1986; Samson *et al.*, 1992), these fluctuations have been generally interpreted in terms of ground signatures of magnetospheric cavity/waveguide compressional modes driven by SW pressure pulses. In this sense, the much clearer statistical evidence in the afternoon sector (Figure 3b) and during higher pressure SW conditions (Figure 3c) suggests to relate the onset of low latitude fluctuations with corotating higher pressure SW structures impinging the postnoon magnetosphere. On the other hand, at high latitudes, discrete frequency fluctuations are mostly observed in the morning and late afternoon (Mathie *et al.*, 1999, Mathie and Mann, 2000). It suggests to relate (Mann *et al.*, 1999) the high latitude morning events with fluctuations driven by magnetopause shear instabilities which superpose on a background of impulsively SW driven events which occur on a wider LT range. Consistently, at low latitudes the polarization pattern of low frequency pulsations, in general, is mostly LH, and a RH polarization becomes dominant from 17 to 21 LT (Figure 3e). Conversely, at subauroral latitudes (Samson, 1972) the polarization reverses around noon, as expected for a wave propagation along the magnetopause flanks (Figure 2a).

On the other hand, we also found several cases in which more irregular, continuous variations of the SW pressure found an impressive correspondence in the low latitude H variations on time scale of several min. (Figure 4). It indicates that the geomagnetic field behavior closely reflects the impact of small amplitude pressure variations on the magnetopause. It implies a highly dynamic magnetopause

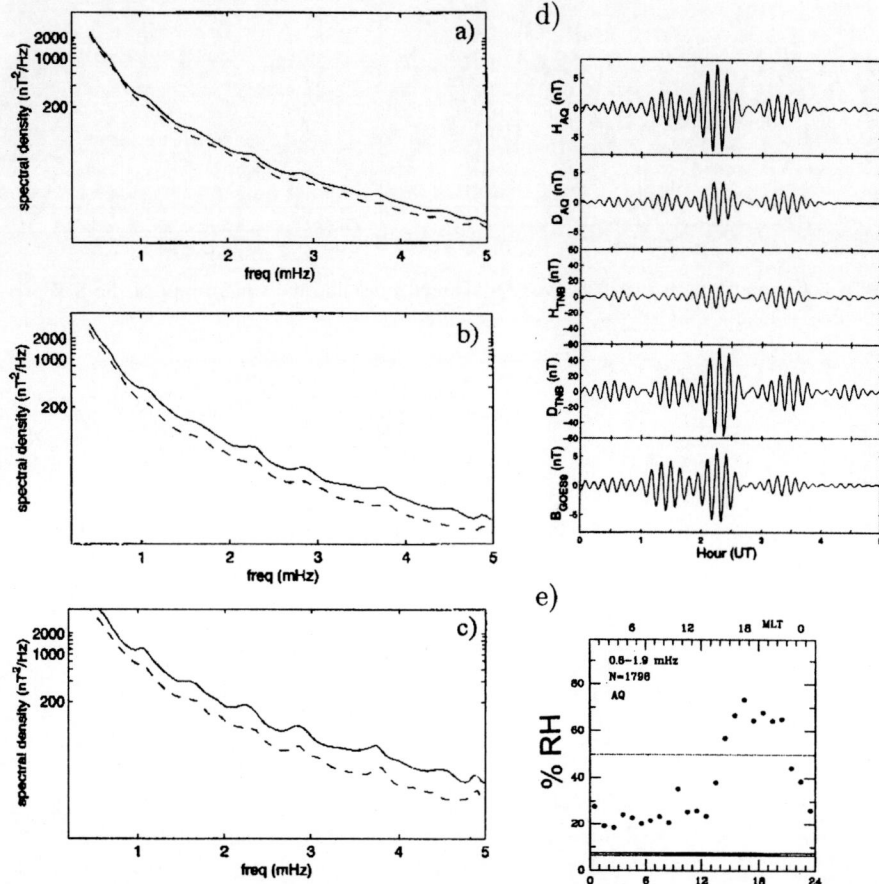

Figure 3. a) Daytime power spectra of the H (solid line) and D component (dotted line) from AQU. b) The same for postnoon intervals. c) The same for high SW pressure. d) The 1.8 mHz filtered data for an event observed at AQU (H and D, top plots), Terra Nova Bay (Antarctica, H and D, middle plots) and at geostationary orbit (field magnitude, bottom plot).(Villante *et al.*, 2001). e) The distribution of the percentage of 3-hr intervals with RH polarization at AQU (Lepidi *et al.*, 1999).

which responds rapidly to external changes, and also suggests SW fluctuations as an additional source also for more regular ground pulsations. Kepko *et al.* (2002) recently reported two events in which enhanced power at discrete frequencies was simultaneously identified in the SW dynamic pressure and in the magnetospheric field (see also Sibeck *et al.*, 1989; Korotova and Sibeck, 1995; Matsuoka *et al.*, 1995). We compared one of these events with low latitude ground measurements and identified the lower frequency enhancements as possible common features of the interplanetary, magnetospheric and geomagnetic field observations (Figure 5a). Nevertheless, other SW and magnetospheric peaks occurring at higher frequencies did not find correspondence in ground measurements. Future investigations will be then important to ascertain the correspondence between external, magnetospheric

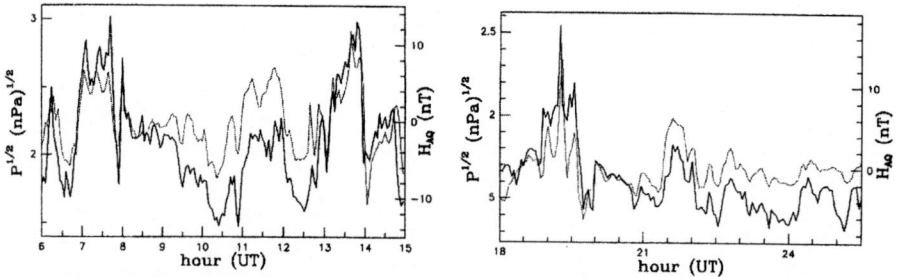

Figure 4. Comparisons between H (at AQU, dotted line) and the square root of the SW pressure (solid line) (Francia *et al.*, 1999).

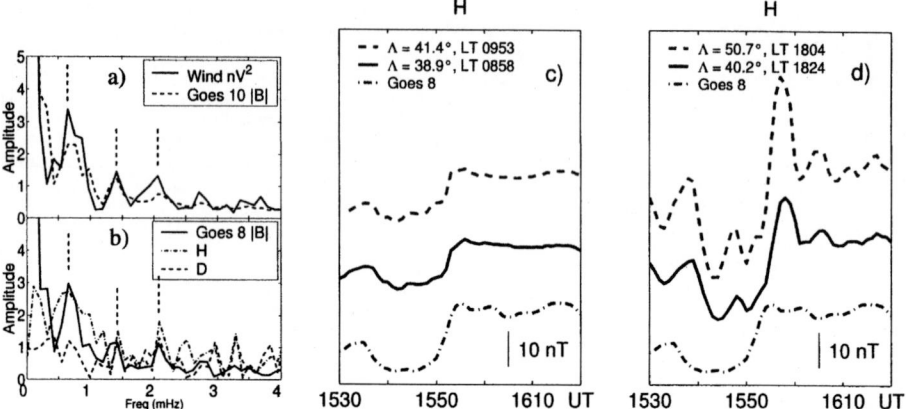

Figure 5. a) Spectral analysis for Goes 10 and Wind. b) Same analysis for Goes 8 and ground observations at AQU. Ground amplitudes have been multiplied by a factor 10 (Villante *et al.*, 2002). c) An example of step like ground response to impulsive variation of the SW pressure at two different stations compared with the magnetospheric response. d) An example of overshoot (Villante and Di Giuseppe, 2002).

and geomagnetic wave modes at different frequencies and LT, as well as for a better understanding of the penetration or triggering mechanisms.

Impulsive variations of the geomagnetic field (SIs) are known to be typically related with the Earth's arrival of impulsive SW pressure variations; they mainly consist of sharp H variations, whose amplitude has been found roughly proportional to the variation of the square root of the SW pressure (Nishida, 1978). In the last several years, the SI characteristics at different latitudes and LTs have been discussed in a number of papers (Le *et al.*, 1993; Russell *et al.*, 1992, 1994; Russell and Ginskey, 1993, 1995; Francia *et al.*, 2001). The emerging consensus suggests that the geomagnetic variation at low latitudes mostly consists of a simple rise of H to a maximum value which is followed by an asymptotic decay to a new steady state (Figure 5c). In addition, an overshoot often occurs (Figure 5d). Its appearance was interpreted in terms of an overcompression of the magnetosphere which occurs when the magnetosphere is more elastic, i.e. when the ring current is less intense

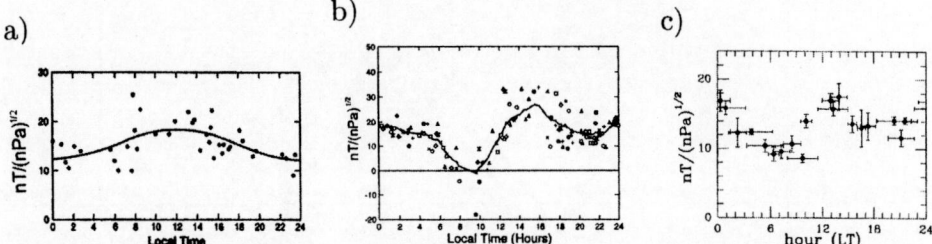

Figure 6. a) The LT dependence of the geomagnetic response at 15°–30° (Russell *et al.*, 1993); b) at 55° (Russell and Ginskley, 1995); c) at 36° (Francia *et al.*, 2001).

(Russell and Ginskey, 1993). The steady state response (i.e. the ratio between the H variation and the square root of the SW pressure) at 15° to 30° latitude shows maximum values around noon (Figure 6a). At ∼55° it shows a different pattern with strongly depressed values in the morning and enhanced values in the afternoon (Figure 6b). Recently, we observed a similar behavior also at ∼36° (Figure 6c). According to Russell and Ginskey (1995), their results might be interpreted in terms of an additional 12-hr disturbance centered at noon and related to a steady polar double-cell convection system due to high latitude reconnection with northward IMF. Similarly, Tsunomura (1998), analyzing SI data at low and middle latitudes, found a morning depression (more pronounced with increasing latitude) that was interpreted in terms of ionospheric current systems of polar origin.

A clear example of the LT dependence of the geomagnetic response was provided by analysis of an event occurring during the recovery phase after the day the SW almost disappeared (May 10–12, 1999; Villante and Di Giuseppe, 2002). Indeed, as shown in Figure 7, the correlation coefficient ρ between the magnetospheric field (B) and H reveals, in the morning hemisphere, a sharp variation across an oblique line that separates a region where ρ is highly significant, from a region where H is not correlated with B. It suggests that in the prenoon sector the ionospheric contributions play a major role also at low latitudes, where they suppress, in a wide quadrant, the correspondence between B and H. On the other hand ρ attains high values in the night sector, suggesting an explicit correspondence of the H variations with magnetospheric sources. Figure 8 provides a summary of the SI characteristics at different LTs, at ∼37° −39° latitude. On the left box we also show the SI characteristics at ∼49°, in the prenoon sector. As can be seen, in the night and afternoon sector, as well as below the separation line in the morning sector, H basically consists of a step like variation. Conversely, above the line, it is better represented by a two pulse structure, in which a positive variation precedes a negative variation. When compared with the Dsc = DL + DP decomposition (Dsc being the global SI disturbance; DL, the step like variation due to the magnetospheric compression; DP, the two pulse structure due to the ionospheric current system, Araki, 1994), the H behavior might be generally interpreted in terms of a dominant effect of the DL field. Above the separation line, the two pulse structure

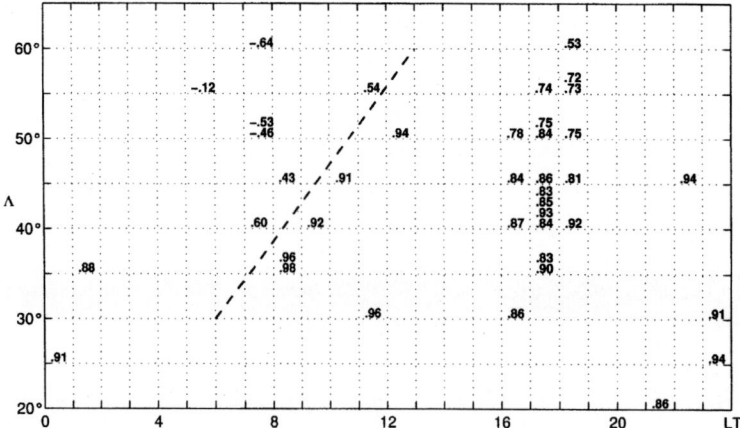

Figure 7. The correlation coefficient between the magnetospheric field and H at different latitudes and LT. (Villante and Di Giuseppe, 2002).

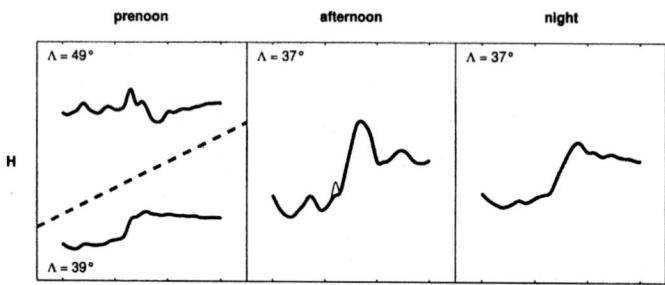

Figure 8. SI characteristics in different LT sectors (Villante and Di Giuseppe, 2002).

might be better interpreted in terms of a dominant DP contribution. Theoretical models predict a dominant DP contribution at higher (auroral) latitudes, and later LTs (Araki, 1994). However, in the present case, the DP field is observed at least up to ~43° and in the 7–8 LT sector. This reinforces the conclusion in favor of a major effect of the high latitude ionospheric currents on the low latitude observations in the prenoon quadrant.

According to early investigations (Wilson and Sugiura, 1961), from low to high latitudes in the northern hemisphere, the polarization of the SI field is elliptical and the polarization sense is RH between 22–10 LT, and LH between 10–22 LT. This conclusion was criticized by Matsushita (1962, see also Araki and Allen, 1982), who reported that only two fifths of SIs indicated elliptical polarization, and only half of those agree with the proposed pattern. We found approximately the same polarization for the SI field as for Pc5 pulsations (Figure 3e, Francia *et al.*, 2001), i.e. consistent with corotating interplanetary discontinuities impacting the postnoon magnetosphere, an aspect that might also interpret the more frequent and explicit occurrence of the overshoot enhancements in the afternoon sector (Francia *et al.*, 2001, Villante and Di Giuseppe, 2002).

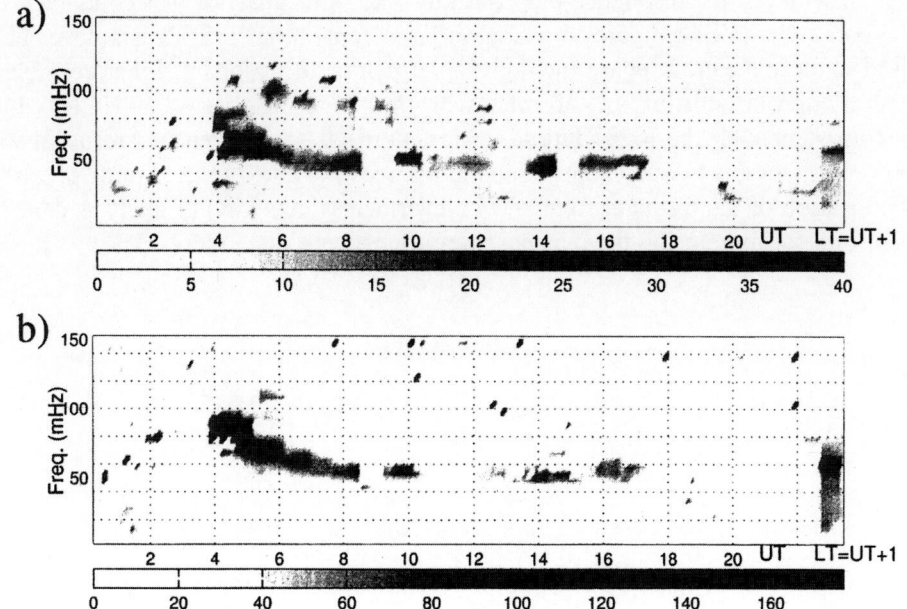

Figure 9. Dynamic cross-spectral analysis of the H component. a) The phase difference between WFB and NCK. b) The same for AQU and WFB (Vellante *et al.*, 2002a).

4. SEGMA: A New Meridional Magnetometer Array

Geomagnetic arrays are important for several aspects of the magnetospheric dynamics. In particular, they allow to conduct high resolution investigations of different scale phenomena, discriminating among temporal and spatial variations and also represent an important ground support to space missions. Dense arrays of stations presently exist at high, auroral, and subauroral latitudes while at lower latitudes the only existing array is the '210° magnetic meridian chain' which also includes the eastern Australian chain. In this sense the new South Europe Geo-Magnetic Array (SEGMA), recently installed by our group in cooperation with the Space Research Institute (IWF) of Graz (Austria), represents a useful opportunity for developing accurate research activity at low latitudes. Presently, SEGMA consists of four stations: Nagycenk/Hungary (NCK, L = 1.86), Wolfgruben/Austria (WFB, L = 1.82), Castello Tesino/Italy (CST, L = 1.75), and AQU. An example of the preliminary results is in Figure 9a which shows the cross-phase dynamic spectrum for the pair WFB-NCK. As can be seen, a clear resonant signature is identified almost continuously in the daytime hours and a decrease of the resonance frequency f_r (from ~80 mHz to ~50 mHz) can be noted in the early morning hours. Likely, it reflects the dynamic ion mass loading effect which takes place in the ionosphere at sunrise (Poulter *et al.*, 1988; Waters *et al.*, 1994). A signature of the 2nd harmonic is also noticeable (0600–0900 LT). Maximum phase differences are of ~20° which, given the latitudinal separation of ~0.6°, correspond to a spatial width of the resonance of ~200 km. The same analysis was conducted for the pair AQU-WFB (Figure 9b). Such a large separation (~5.9°) has never been used so far in the cross-phase method. Nevertheless, the resonant signature is clear, with an upward shift of ~5–10 mHz with respect to the pair WFB–NCK, which is consistent with the lower latitude. Maximum phase differences are usually of 60°–90°. Obviously, such a large separation does not allow to determine precisely the location of the resonant frequency. Nevertheless, this kind of analysis can still be useful for monitoring the plasmasphere-ionosphere dynamics.

References

Ansari, I. A. and Fraser B. J.: 1986, *Planet. Space Sci.* **34**, 519.

Araki, T.: 1994, *Geophys. Monogr. Ser., AGU* **81**, 183.

Araki, T. and Allen, J. H.: 1982, *J. Geophys. Res.* **87**, 5207.

Engebretson, M. J., Cahill Jr., L. J., Arnoldy, R. L., Anderson, B. J., Rosenberg, T. J., Carpenter, D. L., Inan, U. S. and Eather, R. H.: 1991, *J. Geophys. Res.* **96**, 1527.

Francia, P., Lepidi, S., Di Giuseppe, P. and Villante, U.: 2001, *J. Geophys. Res.* **106**, 21,231.

Francia, P., Lepidi, S., Villante, U., Di Giuseppe, P. and Lazarus, A. J.: 1999, *J. Geophys. Res.* **104**, 19,923.

Francia, P. and Villante, U.: 1997, *Annales Geophysicae* **15**, 17.

Hughes, W. J.: 1994, *Geophys. Monograph Ser., AGU* **81**, 1.

Kepko, L., Spence, H. E. and Singer, H. J.: 2002, *Geophys. Res. Lett.* **29**, 1197.

Kivelson, M. and Southwood, D.: 1985, *Geophys. Res. Lett.* **12**, 49.

Kivelson, M. and Southwood, D.: 1986, *J. Geophys. Res.* **91**, 4345.

Korotova, G. I. and Sibeck, D. G.: 1995, *J. Geophys. Res.* **100**, 35.

Lanzerotti, L. J., Medford, L. V., Mclennan, C. G. and Hasegawa, T.: 1981, *J. Geophys. Res.* **86**, 5500.

Le, G., Russell, C. T., Petrinec, S. M. and Ginskey, M.: 1993, *J. Geophys. Res.* **98**, 3983.

Lepidi, S., Francia, P., Villante, U., Lanzerotti, L. J. and Meloni, A.: 1999, *J. Geophys. Res.* **104**, 305.

Li, Y., Fraser, B. J. and Menk, F. W.: 1998: *J. Geophys. Res.* **103**, 2343.

Mann, I. R., Wright, A. N., Hills, K. and Nakariakov, V. H.: 1999, *J. Geophys. Res.* **104**, 333.

Mathie, R. A., Mann, I. R., Menk, F. W. and Orr, D.: 1999, *J. Geophys. Res.* **104**, 7025.

Mathie, R. A. and Mann, I. R.: 2000, *J. Geophys. Res.* **105**, 10,713.

Matsuoka, H., Takahashi, K., Yumoto, K., Anderson, B. J. and Sibeck, D. G.: 1995: *J. Geophys. Res.* **100**, 12,103.

Menk, F. W., Fraser, B. J., Waters, C. L., Ziesolleck, C. W. S., Feng, Q., Lee, S. H. and McNann, P. W.: 1994, *Geophys. Monogr. Ser., AGU* **81**, 299.

Nishida, A.: 1978, *Geomagnetic Diagnosis of the Magnetosphere*, Springer Verlag, Berlin.

Poulter, E. M., Allan, W. and Bailey, G. J.: 1988, *Planet. Space Sci.* **36**, 185.

Radoski, H. R.: 1974, *J. Geophys. Res.* **79**, 595.

Ruohniemi, J. M., Greenwald, R. A., Baker, K. B. and Samson, J. C.: 1991, *J. Geophys. Res.* **96**, 15,697.

Russell, C. T. and Ginskey, M.: 1993, *Geophys. Res. Lett.* **20**, 1015.

Russell, C. T. and Ginskey, M.: 1995, *J. Geophys. Res* **100**, 23,695.

Russell, C. T., Ginskey, M. and Petrinec, S. M.: 1994, *J. Geophys. Res.* **99**, 253.

Russell, C. T., Ginskey, M., Petrinec, S. M. and Le, G.: 1992, *Geophys. Res. Lett.* **19**, 1227.

Russell, C. T., Luhmann, J. G., Odera, T. J. and Stuart, F. W.: 1983, *Geophys. Res. Lett.* **10**, 663.

Saka, O. and Kim, J. S.: 1985, *Planet. Space Sci.* **33**, 1073.

Samson, J. C.: 1972, *J. Geophys. Res.* **77**, 6145.

Samson, J. C., Greenwald, R. A., Ruohoniemi, J. M., Hughes, T. J. and Wallis, D. D.: 1991, *Can. J. Phys.* **69**, 929.

Samson, J. C., Harrold, B. G., Ruohoniemi, J. M., Greenwald, R. A. and Walker, A. D. M.: 1992, *Geophys. Res. Lett.* **19**, 441.

Sibeck, D. G., Baumjohann, W., Elphic, R. C., Fairfield, D. H., Fennell, J. F., Gail, W. B., Lanzerotti, L. J., Lopez, R. E., Luehr, H., Lui, A. T. Y., Mclennan, C. G., McEntire, R. W., Potemra, T. A., Rosenberg, T. J. and Tkahashi, K.: 1989, *J. Geophys. Res.* **94**, 2505.

Takahashi, K.: 1998, *Ann. Geophysicae* **16**, 787.

Tsunomura, S.: 1998, *Earth Planets Space* **50**, 755.

Vellante, M.: 1993, *Annali di Geofisica* **5**, 36.

Vellante, M., De Lauretis, M., Villante, U., Adorante, N., Piancatelli, A., Schwingenschuh, K., Magnes, W., Koren, W. and Zhang, T. L.: 2002a, *Solar Cycle and Space Weather Conference*, European Space Agency, SP-477, p. 491.

Vellante, M., Villante, U., De Lauretis, M. and Barchi, G.: 1996, *Geophys. Res. Lett.* **23**, 12.

Vellante, M., Villante, U., Santarelli, L. and De Lauretis, M.: 2002b, *Solar Cycle and Space Weather* Conference, European Space Agency, SP-477, p. 487.

Villante, U., and Di Giuseppe, P.: 2002, sub. *Ricerca Solare Italiana*.

Villante, U., Di Giuseppe, P. and Francia, P.: 2002, sub. *Ricerca Solare Italiana*.

Villante, U., Francia, P. and Lepidi, S.: 2001, *Ann. Geophysicae* **19**, 321.

Villante, U. and Vellante, M.: 1997, *Solar System Plasma Physics*, Società Italiana di Fisica **56**, 189.

Walker, A. D. M., Ruohoniemi, J. M., Baker, K. B., Greenwald, R. A. and Samson, J. C.: 1992, *J. Geophys. Res.* **97**, 12,187.

Waters, C. L., Menk, F. W. and Fraser, B. J.: 1994, *J. Geophys. Res.* **99**, 17,547.

Wilson, C. R. and Sugiura, M.: 1961, *J. Geophys. Res.* **66**, 4097.

Yedidia, B. A., Vellante, M., Villante, U. and Lazarus, A. J.: 1991, *J. Geophys. Res.* **96**, 3465.

Yumoto, K.: 1986, *J. Geophys.* **60**, 79.

Ziesolleck, C. W. S. and Chamalaun, F. H.: 1993, *J. Geophys. Res.* **98**, 13,703.

Ziesolleck, C. W. S. and McDiarmid, D. R.: 1994, *J. Geophys. Res.* **99**, 5817.

SHOCK WAVE INTERACTION WITH THE MAGNETOPAUSE

C. C. WU

Institute of Geophysics and Planetary Physics, University of California,
Los Angeles, CA 90095, U.S.A.

Abstract. The magnetopause is in continuous motion and shock waves and impulsive acceleration events can occur. As an example, we show that the interaction of an interplanetary shock with the bow shock can generate a shock wave that after passing through the magnetosheath can interact with the magnetopause. In fluid dynamics, when a shock wave encounters a fluid discontinuity, the interface may become unstable and form bubbles and spikes. We consider this Richtmyer–Meshkov instability in magnetohydrodynamics. At the dayside magnetopause, the instability tends to be stabilized by the magnetic field. However, the shock wave interaction can initiate magnetic field reconnection for the southward IMF, which may be important in strong interplanetary shock events.

Key words: magnetopause, MHD, Richtmyer-Meshkov instability, shock waves

1. Introduction

In fluid dynamics, when a shock wave encounters a fluid discontinuity, small perturbations at the fluid interface can grow into nonlinear structures in the form of 'bubbles' and 'spikes'. This Richtmyer–Meshkov (RM) instability (Richtmyer, 1960; Meshkov, 1970) is closely related to the familiar Rayleigh–Taylor (RT) instability (Lord Rayleigh, 1990; Taylor, 1950), which occurs when a heavy fluid is supported by a lighter fluid. However, the RM and RT instabilities are qualitatively different. In the RT instability a sustained acceleration causes the perturbation amplitude to grow exponentially in time, whereas in the RM instability the shock acceleration is impulsive and causes the amplitude to grow linearly in time. In addition, an RT instability can occur only if light fluid is accelerated into heavy fluid, but an RM instability can work in either direction.

Recently, Wu and Roberts (1999) and Wu (2000) considered the RM instability in magnetohydrodynamics (MHD) and discussed its possible role in the dynamics of the magnetosphere. In MHD there are two types of fluid discontinuities: a tangential discontinuity (TD) if the component of the magnetic field normal to the interface (B_n) vanishes and a contact discontinuity (CD) if B_n is nonzero. These interfaces exist in many astrophysical and space situations, such as the magnetopause and the tail current sheet in the magnetosphere.

The magnetic field plays an important role in the RM instability. When the magnetic field is transverse to the surface perturbations, the RM instability can lead to the formation of spikes and bubbles as in the hydrodynamic situation, and when the magnetic field is along the direction of surface perturbations, the magnetic field tends to stabilize the instability. For the latter situation, the evolution depends on whether or not the magnetic field changes sign at the fluid interface. If the sign change occurs, a magnetic field reconnection process can be initiated at the interface upon the interaction with the shock wave.

Thus, for the magnetopause, one may expect the following: The RM instability is stronger in the east-west direction than it is in the north-south direction. The instability can have a large effect at the flanks. In addition, in the southward IMF (interplanetary magnetic field) condition, the instability can cause magnetic field reconnection at the magnetopause.

In Section 2, as a review, a set of numerical examples is presented to illustrate the effects of the magnetic field on the RM instability. In Section 3, we show that the interaction of an interplanetary shock with the bow shock can generate a shock wave that passes through the magnetosheath and then interacts with the magnetopause.

2. Effects of Magnetic Field on RM Instability

As in Wu and Roberts (1999), we model the magnetopause as a TD, across which the total pressure (p_T) is constant. Plasmas are assumed to obey perfect gas laws with adiabatic constant $\gamma = 5/3$. The discontinuity in pressure p and magnetic field B is located at $x = 0$. The density ρ is initially perturbed so that the interface for ρ is $x_0 = a_0 \sin ky$ with $a_0 = 2$, $k = 2\pi/L$, and $L = 10$. Table 1 shows the initial configurations. The region $x > 0$ (for p and B) or $x > x_0$ (for ρ) is denoted by R (right), the region in $-10 < x < 0$ (p and B) or $-10 < x < 0$ (ρ) is denoted by L (left), and the area where $x < -10$ is denoted by D (downstream of the shock). In Table 1, for the transverse configuration, $B = B_z$ and for the parallel configuration, $B = B_y$. For the anti-parallel configuration, $B_y = \sqrt{2}$ in R, -0.8 in L, and -1.79 in D. At time $t = 0$, a shock wave is initiated at $x = -10$ with a fast Mach number $M_f = 2$, which moves with a shock speed of 2.62 towards the magnetopause. (Here we use a stronger shock than that in the calculations of Wu and Roberts (1999), since, as we'll show in Section 3, a strong shock can indeed interact with the magnetopause.) This is to simulate a fast shock crossing from the magnetosheath on to the magnetosphere. The states L and R define the relative physical quantities such as sound speed, Alfvén speed and temperature. The 5th-order upwind MHD code of Jiang and Wu (1999) was used in the calculations, which cover a domain $-25 \leq x \leq 75$ and $0 \leq y \leq 10$ with 400 by 40 grid points. Periodicity is assumed in y, and values at the x boundaries are held constant because no perturbations can reach them during the runs.

If there is no perturbation in the density interface, the interaction of the shock wave with the magnetopause is a typical Riemann problem. It results in the form-

TABLE I

Initial configurations

Variables	D	L	R
ρ	4.50	2	0.1
p	8.02	1.68	1
B	1.79	0.8	$\sqrt{2}$
v_x	1.46	0	0

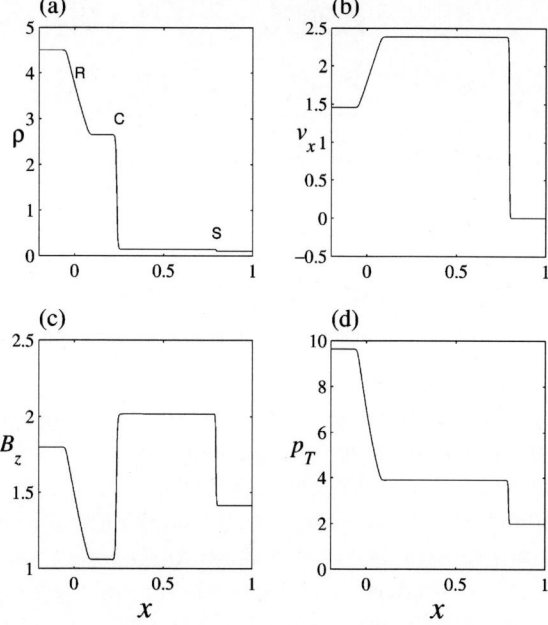

Figure 1. Riemann solution of the shock-magnetopause interaction at $t = 0.1$ after the interaction. Plot (c) is for the transverse configuration. Replace B_z by B_y for the parallel case, and replace B_z by B_y for $x > 0.238$ and by $-B_y$ for $x < 0.238$ for the anti-parallel situation.

ation of a transmitted shock wave moving into the magnetosphere with a speed of 7.96, a rarefaction wave moving back towards the magnetosheath, and the magnetopause moving with a speed, v_C, of 2.38. The solution is shown in Figure 1. This is the 'average' solution even when perturbation is included. The transmitted shock is very stable; however, the tangential discontinuity and the rarefaction can subject to instability.

With the perturbation in the density interface, the shock wave encounters the interface at different times at different y positions. Figure 2 shows the results for the transverse configuration where the initial magnetic field is in the z direction.

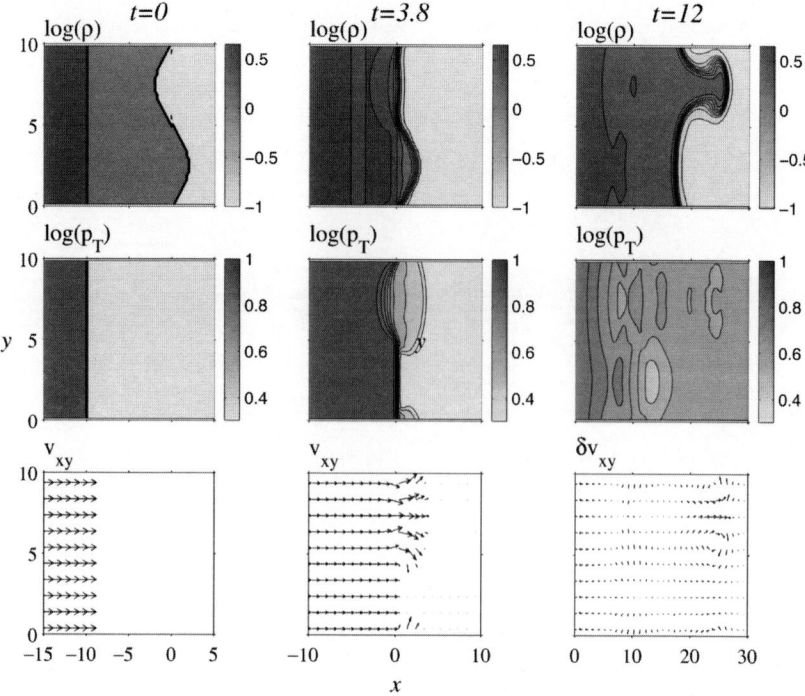

Figure 2. RM instability – transverse configuration

In this configuration, the MHD calculations are similar to that of hydrodynamical calculations and the magnetic field does not play any role. At $t = 3.8$, the shock reaches $x \sim 0$, but it has already interacted with the interface at $[x, y] \sim [-2, 7.5]$. Since the transmitted shock has a faster speed than the initial impinging shock, the transmitted shock diverges along $y \sim 7.5$ and forms a vortex flow which converges near $[x, y] \sim [0, 2.5]$. This creates a phase inversion: the low density plasma in the magnetosphere is accelerated into the high density plasma in the magnetosheath near $y \sim 2.5$ and the other way around near $y \sim 7.5$, which is out of phase with the initial perturbation. The results at $t = 12$ show the instability grows to form spikes and bubbles and the initially uniform current sheet at the interface is distorted. At that time, the transmitted shock has reached $x \sim 65$ and is not shown in the plot. The interface moves with an average velocity of v_C; the relative velocity $(v_x - v_C, v_y)$ is plotted in the lower panel. Because of the reflected rarefaction waves, the magnetosheath region is highly non-uniform; p_T varies as much as 30%. This suggests that the instability can contribute to the turbulence observed in the magnetosheath.

Figure 3 shows two cases where the magnetic field is along the y direction. Figure 3a is for the parallel configuration, where the magnetic fields in the magnetosphere and the magnetosheath point to the same y direction (for northward IMF condition), while Figure 3b is for the anti-parallel configuration, where the

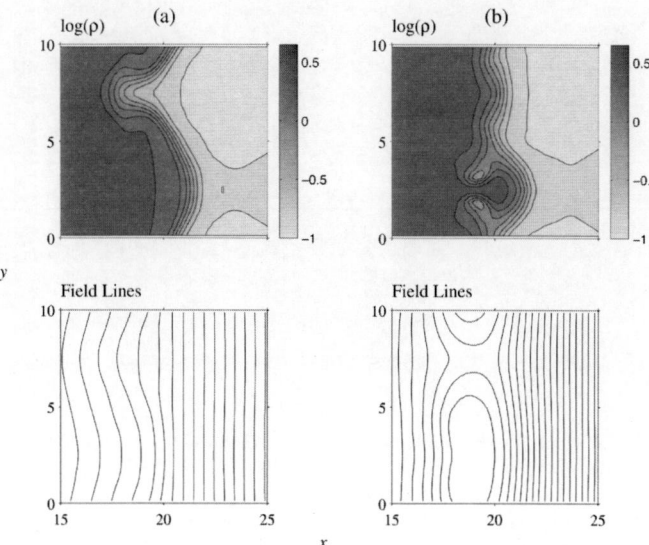

Figure 3. RM instability – results at $t = 12$ for parallel (a) and anti-parallel (b) configurations

magnetic field in the magnetosheath and that in the magnetosphere point to opposite y directions (for southward IMF condition). Because of the field line tension, the parallel magnetic field tends to stabilize the RM instability. This means that at the magnetopause, the RM instability may be less likely in the northward IMF situation.

For the anti-parallel configuration, the vortex motion at the interface tends to contract the interface (current sheet) at one place and to expand it at the other position. Since the magnetic field changes its direction through the interface, magnetic field reconnection can then occur. As shown in Figure 3b, the reconnection takes place at $(x, y) \sim (19, 7)$, where a thin current sheet is formed.

In these cases, the initial perturbation consists of a single mode. In hydrodynamics it has been shown that if the initial perturbation includes a mixture of different wavelength modes, the RM instability can lead to turbulent mixing (Youngs, 1984). This should be true at least for the transverse case.

3. Interaction of Interplanetary Shock with Bow Shock

In this section, by using a model magnetosphere, we show that the interaction of an interplanetary shock wave with the bow shock can generate a transmitted shock that increases its strength when passing through the magnetosheath and then interacts with the magnetopause. The calculation is based on hydrodynamics and is carried out on the $x - z$ plane. The magnetosphere is modeled as a circular cylinder of radius 1. The solar wind parameters are $\rho = \gamma$, with the adiabatic constant $\gamma = 1.4$, $p = 1$, and $v_x = -2$. This is a supersonic flow of Mach number $M = 2$

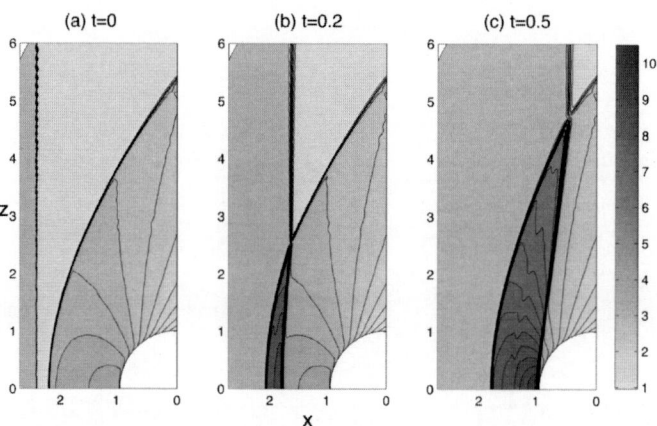

Figure 4. Interplanetary shock–magnetosphere interaction: density contours at various times.

over a circular cylinder. Figure 4a shows the resulting flow configuration using the WENO code (Jiang and Wu, 1999). The magnetosheath extends from $x = 1$ to $x = 2.26$ along the x axis. Across the bow shock, ρ jumps from $\rho = \gamma$ to 3.733, v_x from 2 to 0.75, and p from 1 to 4.5.

At $t = 0$ an $M = 2$ shock is introduced at $x = 2.5$, which is shown in Figure 4a. The downstream values of the shock are $\rho = 3.733$, $v_x = -3.25$, and $p = 4.5$. In the magnetospheric frame, the shock speed is -4. At $t = 0.04$, the shock interacts with the bow shock at $[x, z] = [2.26, 0]$. It can be regarded as a Riemann problem. The interaction produces a transmitted shock of $M = 1.73$, moving with a velocity -3 towards the cylinder, a contact discontinuity (with no jumps), moving with a velocity -2, and a reflected shock of $M = 1.73$, moving with a velocity -1 in the magnetospheric frame.

Figures 4b and 4c show further evolution of the interaction. The transmitted shock continues to move towards the magnetosphere. The reflected shock (namely the new bow shock boundary) also moves inwards. This is shown clearly in Figure 5, where the density distributions at $t = 0, 0.2$, and 0.4 along the x axis are plotted. The shock strength increases as the transmitted shock moves closer to the magnetosphere. At $t = 0.04$, the transmitted shock has $M = 1.73$ (see above). At $t = 0.2$ and 0.4, the shock strengths are $M = 1.83$ and 1.85, respectively.

In addition to interplanetary shocks, solar wind contact discontinuities may interact with the bow shock. If the solar wind density increases at the discontinuity, the interaction of it with the bow shock can generate a transmitted shock that can then interacts with the magnetopause. On the other hand, if the solar wind density decreases at the discontinuity, the interaction of it with the bow shock can create a transmitted rarefaction wave and produces no shock wave–magnetopause interaction. The inclusion of magnetic field does not alter much from these results. The MHD results will be reported elsewhere.

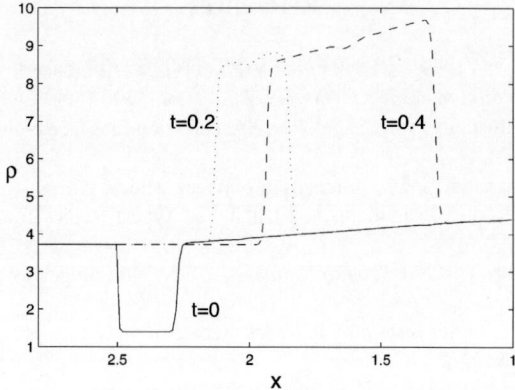

Figure 5. Interplanetary shock–magnetosphere interaction: density distributions at various times.

4. Conclusions

In this paper, we have considered the interaction of shock waves with the magnetopause and the RM instability. When the magnetic field is transverse to the surface perturbations, the RM instability can lead to the formation of spikes and bubbles as in the hydrodynamic situation. When the magnetic field is along the direction of surface perturbations, the magnetic field tends to stabilize the instability. However, the interaction has a significant effect. While the interaction does not produce much effect in a northward IMF configuration, it can initiate magnetic reconnection in a southward IMF configuration. The difference in the response between the two situations may be important in the understanding of the dynamics of the magnetosphere.

In addition, we show that as the interplanetary shock interacts with the magnetosphere, the shock wave–magnetopause interaction can occur. We expect that the stronger the interplanetary shock is, the stronger is the transmitted shock that interacts with the magnetopause. A solar wind contact discontinuity that increases the plasma density can also create the shock wave–magnetopause interaction.

The shock wave induced magnetic field reconnection process at the current sheet is likely to be applicable to other situations. For example, it may occur at the magnetotail current sheet if a shock wave can somehow interact with it. Odstrcil and Karlicky (1998) have considered such a reconnection process for current sheets in solar corona.

Acknowledgements

I wish to thank Prof. Abraham Chian for the invitation to WSEF2002. This work is supported in part by a NASA grant NAG 5-9111.

C. C. WU

References

Jiang, G. S. and Wu, C. C.: 1999, 'A High Order WENO Finite Difference Scheme for the Equations of Ideal Magnetohydrodynamics', *J. Comput. Phys.* **150**: 549–560.

Meshkov, E. E.: 1970, 'Instability of a Shock Wave Accelerated Interface between Two Gases', *NASA Tech. Trans.* **F-13**, 74.

Odstrcil, D. and Karlicky, M.: 1998, 'Interaction of Weak Shock Waves with Current Sheets in an Active Region to Produce Nanoflares and Chains of Type I Radio Bursts', *Solar Physics* **177** (1–2): 415–425.

Rayleigh, Lord (J. W. Strutt): 1990, *Scientific Papers*, Vol. 2, Cambridge University Press, New York. p. 200.

Richtmyer, R. D.: 1960, 'Taylor Instability in Shock Acceleration of Compressible Fluids', *Commun. Pure Appl. Math.* **13**: 297.

Taylor, G. I.: 1950, 'The Instability of Liquid Surfaces when Accelerated in a Direction Perpendicular to Their Planes, I.', *Proc. R. Soc. London, Ser. A* **201**: 192.

Youngs, D. L.: 1984, 'Numerical Simulation of Turbulent Mixing by Rayleigh–Taylor Instability', *Physica* **12D**: 32.

Wu, C. C. and Roberts P. H.: 1999, 'Richtmyer-Meshkov Instability and the Dynamics of the Magnetosphere', *Geophys. Res. Lett.* **26**: 655–658.

Wu, C. C.: 2000, 'Shock Wave Interaction with the Magnetopause', *J. Geophys. Res.* **105**: 7533–7543.

III: IONOSHERE/ATMOSPHERE

OPTIMAL ASSIMILATION FOR IONOSPHERIC WEATHER
Theoretical aspect

JIAN-SHAN GUO, SHE-PING SHANG, JIANKUI SHI, MANLIAN ZHANG,
XIGUI LUO and HONG ZHENG

Laboratory for Space Weather, Center for Space Science and Applied Research, Chinese Academy of Sciences, P.O.Box 8701, Beijing 100080, P.R.China (e-mail: guojs@center.cssar.ac.cn)

Abstract. Observation, specification and prediction of ionospheric weather are the key scientific pursuits of space physicists, which largely based on an optimal assimilation system. The optimal assimilation system, or commonly called data assimilation system, consists of dynamic process, observation system and optimal estimation procedure. We attempt to give a complete framework in this paper under which the data assimilation procedure carries through. We discuss some crucial issues of data assimilation as follows: modeling a dynamic system for ionospheric weather; state estimation for static or steady system in sense of optimization and likelihood; state and its uncertainty estimation for dynamic process. Meanwhile we also discuss briefly the observability of an observation system; system parameter identification. Some data assimilation procedures existed at present are reviewed in the framework of this paper. As an example, a second order dynamic system is discussed in more detail to illustrate the specific optimal assimilation procedure, ranging from modeling the system, state and its uncertainty calculation, to the quantitatively integration of dynamic law, measurement to significantly reduce the estimation error. The analysis shows that the optimal assimilation model, with mathematical core of optimal estimation, differs from the theoretical, empirical and semi-empirical models in assimilating measured data, being constrained by physical law and being optimized respectively. The data assimilation technique, due to its optimization and integration feature, could obtain better accurate results than those obtained by dynamic process, measurement or their statistical analysis alone. The model based on optimal assimilation meets well with the criterion of the model or algorithm assessment by 'space weather metrics'. More attention for optimal assimilation procedure creation should be paid to transition matrix finding, which is usually not easy for practical space weather system. High performance computing hardware and software studies should be promoted further so as to meet the requirement of large storage and extensive computation in the optimal estimation. The discussion in this paper is appropriate for the static or steady state or transition process of dynamic system. Many phenomena in space environment are unstable and chaos. So space environment study should include and integrate these two branches of learning.

Key words: data assimilation system, dynamic process, model and measurement statistic properties, optimal estimation, state space

Abbreviations: AMIE – Assimilative Mapping of Ionosphere Electrodynamics; CMP – Chinese Meridian Project; DAS – data assimilation system; ISTP – International Solar Terrestrial Program; NSWP – National Space Weather Program; OI – Optimal Interpolation; OSSE – observation system simulation experiment

Space Science Reviews **107**: 229–250, 2003.
© 2003 *Kluwer Academic Publishers.*

1. Introduction

Currently the main objective of space environment research is to understand the physical law that governs spatial and temporal variation of disturbed processes in the Sun-Earth system in view of science and to realize space weather specification and prediction in practical application. This objective is embodied adequately in some ambitions scientific programs, such as, the National Space Weather Program (NSWP), the International Solar Terrestrial Program (ISTP), and the Chinese Meridian Project (CMP) (Guo *et al.*, 1996). Some space- and ground-based exploration campaign have successfully recorded some entire process from corona mass ejections to the responses in near earth space, provided more or less ability of forecast for a period of a few days or hours (Peredo *et al.*, 1997). On the other hand numerous efforts have been made over decades in theoretical modeling in heliosphere, magnetosphere and ionosphere-thermosphere, and significant progresses have been made in understanding of the Solar-terrestrial process. However, the data procession technique for such costly experiments seems to be far from optimized and integrated. And none of the theoretical or empirical models alone seems to be able to provide specifications and predictions with practically required accuracy. Optimal assimilation technique could be a potential approach to combine the theoretical models or physical law with the coordinately measured data to provide optimized specification and forecast for space environment system. This technique has been applied in meteorology and oceanography in varies forms as early as in fifties. Currently two types of four dimensional data assimilation procedures have already been successfully used operationally and in research (Kuo *et al.*, 2001; Grell *et al.*, 1994). For space environment application the data assimilation originated from the requirements of research in responses of convection electric field in polar region to the interplanetary magnetic field disturbances. Kamide *et al.* (1981), devised to deduce electromagnetic features in polar region from ground based magnetometer data alone for the first time and got a convection electric field pattern basically consistent with that obtained by the statistical procession from the measurement. But the equivalent electric currents remain under investigation. Seven years latter Richmond and Kamide (1988) took account of of different data sets from the ground- and space-based measurements, statistical properties of the data and the physical law relating to the observables, assimilated and formed a procedure, called Assimilative Mapping of Ionosphere Electrodynamics (AMIE). Significant results have been obtained afterwards (Schunk and Sojka, 1996). One of them is the analysis of global energy deposition during the January 1997 magnetic cloud event (Lu *et al.*, 1998; Richmond and Lu, 2000). The AMIE is a steady state estimation procedure of ionospheric electrodynamics process that was named as Optimal Interpolation (OI) mathematically by Akemaev (Akmaev, 1999a, 1999b). Recently data assimilation technique has been used in ionospheric specification modeling. More and more types of data and simple dynamic transition property have been assimilated together to get the optimal state estimate and its uncertainty.

Currently some of the results have been obtained based on observation system simulation experiment (OSSE) only, it is hopefully that some of them could be possible to put into operation in the near future (Howe *et al.*, 1998; Sojka *et al.*, 2001; Kuo *et al.*, 2001; Hajj *et al.*, 2001; Ganguly and Brown, 2001). Many data assimilation offices, centers and projects begin to dedicate to the study of the technique and theory in recent years. Therefore understanding and mastering the optimal assimilation technique might be able to play a crucial role for space environment specification and prediction.

Although space physicists are attracted very much to the data assimilation technique, the technique itself is still far from mature at moment, with regard to both its theoretical aspect and practical application. Even in meteorology, operational data assimilation procedure is still very few (Schlatter, 2000). In ionospheric weather field, the effective data assimilation technique seems to be limited in the static optimal estimation in electrodynamics, the consideration of dynamic transition property just proceeds at a very first step. The theory is neither systematic nor perfect, even the nomenclatures are not satisfactorily consistent in the literature.

Observation, specification and prediction of the ionospheric weather are three crucial scientific activities for space weather community. These activities base on an estimation system that consists of three elements: ionospheric dynamic process, its observation system and an estimation procedure. We call this system the data assimilation system to keep the consistency with the previous paper. In order to estimate the state and its uncertainty of the dynamic process, a mathematical model of the dynamic system and its observation should be defined. We will discuss the elements of the data assimilation system and give a mathematical model of a specific ionospheric process in Section 2. The estimation procedure or algorithm, which provides the specification and prediction for the states and its uncertainty of the dynamic process, bases on the knowledge of dynamic law, measurement, and the statistical property of them. The dynamic process could be at static, steady or dynamic states. The estimation for the static or steady state corresponds to the estimation for time-invariant parameter and the estimation for dynamic process the state evolution. Actually, the static or steady estimation is a special case of dynamic estimation (Section 3). Second order dynamic system is used very often in ionospheric dynamics and may be familiar to most readers. It is used as an example to illustrate the important aspects of the dynamic estimation procedure (Section 4). We will summarize and discuss some issues in the Section 5. This paper has a tutorial style, some of the existed data assimilation procedures are going to be briefly reviewed on the framework provided in the paper.

2. Modeling the Data Assimilation System

The data assimilation system (DAS) consists of dynamic process subsystem, its observation subsystem and system estimation procedure. The object of the es-

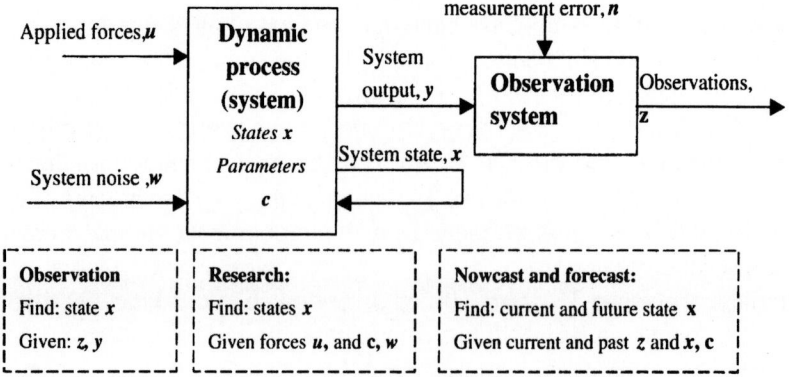

Figure 1. Dynamic system and scientific assignments.

timation is the state, its uncertainty, and sometimes the system parameter of the ionospheric dynamic process. The estimation bases on a dynamic system, which consists of dynamic process and its observation (Figure 1). The dynamic system can be mathematically expressed as a equation set in state space, which is called the mathematical model or dynamic model. These equations along with its initial and boundary conditions form an initial or boundary problem, which govern the variation of the system states in the state space. In the case of noisy internal and external environment, the state trajectory in the state space becomes uncertain. The optimal trajectory should be estimated. We will concentrate on the discussion of the modeling the dynamic system in this section, and the estimation of dynamic process in the next two sections.

The optimal assimilative estimation carries through according to the dynamic model. The system states or parameters are usually a matrix, as shown by bold letter x in Figure 1. The system behavior can also be governed by imposed forces u and dynamic system noise w. The w could include unknown physical process, truncation error and representativeness error etc. Sometimes the system state x is not necessarily the same as system output y or system parameter c. The y is the measurement undergraded by instrument noise n. System states x feedback may provide a way to modify the system and change its output to meet a system control objective requested. This is another ambitious topic for ionospheric weather we do not discuss it in this paper. Observation subsystem measures the output of the dynamic process for the usage of estimation. The goal of ionospheric weather study is to estimate system states, their evolution and uncertainty, or their parameter based on a given set of measurements z, statistics of w, n and under constraint of the dynamic equation set. The purpose of estimation is to give specification of ionospheric dynamic condition at the present (specification), the past (research) and the future (prediction). Three main assignments for ionospheric weather study, observation, research and prediction, are also shown on Figure 1. The system concepts above are very general in nature, and may be used in some other fields of nature science, even economics. It would be also very important for providing quantitatively description of extensive space weather processes in magnetosphere and heliosphere even the entire Solar-terrestrial system. In the next we are going to give a specific ionosphere dynamic process expressed in the state space for understanding the mathematical model in the concrete.

Now let us express a specific ionospheric dynamic process as a mathematical or dynamic model of the DAS. Rishbeth *et al.* (1978) gave a F_2 layer peak model to associate the electron density peak N_m with the perturbing force, the vertical drift velocity U_z, which imposed by neutral wind or electric field. The model starts with the continuity equation, results in following servo equations.

$$\frac{dz_m}{dt} = (1-a)\frac{q_m}{N_m} + \frac{kac-1}{k}\beta_m - \frac{D_m}{\sin^2 I}2H' + \frac{U_z}{H'}$$

$$\frac{dN_m}{dt} = q_m - c\beta_m N_m - \frac{\Phi_\infty}{aH'}$$

(1)

This set of equations involves processes of optical ionization, loss action and neutral and electric dynamics in the ionosphere. Here suffix m represents the para-meter evaluated at the position of F_2 peak. N-electron density, $q = e^{-z}$-production rate variation with height z, $\beta = e^{-kz}$-loss coefficient variation with height z, c-empirical constant, D-plasma diffusion coefficient, H'-scale height of neutral ionizable gas, Φ_∞-vertical upward plasma flux at the top of the ionosphere, a-ionosphere shape-related constant. After peak electron density are expressed by the background field N_{m0} plus perturbation ΔN_m, Equation (1) becomes

$$\dot{x} = Fx + Gu + Lw$$

(2)

$$z = Hx + n$$

(3)

where

$$x = \begin{bmatrix} z_m \\ (\Delta N_m) \end{bmatrix}, \quad u = \begin{bmatrix} q_m \beta_m D_m U_z \Phi_\infty \end{bmatrix}^T, \quad F = \begin{bmatrix} (1-a)q_m \\ -c\beta_m \end{bmatrix},$$

$$G = \begin{bmatrix} (1-a) & \frac{kac-1}{k} & -\frac{\sin^2 I}{2H'} & \frac{1}{H'} & 0 \\ 1 & -c & 0 & 0 & -\frac{1}{aH'} \end{bmatrix}, \quad H = \begin{bmatrix} 1 & 0 \\ 0 & 1 \end{bmatrix}, \quad w = 0$$

This is a typical dynamic system that consists of dynamic process (2), obser-vation system (3) with $H=I$, unit matrix, meaning the output vector is just the state vector, and the system noise $w = 0$. The system noise w could be non zero, and include all unknown physical processes, inappropriate assumptions, etc. From measurement z (z_m, ΔN_m), which is degraded by measurement error n, we can determine U_z if the forces q_m, β_m, D_m and Φ_∞ are known. Although the model (1) is simple from view of ionospheric dynamics, it is likely rather complex in finding the solution as it is treated as a mathematical model (2) and (3) of the DAS. Up to now an ionospheric weather process has been expressed as dynamic model of (2)

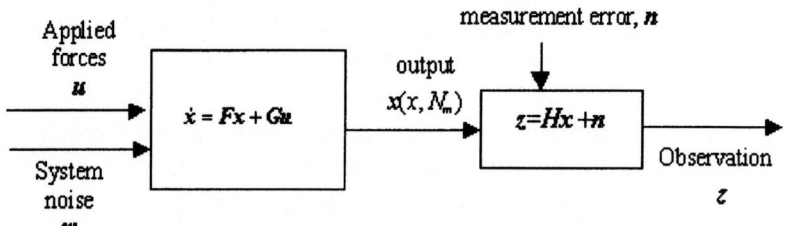

Figure 2. F_2 layer peak dynamic model.

and (3), and schematically shown in Figure 2 in framework of Figure 1. This is the important parts of data assimilation works we have to be done before we create an assimilation procedure. Based on the dynamic model, the optimal estimation, the main topic of this paper, carries through. This is what we are going to discuss in Section 3 (see (6) and (7)) and Section 4 (see (41) and (47)). However we keep the discussion in a more general form rather than an specific optimal assimilation procedure. Interested readers in the specific procedures are referred to the papers by Sojka *et al.* (2001), Ganguly and Brown (2001) and the example in Section 4.

We now define the data assimilation procedure, the third element of DAS. By data assimilation procedure (optimal assimilation procedure has the same connotation in this paper), we mean a coordinated data processing procedure based on dynamic and observation subsystems in DAS. The data assimilation procedure assimilates various observations constrained by dynamic process, and priori statistic knowledge of both dynamic process and measurement, in order to obtain an optimized estimate of the system states or parameters. This definition means that the data assimilation technique is not only the data processing but also the dynamic modeling. The data assimilation technique includes three parts of work: modeling dynamic process, measuring the states of the process, and estimating the states of the process. Data assimilation mainly refers the technique, sometimes, the estimation procedure only in this paper. It is worthwhile to note that it is not necessarily to measure the complete states, but just part of them could make sense on reconstruction of the complete states of the system. However this part of measurements should make the system to have observability (this is another issue).

3. Data assimilation for Dynamic System

To make our discussion have more general applicability, we starts with a nonlinear, time-varying system that can be expressed by n-dimensional nonlinear dynamic equations and k-dimensional observation equations

$$\dot{x}(t) = f\left[x(t), u(t), w(t), p(t), t\right] \tag{4}$$

$$z(t) = h\left[x(t), u(t), w(t), p(t), t\right] \tag{5}$$

where $u(t)$ is an imposed force, for example, could be k_p index, neutral wind U_z or penetrating electric field E etc. as common in the ionospheric forces. $p(t)$ the system parameter, e.g. resonate frequency for second order system. The nonlinear Equations (4), (5) could be approximated by linearization in a leaner random state space. The corresponding dynamic and observation subsystems describing internal and external behavior of the linear system are as follows

$$\dot{x}(t) = Fx(t) + Gu(t) + Lw(t), \quad x(0) = x_0 \tag{6}$$

$$z(t) = Hx(t) + n(t), \quad z(0) = z_0, \quad t \geq t_0 \tag{7}$$

where $F(t), G(t), L(t), H(t)$ are Jacobian matrix of functions of f and h in (4), (5). This is more general procedure of linearization than that we have done for (1). However the results of (6) and (7) are the same as the model of (2) and (3) mathematically. Note that x,u in (6) and (7) are the small perturbations from nominal trajectory of (4) and (5). When the f and h are relatively slow varying function within sampling interval $\Delta t, f$ (or h) can be expressed as $f=f_0+\Delta f$, and f_0 is considered as constant during sampling interval, and Δf a random sequence for which linear estimation could be used (Howe et al., 1998).

Sometimes other easier way, 'variable conversion', could express some complex equation as the linear. For example, for a function $z(x)$ expressed by the Fourier series truncated the terms with index higher than 3:

$$z = a_0 + \sum_{i=1}^{2} (a_i \cos ix + b_i \sin ix) \tag{8}$$

Let a_0, a_1, a_2, b_1, b_2 be expressed by c_0, c_1, c_2, c_3, c_4 and

$$z_1 = \sin 1x, \quad z_2 = \sin 2x, \quad z_3 = \cos 1x, \quad z_4 = \cos 2x$$

We have

$$y = c_0 + c_1 z_1 + c_2 z_2 + c_3 z_3 + c_4 z_4 \tag{9}$$

This is a typical linear equation, and the linear estimation procedure can be carried through based on optimization method. Actually the basic function could be others, like associate Legendre function (Richmond and Kamide, 1988), which should be a complete collection of basic functions. For asymptotic model solution, expressed like (9), usually has limited terms and remains truncate error, which is usually considered as part of model error (Richmond, 1992; Richmond and Kamide 1988; Howe et al., 1998; Akmaev, 1999a). From now on we think that the dynamic equation has been linearized as (6) and (7), and concentrate on the linear estimation study. In the stationary or static case the right side of (6) vanishes, the treatment could be easier than a time-varying system. This is what we are going to discuss next.

3.1. STATIC STATE ESTIMATION

We start with a very simple example to illustrate the concept of the optimal estimation and the relationship between optimal estimation and maximum-likelihood calculation in a static system. Suppose we have two instruments to measure the same physical parameter in the same station at the same time. In the sense of data assimilation the parameter is always estimated from redundant observations, and we have here two measurements but one state (parameter) to estimate. Although the true value of the parameter is theoretically existed, what we can get from redundant measurements is an optimized estimation $\hat{x}$, that is, there is an error between true value and estimation. This is because the measurements are noisy in practice. Suppose the observed values are z_1, z_2, which are not perfect but degraded by noise n_1, n_2. We have no choice but arithmetic average of state x and its variance p as

$$\bar{x} = (z_1 + z_2)/2 \tag{10}$$

$$p = \left(n_1^2 + n_2^2\right)/2 \tag{11}$$

The situation is different when the noise statistical characteristics of the two instruments have already been known, for instance, it has a Gauss probability distribution, variances of σ_1^2, σ_2^2 and zero assembly average $E(n_1) = E(n_2) = 0$; $E(n_1 n_2) = 0$. After considering the statistics of the measurement noise, the optimal value $\hat{x}$ with maximum combination probability for z_1, z_2 can be found from probability calculation

$$\hat{x} = \alpha_1 z_1 + \alpha_2 z_2 \tag{12}$$

$$\alpha_1 = \sigma_2^2/\left(\sigma_1^2 + \sigma_2^2\right); \qquad \alpha_2 = \sigma_1^2/\left(\sigma_1^2 + \sigma_2^2\right) \tag{13}$$

$$\alpha_1 + \alpha_2 = 1 \tag{14}$$

We can see that this estimated value is a weighted average of the measurements. The right side of (12), (13) show that the weight is reversely proportional to the measurement variance. In other words the less accurate measurement has the less contribution to the estimated value or vice versa. This is more reasonable and meaningful than the arithmetic average of (10) (Guo *et al.*, 2000; Schlatter, 2000). This shows the roles of statistical knowledge of measurement in the data assimilation.

We will see that the same result can be obtained by minimizing the quadratic cost function of normalized measurement residual $(z - \hat{x})$. This optimization with criterion of least square error is what we concerned mostly in the paper. Now we discuss it in a more rigorous way and more general case in the view of data assimilation system. Suppose the system state with n-dimension x $(x_1, x_2 \ldots x_n)$ that is to be estimated from k-dimension measurements z $(z_1, z_2 \ldots z_k)$. The measurement is a linear combination of the system output y $(y_1, y_2 \ldots y_k)$ which is degraded by measurement noise n $(n_1, n_2 \ldots n_k)$. Suppose $k \geq n$ that means redundant measurements. All or some elements of y $(y_1, y_2 \ldots y_k)$ can be direct or indirect measurements of the system states, for example, the indirect one can be converted from another type of measurements or coordinate system by a conversion observation matrix H. So output $y = Hx$. Each measured value has different weights forming weighting matrix S.

$$S = E\left[(z - y)(z - y)^T\right] = E\left[nn^T\right] = R$$
$$n_i = z - y = z - H_i x \tag{15}$$

We use n instead of σ in (12), which is defined as measurement standard deviation. H_i is the i-th row of $n \times k$ matrix H. We wish to estimate the system states $\hat{x}$ from measurement z plus noise n by observation system equation.

$$z = Hx + n = y + n \tag{16}$$

Creating a weighting quadratic cost function and minimizing it, an optimal weighted estimation of state and state residual covariance can be obtained. We just list the results here, interested reader can refer to the books by Stengel (1986), or Bryson and Ho (1975).

$$\hat{x}_w = \left(H^T S^{-1} H\right)^{-1} H^T S^{-1} z = x + \left(H^T S^{-1} H\right)^{-1} H^T S^{-1} n \tag{17}$$

In the case of S=I (unity matrix) i.e. without weighting, optimal state estimation becomes

$$\hat{x} = \left(H^T H\right)^{-1} H^T z \tag{18}$$

Now we go back to the simple example for the case of $k = n = 2$ at the beginning of this subsection, we have $H = \begin{bmatrix} 1 & 1 \end{bmatrix}^T$, $S = \begin{bmatrix} n_{11} & 0 \\ 0 & n_{22} \end{bmatrix}$, weighting state estimate of (17) is

$$\hat{x}_w = \left([1 \quad 1] \begin{bmatrix} 1/n_{11} & 0 \\ 0 & 1/n_{22} \end{bmatrix} \begin{bmatrix} 1 \\ 1 \end{bmatrix}\right)^{-1} [1 \quad 1]$$

$$\begin{bmatrix} 1/n_{11} & 0 \\ 0 & 1/n_{22} \end{bmatrix} \begin{bmatrix} z_1 \\ z_2 \end{bmatrix} = \left[\varsigma_1^2 z_1 + \varsigma_2^2 z_2\right] \tag{19}$$

Where $\zeta_{ii} = 1/n_{ii}$. In the simple case the estimated state x is one parameter, a scalar, measured by two instruments at the same site, so the sum of the weighting coefficients ζ_i have to be normalized to be 1, i.e. divided by $\Sigma \zeta_i$ respectively for each i terms of (19). Then we have the estimation $\hat{x}_w$ of (19)

$$\hat{x}_{wn} = \frac{\varsigma_1^2 z_1 + \varsigma_2^2 z_2}{\varsigma_1^2 + \varsigma_2^2} = \frac{n_{22}^2}{n_{11}^2 + n_{22}^2} z_1 + \frac{n_{11}^2}{n_{11}^2 + n_{22}^2} z_2 = \alpha_1 z_1 + \alpha_2 z_2 \tag{20}$$

with

$$\alpha_1 + \alpha_2 = 1 \tag{21}$$

The (20) and (21) are the same as (12), (13) and (14). The weighted estimation $\hat{x}_{wn}$ of (20) obviously has a constraint of (21). The state estimation $\hat{x}$ of (19) without weighting is

$$\hat{x} = \left(H^T H\right)^{-1} H^T z = \left(\begin{bmatrix} 1 & 1 \end{bmatrix} \begin{bmatrix} 1 \\ 1 \end{bmatrix} \right)^{-1} \begin{bmatrix} 1 & 1 \end{bmatrix} \begin{bmatrix} z_1 \\ z_2 \end{bmatrix} = \frac{z_1 + z_2}{2} \tag{22}$$

The (22) is the same as (10). If the two measurements have same errors i.e. $n_{11} = n_{22}$ in (21) and note when $\alpha_1 = \alpha_2 = 1/2$, which means without weighting, (20) is the same as (22).

Now let us consider the estimation errors for the simple example by the sum of quadratic errors, i.e. the cost function. For the case of without weighting and knowing n, quadratic state residual error $J(x)$ and measurement residual error $J(z)$ can be get by substituting (18) in expressions of $J(x)$ and $J(z)$ below and using (16)

$$J(x) = \tfrac{1}{2} \left(x - \hat{x}\right)^T \left(x - \hat{x}\right) = \tfrac{1}{2} \left[n^T H \left(H^T H\right)^{-1} \left(H^T H\right)^{-1} H^T n \right] \\ = \tfrac{1}{8} \left(n_{11} + n_{22}\right)^2 \tag{23}$$

$$J(z) = \tfrac{1}{2} \left(z - H\hat{x}\right)^T \left(z - H\hat{x}\right) = \tfrac{1}{2} \left[n^T \left[I_k - H \left(H^T H\right)^{-1} H^T \right] n \right] \\ = \tfrac{1}{4} \left(n_{11} - n_{22}\right)^2 \tag{24}$$

In case of $n_{11} = n_{22}$, $J(z)=0$ means that it is minimized, however $J(x)=n_{11}^2/2$ indicating there is a bias in the estimate. If the measurements are much more than two, which meet the requirement of statistic estimation, much more precise result can be obtained. This simple problem shows that the important roles of much more redundant measurement in case of statistic knowledge of measurement unknown in the data assimilation.

In case of static system, the estimated states or parameters are time–invariant. The estimation can be proceeded recursively to get more and more accurate result. After getting estimation from k measurements at time step $k - 1$, some new k_1 measurements at step k are made. In this case we need to do recursive estimation to include the new measured information. Conducting a optimization for all k and k measurements, a quite simple estimation procedure can be found

$$\hat{x}_k = \hat{x}_{k-1} + K_k \left(z_k - H_k \hat{x}_{k-1}\right) \tag{25}$$

$$K_k = P_{k-1} H_k^T \left(H_k P_{k-1} H_k^T + R_k\right)^{-1} \tag{26}$$

$$P_k = \left(P_{k-1} + H_k^T R_k^{-1} H_k\right)^{-1} \tag{27}$$

Where P is the $n \times k$ state residual covariance matrix, $R=[nn^t]$ is a $k_1 \times k_1$ squared error matrix, K is $n \times k$ estimate gain matrix, and k is dimensions of observation vector at step k-1. Equation (25) means that the old estimation plus K times the residual can determine new estimation. It is also clear from (27) that since $H^t R^{-1} H$ is at least a positive semi-definite matrix, the error covariance always decrease when the additional measurements are made. The $\hat{x}_{k-1}$ can also be output of a theoretical model at the time step k. If we known statistic knowledge $E\left[(x - x_m)(x - x_m)^T\right] = M$ of the output x_m of the model. We can minimize the combined cost function below for optimal estimation

$$J = \frac{1}{2}\left[(x - x_m)^T M^{-1}(x - x_m) + (z - Hx)^T R^{-1}(z - Hx)\right] \tag{28}$$

The cost function J here essentially is the same as what AMIE procedure used (Richmond, 1992) in which the coefficients of a statistical electric potential model were used instead of state x_m here. This is a general treatment for the model, which can be expanded as a linear expression of a periodic basic function and the coefficients of the basic function are estimated optimally [see (8) and (9)]. A minimum of variance was sought for both points of measurement and estimate, and the statistic characteristics of states at estimate points are supposed to be known as a priori (Akmaev, 1999a, 1999b). So the cost function J here is also essentially the same as the estimation error squared in Akmaev's paper. If we use second term alone in the right side of (28), inverting a singular matrix may be resulted in. If both terms are considered, the difficulty would be overcome. Moreover a reasonable estimation in region of devoid of measurement could be made.

Now we turn to discuss the estimate of (28)

$$\hat{x} = \hat{x}_m + PH^T R^{-1}\left(z - H\hat{x}_m\right) \tag{29}$$

$$K = PH^T R^{-1} \tag{30}$$

$$P^{-1} = M^{-1} + H^T R^{-1}H \quad or \quad P = (I - KH)M \tag{31}$$

This also shows that the error covariance always decreases when additional measurements are made. Of course, these cases are correct only if the system is at the static or steady, i.e. the state values measured at k step and $k + 1$ step are temporally constant. In the dynamic system, the state and its uncertainty propagate from one time step to the next. The propagation property relies on the dynamic law and the new measurement. That situation is quite different, we are going to discuss the dynamic system in the next subsection.

3.2. ESTIMATION OF DYNAMIC SYSTEM

In the static estimation the gain matrix K gets vanishingly smaller as more measurements are made, so that the state estimation gets closer to the true value and its uncertainty gets to be zero at last. However the state or parameter in a time-varying dynamic system is generally not a temporally constant. The dynamic law and the forced inputs are going to control the variations of state and its uncertainty. Therefore a mechanism which includes the effects of both dynamic model and measurement system should be considered. The role of dynamic model in the estimation procedure is to propagate the state and its uncertainty from one time step to another. The estimation procedure, combining dynamics and measurement, gives lower uncertainty than that estimated by either of dynamics or measurement procedures alone. This will be quantitatively shown in the example of Section 4. For ionospheric weather dynamic system we need to estimate the states $x(t)$ and their uncertainty $P(t)$ based on given measurement z and the transit property of dynamic system in the period of time from t_0 to t_f. We call these estimations as interpolation as $t < t_f$, specification as $t = t_f$ and prediction as $t > t_f$. The criterion of least square error, as in the static case, is again used as optimal dynamic estimation. The idea of optimal assimilation system for ionospheric weather could relate in some way to optimal control theory. Interested readers can refer to relevant books (Stengel, 1986; Goodwin and Sin, 1984; Bryson and Ho, 1975).

We now discuss the propagation features by discussing the solution of the dynamic Equation set (6). The discussion also provides the prerequisites for the example in Section 4. As a common problem of first order differential equation set, we just list the solution of dynamic Equation (6) in the sampling interval $(t_{k-1} - t_k)$ below without derivation

$$x(t_k) = \Phi(t_k, t_{k-1}) x(t_{k-1}) \\ + \int_{t_{k-1}}^{t_k} \Phi(t_k, \tau) [G(\tau) u(\tau) + L(\tau) w(\tau)] d\tau, \quad x_0 \text{ given} \tag{32}$$

This is total responses of the initial condition (the first term of right side of (32)), input disturbance $u(\tau)$ and system noise $w(\tau)$ (the second term of right side of (32)). Transition matrix Φ satisfies the equation $\dot{\Phi}(t, t_i) = F(t) \Phi(t, t_i)$. If F, G, L, Δt, Δu, Δw can be considered as constant during the sampling interval, and use k index instead of t index, the state propagating equation becomes

$$x_k = \Phi_{k-1} x_{k-1} + \Gamma_{k-1} u_{k-1} + \Lambda_{k-1} w_{k-1}, \quad x_0 \text{ given} \tag{33}$$

$$\Gamma(\Delta t) = [\Phi(\Delta t) - I_n] F^{-1} G \\ \Lambda(\Delta t) = [\Phi(\Delta t) - I_n] F^{-1} L$$

As we know many natural and man-made dynamic phenomena may be approximated quite accurately by Gauss–Markov random sequence. In practice w and n are always existed in dynamic and measurement systems and with certain statistical property, for instance, the Gaussian distribution. Because a linear transformation of a Gaussian vector preserves its Gaussian character, a Gauss–Markov random sequence can always be represented by the state vector of a linear dynamic system forced by a Gaussian random sequence in which the initial state vector is Gaussian. Therefore solution (33) describes a state evolution of a Gauss–Markov random sequence. The sequence has a conditional probability density distribution, and the condition is that the past and current states are known. So the conditional probability density function of the states x can be expressed as

$$p\left(x_k/x_{k-1} \cdots x_0\right) = p\left(x_k/x_{k-1}\right) \tag{34}$$

The purpose of optimal estimation is to find optimal state $\hat{x}_k$ with minimum uncertainty, this is equivalent to finding the minimum spread of the estimate-error probability density of the sate. The optimal estimation for dynamic system propagates the conditional probability from one time step to another and incorporates system dynamics, its input, measurements and uncertain properties in the estimation. What we need to do in the estimation procedure are three steps (Kalman filter) (Stengel, 1986; Goodwin and Sin, 1984; Bryson and Ho, 1975; Howe *et al.*, 1998).

First propagate state x_k $(-)$ and covariance p_k $(-)$ estimates from one time step to the next according to the recursive formulae

$$\hat{x}_k\left(-\right) = \Phi_{k-1}\hat{x}_{k-1}\left(+\right) + \Gamma_{k-1}u_{k-1} \tag{35}$$

$$P_k\left(-\right) = \Phi_{k-1}P_{k-1}\Phi_{k-1}^T + \Lambda_{k-1}Q'_{k-1}\Lambda_{k-1}^T \tag{36}$$

Here $(-)$ stands for without considering new measurement of step k, the $(+)$ for the opposite, the similar for other parameters bellow. And $Q'_k = E\left(w_k w_k^T\right)$.

Then propagate filter gain K_k $(-)$ from one time step to the next according to

$$K_k\left(+\right) = P_k\left(-\right) H_k^T \left[H_k P_k\left(-\right) H_k^T + R\right]^{-1} \tag{37}$$

Finally update state x_k $(-)$, and covariance p_k $(-)$ estimates by new measurements at step k according to

$$\hat{x}_k\left(+\right) = \hat{x}_k\left(-\right) + K_k\left[z_k - H_k\hat{x}\left(-\right)\right] \tag{38}$$

$$P_k\left(+\right) = \left[P_k\left(-\right)^{-1} + H_k^T R_{k-1}H_k\right]^{-1} \tag{39}$$

The statistic property of the system noise w, observation system noise n and the initial condition should be known before the estimation procedure can proceed. They are

$$E\left\{\left|\begin{matrix} w\left(t\right) \\ n\left(t\right) \end{matrix}\right| \left[w\left(\tau\right)n\left(\tau\right)\right]\right\} = \left|\begin{matrix} Q & S \\ S^T & R \end{matrix}\right| \delta\left(t - \tau\right) \tag{40}$$

$$E\left(x_0\right) = \hat{x}_0 \quad E\left[\left(x_0 - \hat{x}_0\right)\left(x_0 - \hat{x}_0\right)^T\right] = P_0 \tag{41}$$

This estimation procedure is expressed as flow charts in Figure 3 and Figure 4, respectively.

In the case of $\Phi = I, \Gamma = 0$ and $Q = 0$, the system becomes an observation system for static or steady case

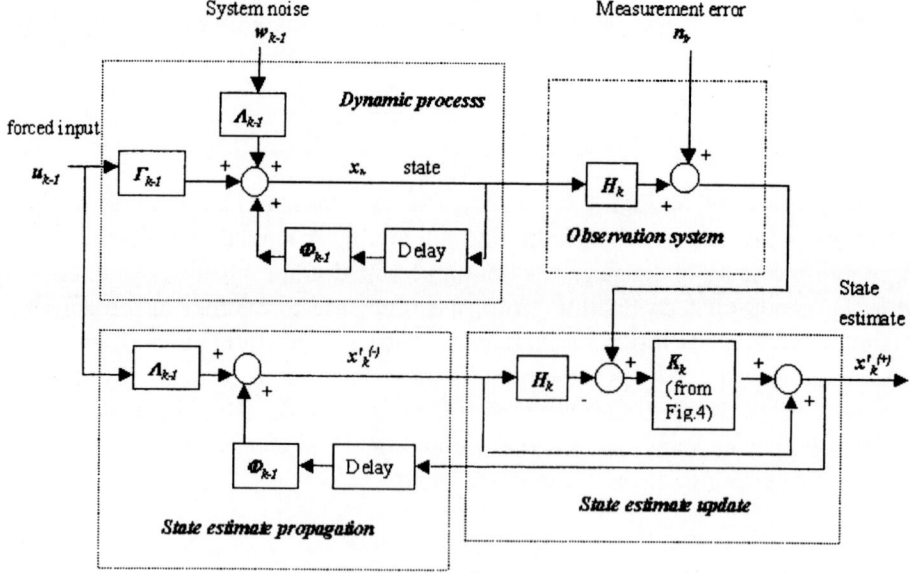

Figure 3. Dynamic system and state estimate (Stengel, 1986).

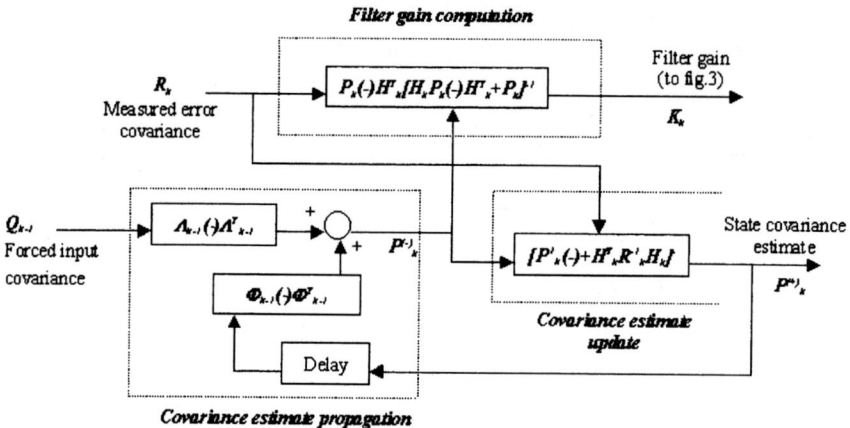

Figure 4. Covariance estimator and gain computation (Stengel, 1986).

$$x_{k+1} = x_k$$
$$z_k = H_{k-1}^T x_k + n_k \qquad (42)$$

and the estimator becomes least square estimator as we have discussed in the previous subsection.

The static or steady estimator calculates state and its uncertainty on the basis of knowing all information, but without measurement updating. The predictor uses optimal initial condition provided by the estimator and the dynamic system model to extend the state and its uncertainty estimate beyond the time of the last available measurement. Suppose present sampling time as t_k, the state prediction at t_f is

$$\hat{x}_f = \Phi_k \left(t_f - t_k\right) \hat{x}_k \left(+\right) + \Gamma_k \left(\cdot\right) u_k \tag{43}$$

Where $\Phi_k \left(t_f - t_k\right)$ is state transition matrix for prediction interval $(t_f - t_k)$, $\Gamma_k(\cdot)$ is the control effect matrix, and $x(+)$ is the current optimal estimation, u_k is the current control, which is assumed as a constant within the prediction interval. The covariance prediction is

$$P_f = \Phi_k \left(t_f - t_k\right) P_k \left(+\right) \Phi_{k-1}^T + Q_k \left(t_f - t_k\right) \tag{44}$$

Where $Q_k(\cdot)$ is covariance of system noise inputs during the prediction interval. We can see that the predicted uncertainty must be at least as large as the present one.

4. Second Order Dynamic System

In this section we will give an example, which clearly expounds the state and its uncertainty evolution in the state space. And it also shows us the roles of the observation system and the dynamic system in terms of reduction of estimation error. This optimal assimilation system is depicted by a second order ordinary differential equation and an observation system. As we mentioned before the second order differential equation play an important role in ionospheric dynamics, since the momentum transport equation is the second one. Suppose a second order dynamic system forced by a white noise w as follows (refer to Stengle, 1986; Bryson and Ho, 1975; Wax, 1954, and application for ionospheric process; Howe *et al.*, 1998; Sojka *et al.*, 2001)

$$\ddot{x} + 2\zeta \omega \dot{x} + \omega^2 x = \omega^2 w\left(t\right) \tag{45}$$

with

$$E\left[w\left(t\right)\right] = 0, \quad E\left[w\left(t\right) w\left(\tau\right)\right] = q\delta\left(t - \tau\right) \tag{46}$$

Let $x_1 = x$, $x_2 = \dot{x}$, the (45) can be written in form of matrix in state space

$$\begin{bmatrix} \dot{x}_1 \\ \dot{x}_2 \end{bmatrix} = \begin{bmatrix} 0, & 1 \\ -\omega^2, & -2\zeta\omega \end{bmatrix} \begin{bmatrix} x_1 \\ x_2 \end{bmatrix} + \begin{bmatrix} 0 \\ \omega^2 \end{bmatrix} w \tag{47}$$

or

$$\dot{X}\left(t\right) = F X\left(t\right) + G w\left(t\right) \tag{48}$$

with initial conditions for expected value

$$E\left[x_1\left(0\right)\right] = E\left[x_2\left(0\right)\right] = 0 \tag{49}$$

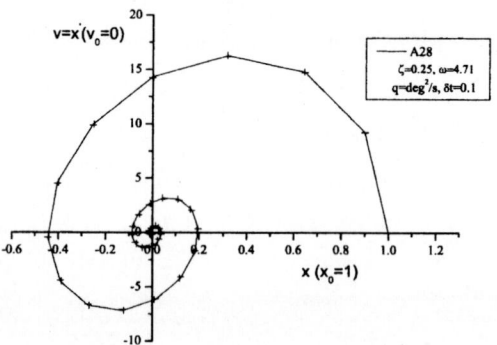

Figure 5. System responses to initial condition with position and velocity variance.

From (45) we can see that the system input is just the system noise w so $P(0) = (x - \bar{x})(x - \bar{x})^T |_0 = xx^T |_0 = X(0)$, and we write the initial covariance as

$$E \begin{bmatrix} x_1 x_1 & x_1 x_2 \\ x_2 x_1 & x_2 x_2 \end{bmatrix}_{t=0} = \mathbf{X}(0) \tag{50}$$

The transition matrix subjects to

$$\dot{\Phi}(t, t_0) = F(t)\,\Phi(t, t_0), \quad \Phi(t_0, t_0) = I \tag{51}$$

The transit matrix can be found more generally by Cayley-Hamilton theorem

$$\Phi(t) = \frac{1}{\lambda_2 - \lambda_1} \begin{bmatrix} \lambda_2 e^{\lambda_1 t} - \lambda_1 e^{\lambda_2 t} & e^{\lambda_1 t} - e^{\lambda_2 t} \\ -\omega\left(e^{\lambda_1 t} - e^{\lambda_2 t}\right) & (\lambda_2 + 2\varsigma\omega)\,e^{\lambda_1 t} - (\lambda_1 + 2\varsigma\omega)\,e^{\lambda_2 t} \end{bmatrix} \tag{52}$$

$$\lambda_1 = -\varsigma\omega + \omega\sqrt{\varsigma^2 - 1}, \quad \lambda_2 = -\varsigma\omega - \omega\sqrt{\varsigma^2 - 1}$$

Here $\lambda_{1,2}$ are eigenvalues of matrix $\mathbf{F}$. In case of $\zeta < 1$ with $\beta = \omega\sqrt{1 - \varsigma^2}$, the general transit matrix (52) is the same as the result of Bryson and Ho (1975). It is obviously that the transition function quantitatively relates to parameters ω, ς of the dynamic system and the sampling interval Δt. For the example of Stengel (1986), ω=6.28 rad/s, $\varsigma = 0.3$

$$\phi(0.1) = \begin{bmatrix} 0.8309 & 0.078 \\ -3.075 & 0.5372 \end{bmatrix}, \phi(0.01) = \begin{bmatrix} 0.9981 & 0.0098 \\ -0.3868 & 0.9611 \end{bmatrix}.$$

Therefore the readers should be cautioned when you make choice of the sampling interval for a known dynamic system in case of discrete equation, since the transition matrix varies with sapling interval. We will also see below that the transition function has an effect on covariance propagation. Integrating (48) we have covariance expressed by transition matrix

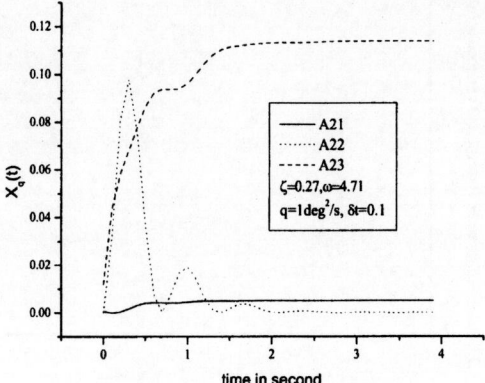

Figure 6. Variation of covariance.

$$\mathbf{X}(t) = \Phi(t) X(0) \Phi^T(t) + \int_0^t \Phi(t-\tau)\mathbf{GqG}^T \Phi^T(t-\tau) \, d\tau$$
$$= X_0 + X_q$$

(53)

where $G = \begin{bmatrix} 0 \\ \omega^2 \end{bmatrix}$, and q as defined by (46). X_0 varies under the control of dynamics, and x_q is system response to the noise q.

Substitution of (52) in (54) and note (46), (47), the responses x_{0ij}, for initial condition and responses x_{qij}, covariance of uncertainty, due to system noise q, can be found. The x_{011} and x_{022} trajectory in state space with error marks x_{q11} and x_{q22} is shown in Figure 5, and the variation of x_{qii} and x_{q12} themselves are shown in Figure 6. The true state trajectory is a spiral, which ends at the focus of zero values of x_{11} and x_{22}, the stable focus. This is the typical property for the under-damped system. The errors are coupled to each other due to dynamic property of the system, though the initial position error is zero, it is stable at a non-zero value after about 0.6 second. The covariance x_{q12} becomes zero after about 1.5 seconds. This means the x_{q11} and x_{q22} are not correlated after the system getting stable.

That measurements can reduce the system errors can be quantitatively explained as bellow. Suppose a continuous measurement of the velocity variable, x_2, is made that forms a observation system

$$z(t) = Hx(t) + n(t), \quad E(n(t)) = 0,$$
$$E(n(t)n(\tau)) = n\delta(t-\tau)$$

(54)

The variance and covariance, which is designed as P to distinguish from X without measurement in (46), can be found from measured data. For simplicity but without loss of generality we just exam the case of steady state during the time period (t) great than about 1.5 second in Figures 5 and 6. From optimal estimation formula for continuous dynamic system

$$\dot{P} = FP + PF^T + GQG^T - PH^TR^{-1}HP, \quad P(0) = P_0$$

(55)

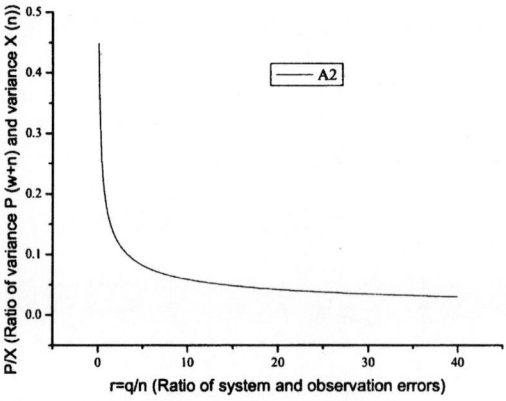

Figure 7. R variation with r.

we can find covariance matrix with measurements P, and in turn, the ratio of P and the covariance X_q without measurements, using $H = [0, 1]$, $Q = q$, $R = n$ and $r = q/n$. Here $H = [0, 1]$ means velocity measurement only. We have

$$R = \frac{P_{11}}{X_{11}} = \frac{P_{22}}{X_{22}} = \frac{8\varsigma^2}{r\omega^2} \left[\sqrt{1 + \frac{\omega^2 r}{4\varsigma^2}} - 1 \right], \qquad r = \frac{q}{n} \qquad (56)$$

Variation of R with r is shown in the Figure 7. We can see that for any values of r the ratio of covariance P with measurements and the covariance of dynamic system X_q is always less than one and decreases very fast. In other wards P_{11} and P_{22} are less than the case without measurement, or the measurement always makes covariance getting smaller and decreases very fast as measurement error getting smaller. This is evidence to demonstrate the importance of observation system in space weather system.

The system parameters, like ς, ω, can be also estimated by a history measurement of position, velocity and acceleration. This kind of job is so called 'system identification', and also another issue, we do not discuss it here.

5. Discussion and Conclusion

We attempt to present a framework of optimal assimilation technique for ionospheric weather, as a branch of space science, which coordinately integrates the observation, dynamics research and prediction of the ionospheric weather. We considered that the object of ionospheric weather study is the physical, chemical and optical processes taking place in the ionosphere, and the scientific and practical objective of the study is to obtain an optimal estimation of the state evolution of these processes. Dynamic law, measured data and their statistic knowledge are three important factors for the optimal estimation. 'Optimal assimilative modeling' representing essence of the framework seems more precise than 'data assimilation' itself. But we still use 'data assimilation' in the paper to keep consistency with the nomenclature used in previous papers. The mathematical core of the optimal assimilation is optimal estimation, which is also the underlay of extensive disciplines, ranging from automatization, communication and artificial intelligence to decision making for economic and social system. One of the popular optimal estimation techniques is constrained weighted least square fitting. Constraint refers to that the dynamic law and measured data are constrained to each other, weighting the heavy-duty of the data at different spatial-temporal grids, least square the criterion to fit the estimated value (or function) in with the objective value (or function). Optimal assimilation differs from theoretical model in that it assimilates different types of and spatial-temporally measured data, empirical model in that the measurement is constrained by dynamic law, semi-empirical model in that it is optimized. Therefore the optimal assimilation technique could obtain more accurate results than those obtained by theoretical, empirical or semi-empirical models alone. The optimal assimilation technique quantitatively integrates the elements of 'space weather metrics' into a estimation procedure, so could create a model or algorithm with optimal ability in terms of space weather metrics (Committee for Space Weather, The, 2000). The consideration here is very general, so the optimal assimilation theory might have a potential to become a fundamental theory for constructing a space weather system for specification and prediction.

A dynamic system is not certainly observable, in turns, predictable. System transition matrix Φ, observation matrix H and the types of the measured parameters determine the observability of an optimal assimilation system. A specific physical process needs a specific observation system, which can not be an arbitrary combination of a set of instruments. This is an important point when designing an observation system, but another interesting topic we do not discuss in detail in the paper. In practice, understanding the observability of a system is easier. For example, the direction of a traveling wave can only be determined by phase measurement at three-station system. For system with two stations the wave direction is not observable. As we mentioned before prediction is based on the dynamic model and the last available measurement to provide estimation beyond the present time. The future measurement can not be obtained at present time, so the dynamic model is the basis of prediction. Of course the model can be the theoretical, empirical or semi-empirical. When finding transition matrix is impossible, nudging relaxes or Newtonian relaxation can be applied (Grell et al., 1994), this is also another topic.

An optimal assimilative estimation procedure of space weather system involving all regions in solar-terrestrial system is quite too complex to realize at moment, but it is the ultimate goal of space weather program. Learning and somehow linking estimation procedures of different regions or processes may be the best practical way to go ahead at present stage. As for theoretical aspects, designing observable system, finding dynamic transition matrix and learning statistic property of dynamic model and measurement are important topics for optimal estimation procedure creation for ionospheric weather. They should be but sometimes not easily to be realized in practice even for a regional estimation procedure. So further study of both theory and application of optimal assimilation should be simultaneously promoted.

Linear estimation in state space is a relatively mature subject. Though some non-linear system can be linearized, linear estimation becomes incapable as system is in chaos or unstable state, which is often encountered in the upper atmosphere. Nonlinear analysis should be carried through in these cases. The discussion in this paper is only appropriate when the dynamic system is stable or steady and the noise is statistically stationary. The dynamic system for ionospheric environment is usually characterized by the spatially distributed parameter. This is different from the optimal control system in industry so that some specific features should be studied further also.

Mathematically, problem of dynamic optimal estimation is to find solution of the differential equation set. Unfortunately only few types of that can be solved analytically. Linear ordinary differential equation set is the one that can be solved by the mature method. So high order ordinary or some partial differential equations is to be somehow treated as a first order differential equation set, and solved in so-called state space. Solution of second ordinary differential equation is a common problem in the ionospheric dynamics (Schunk and Nagy, 2000; Guo *et al.*, 1995), that is why we discussed it in more detail in the paper. It is easy to visualize that varying coefficients, nonlinear and partial differential equations are difficult to treat analytically, one understands only the property and structure of the solution. So the numerical solution plays important role in this scenery.

For operational space weather purpose, high performance computing is required for the larger, faster and finer resolution simulation, and retrieval, statistic analysis, numerous matrix computing, intermediate data storage, in the optimal assimilation procedure. Mass storage and fast CPU speed are needed for the computer hardware system. The efficient computing technique, such as parallel and recursive computing are needed for software design. For example, if a series of orthogonal polynomial in a recursive form is used for least square fitting, the coefficients of each terms of the fitting function can be found at the same time when the polynomial computed. Then accumulate the new terms step by step, the final fitting function can be obtained at last without solving matrix equation. These recursive techniques enormously save computing time and yet have already some common programme existed. The basic formulae (25), (29) for the static or steady state, (35), (36), (38), (39) for dynamic state and uncertainty estimation in this paper are all in recursive forms, the Kalman filter gain (37) is requested to be computed just once if the coefficients of the dynamic equation are constant. For a practice purpose these types of formulae, which could be expressed in other forms, are effective for saving computing time.

Acknowledgements

This work was performed as part of innovation program 2000–2001 and China-Russian space weather study with financial support by Chinese academy of Sciences. Partially supported by NSFC-49944008, NSFC-49990450, NSFC-49704057. The author GJS thanks Professor Abraham C.-L. Chian, Director of WISER to give him support and an opportunity to present part of material of the paper as an invited talk and mini-tutorial lecture on WSEF2002 and HPC2002, and professors F.S. Wei and C. Wang of Laboratory for Space Weather for the recommendation for the meetings and polishing the paper in language aspect respectively. Thanks referee for the very helpful comments.

References

Akmaev, R. A.: 1999a, 'A Prototype Upper-atmospheric Data Assimilation Scheme Based on Optimal Interpolation: 1. Theory', *J. Atmos. Solar-Terr. Phys.* **61**, 491–504.

Akmaev, R. A.: 1999b, 'A Prototype Upper-atmospheric Data Assimilation Scheme Based on Optimal Interpolation: 2. Numerical Experiment', *J. Atmos. Solar-terr. Phys.* **61**, 505–517.

Bryson, Jr. A. E. and Yu-Chi, Ho: 1975, *Applied Optimal Control: Optimization, Estimation, and Control*, Hemisphere Publishing Corporation.

Committee for Space Weather The: 2000, *The National Space Weather Program, The Implementation Plan*, 2nd edition, July.

Ganguly, S. and Brown, A.: 2001, 'Real-time Characterization of the Ionosphere Using Diverse Data and Models', *Radio Sci.* **36**, 1181–1197.

Grell, G. A., Dudhia, J. and Stauffer, D. R.: 1994, 'A description of the Fifth-generation Penn State/NCAR Mesoscale Model (MM5)', *NCAR/TN 398+STR*, June.

Goodwin, G. C. and Sin, K. S.: 1984, *Adaptive Filtering, Prediction and Control*, Prentice Hall.

Guo, J. -S., Bao, D. -H. and Zheng, H.: 1996, in China Geophysics Union (ed.), *space weather, Annual of Chinese Geophysics, China Construction and Finance Industry Publishing House*, p. 274 (in Chinese).

Guo, J. -S., Shang, S. -P., Zhang, M., Lou, X., Zheng, H., Shi, J. -K. and Zhang, Q. J.: 2000, 'Data Assimilation in Space weather study', *Science in China (Series A)* **30**, Supplement, 115–118 (in Chinese).

Guo, J. -S., Zhang, Q. -W., Zhang, G. -L. and Zheng, H.: 1995, 'Mid- and Low-latitude Ionospheric Responses to Different Types of Magnetic Storms', *Chinese J. Geophys.*, 38341–349.

Hajj, G., Lee, L. Q., Pi, X., Romans, L., Schreiner, W., Straus, P., and Wang, C.: 2001, in Lee L. Q., R. Christian, K. Robert (eds.), *CONSMIC GPS Ionospheric Sensing and Space Weather, Application of Constellation Observing System for Meteorology, Ionosphere & Climate*, Springer-Verlag, pp. 235–272.

Howe, B. M., Runciman, K. and Secan, J. A.: 1998, 'Tomograghy of the Ionosphere: Four-dimensional Simulation', *Radio Sci.* **33**, 109–128.

Kuo, Y. H., Sokolovsky, S., Anthes, R. and Vandenberghe, F.: 2001, in Lee L. Q., R. Christian, K. Robert (eds.), Assimilation of GPS Occultation Data for Numerical Weather Prediction, *Applicaton of Constellation Observing System for Meteorology, Ionosphere & Climate*, Springer-Verlag, pp. 57–186.

Kamide, Y., Richimond, A. D. and Matsushita, S.: 1981, 'Estimation of Ionospheric Electric Fields, Ionospheric Currents, and Field Aligned Currents from Ground Magnetic Records', *J. Geophys. Res.* **86**, 801–813.

Lu, G., Baker, D. N., Farrugia, C. J., Lummerzheim, D., Ruohoniemei, J. M., Rch, F. j., Evans, D. S., Brittnacher, M., Li, X., Greenwald, R., Sofko, G., Villain, J., Laster, M., Thyer, J., Moretto, T., Milling, D., Troshichev, O., Zaitzev, A., Makarov, G. and Hayashi, K.: 1998, 'Global Energy Deposition During the January 1997 Magnetic Cloud Event', *J. Geophys. Res.* **103**, 11,685– 11,694.

Peredo, M., Fox, N. and Thompson, B.: 1997, 'Scientists Track Solar Event All the Way to Earth', *Adv. Space Res.* **78**, 477–483.

Richmond, A. D.: 1992, 'Assimilative Mapping of Ionosphere Electrodynamics', *EOS* **12**, (6)59– (6)68.

Richmond, A. D. and Kamide, Y.: 1988, 'Applying Electrondynamic Features of the High-altitude Ionosphere from Localized Observations: Technique', *J. Geophys. Res.* **93**, 5741–5759.

Richmond, A. D. and Lu, G.: 2000, 'Upper-atmospheric Effects of Magnetic Storms: A Brief Tutorial', *J. Atmos. Solar-Terr. Phys.* **62**, 1115–1127.

Rishbeth, H., Ganguly, S. and Walker, J. C. G.: 1978, 'Field-aligned and Field-perpendicular Velocities in the Ionospheric F2-Layer', *J. Atmos. Solar-Terr. Phys.* **40**, 767–784.

Schlatter T. W.: 2000, 'Variational AssiCommittee for milation of Meteorological Observation in the Lower Atmosphere: A Tutorial on How it Works', *J. Atmos. Solar-Terr. Phys.* **62**, 1057–1085.

Schunk, R. W. and Nagy, A. F.: 2000, *Ionosphere: Physics, Plasma physics, and Chemistry*, Cambridge university press.

Schunk, R. W. and Sojka, J. J.: 1996, 'Ionosphere-thermosphere Space Weather Issues', *J. Atmos. Solar Terr. Phys.* **58**, 1527–1574.

Sojka, J. J., Thompson, D. C., Shunk, R. W., Bullett, T. W. and Makela, J. J.: 2001, 'Assimilation Ionospheric Model: Development and Testing with Combined Ionospheric Campaign Caribbea Measurements', *Radio Sci.* **36**, 247–259.

Stengel, R. F.: 1986, *Stochastic Optimal Control: Theory and Application*, John Wiley & Sons.

Wax, N. (ed.): 1954, *Collected Papers on Noise and Stochastic Processes*, Dover, New York.

PLASMA INSTABILITIES AND THEIR SIMULATIONS IN THE EQUATORIAL *F* REGION – RECENT RESULTS

R. SEKAR

Physical Research Laboratory, Navrangpura, Ahmedabad 380 009, India

Abstract. In this paper the developments made in the last five years on numerical simulation/modeling studies of a complex nighttime equatorial spread *F* phenomenon are reviewed. Emphasis is given to the Indian work and necessary comparisons are done with other international works on this field. Investigations involving the important aspects, namely the confinement of the plasma bubble in the bottomside of the ionosphere, linear and nonlinear effects of molecular ions in the development of plasma bubbles, interaction of two modes as a seed perturbation are discussed in detail.

Key words: equatorial ionosphere, plasma bubble, plasma irregularities

1. Introduction

The plasma irregularities in the *F* region of the equatorial ionosphere manifest as diffused echoes on the ionograms which are generally known as equatorial spread *F* (ESF) phenomenon. These plasma irregularities generally occur during nighttime and are observed (Tsunoda, 1980) to be aligned to Earth's magnetic field. The scale sizes of these irregularities are ranging from a few hundreds of kilometers to a few centimeters. The ESF irregularities extend around ±20° about the magnetic equator with a zonal dimension of about a few hundred kilometers to a few thousand kilometers. The altitude extent starts from 200 km to go well beyond 1000 km. These irregularities generally move eastward with velocities of a few hundred ms^{-1} while the upward movement varies in general from a few tens of ms^{-1} to a few hundreds of ms^{-1}.

Though the phenomenon of ESF was discovered as early as 1938, a suitable physical mechanism was proposed only in early seventies. Earlier studies till then were essentially morphological to describe the characteristic of ESF. A linear theory involving the generation of large scale irregularities by Collisional Rayleigh–Taylor instability mechanism was proposed (Haerendel, 1974) based on radar measurement made at Thumba (Balsley *et al.*, 1972). Hierarchy of plasma instability processes, induced by the primary collisional Rayleigh–Taylor mode give rise to the wide spectrum of irregularities. The primary mode evolves nonlinearly in the bottomside of the ionosphere to form into a plasma-depleted region commonly known as a plasma bubble. Pioneering numerical simulations demonstrating the development of irregularities on the topside of the ionosphere, a region which oth-

Space Science Reviews **107**: 251–262, 2003.
© 2003 *Kluwer Academic Publishers.*

erwise remains linearly stable were made by Naval Research Laboratory (Ossakow, 1981). With the advent of rocket, radar satellite and optical techniques, it is rather well established that plasma fluid type Collisional Rayleigh–Taylor instability is the prime causative mechanism for the development of long wavelength plasma irregularities (see review of Kelley and McClure, 1981). Later on, based on the rocket measurements (Raghavarao et al., 1987) and theory (Sekar and Raghavarao, 1987; Raghavarao et al., 1992; Sekar et al., 1994), the importance of neutral wind was realized.

The important background scenario in the form of background electron density profiles have been obtained over an equatorial station using Jicamarca incoherent scatter radar (McClure, 1965). These results revealed that the electron density peak values reduce by an order of magnitude during nighttime compared to daytime values and steep electron density gradients in the bottomside of the ionosphere is found in the nighttime profiles. After local sunset, the ionization in the E region begins to recombine while the F region moves up due to electro-dynamical effects. Owing to these processes, the altitude profile of electron density is restructured and a steep electron density gradient is developed in the bottomside of the F region of the ionosphere. Under this condition, the heavier fluid (F region) supported by the lighter fluid (base of the F region) is susceptible to become unstable under the action of gravity. Some amount of free energy is dissipated through ion-neutral collisions. The background ionospheric conditions modify the evolution of this instability and results obtained in this regard are reviewed in the present paper.

2. Models

The growth of the Collisional Rayleigh–Taylor instability in the context of the ESF phenomenon has been investigated by means of linear theories (Haerendel, 1974; Ossakow, 1981; Hanson et al., 1986; Sekar and Raghavarao, 1987; Sekar and Kherani, 1999) and nonlinear numerical simulation models (Scannapieco and Ossakow, 1976; Ossakow, 1981; Raghavarao et al., 1992; Sekar et al., 1994; Huang and Kelley, 1996a, 1996b).

3. Linear Models

The linear models provide growth rate and zero order conditions for the development of instability. The growth rate (γ) for the Collisional Rayleigh–Taylor (Haerendel, 1974) is given by the following equation

$$\gamma = \frac{1}{H}[\frac{g}{v_{in}}],\tag{1}$$

where H is plasma scale length, g is acceleration due to gravity and ν_{in} is ion neutral collision frequency. After local sun set time, when the plasma scale length decreases, the bottomside *F* region is susceptible to become unstable under the action of gravity when it can overcome the recombination effects. The contribution from gravity is significant when *F* region peak is at higher altitude as ν_{in} decreases exponentially with altitude due to exponential decrease of neutral density. However, the occurrence variability of ESF phenomenon cannot be explained with the omnipresent gravitational term. The importances of zonal electric field in modifying the linear growth rate are discussed in the literature (Ossakow and Chaturvedi, 1978; Hanson *et al.*, 1986). Based on the results obtained from co-ordinated measurements conducted at the onset time of ESF (Raghavarao *et al.*, 1987), the importance of vertical wind in the Collisional Rayleigh–Taylor instability was brought out by Sekar and Raghavarao (1987). Recently, based on the ion composition measurements in the nighttime equatorial regions (Narcisi and Szuszczewicz, 1981; Sridharan *et al.* 1997), the effect of molecular ions on the collisional Rayleigh–Taylor is examined (Sekar and Kherani, 1999) and a growth rate expression in the presence of double ion species is obtained. This exercise reveals that the growth rate depends on the number densities of both the ions species. This is in contrast to a single ionic constituent wherein the growth rate is independent of number density. Thus the above discussed linear models provide preconditions for the development of instability processes but not the evolutionary stages of the instability. Nonlinear numerical simulation models have been used to study the evolutionary stages.

4. Numerical Simulation Model

The equatorial ionosphere is considered in a slab geometry with Earth's magnetic field (**B**) directed along **Z** axis. **X**, **Y** and **Z** are the unit vectors directed along westward, upward and northward respectively. A set of plasma fluid equations describing the conservation of momentum, mass and current are considered. These equations which govern the motion of the species α [ions (i) and electrons (e) whose masses are represented as m_i and m_e respectively] are

$$\frac{\partial n_\alpha}{\partial t} = P - L - \nabla \cdot (n_\alpha \mathbf{V}_\alpha)(\alpha = i, e) \tag{2}$$

$$(\frac{\partial}{\partial t} + \mathbf{V}_\alpha \cdot \nabla)\mathbf{V}_\alpha = \frac{e_\alpha}{m_\alpha}[\mathbf{E} + \mathbf{V}_\alpha \times \mathbf{B}] + \mathbf{g} - \nu_{\alpha n}(\mathbf{V}_\alpha - \mathbf{W_n}) \tag{3}$$

$$\nabla \cdot \mathbf{J} = 0 \tag{4}$$

$$\mathbf{J} = |e|(n_i \mathbf{V}_i - n_e \mathbf{V}_e) \tag{5}$$

During nighttime, the production (P) in the continuity Equation (2) is taken as zero. The loss term (L) in the same equation is given by $\nu_R N_n$ where ν_R is the recombination co-efficient and N_n is neutral number density.

The equation of momentum (3) governs the movement of the species (α). The influences of neutral wind (W_n), gravity (g), electric field (E) and the Lorentz force due to the movement of the species (V_α) are included for describing the motion. Here e and $\nu_{\alpha n}$ represent the charge of the species and the collisional frequency of the species with neutral respectively.

Equation (4) represents the current conservation where **J** is the total current density given by (5).

The steady state values for V_i and V_e are obtained by solving momentum equations with valid F region approximations. Substituting these values in current conservation equation with an assumption of electrostatic field ($E = -(\nabla\phi)$, where ϕ is potential) and performing the first order perturbation analysis, the following differential equation for the perturbation potential (ϕ_1) is obtained (see for details Sekar et al., 1994).

$$\nabla.(\nu_{in} N \nabla \phi_1) = B[-g + W_y \nu_{in} + (E_{xo}/B)\nu_{in}]\frac{\partial N}{\partial x} \tag{6}$$

The continuity equation gets modified into

$$\frac{\partial N}{\partial t} - \frac{\partial}{\partial x}[(N/B)(E_{yo} + \frac{\partial \phi_1}{\partial y})] + \frac{\partial}{\partial y}[(N/B)(-E_{xo} + \frac{\partial \phi_1}{\partial x})] = -\nu_R N \tag{7}$$

Here the E_{xo} and E_{yo} are the ambient electric field in zonal (x) and vertical (y) directions.

Equation (6) describes the spatial distribution of the perturbation potential generated by the generalized Rayleigh–Taylor in a slab geometry, while (7) describes the temporal evolution of total plasma density. The inclusion of the background electric field in (7) enables the investigation of the transport effects due to the altitude variation of the vertical electric field.

As the problem of interest is to study the growth of the periodic perturbations in plasma densities in zonal direction, periodic boundary conditions are imposed on N and ϕ_1 in the zonal direction. In vertical direction, transmittive boundary conditions (i.e. $\frac{\partial N}{\partial y} = 0$) for N and Neumann condition(i.e. $\frac{\partial \phi_1}{\partial y} = 0$) for perturbation potential are imposed. These are closer to realistic situation.

These coupled differential equations are solved numerically over a plane orthogonal to Earth's magnetic field over the dip equator. The region encompasses ±200 km in zonal direction (positive x axis is along westward) and from 182 to 534 km in vertical (positive y axis is upward) direction. The region is divided into 80 (zonal) × 176 (altitude) grid points with the grid size of $\Delta x = 5$ km and $\Delta y = 2$ km. The above parameters are used in most of the simulations presented in this paper except a simulation involving a very long wavelength wherein the grid size was suitably altered in zonal direction to accommodate one full cycle of the wave. The numerical methods were adopted as described in Sekar et al. (1994).

The physics that governs the model is as follows. Consider a sinusoidal perturbation (n_1) over a steady state (n_o) in plasma density along zonal direction (see

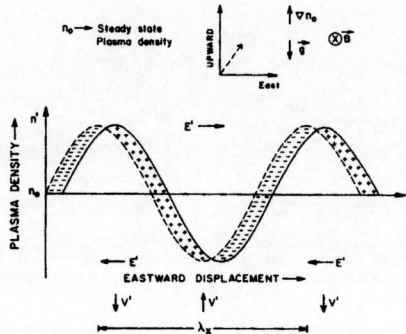

Figure 1. Simplified physical picture of Rayleigh–Taylor instability. The ion and electron densities are represented by continuous and dashed curves.

Figure 1). The magnetic field is directed into the plane of the diagram and the geometry is given in Figure 1. The dominant movement of the plasma in the *F* region of the equatorial ionosphere is essentially due to Hall drift which arises from the ambient electric field. The Hall drift is independent of charge and mass. However, there exists a differential drift between ions and electrons along the zonal direction owing to other agencies like gravity, vertical wind and Pedersen drift component due to the zonal electric field. Owing to the differential drift between ions and electrons, there are regions of space charge (denoted by $+$ and $-$ in Figure 1) which induces polarization electric fields which are directed eastward in the depleted region and westward in the plasma enhanced region. The drifts associated with these perturbation electric fields bring the plasma from higher altitude region and dumps over an enhanced region. Similarly, the plasma is excavated from the depleted region due to upward polarization drift. Thus the instability grows with time. As it grows nonlinearly, western edge becomes more steep and at the expense of enhancement region the depleted region grows further. The wave gets de-shaped and a plasma bubble is formed in the bottomside of the *F* region.

Many investigations have been carried out using this model, some of them to explain the observations under varying geophysical conditions (Sekar *et al.*, 1994, 1997) while the other investigations are concentrated to predict the nature of irregularities (Sekar and Raghavarao, 1995). Further, the important problems such as seeding mechanism (Sekar *et al.*, 1995), topside electron density control (Sekar and Raghavarao, 1995) and the bottomside confinement of the plasma bubble (Sekar and Kelley, 1998) have been addressed using this model. Recently, the observation of the down drafting of the plasma structures by the Indian MST Radar is explained using this model (Sekar *et al.*, 2001). In addition to the above, the effects of molecular ions are examined by nonlinear model (Sekar and Kherani, 2002). The significant results obtained in the last five years are highlighted in this paper.

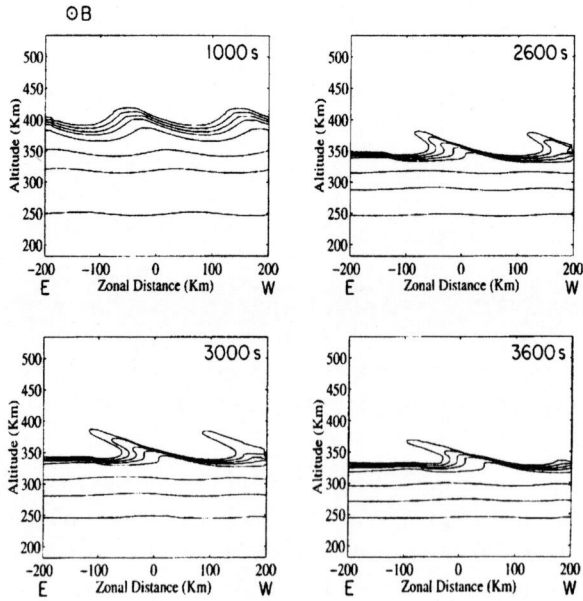

Figure 2. Temporal sequence of the isoelectron density contours in the simulation plane for a zonal drift pattern as per case 1.

5. Results

5.1. CONFINEMENT OF ESF IRREGULARITIES IN THE BOTTOMSIDE OF THE IONOSPHERE

The Plasma irregularities in the F region of the ionosphere as revealed by the VHF Radars, manifest themselves into many different forms ranging from spectacular rising plasma plumes to moderate bottomside spread F events. Though the backscatter map corresponds to scale sizes of a few meters, earlier co-ordinated measurements (Kelley and McClure, 1981) reveal that the radar plumes are collocated with large scale plasma bubble. Plumes extending from bottomside to topside of the ionosphere were identified by earlier studies to be due to the development of plasma bubble. However, the confinement of plumes to the bottomside of the ionosphere was not understood comprehensively. In order to understand the suppression mechanism of large scale plume events on some nights, an investigation was carried out with various background ionospheric conditions (Sekar and Kelley, 1998). Figure 2 depicts the temporal sequence of the isoelectron density contours in the zonal and vertical plane for a case 1 in which the initial conditions are set with shears in zonal plasma drift with maximum value of 1.8×10^{-3} s^{-1} at 375 km altitude along with a vertical upward drift of 10 ms^{-1} up to initial 800 s, followed by a downward drift of 20 ms^{-1} after 900 s, which remains for an hour. The rising contours in the figures represent the development of plasma bubble.

Figure 2 reveals that the growth of the instabilities due to Rayleigh–Taylor instability mechanism is initially accelerated at the lower altitude by eastward electric field and is then decelerated by the westward electric field. Equally important is the fact that the higher altitude structures are tilted by shears towards the zonal direction. Individual effects of the above two agencies were unable to confine the irregularities (see Sekar and Kelley, 1998). However, this investigation revealed that the development of irregularities can only be confined to a localized region in the bottomside of the ionosphere with the combined effects of shears in zonal plasma flow and westward electric field associated with a certain temporal pattern of the zonal electric field. In order to understand the variabilities in the occurrence of the plume events, investigations were carried out with the same (shears) zonal plasma drift however with different temporal zonal electric field patterns. In case 2, the upward drift of 30 ms^{-1} is assumed until 800 s, which was allowed to change linearly to a 20 ms^{-1} downward drift at 900 s; and remains the same after 900 s. In another case (case 3), the upward drift of 20 ms^{-1} is assumed until 1200 s, which decreases linearly to 20 ms^{-1} downward by 1400 s and remains the same afterward. Figure 3 depicts the isoelectron density contours in the simulation plane for the cases 2 and 3. Both the cases show the development of plasma bubbles which penetrated beyond the *F* region peak altitude. This result shows the importance of the magnitude and the time duration of eastward electric field after the initiation of the plasma instability process in the development of plasma bubble events. Thus the variabilities in the occurrence characteristic of bottomside ESF are shown to depend on the variabilities in the pre-reversal enhancement in the zonal electric field. Further, the resurgence of plume event at a later time, is explained (Sekar and Kelley, 1998) on the basis of a pre-seeded structure subject to the presence of a fresh source of instability like an electric field reversal during nighttime associated with storm, sub-storm etc.

Another possible mechanism of confining the plasma bubbles in the bottomside of the ionosphere by means of a possible neutral wind jet (in zonal direction) stream was proposed by Kuo *et al.* (1998). However, neutral wind measurements (Raghavarao *et al.*, 1987; Sridharan *et al.*, 1989) did not show strong shears in the equatorial thermosphere (equatorial *F* region) associated with the presence of neutral wind jet stream. No such measurement is reported so far. Further, it is rather difficult to maintain a strong wind shear in thermosphere in the presence of strong kinematic viscous force (Rishbeth, 1972).

5.2. INTERACTION OF TWO LONG-WAVELENGTH MODES IN THE DEVELOPMENT OF ESF

Gravity waves in the neutral atmosphere is considered to be a potential source of seeding plasma instability (Huang and Kelley, 1996a). The exact mechanism that couples the transfer of perturbation from neutral atmosphere to plasma is yet to be understood comprehensively. The spectra associated with gravity waves are

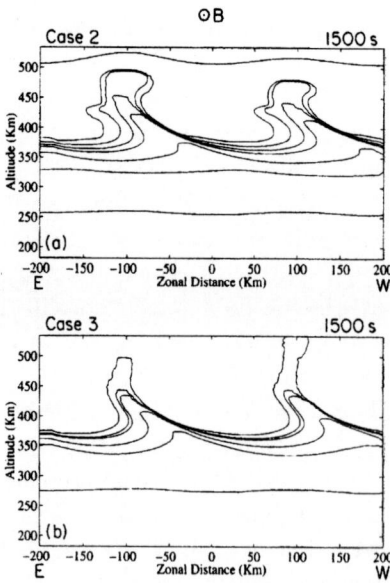

Figure 3. Isoelectron density contours in the simulation plane for two different initial conditions as per cases 2 and 3.

diverse. However, on many occasions, the radar observations reveal that the presence of more than one mode as seed perturbations. The earlier simulation attempts (Huang and Kelley, 1996a) using gravity wave as seed perturbations were restricted to modes with wavelength not exceeding 200 km. In addition, in their simulations, different combination of wave parameters were not examined. Further, in order to understand a unique structural pattern with down drafting of ESF structures obtained by Indian MST Radar (Patra *et al.*, 1995), an investigation was carried out with seed perturbations having two long wavelengths modes, as a first step to understand the effects of gravity waves. The investigation (Sekar *et al.*, 2001) revealed that the superposition of two modes gives rise to a low level plasma depletion which moves downward even when the polarity of the ambient plasma drift is upward (see Figure 4). The evolution of well developed plasma bubble (Figure 4) with the scale size corresponding to smaller wave length with 0.5% amplitude is possible if it rides over a longer-wavelength mode with large amplitude (5%). Further, longer-wavelength mode develops into bottomside structures, while the shorter wavelength mode develops into multiple plumes similar to earlier investigations (Huang and Kelley, 1996a) however, subject to the conditions that the amplitude of the former is larger in comparison with latter. The interaction between the modes was shown to modify the electric field structures. The rising multiple plumes and the descending structure along with a downward moving streak as observed by the Indian MST Radar can qualitatively be understood on the basis of the interaction between the two modes which gives rise to depletions with varying degrees and a rapidly descending enhancements followed by a downward

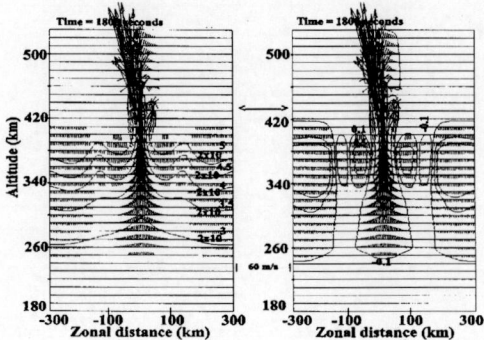

Figure 4. Isoelectron density and depletion contours along with vertical drift pattern at 1800 s over the simulation plane for the superposition of two modes with amplitudes 0.5% and 5% of the ambient for the wavelengths 150 km and 600 km respectively.

moving weakly depleted region, depending upon the wave parameters of the two modes (Sekar *et al.*, 2001).

5.3. EVOLUTION OF ESF IN THE PRESENCE OF MOLECULAR ION

Satellite observations of ESF (McClure *et al.*, 1977) revealed the presence of molecular ion (NO^+) even in the topside of the ionosphere and also enhancements in NO^+ were found to be collocated with the depletions in atomic oxygen ion (O^+). The presence of short-lived molecular ions at higher altitude was an enigmatic problem. In order to understand these results, the model was suitably extended to two fluid type (involving O^+ and NO^+) to investigate the nonlinear effects of plasma bubble on molecular ions (NO^+) (Sekar and Kherani, 2002). During initial phase (<500 s), instability grows in O^+ densities while the NO^+ densities are controlled by the chemical process. As the instability grows with time in O^+, the nonlinear evolution of perturbation velocities makes the transport process to dominate even in NO^+ densities as the polarization (Hall) drift is independent of charge and mass. The investigation revealed that the plasma transport process associated with the plasma instability brings the plasma from the base of F region. The experimental observations conducted during evening hours (Sridharan *et al.*, 1999) and nighttime (Narcisi and Szuszczewicz, 1981) revealed that NO^+ was dominant up to the base of the F region and the concentration progressively reduces with altitude and the cross-over from NO^+ to O^+ takes places during evening hour around the base of the F region which is also theoretically supported (Andersen and Rusch, 1980). Thus, if the cross-over of O^+ and NO^+ takes places at the base of the F region due to pre-reversal enhancement in zonal electric field, the plasma bubble development can take over and redistribute over topside of F region. As the NO^+ densities are decreasing with altitude beyond the base of F region, the transport process associated with plasma bubble development makes enhancement in NO^+ densities. Further, depletion of atomic oxygen is found to

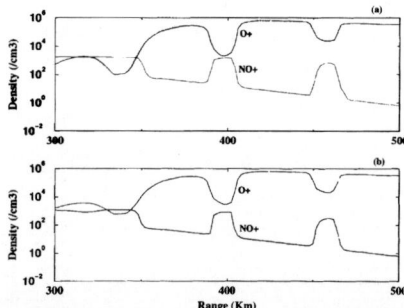

Figure 5. The simulation of NO^+ and O^+ densities along an assumed trajectory similar to an equatorial orbiting satellite for two different ratios of NO^+ and O^+ at the base of the F region.

be collocated with the enhancement in molecular ion similar to that of satellite measurements. Figure 5 depicts the simulation of NO^+ and O^+ determined along the trajectory some what similar to equatorial orbiting satellite revealing simultaneous occurrences of enhancement in NO^+ and depletions in O^+. The variations in the occurrences of NO^+ at higher altitudes are shown to be associated with the variations of the relative concentration of NO^+ and O^+ at the base of F region due to the variability associated with post sun set enhancement in zonal electric field (Sekar and Kherani, 2002).

5.4. RELATION BETWEEN NIGHTTIME E AND F REGION STRUCTURES

In order to understand the radar observation (Woodman and Chau, 2001) of rising structures in the plasma density in the upper E region during nighttime, an investigation was carried out to explore the possible relation between E and F region structures. The investigation (Kherani *et al.*, 2002) revealed that the fringe fields associated with the development of equatorial spread F structures initiated by large scale wave in zonal direction can penetrate well below the E region. These fringe fields pull the structures upward and tilt them left or right hand side to generate rising tilted structures in the E region. The depth of penetration of the fringe fields from E region altitudes mainly depends on the wavelength of initial penetration. The fringe fields can move the E region structures upward with varying speeds, even when the background drift during nighttime is downward, depending on the scale size and the strength of ESF development (Kherani *et al.*, 2002).

6. Summary

Some of the recent developments in last five years in the field of nonlinear simulation of equatorial spread F are reviewed. The important problems such as the bottomside confinements of the plasma bubble and their day-to-day variabilities, evolution of plasma bubble in the presence of molecular ions, interaction of two long-wavelength modes and a possible relation between E and F region structures during nighttime are discussed in detail. Future work must be directed in understanding the aspects like, relation between the plasma fountain and ESF, generation mechanism of meter scale size irregularities, evolution of supersonic plasma bubble etc. More quantitative investigation between the coupling of E and F regions is needed. Three dimensional modeling or magnetic flux tube integrated evolution similar to Keskinen *et al.* (1998) is needed to investigate topside structures.

Acknowledgement

The author thanks Prof. Raghavarao for constant encouragement. The author acknowledges Dr R. Suhasini Rao for her advice during the development of the numerical model. This work is supported by Department of Space, Government of India.

References

Anderson, D. N. and Rusch, D. W.: 1980, 'Composition of the Nighttime Ionospheric F Region Near Magnetic Equator', *J. Geophys. Res.* **85**, 569–574.
Balsley, B. B., Haerendel, G. and Greenweld, R. A.: 1972, 'Equatorial Spread F: Recent Observation and a New Interpretation', *J. Geophys. Res.* **77**, 5625–5628.
Hanson, W. B., Cragin, B. L. and Dennis, A.: 1986, 'The Effect of Vertical Drift on the Equatorial F Region Stability', *J. Atmos. Terr. Phys.* **48**, 205–212.
Haerendel, G.: 1974, *Theory of Equatorial Spread F*, Max Planck Inst. fur. Phys. und Astrophys., Garching, Germany.
Huang, C. S. and Kelley, M. C.: 1996a, 'Nonlinear Evolution of Equatorial Spread F, 2. Gravity Wave Seeding of Rayleigh–Taylor Instability', *J. Geophys. Res.* **101**, 293–302.
Huang, C. S. and Kelley, M. C.: 1996b, 'Nonlinear Evolution of Equatorial Spread F, 4. Gravity Waves, Velocity Shear and Day-to-day Variability', *J. Geophys. Res.* **101**, 24,521–24,532.
Kelley, M. C. and McClure, J. P.: 1981, 'Equatorial Spread F: A Review of Recent Experimental Results', *J. Atmos. Terr. Phys* **3**, 427–435.
Keskinen, M. J., Ossakow, S. L., Basu, S. and Sultan, P. J.: 1998, 'Magnetic Flux Tube Integrated Evolution of Equatorial Ionospheric Plasma Bubbles', *J. Geophys. Res.* **103**, 3957–3967.
Kherani, E. A., Raghavarao, R. and Sekar, R.: 2002, 'Equatorial Rising Structure in Nighttime Upper E Region: A Manifestation of Electro-dynamical Coupling of Spread-F', *J. Atmos. Solar Terr.Phys.* **64**, 1505–1510.
Kuo, F. S., Chou, S. Y. and Shan, S. J.: 1998, 'Comparison of Topside and Bottomside Irregularities in Equatorial F Region Ionosphere, *J. Geophys. Res.* **103**, 2193–2199.
McClure, J. P.: 1965, *Electron Density Studies at Jicamarca*, in Proceeding of the Second International Symposium on Equatorial Aeronomy, Brazil, pp. 170–177.
McClure, J. P., Hanson, W. B. and Hoffman, J. H.: 1977, 'Plasma Bubbles and Irregularities in the Equatorial Ionosphere', *J. Geophys. Res.* **82**, 2650–2656.
Narcisi, R. S. and Szuszczewicz, E. P.: 1981, 'Direct Measurements of Electron Density, Temperature and Ion Composition in an Equatorial Spread F Ionosphere', *J. Atmos. Terr. Phys.* **43** 463–471.
Ossakow, S. L.: 1981, 'Spread F Theories – A Review', *J. Atmos. Terr. Phys.* **43**, 437–452.

Ossakow, S. L. and Chaturvedi, P. K.: 1978, 'Morphological Studies of Rising Equatorial Spread F Bubbles', *J. Geophys. Res.* **83** 2085–2090.

Patra, A. K., Anandan, V. K., Rao, P. B. and Jain A. R.: 1995, 'First Observation of Equatorial Spread F Irregularities from Indian MST Radar', *Radio Sci.* **30**, 1159–1165.

Raghavarao, R., Gupta, S. P., Sekar, R., Narayanan, R., Desai, J. N., Sridharan, R., Babu, V. V. and Sudhakar, V.: 1987, 'Insitu Measurements of Winds, Electric Fields and Electron Densities at the Onset of Equatorial Spread F', *J. Atmos. Terr. Phys.* **49**, 485–492.

Raghavarao, R., Sekar, R. and Suhasini, R.: 1992, 'Nonlinear Numerical Simulation of Equatorial Spread F: Effects of Winds and Electric Fields', *Adv. Space Res.* **12**, 227–230.

Rishbeth, H.: 1972, 'Thermospheric Winds and the F Region: A Review', *J. Atmos. Terr. Phys.* **34**, 1–47.

Scannapieco, A. J. and Ossakow, S. L.: 1976, 'Nonlinear Equatorial Spread F', *Geophys. Res. Lett.* **3**, 451–454.

Sekar, R. and Raghavarao, R.: 1987, 'Role of Vertical Winds on the Rayleigh–Taylor Mode Instabilities of the Nighttime Equatorial Ionosphere', *J. Atmos. Terr. Phys.* **49**, 981–985.

Sekar, R. and Kherani, E. A.: 1999, 'Effects of Molecular Ions of the Rayleigh–Taylor Instabilities in the Nighttime Equatorial Ionosphere', *J. Atmos. Terr. Phys.* **61**, 399–405.

Sekar, R. and Kelley, M. C.: 1998, 'On the Combined Effects of Vertical Shear and Zonal Electric Field Patterns on Nonlinear Equatorial Spread F Evolution', *J. Geophys. Res.* **103**, 20,735–20,747.

Sekar, R. and Kherani, E. A.: 2002, 'Effects of Molecular Ions on the Collisional Rayleigh-Taylor Instability: Nonlinear evolution', *J. Geophys. Res.* **107**, No A7, 10.1029/2001JA000167. SIA 16–1 to 16–9.

Sekar, R. and Raghavarao, R.: 1995, 'Critical Role of the Equatorial Topside F Region on the Evolutionary Characteristic of the Plasma Bubbles', *Geophys. Res. Lett.* **22**, 3255–3258.

Sekar, R., Suhasini, R. and Raghavarao, R.: 1994, 'Effects of Vertical Winds and Electric Fields in the Nonlinear Evolution of Equatorial Spread F', *J. Geophys. Res.* **99**, 2205–2213.

Sekar, R., Suhasini, R. and Raghavarao, R.: 1995, 'Evolution of Plasma Bubbles in the Equatorial F Region with Different Seeding Conditions', *Geophys. Res. Lett.* **22**, 885–888.

Sekar, R., Sridharan, R. and Raghavarao, R.: 1997, 'Equatorial Plasma Bubbles Evolution and its Role in the Generation of Irregularities in the Lower F-region', *J. Geophys. Res.* **102**, 20,063–20,067.

Sekar, R., Kherani, E. A., Rao, P. B. and Patra, A. K.: 2001, 'Interaction of Two Long-wavelength Modes in the Nonlinear Numerical Simulation Model of Equatorial Spread-F', *J. Geophys. Res.* **106**, 24,765–24,775.

Sridharan, R. *et al.*: 1997, 'Ionization Hole Campaign- a Co-ordinated Rocket and Ground Based Study at the Onset of Equatorial Spread -F: First Results', *J. Atmos. Terr. Phys.* **59**, 2051–2067.

Sridharan, R., Raghavarao, R., Suhasini, R., Narayanan, R., Sekar, R., Babu, V. V. and Sudhakar, V.: 1989, 'Winds, Wind Shears and Plasma Densities During the Initial Phase of Magnetic Storm from Equatorial Latitude', *J. Atmos. Terr. Phys.* **51**, 169–177.

Tsunoda, R. T.: 1980, 'Magnetic Field Aligned Characteristic of Plasma Bubbles in the Nighttime Equatorial Ionosphere', *J. Atmos. Terr. Phys.* **42**, 743–752.

Woodman, R. F. and Chau, J. L.: 2001, 'Equatorial Quasiperiodic Echos from Field-aligned Irregularities Observed over Jicamarca', *Geophys. Res. Lett.* **28**, 207–210.

SPACE WEATHER IN THE EQUATORIAL IONOSPHERE

R. J. STENING

School of Physics, University of New South Wales, Sydney 2052, Australia
(e-mail: r.stening@unsw.edu.au)

Abstract. The 'scintillations' observed on signals received in the equatorial region from GPS satellites are due to plasma instabilities in the F region of the ionosphere, also detected as spread F. These instabilities give rise to depletions of ionisation or 'bubbles'. The occurrence of these events and their relation to the equatorial electrojet are reviewed. Possibilities of short-term forecasting are examined with particular attention to problems encountered in modelling the equatorial electrojet.

Key words: equatorial electrojet, equatorial ionosphere, plasma bubbles, spread F

1. Introduction

Whereas the origin of most of the 'weather' in the ionosphere at mid and high latitudes lies with events on the Sun, much of the variability of the equatorial ionosphere is due to changes in the Quiet Day Sq current system which is driven by winds associated with various atmospheric tides. These changes are much harder to predict except for maybe a few hours ahead.

Although magnetic disturbances originating at the Sun do produce effects in the equatorial ionosphere, this paper will focus on changes occurring on quiet days.

Some of the phenomena of interest in the equatorial region are outlined in Figure 1. The tidal winds in the 'dynamo region' of the ionosphere (altitudes about 100–150 km) move the ions across the Earth's magnetic field to produce emfs. The emfs drive electric currents in the conducting E region and these give rise to daily variations in the magnetic field measured on the ground. This mechanism operates over the whole earth to give the well-known Sq current system with its two current whorls, counter-clockwise in the northern hemisphere and clockwise in the southern hemisphere. At the magnetic equator, where the Earth's magnetic field is horizontal, the electric current cannot flow far in the vertical direction because the conductivity is poor both above and below this region. This gives rise to a strong electrostatic field in the vertical direction, which drives an eastward Hall current known as the equatorial electrojet.

There remains some controversy over the equatorial electrojet as some workers maintain that it is a current system separate from the rest of the worldwide Sq system and is controlled primarily by local winds and emfs (Onwumechili, 1992). Other authors see the electrojet as an integral part of the worldwide Sq system (e.g.

Space Science Reviews **107**: 263–271, 2003.
© 2003 *Kluwer Academic Publishers.*

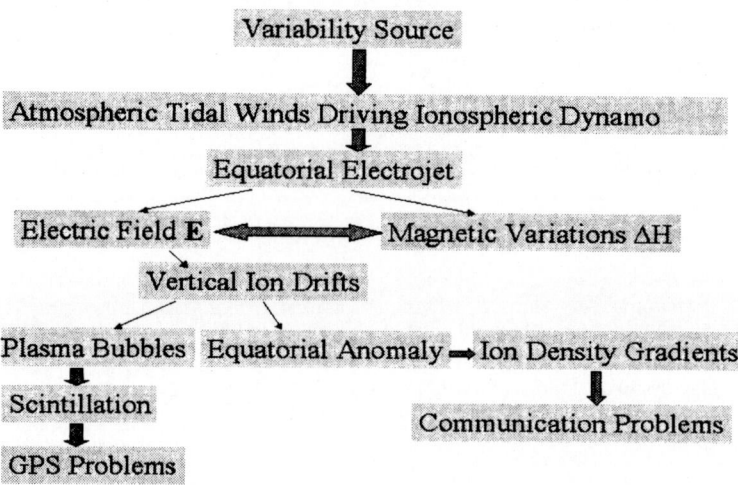

Figure 1. Relationships in the region of the magnetic equator.

Stening, 1995). It is important that this question be resolved, as we would need to look in different places for the 'Variability Source' shown in Figure 1 according to which is correct. In one case the variability will be caused by changes in local winds at the equator. In the other case changes in world-wide tidal wind systems will be responsible. There is a third possibility that both ideas are correct and the two mechanisms contribute to the electrojet different amounts on different days.

A horizontal electrostatic field is also associated with the electrojet and this interacts with the Earth's magnetic field to generate vertical $\mathbf{E} \times \mathbf{B}$ ion drifts. These drifts move plasma upwards across magnetic field lines at the equator. This plasma then diffuses downwards under the influence of gravity along magnetic field lines to form two 'crests' of enhanced ionisation in the F region at about 18° north and south of the magnetic equator. The result is a depletion of ionisation in the F region at the magnetic equator and two peaks on either side at these crests. This is known as the equatorial anomaly.

Around sunset the sudden decrease in the conductivity of the E region gives rise to a post-sunset enhancement of the electrostatic fields. The enhanced field gives rise to increased vertical ion drifts, which lead to 'bubbles' or depletions in ion density due to the Rayleigh–Taylor instability. During the existence of these bubbles, the ionosphere is changing very fast both in time and in space. When the bubbles form at a high altitude, they extend along the field lines to the anomaly crests. If a signal from a satellite passes through a region disturbed like this, there is much fluctuation in the signal strength received at a ground station and this is known as 'scintillation'. Such large changes in signal strength may cause problems to GPS receivers, limiting the accuracy with which they can pinpoint their position.

2. Equatorial space weather effects

The best known problems arising from ionospheric scintillation relate to the use of the GPS system. Wanninger (1993) has summarised these effects together with their spatial, seasonal and solar cycle variations. Recent observations in the Australian and Southeast Asian region have been given by Thomas *et al.* (2001) and Cervera *et al.* (2001). Figure 2 shows the observed seasonal and local time dependence as recorded by an ionospheric scintillation monitor at Parepare ($-12.6°$ magnetic latitude, $-26.2°$ dip) on the island of Sulawesi in Indonesia. Also shown is the predicted scintillation according to the empirical model WBMOD (Secan *et al.*, 1995). Here the occurrence is clearly centred around 21 hours local time and the equinoctial months. The S4 index is the normalized standard deviation of the detrended signal power received from a satellite. With no scintillation present $S4 < 0.05$.

The use of GPS satellites for positioning has increased importance as they are used with the Wide Area Augmentation System (WAAS), a connected system of satellites and receivers operated by the Federal Aviation Authority in the U.S.A., which seeks to use real time information on the ionosphere to correct GPS positioning to an accuracy of a few centimetres.

Other effects in the equatorial region relate to the operation of over-the-horizon radars (Dandekar *et al.*, 1998) and to the use of the trans-equatorial propagation mode for communication. There is one propagation mode in the evening which appears to use the depleted ionisation (bubble) as a waveguide between the northern and southern hemispheres (Heron and McNamara, 1979).

3. What Conditions are Necessary for Bubble Formation?

Fejer *et al.* (1999), using observations from the American sector, showed that the magnitude of the vertical plasma drift during the evening prereversal enhancement was the main factor in determining the occurrence of equatorial spread F. As solar activity increases (as measured by the F10.7 solar flux), a larger drift velocity is required for the generation of spread F. During the June solstice the drift velocities are generally smaller and the reversal occurs earlier, so spread F occurrence is a lot less in this season.

Whalen (2002), using South American ionosonde data from 1958, found that strong (obscuring f_oF2) range spread F, lasting at least 2 hours and seen at observatories near the dip equator at least $2°$ of latitude apart, was a necessary but not sufficient criterion for bubble formation. He also showed that bubble occurrence decreases with increasing magnetic index Kp value till there is virtually no occurrence with Kp greater than 4 during the 6 hours preceding sunset. This relates to the fact that the equatorial electrojet, the eastward electric field, the vertical drifts and the development of the equatorial anomaly are all inhibited during magnetic

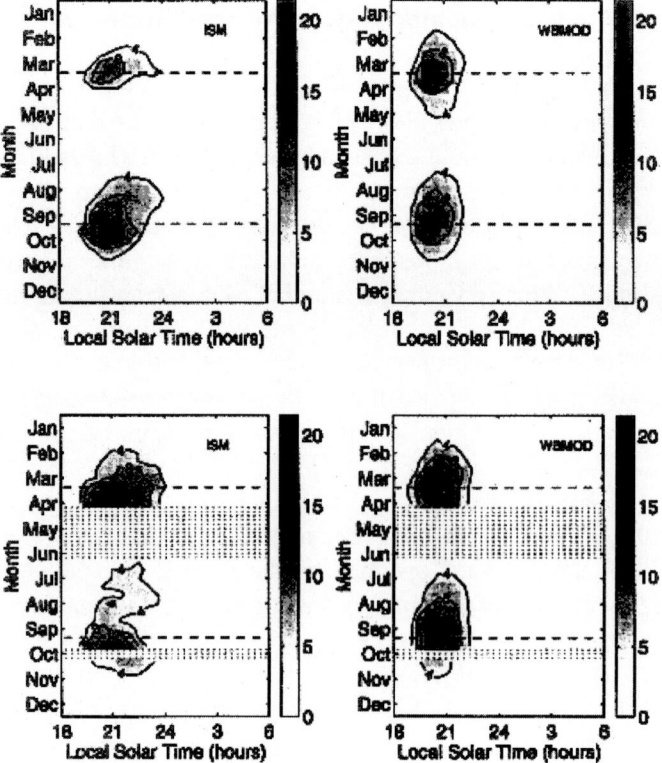

Figure 2. Percentage of time that the S4 scintillation index exceed 0.3 as a function of local time and season at Parepare. Top panels are for 1998 and bottom for 1999. Left panels show data gathered from the ionospheric scintillation monitor while right panels are predictions from the WBMOD model. From Cervera *et al.* (2001). Copyright (2001) American Geophysical Union. Reproduced by permission of American Geophysical Union.

disturbance. The disturbance effect may operate either by means of the 'disturbance dynamo' (Blanc and Richmond, 1980) or by the transmission of westward electric fields from the auroral zone down to the equator.

How long the 'bubble disturbance' lasts depends also on the behaviour of the electric field driving the drifts. If the field reverses to a strong westward value, then the disturbance ceases within an hour or so, but if the westward reversal is weak, as is often the case in December in the American sector, then the disturbance persists for several hours after midnight. Sometimes the disturbance dynamo may act to increase bubble generation after midnight (Hysell and Burcham, 2002). Devasia *et al.* (2002) give evidence of the influence of meridional winds on whether spread F forms or not. They find that when the F region maximum is below 300 km an equatorward wind of 'significant amplitude' is necessary to trigger spread F. When the F region has been lifted above 300 km by the electrodynamic drift, the spread F is triggered regardless of the direction of the meridional wind.

On the theoretical side there have been investigations into factors influencing the growth rate of the Rayleigh–Taylor instability. Sultan (1996) emphasises that it is the magnetic flux tube integrated values of quantities, such as ion density gradients, which are important in determining the growth rate. Some 'seeding mechanism' is also required to initiate the instabilities. Abdu (2001) suggests that tropospheric convective activity associated with the Inter-tropical Conversion Zone might be responsible while other authors believe that some seeding mechanism is virtually always there.

There is also a longitude-seasonal control. When the evening solar terminator lines up with the magnetic meridian, then magnetically conjugate points move from light to shadow at the same time and the east-west gradient in field line integrated conductivity is largest. Then the pre-reversal enhancement of the vertical drift is also largest, giving maximum spread F occurrence (Tsunoda, 1985; Abdu, 2001).

4. How Can we Measure Those Parameters which could Help Us Forecast Bubble Formation?

The vertical ion drift velocities are the most important factor in determining post-sunset F region behaviour. These may be measured by an incoherent scatter radar but these radars are not run continuously. Furthermore, when spread F and bubbles are present, the radars are often unable to determine the drift velocity. We should therefore look for a proxy for the drift velocity and the strength of the equatorial electrojet is an obvious candidate.

Anderson *et al.* (2002) have investigated the relation between the drift velocity, as measured by the Jicamarca radar, and the electrojet strength, as measured by a ground-based magnetometer. They have shown that a linear relationship between these two parameters can be established but they have not yet determined how the proportionality varies with season, solar cycle etc. They find that the proportionality does change significantly if the direction of the electrojet is reversed.

We should be able to determine the proportionality by means of a theoretical model and some work has been done towards this (Richmond, 1973a; Stening, 1985). However the models have had problems in correctly simulating the latitude profile of the electrojet as measured by a chain of magnetometers on the ground (Fambitakoye *et al.*, 1976). In particular there is a problem with the relation of the horizontal component of the magnetic field at the electrojet centre, ΔH, to that measured about $5°$ latitude away from the centre (the so-called planetary Sq component). The models give a larger ratio of these two values than is generally observed.

In Figure 3 the model used is the equivalent circuit model of Du and Stening (1999) with a simple diurnal wind with no variation with height and a latitude dependence corresponding to the $(1,-2)$ tidal mode. The fit looks good except 86 nT

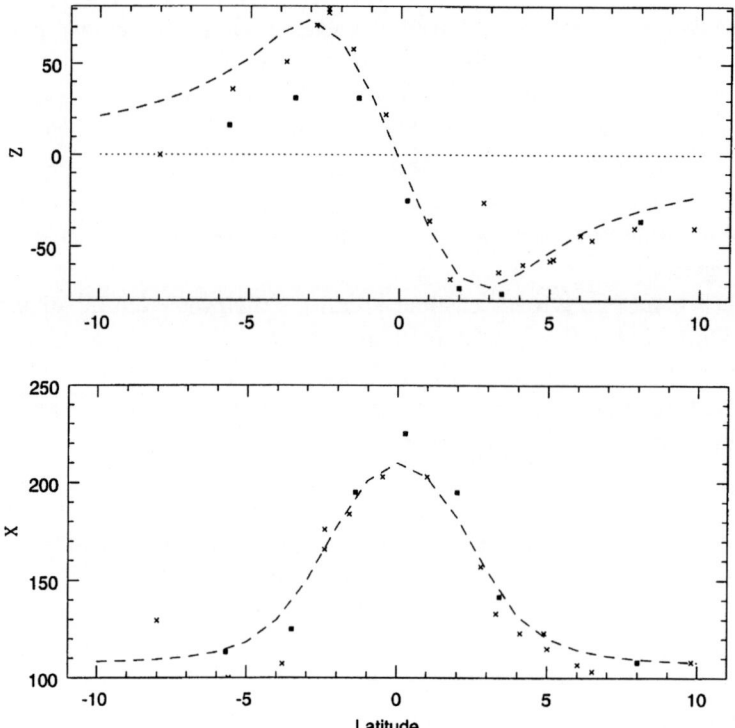

Figure 3. Latitude profile of magnetic fields due to the equatorial electrojet. Top panel: vertical component. Bottom panel: northward component. Dashed curve: simulation. The bottom simulation curve has 86 nT added to it. Crosses: data from Peru (Richmond, 1973b). Squares: African data (Fambitakoye and Mayaud, 1976). All fields have been scaled to 205 nT at Huancayo in Peru.

has been added to the simulated curve to bring it into agreement. The observed ΔH variation falls to half its maximum value at about 5° latitude while the simulated variation falls to half at 3° latitude.

Fambitakoye *et al.* (1976) give three possible causes for this discrepancy:

1. There are local wind systems which enhance the equatorial electrojet without affecting the worldwide Sq current system. This idea is part of Onwumechili's picture of an electrojet current system separate from the worldwide system. It is supported by such observational evidence as the poor correlation between ΔH at equatorial electrojet observatories and ΔH from low latitude observatories away from the dip equator (Mann and Schlapp, 1988). Yet questions remain as to the nature of such local wind systems. If they are truly local, then they are likely to have the properties of atmospheric internal gravity waves with fairly short vertical wavelengths and intermittent occurrence and variable phase. If they are part of an atmospheric tidal system, then they are not truly local and the tide will have affects on current systems far away from the equator as well.

Note that these affects may not be seen at stations just off the equator but rather at higher latitudes (e.g. Stening, 1977).

2. Non-ionospheric currents such as those generated on the magnetopause (Olson, 1970). These yield only a few nanotesla at the equator and so are an unlikely answer to the problem.

3. Instability effects. Richmond (1973a) suggested that, when the electrojet becomes intense, a gradient-drift instability occurs in the dynamo region, decreasing the effective conductivity and so producing a saturation effect: the electrojet current is unable to increase past a certain value. This is an attractive suggestion except that the discrepancy between observed and model electrojet widths exists also at low electrojet intensities.

Another two possibilities may be considered:

1. A secondary wind system acts to deplete the electrojet more than nearby currents. This might be achieved perhaps with some semidiurnal tide added to the dominant diurnal tide usually thought to be responsible for the Sq system. However it seems unlikely that such an added tidal effect would always have the right phase to yield what is observed.

2. The latitude profile of the currents, induced in the ground by the electrojet, is much wider than the ionospheric current profile. It is indeed likely that the induced current profile will be wider as, unlike the ionosphere, the ground has a fairly uniform conductivity. The magnitude of this effect remains to be calculated.

Once the relationship between ΔH and the vertical drifts has been established, then it would be possible to look for signs in the morning behaviour of ΔH which might signal a larger or smaller post-sunset drift and so forecast the likelihood of spread F bubbles and scintillation effects.

There is some evidence that the 'regular' electrojet is modulated to some extent by the influence of semidiurnal atmospheric tides. Some effect comes from lunar semidiurnal tides (Stening and Fejer, 2001) but the solar tides are likely to be larger. If the phase of the tide is suitable, the electrojet may reverse direction in the afternoon. This occurrence is often accompanied by an increased amplitude in the forward electrojet in the morning, thus permitting a few hours of forecast capability (Stening, 1977). However the relation to the vertical drift speeds has not yet been firmly established. The reversed afternoon electrojet would also lead to an earlier reversal in drift speed direction and so to a decreased likelihood of scintillation effects. Millward *et al.* (2001) have modelled the prereversal drift enhancement and also conclude that the magnitude and phase of the semidiurnal atmospheric tide are important factors in determining the extent of the enhancement. Fesen *et al.* (2000) have also modelled the equatorial ion drifts using the Thermosphere-Ionosphere-Electrodynamic General Circulation Model and they likewise established the importance of semidiurnal tides.

Planned for launch in 2003, the C/NOFS (Communication/Navigation Outage Forecasting System) satellite will measure ion densities, electric fields, neutral winds, the electron density profile and ionospheric scintillations. The development of this mission indicates the importance of these phenomena in the equatorial region.

5. Conclusions

As greater and greater accuracy is demanded from GPS systems, the ionosphere is found to give the most serious limitations. Forecasting the behaviour of the evening equatorial ionosphere may become possible with greater understanding of the mechanisms which control day-to-day changes in the equatorial electrojet. While space weather associated with disturbances originating on the Sun do affect this equatorial region, the scintillation phenomena discussed here are often diminished during increased magnetic activity. They are different from other geographical regions in that the mechanisms controlling the occurrence of scintillations originate in the lower atmosphere.

Acknowledgement

The assistance of Dr J Du in preparing this report is gratefully acknowledged.

References

Abdu, M. A.: 2001, 'Outstanding Problems in the Equatorial Ionosphere-Thermosphere Electrodynamics Relevant to Spread F', *J. Atmos. Solar Terr. Phys.* **63**, 869–884.

Anderson, D. N., Anghel, A., Yumoto, K., Ishitsuka, M. and Kudeki, E.: 2002, 'Estimating Daytime Vertical E × B Drift Velocities in the Equatorial F-region Using Ground-based Magnetometer Observations', *Geophys. Res. Lett.* 10.1029/2001GL014562.

Blanc, M. and Richmond, A. D.: 1980, 'The Ionospheric Disturbance Dynamo', *J. Geophys. Res.* **85**, 1669.

Cervera, M. A., Thomas, R. M., Groves, K. M., Ramli, A. G. and Effendy: 2001, 'Validation of WBMOD in the Southeast Asian Region', *Radio Sci.* **36**, 1559–1572.

Dandekar, B. S., Sales, G. S., Weijers, B. and Reynolds, D.: 1998, 'Study of Equatorial Clutter Using Observed and Simulated Long-range Backscatter Ionograms', *Radio Sci.* **33**, 1135–1157.

Du, J. and Stening, R. J.: 1999, 'Simulating the Ionospheric Dynamo - I. Simulation Model and Flux Tube Integrated Conductivities', *J. Atmos. Solar Terr. Phys.* **61**, 913–923.

Devasia, C. V., Jyoti, N., Subbarao, K. S. V., Viswanathan, K. S., Diwakar Tiwari and Sridharan, R.: 2002, 'On the Plausible Linkage of Thermospheric Meridional Winds with the Equatorial Spread F', *J. Atmos. Solar. Terr. Phys.* **64**, 1–12.

Fambitakoye, O. and Mayaud, P. N.: 1976, 'Equatorial Electrojet and Regular Daily Variation S_R – I. A Determination of the Equatorial Electrojet Parameters', *J. Atmos. Terr. Phys.* **38**, 1–17.

Fambitakoye, O., Mayaud, P. N. and Richmond, A. D.: 1976, 'Equatorial Electrojet and Regular Daily Variation S_R – III. Comparison of Observations with a Physical Model', *J. Atmos. Terr. Phys.* **38**, 113–121.

Fejer, B. G., Scherliess, L. and de Paula, E. R.: 1999, 'Effects of the Vertical Plasma Drift Velocity on the Generation and Evolution of Equatorial Spread F', *J. Geophys.Res.* **104**, 19,859–19,869.

Fesen, C. G., Crowley, G., Roble, R. G., Richmond, A. D. and Fejer, B. G.: 2000, 'Simulation of the Pre-reversal Enhancement in the Low Latitude Vertical Ion Drifts', *Geophys. Res. Lett.* **27**, 1851 –1854.

Heron, M. L. and McNamara, L. F.: 1979, 'Transequatorial VHF Propagation Through Equatorial Plasma Bubbles', *Radio Sci.* **14**, 897–910.

Hysell, D. L. and Burcham, J. D.: 2002, 'Long Term Studies of Equatorial Spread F using the JULIA Radar at Jicamarca', *J. Atmos. Solar Terr. Phys.*, in press.

Mann, R. I. and Schlapp, D. M.: 1988, 'The Equatorial Electrojet and Day-to-day Variability of Sq', *J. Atmos. Terr. Phys.* **50**, 57.

Millward, G. H., Müller-Wodarg, I. C. F., Aylward, A. D., Fuller-Rowell, T. J., Richmond, A. D. and Moffett, R. J.: 2001, 'An Investigation into the Influence of Tidal Forcing on F Region Equatorial Vertical Ion Drift using a Global Ionosphere-thermosphere Model with Coupled Electrodynamics', *J. Geophys. Res.* **106**, 24733–24744.

Olson, W. P.: 1970, 'Contribution of Nonionospheric Currents to the Quiet Daily Magnetic Variation at the Earth's Surface', *J. Geophys. Res.* **75**, 7244–7249.

Onwumechili, C. A.: 1992, 'Study of the Return Current of the Equatorial Electrojet', *J. Geomagn. Geoelectr.* **44**, 1.

Richmond, A. D.: 1973a, 'Equatorial Electrojet, Part 1: Development of a Model Including Winds and Instabilities', *J. Atmos. Terr. Phys.* **35**, 1083–1103.

Richmond, A. D.: 1973b, 'Equatorial Electrojet – II. Use of the Model to Study the Equatorial Ionosphere', *J. Atmos. Terr. Phys.* **35**, 1105–1118.

Secan, J. A., Bussey, R. M., Fremouw, E. J. and Basu, S.: 1995, 'An Improved Model of Equatorial Scintillation', *Radio Sci.* **30**, 607–617.

Stening, R. J.: 1977, 'Magnetic Variations at Other Latitudes During Reverse Equatorial Electrojet', *J. Atmos. Terr. Phys.* **39**, 1071–1077.

Stening, R. J.: 1985, 'Modeling the Equatorial Electrojet', *J. Geophys. Res.* **90**, 1705–1720.

Stening, R. J.: 1995, 'What Drives the Equatorial Electrojet?', *J. Atmos. Terr. Phys.* **57**, 1117–1128.

Stening, R. J. and Fejer, B. G.: 2001, 'The Lunar Tide in the Equatorial F Region Vertical Ion Drift Velocity', *J. Geophys. Res.* **106**, 221–226.

Sultan, P. J.: 1996, 'Linear Theory and Modelling of the Rayleigh–Taylor Instability Leading to the Occurrence of Equatorial Spread F', *J. Geophys. Res.* **101**, 26875–26891.

Thomas, R. M., Cervera, M. A., Eftaxiadis, K., Manurung, S. L., Saroso, S., Effendy, Ramli, A. G., Salwa Hassan, W., Rahman, H., Dalimin, M. N., Groves K. M. and Wang, Y.: 2001, 'A Regional GPS Receiver Network for Monitoring Equatorial Scintillation and Total Electron Content', *Radio Sci.* **36**, 1545–1557.

Tsunoda, R. T.: 1985, 'Control of the Seasonal and Longitudinal Occurrence of Equatorial scintillation by Longitudinal Gradient in Integrated Pedersen Conductivity', *J. Geophys. Res.* **90**, 447– 456.

Wanninger, L.: 1993, 'Effects of the Equatorial Ionosphere on GPS', *GPS World*, 48–54, July 1993.

Whalen, J. A.: 2002, 'Dependence of Equatorial Bubbles and Bottomside Spread F on Season, Magnetic Activity, and E × B Drift Velocity during Solar Maximum', *J. Geophys. Res.* **107**, 10.1029/2001JA000039.

LIGHTNING INDUCED OPTICAL EMISSIONS IN THE IONOSPHERE

J. A. VALDIVIA

Departamento de Fisica, Universidad de Chile, Santiago, Chile and World Institute for Space Environment Research (WISER), University of Adelaide, Adelaide, SA 5005, Australia

Abstract. A discussion of lightning induced optical emissions in the ionosphere is presented. Emphasis is placed on accounting for the puzzling observation of the spatial structure in the optical emissions and the Sprite 'seeding' before the development of the 'tendrils' (or streamers). In this context we discuss the generation of spatial brightness variations, within the required lightning parameter thresholds, due to spatio-temporal electric fields and spatial neutral density perturbations.

Key words: fractal antenna, ionosphere, optical emissions, Sprites

1. Introduction

The problem of the generation, propagation and interaction of the lightning induced electric field in the upper atmosphere and ionosphere is a very interesting subject that requires bridging together many fields of physics. Of special current interest is the issue of the spatially structured optical emissions and the 'seeding' of Sprites.

High-altitude optical flashes were detected more than 100 years ago (Kerr, 1994). Interest in the subject was renewed recently following the observations of such flashes by the University of Minnesota group (Franz *et al.*, 1990). Since then, observations of the optical emissions at altitudes between 50 and 90 km associated with thunderstorms have been the focus of many ground and aircraft campaigns (to name a few, Boeck *et al.*, 1992; Vaughan *et al.*, 1992; Winckler *et al.*, 1993; Lyons, 1994; Sentman *et al.*, 1995; Lyons, 1996; Wescott *et al.*, 2001). Sentman and Wescott (1993) and Winckler *et al.* (1996) described the phenomenon, called 'red Sprites', as a luminous column that stretches between 50 and 90 km, with peak luminosity in the vicinity of 70–80 km. The flashes have an average lifetime of a few milliseconds and an optical intensity of about 100 kR. Red Sprites are associated with the presence of massive thunderstorm clouds, and among the most puzzling aspects of the observations is the presence of spatial structure in the emissions reported by Winckler *et al.* (1996). Vertical striations, denominated 'tendrils', with horizontal size of 1 km or smaller, often limited by the instrumental resolution, are apparent in the red Sprite emissions (Stanley *et al.*, 1999; Gerken *et al.*, 2000). These 'tendrils', which seem to be a strong component of the spatial structure present in Sprites, have been modeled as streamers (Raizer *et al.*, 1998; Pasko *et al.*, 1998). The source of the 'seeds' that start the streamers remains one

Space Science Reviews **107**: 273–291, 2003.
© 2003 *Kluwer Academic Publishers.*

of the open questions in Sprite research, and is one of the topics of the present paper. High speed images have revealed that the optical emissions appear in a variety of situations such as: (a) large regions with spatially homogeneous emissions, (b) regions with spatially structured emissions, (c) Sprites initiating from areas of considerable optical emissions, (d) Sprites initiating in regions with no previous optical emissions, (e) Sprites initiating from long-lived beads, (f) optical emissions appearing in the same spot after many milliseconds, (g) Sprites stimulated by upward propagating discharges, (h) etc. (see for example Stanley *et al.*, 1999; Gerken *et al.*, 2000; Moudry *et al.*, 2002; Pasko *et al.*, 2002). Still, it is traditional to try to characterize these optical emissions as either Elves, Halos or Sprites, even though this may be difficult in some situations, e.g., optical emissions that can be considered as spatially varying Halos or as Sprites that don't develop streamers (see figures in Moudry *et al.*, 2002).

In this paper, we address the very important question that is related to the generation of the spatio-temporal optical emission patterns, and the seeding of 'Sprites'. These issues become relevant before the development of the 'tendrils' (or streamers) and should be discussed in detail. In particular, we will consider two possible candidates for producing spatio-temporal optical emissions and 'seeding' of Sprites: (a) the spatio-temporal electric field pattern produced by the intracloud portion of the lightning discharge, and (b) the neutral density perturbations. Of course other candidates must also be considered in detail.

The first published theoretical models of these optical emissions associated their generation with transient electric fields induced by lightning discharges. In a broad sense, the electric field produced by a current structure, is given by Appendix A,

$$4\pi\epsilon_o E \sim \frac{M_1}{r^3} + \frac{1}{cr^2}\frac{dM_2}{dt} + \frac{1}{c^2 r}\frac{d^2 M_3}{dt^2} \tag{1}$$

where M_i are the moment array factors. The first term represents the quasi-electrostatic (QE) field generated by the charge separation, while the last term, due to the charge motion, is usually associated with the radiative electromagnetic pulses (EMP) which may produce spatially structured field patterns in the far field. Due to the large variability in the lightning discharge parameters (see book by Uman, 1984), it is expected that the relative importance of each term (hence QE, EMP, etc.) depends on the specific discharge case. It has been traditional to try to associate a specific type of field with a particular optical emission, such as Elves, Halos or Sprites. It is important to understand that in the convoluted environment of the ionosphere, a nonlinear media, such association may turn out to be difficult, with some of the observations including all the effects together: QE fields, EMPs, pressure/compositional atmospheric variations, etc.

We will discuss the general framework for understanding the coupling of the lightning induced fields (Appendix A), their nonlinear propagation through the upper atmosphere (Appendix B), and the interaction with the ionospheric media responsible for the observed optical emissions (Appendix C). Particular attention will be given to the issue of the observed spatial structure in the optical emissions and the 'seeding' of Sprites, which due to the large variability in the optical emission phenomena mentioned above, it is clearly a nontrivial problem. We will describe the spatially structured field pattern produced by the intracloud lightning component, as one of the possible contributions to the spatially structured optical emissions, and a possible 'seed' for Sprites. Another component is the background pressure or compositional spatial structure of the ionosphere, as described by density/pressure/compositional fluctuations, such as gravity wave fluctuations (Yamada *et al.*, 2001).

2. Ionospheric Electrodynamics in the Lightning Context

The charge separation and the charge motion during a lightning discharge generates electromagnetic fields that propagate upward into the upper atmosphere. In Appendix A, Equation (5), we discuss how to calculate the electric field produced by a distribution of line elements. As the field propagates into the upper atmosphere, it energizes the ionospheric electrons, inducing a self-absorption effect that requires the self-consistent computation of the propagation of the lightning-induced fields in the upper atmosphere. This involve solving a nonlinear wave equation as described in Appendix B, given by Equation (7), where the transport coefficients, such as the conductivity, depend implicitly on the field amplitude. As the fields propagate through the ionosphere, the ambient electrons get energized generating non-Maxwellian distribution functions, that could be described, for example, by the Fokker Planck approach of Appendix C. Energizing the electrons is a two step process. First, the energization of the electrons is done at the expense of the field energy affecting the field transport coefficient of Equation (7). And second, the electrons give their increased energy through collisional excitation of neutral and ion particles, including ionization. In this manner we have a closed coupled system that will guide us in understanding the observations. All the transport coefficients are described in Appendix C. It is possible skip the calculation of Appendix A, by including directly in Equation (7) the current and charge densities as source terms. This may be difficult in the case of a complex current source, but some simplifications can be accomplished with the help of the Hertz vector. A further simplification can be done by assuming QE fields and by using a semi-empirical approach for the transport coefficients (see Pasko et al., 1997).

Elves are described as diffuse optical flashes that last < msec with a horizontal diameter > 100 km and that appear at a height ~ 100 km. Elves, due to their direct timing relation to the parent positive cloud to ground (+CG) discharge, seem to be causally connected with the EMPs produced directly by the +CG parent return stroke (Inan *et al.*, 1997; Rowland, 1998). This direct causal connection is not necessarily clear for the other optical phenomena. The EMP produced by a 'vertical' dipole current (a simple model of a +CG discharge) will reach the ionosphere is less than a msec, and will produce a very homogeneous field pattern reminiscent of an Elve. Similarly, Pasko *et al.* (1997) estimated the relevance of the QE fields in the ionosphere produced by the neutralization of charge during the same +CG discharge, which produces a homogeneous (characteristic r^{-3} QE field) emission pattern, reminiscent of a Halo (see also Barrington-Leigh *et al.*, 2001). Halos (or Sprite Halos) are described as diffuse optical flushes that seem to occur at an altitude of ~ 80 km and have a diameter of ~ 60 km and may last for a few msec producing more photons that an Elve (Wescott *et al.*, 2001). It is assumed that Halos may occur with or without Sprites.

The first published theoretical model of red Sprites assumed an intracloud discharge, modeled as a horizontal electric dipole at an altitude of 10 km, and the field calculation included quasi-static, intermediate, and far-field components (Milikh *et al.*, 1995). In this paper Milikh *et al.* (1995) demonstrated that energization of the ionospheric electrons by the transient fields could account for several of the observed features of the observed optical emissions, such as color, duration, height, etc. Pasko *et al.* (1995) later reached the same conclusions by modeling the electric fields energizing the electrons as due to a point charge Q (monopole) at an altitude of 10 km. The analysis emphasized the importance of dielectric relaxation of the field in the ionosphere.

All of the above models, while successful in explaining some observed characteristics of the optical emissions, such as the color and the generation altitude of the emissions, suffer from two important drawbacks. First, dipole or monopole charge distributions generate electric fields smoothly distributed at ionospheric heights, thereby failing to account for the persistent spatial structure in the optical emissions. Second, the threshold charge and dipole moment requirements for all three models have been criticized as unrealistically large (based on the lightning parameters cited by Uman (1984).

Even as early as 1996, Lyons (1996) noticed that in some situations there was a time delay between the Sprites and the parent +CG, which could reach values as large as 100 msec. Such time delays may be difficult to reconcile with the QE fields produced by the +CG return stroke. It was then suggested that the complex dynamics (Mazur *et al.*, 1998; Villanueva *et al.*, 1994) of the associated intracloud discharge, and its induced electric field pattern in the ionosphere (EMP and/or QE), may be able to explain, in a natural manner, the time delay observed between the sprite and the parent +CG (Lyons, 1996). Milikh *et al.*, (1995) studied the energetics of the optical emissions produced by a horizontal current element, because it was recognized that it would be more efficient to radiate the EMPs if the lightning discharge had an horizontal, or intracloud discharge, component. Along this line, Lyons (1996) suggested that the +CG discharges correlated with Sprites were associated with 'extremely complex' horizontal intracloud 'spider' or 'dendritic' discharges and that these may be related to the production of Sprites. A number of observations suggest that this may indeed be the case, e.g. (a) time delay between +CG and Sprite, (b) size requirement of convective cell, (c) displacement of sprite with +CG, (d) etc. (see the details in Lyons, 1996). Lyons (1996) concluded that "while +CG seems to be a necessary condition for sprite formation, it may not be a sufficient one". At about the same time, Winckler *et al.* (1996) observed that the optical emissions from Sprites were spatially structured which suggested that the radiation pattern of the lightning induced fields could be spatially structured as well. A horizontal intracloud fractal lightning discharge model (Valdivia *et al.*, 1997) was constructed to try to account for these observations, in the hope of quantifying the spatially structured contribution of the lightning induced EMPs to the associated optical emissions.

Similarly, the Valdivia *et al.* (1998) model hoped to capture some of the very complex dynamics of the unbalanced charge distribution in the cloud (Mazur *et al.*, 1998; Villanueva *et al.*, 1994), just before or after the +CG discharge, using a simple fractal parameterization. It is important to notice that the charge re-organization happening during this complex intracloud discharge may equally well produce a QE field pattern large enough to be observable, but due to their r^{-3} dependence it is expected to be homogeneous in nature. As it became clear that the 'tendrils' seem to be a strong component of the spatial structure present in Sprites, a similar fractal parameterization was used by Pasko *et al.* (2000) to model the spatial structure of the Sprites based on a fractal streamer evolution model in the presence of a QE field pattern, after the initial spatially dependent 'seeds' were in place. In this paper there was no clear discussion of the original 'seed' that starts the streamers (Raizer *et al.*, 1998; Pasko *et al.*, 1998), but seemed to capture some characteristics of their evolution. It is important to remember that these tendrils may reach scales as small as < 50 m (Gerken *et al.*, 2000). Raizer *et al.* (1998) proposed that the streamers would naturally start, in the presence of a QE field, from the spatially structured conductivity profile left in the ionosphere by the spatially structured electric field pattern produced by the intracloud discharge, thus providing the 'seeding' for Sprites. Other postulated 'seeds' must also be considered in detail, such as cosmic rays, meteoric dust (Zabotin and Wright, 2001), etc.

Most of the Sprites seem to be correlated with +CG discharges (Lyons, 1996), but there have been reports of a few Sprites correlated with −CG discharges, and even Sprites with no associated CG discharge (see Westcott *et al., 2001*). These intriguing observations may be relevant when comparing models (see Valdivia *et al.,* (1998) for a possible explanation when considering the intracloud fractal source).

There has been a large amount of work on these, and other issues, such as: the ionization and charge dynamics during Sprites, the electromagnetic waves produced by Sprites, mesospheric coupling and chemistry, magnetospheric coupling, etc. But due to size restrictions we cannot mention them all (e.g., Veronis *et al.,* 1999; Stanley *et al.,* 1999; Wescott *et al.,* 2001; Moudry *et al.,* 2002).

We will now review two mechanisms that may provide possible 'seeds' for the generation of Sprites and streamers, namely (Section 2.1) the spatially structured electric fields and (Section 2.2) the atmospheric spatial variations of the density perturbations.

2.1. THE SPATIALLY STRUCTURED ELECTRIC FIELDS

Since it is expected that a horizontal discharge is a better ionospheric radiator than a vertical current discharge, we assume that we have a horizontal intracloud lightning discharge placed for example at a height $h = 5$ km above the ground. In view of the complex dynamics of the charge imbalance inside the cloud, assuming an intracloud lightning discharge would naturally offer a solution for the observed time delay between the Sprite and the +CG, in the case of a long-delayed Sprite.

It is well known that lightning discharges follow a tortuous path (LeVine and Meneghini, 1978), and that intracloud discharges resemble the well-known Lichten-berg patterns observed in dielectric breakdown (Williams, 1988). These patterns are considered fractal structures with a fractal dimension $D \approx 1.6$ (Niemeyer et al., 1984). A fractal intracloud lightning discharge can be considered as a two-dimensional horizontal phased array antenna that will naturally exhibit a spatially dependent radiation pattern with an effective gain factor. Such a gain factor is extremely important, for it will reduce the lightning energetics, compared with a dipole model, needed to produced the optical emissions in the ionosphere. For an oscillating current $e^{i\omega t}$, each element of the phased array contributes to the total radiated power density at a given point with a vectorial amplitude $\mathbf{A}_n$ and phase ϕ_n,

$$\mathbf{E} \cdot \mathbf{E}^* = \sum_{n,m} (\mathbf{A}_n \cdot \mathbf{A}_m^*) e^{i(\phi_n - \phi_m)}$$

and in the sense of statistical optics, we can consider the ensemble average using an ergodic principle to obtain (Valdivia et al., 1998)

$$G \sim \langle \mathbf{E} \cdot \mathbf{E}^* \rangle \sim N^2 \left(\frac{\langle |\mathbf{A}|^2 \rangle}{N} + \frac{N-1}{N} |\langle \mathbf{A} \rangle|^2 \, |\langle e^{i\phi} \rangle|^2 \right) \tag{2}$$

The first term represents the incoherent radiation component, $G \sim N$, and the second term is the coherent (interference) radiation component that may contribute with significant gain, e.g. $G \sim N^2$ depending on the spatial current distribution that can be characterized by its fractal dimension. A fractal antenna will display partial coherence in some direction(s), and the peak radiated power will be in between these two limits, showing a significant gain over a random distribution of phases (see Figure 1).

In Figure 1a the top drawing shows the dipole, or linear, model of the current channel between two points, and in general it will generate a dipole radiation pattern. The middle drawing shows a tortuous model of the discharge between the same two points, where this tortuous current channel can be considered as a two-dimensional phased array, as shown in the bottom drawing. Note that the tortuous model will have a longer path length that will increase the effective N and the

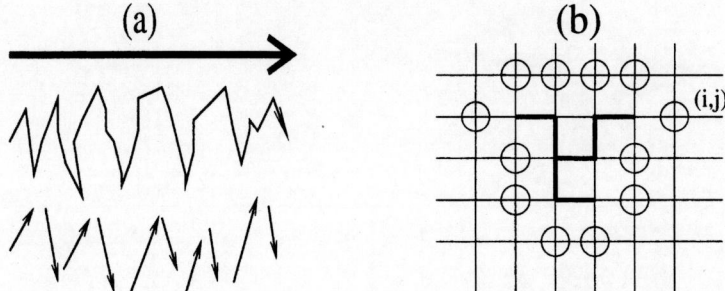

Figure 1. (a) Dipole or straight line model (top), tortuous channel model (middle), and equivalent phased array (bottom). (b) Diagram of the discretized fractal discharge and its adjacent grid points. The solid circles correspond to the discharge, and the open circles correspond to the adjacent points that can be added to the discharge.

individual A_n in Equation (2) with respect to the dipole model. Clearly, there may be positions at which the radiation pattern from the tortuous line elements will add constructively, while at other positions they may add destructively.

Therefore, the radiated electromagnetic pulses from a horizontal intracloud fractal lightning discharge can add coherently in the ionosphere producing a spatially structured radiation pattern with a significant antenna gain compared with a simple dipole radiator. Our intracloud fractal discharge can then be modeled as a nonuniform set of small current line elements as described in Appendix A (see also Vecchi *et al.*, 1994)

The dynamics of the unbalanced charge distribution in the cloud is very complex (Mazur *et al.*, 1998; Villanueva *et al.*, 1994), and in the following we will try to capture some of this complexity through some specific fractal parameterizations. Since Femia *et al.* (1993) found experimentally that a dielectric discharge pattern is approximately an equipotential, we construct a fractal discharge by adapting for our purposes the two-dimensional stochastic dielectric discharge model proposed by Niemeyer *et al.* (1984). This model naturally leads to fractal structures where the fractal dimension can be easily parametrized by a parameter η. We start with a charge Q at the center of our computational box, and the discharge propagates in steps of 100 m. Consider a discharge pattern (thick line in Figure 1b) at a later time t, which we assume to be an equipotential with value $\phi = 1$. The boundary condition $\phi = 0$ is forced at infinity (on a circle at $L = 10$ km). The electric potential outside the discharge structure satisfies Laplace's equation $\nabla^2 \phi = 0$ which can be solved by a relaxation method that consists in iterating

$$\phi_{i,j}^{n+1} = \frac{1}{4}(\phi_{i+1,j}^n + \phi_{i-1,j}^n + \phi_{i,j+1}^n + \phi_{i,j-1}^n)$$

until it converges. The discharge pattern grows in single steps by the addition of an adjacent grid point (open circles in Figure 1b) to the discharge pattern, generating a new bond. We assume here that an adjacent grid point, denoted (i, j), has a probability proportional to the η power of the local electric field to become part

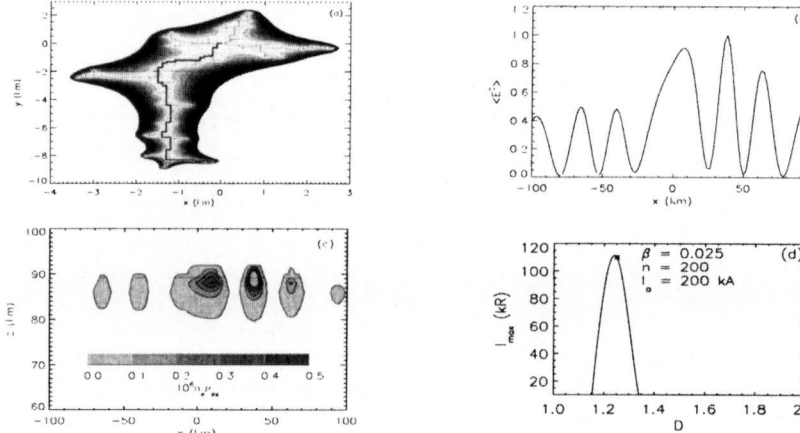

Figure 2. (a) Fractal discharge generated with the stochastic model for $\eta = 3$ (top left). The gray shading corresponds to the electric potential and the thickness is proportional to the current. (b) The radiated pattern at $h = 60$ km for $n_f = 200$ (top right). (c) The emission pattern (bottom left). (d) The maximum emission intensity as a function of fractal dimension (bottom right). The parameters are $\eta = 3$, $\beta = 0.025$, $I_o = 200$ kA.

of the pattern, and we apply the Monte Carlo method to add one of the adjacent points to the discharge. The new grid point, being part of the discharge structure, will have the same potential as the discharge pattern, i.e., $\phi = 1$, changing ϕ over the grid. Therefore, we must solve Laplace's equation for the potential every time we add a new bond. When $\eta = 0$ the discharge will have the same probability of propagating in any direction and the discharge will be a compact structure with a dimension $D \sim 2$. Since the electric field is stronger on the tip, in the other extreme, for $\eta \to \infty$ the discharge will become one dimensional, and hence $D = 1$. In between we have for example Figure 2a for $\eta = 3$, with $D \sim 1.3$, hence the fractal dimension of the discharge is parametrized by η.

In order to compute the radiated fields, i.e. Equation (6), we must describe the current along each of the segments of the fractal discharge. We start with a charge Q at the center of the discharge and allow the current to discharge along each of the dendritic arms. We assume that a series of train pulses $I(t - s/v)$,

$$I(t) = (e^{-\alpha t} - e^{-\gamma t})(1 + \cos(\omega_o t))H(t),$$

propagates with speed v (hence $\beta = v/c$) along the length s of each branch of the horizontal fractal discharge. $\omega_o = 2\pi \alpha n_f$ and $H(t)$ as the step function. Here n_f represent the number of pulses during the decay timescale $1/\alpha$. If we require to have spatial structure in the ionosphere, we need $2\pi \alpha n_f L > 1$, or $n_f > 50$. We chose $\alpha = 10^3$ s^{-1} as the inverse duration and $\gamma = 2 \times 10^5$ s^{-1} as the risetime (see Uman, 1984) of the discharge. We enforce current conservation at each branching point with the current strength partitioned in proportion to the size of each branch (see Valdivia *et al.*, 1998). The initial strengths of the current pulse I_o get divided as the discharge branches, but the total charge discharged is $Q = I_o/\alpha$, which for $I_o = 100$ kA gives $Q = 100$ C. The field Equation (6) is evaluated at $h = 60$ km, and propagation, absorption, heating and emissions are considered selfconsistently for $h > 60$ km as described above.

In general, the electric field and optical emission patterns induced by the fractal discharges in the ionosphere are three-dimensional (3-D), but for the purposes of illustration we take a two-dimensional cross section (e.g., the plane $y = 10$ km from the center of the discharge, but projected in the ionosphere) of this 3-D profile in the ionosphere. Of course other cross-sections will give different spatial patterns, but since we are not comparing with a specific observational situation, any cross section will suffice for the purpose of illustration. The field and optical emission patterns along the cross section $y = 10$ km, averaged over the duration of the discharge, are shown in Figure 2b and Figure 2c respectively for $\eta = 3, \beta = 0.025$, $I_o = 200$ kA. The peak emission intensity for an optimal column integration is about 100 kR. The model not only accounts for spatially structured optical emissions, but it also reduces the required charge Q discharged that is necessary to produce a total optical emission, e.g., of about 100 kR. Since the emission intensity depends on the type of discharge, we have plotted in Figure 2d the maximum intensity in kilorayleighs for an optimal column integration along the x axis is shown as a function of the dimension of the discharge $D(\eta)$. Since the optical emission intensity is extremely sensitive to the power density, a small percentage increase in the electric field strength can have profound effects on the emission pattern of a given fractal discharge. By having a spatially structured radiation pattern, the fractals can increase the electric power density locally in specific regions of the ionosphere and generate considerable optical emissions with relatively low (see scaling below) lightning discharge parameters.

We have seen that the intracloud component of the discharge is a good candidate for generating the observed spatio-temporal optical emissions. Furthermore, the spatio-temporal conductivity profile induced in the ionosphere during this process, is a natural candidate for the 'seeding' of streamers and Sprites, in the presence of the QE field left after the discharge (see Raizer et al., 1998). It is important to realize that we have used a very simple parameterization of the complex behavior of the charge imbalance in the cloud, but there are other ways to include this complexity in the charge dynamics.

2.2. RELEVANCE OF SPATIAL DENSITY/COMPOSITIONAL PERTURBATIONS

It has been suggested that the spatial variations in the density/pressure/compositional perturbations may be responsible for 'seeding' the streamers and Sprites (Pasko et al., 2000; Moudry et al., 2002). We propose to estimate the amplitude of the neutral density perturbation $\Delta n/n$ required to produce an optical emission intensity contrast. Taking n_e as the electron density and v_{ex} as the excitation rate of the 1st excited state of N_2 (see Appendix C), we follow Equation (12) and define the quantity $I[E, n] \sim <v_{ex}n_e>_t$ which is proportional to the optical emission intensity of the 1st excited state of N_2. The averaging is made over a time interval $\Delta t \sim 1$ msec, and we use the neutral atmospheric density shown in Figure 4a. We can do an order of magnitude estimation of $\Delta n/n$ required to generate an optical emission contrast R given by

$$R(E, \frac{\Delta n}{n}, n) = \frac{I[E, n(1 + \frac{\Delta n}{n})] - I[E, n])}{I[E, n]} \sim 1 \tag{3}$$

in a constant electric field. For the purpose of this estimation we will not consider amplification effects due to spatial variations, streamers, focusing, and so on. The value $R \sim 1$, a 100% contrast, represents a situation of spatially structured optical emissions where one region shows optical emissions and a near-by region seems to show no emissions at all (see e.g., Moudry et al., 2002). The evolution of the electron density depends on the value of the effective ionization rate v_i,

$$\frac{dn_e}{dt} = v_i(E, n)n_e \quad - > \quad n_e = n_e(0)e^{<v_i>_t\Delta t} \tag{4}$$

which includes attachment effects. All of these parameters depend on the neutral density n and the electric field strength E, which has been normalized to the ionization threshold field E_i, as described in Appendix C. An order of magnitude estimation can be accomplished by using the quiver energy, which represents the energy of an electron in an oscillating electric field, and using the approach of Valdivia et al. (2002).

Using Equations (3) and (4) and the variation of v_i and v_{ex} with the neutral density for a given electric field value E, we can estimate the optical emission contrast at a given height. For example at $h = 85$ km and electric field $E = E_i$ we plot $R(E, \Delta n/n, n)$ as a function of $\Delta n/n$. Numerically we then find the value of $\Delta n/n$ that satisfies $R = 1$. The same analysis can be repeated for any E, and at any given height. The results of this procedure for $h = 85$ km is shown in Figure 3a. Assuming that the ionization threshold field is the relevant field characterizing Sprites (Milikh et al., 1997) and streamers, we observe from Figure 3a that we need $\Delta n/n \sim 5–10\%$ to induce a spatially structured emission contrast in an homogeneous field, at the scales of interest.

The value of $\Delta n/n$ at the $E = E_i$ is shown as a function of h in Figure 3b. This result may be relevant when discussing spatially structured optical emissions and the 'seeding' of Sprites and streamers.

3. Discussion

We have discussed the modeling of the optical emissions in the upper atmosphere. Due to the large variety of observed optical emissions, it becomes important to discuss the formation of the spatial structure in the optical emissions, and possible

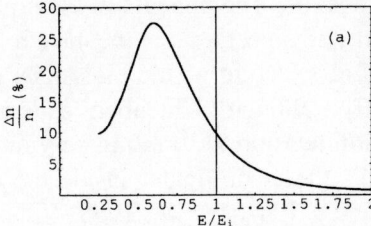 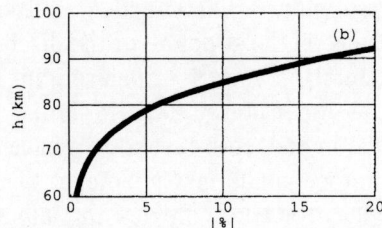

Figure 3. (a) The density perturbation required to produce an emission contrast of 1 at $h = 85$ km as a function of the normalized field. (b) The density perturbation required to produce an emission contrast of 1 at the ionization field as a function of height.

'seeds' for the generation of Sprites and streamers. In this paper we have analyzed two sources: (a) the spatially structured electric fields and (b) the atmospheric spatial variations of neutral density perturbations.

A model for the formation of spatially structured electric field and optical emission patterns, which incorporates the fractal nature of the horizontal lightning discharges was presented. The fractal structure of the discharge is reflected in the subsequent optical emission pattern, and fits the qualitative model suggested by Lyons (1996), which is based on the fact that complex horizontal discharges of the order of 100 km have been observed in connection with positive cloud-to-ground (+CG) events and Sprites. Suggestions that these discharges are relevant for the production of spatially structured optical emissions are discussed throughout the text. For an optimal configuration, so that the fields get projected upward, the lightning discharge must be horizontal, i.e., the so-called intracloud lightning or 'spider lightning' (Lyons, 1996). The model starts with a positive leader toward the ground, which in turn is followed by the positive return stroke, i.e., the +CG discharge. The charge imbalance left in the cloud after the +CG discharge may induce a complex charge dynamics. We have tried to capture some of this complexity by placing a charge Q at the center of a Lichtenberg-like figure, with the intracloud current propagating along the dendritic arms of the horizontal fractal. According to Lyons (1996), +CG lightning discharges having a peak current of up to 200 kA are associated with Sprites, and we expect that a fraction of this current, I_o, propagates along the dendritic arms of the intracloud discharge. Statistics of the speed of propagation for intracloud lightning are incomplete. The propagation speed during a cloud-to-ground return stroke can reach speeds of about $\beta \approx 0.1-0.5$ (Uman, 1984), while the propagation speed of intracloud discharges is at least an order of magnitude lower, and hence we took $\beta \approx 0.01-0.05$.

Certain fractals can radiate more effectively than others, and even though this problem is very complicated, we can take that the electric power density, and thus the emission pattern and intensity, scales as $E^2 \sim \beta^2 I_o^2 g(\theta, \beta, D)$ in the far field. For the parameters we have used, we obtained an emission intensity of about 100 kR for $D(\eta = 3) \approx 1.25$, with $\beta = 0.025$ and $I_o = 200$ kA. However, using the above scaling for the radiated field, we can generate a similar radiation pattern with $\beta = 0.05$ and $I_o = 100$ kA, and so on. The optical emission pattern depends on the structure of the discharge, and we conjecture that the most relevant structural parameter in determining the spatial structure of the emissions is the fractal dimension of the self-similar fractal. Even though Lichtenberg patterns observed in dielectric breakdown have a fractal dimension $D \approx 1.6$ in very uniform laboratory situations (Niemeyer et al., 1984), the environment inside the cloud provides for very complex intracloud discharges (Mazur et al., 1998; Villanueva et al., 1994), for which a range of dimensions may have to be considered. In the present model, the optimal emission intensity is obtained for dimensions $D \simeq 1.25$ for the parameter used above.

The spatial structure of the conductivity in the ionosphere, as suggested by the spatial structure in the optical emissions, produced by the intracloud fractal discharge is a natural candidate for 'seeding' the streamers and Sprites. Therefore, a comprehensive model of Sprites with total electric field (similar to Veronis et al. (1999) but allowing for a complex lightning discharge) which includes the seedings and the subsequent streamer development may be of relevance. Figure 2 provides a good indication for the height we expect the spatially dependent optical emissions to occur, for the electron density profile assumed. This is the natural place for the 'seeding' of the streamers, and correlates well with the height from which 'tendrils' start (see for example the pictures in Gerken et al., 2000; Wescott et al., 2001). The analysis of the optical spectrum that this model would generate is discussed in Milikh et al. (1997), and it agrees well with spectrum measurements (e.g., Mende et al., 1995).

It has been suggested that density/pressure/compositional perturbations can also be responsible for the spatial structure in the optical emissions (Pasko et al., 2000; Moudry et al., 2002). But if density perturbations are responsible for the observed spatial structure in the optical emissions and the 'seeding' of streamers and Sprites, then Figure 3b puts a very strong constraint on the required $\Delta n/n \sim 5\text{--}10\%$. We have to determine if these large background neutral density perturbations are present at the scales of interest (e.g., < 10 km). If density perturbations are to be relevant in shaping the spatial structure of the optical emissions and the 'seeding' of Sprites, then we may need to require gravity wave steepening and breaking to reach such large values of $\Delta n/n \sim 10\%$. Furthermore, if these fluctuations are to 'seed' the initiation of streamers, we need to remember that they have a scale smaller than a km and may reach scales as small as a 100 meters (Raizer et al., 1998). Therefore, for such a small scale it is presumed that the turbulent cascade generated by gravity wave-breaking may be necessary. Suggestions of a turbulent cascade have been observed in the OH airglow, where small scale structure appeared after an initial wave of 27 km wavelength (Yamada et al., 2001). Even though it may turn out to be difficult to estimate $\Delta n/n$ from these images at the heights of interest, it suggests a way to monitor the presence of spatial pressure/compositional fluctuations that exist at the time of Halos or Sprites.

On the theoretical side, it is also important to keep in mind that sometimes the specific parameters chosen to illustrate a physical process or model may turn out not to be the best representation of a specific observational case in this rapidly advancing area of research. But this does not disqualify the merits of the specific physics, when used in a different region of parameter space. This is especially true in view of the large variability in the optical emission phenomena, and in view of the large variability in the lightning discharge parameters (see book by Uman, 1984).

In the convoluted environment of the ionosphere there is such a large variability in the optical phenomena, that it may turn out to be difficult to associate a unique type of electric field (QE, EMP, etc), and/or ionospheric effect, to a particular situation. Some of the observations seem to include all the effects together: QE fields, EMPs, pressure/compositional atmospheric variations, etc. This is specially relevant in view of the nonlinear nature of the ionospheric environment, where these effects would couple in a nontrivial manner. Therefore, we must be careful if we intend to keep insisting in associating a specific optical phenomenon with a specific type of field, and in trying to separate, in a clear-cut manner, the different effects. At the end it is important to realize that the observations may include all the effect together: QE fields, EMPs, compositional atmospheric effects, etc. More than one effect may be relevant in some situations, such as spatially structured halos, Sprites starting away from halos, Sprites having a time delay with the parent +CG, and Sprites starting from long lived beads (Moudry *et al.*, 2002), among others. These situations may represent cases in which a clear-cut association with a specific type of fields may not be useful. These observations definitely require a detailed analysis and consideration, both theoretically and experimentally.

We therefore pose the following questions: Why do these neutral density perturbations sometimes don't seem to affect the images of homogeneous Halos and/or Elves, that may precede localized Sprites? We have seen that if these neutral density perturbations are capable of seeding the Sprites, they should cause a strong effect on the spatial structure of homogeneous Halos and/or Elves as well. Who is responsible for the spatially structured optical emissions and the 'seeding' of Sprites: the QE fields, the EMPs, the composition of the atmosphere? All of them? Or something else?

4. Appendix: Theoretical Ionospheric Electrodynamics

The basis for modeling the upper atmospheric electrodynamics, as related to lightning induced optical emissions, can be described by a coupled model that contains: (Appendix A) the generation of lightning induced fields from a lightning current structure, (Appendix B) the nonlinear propagation of the electric field through the upper atmosphere, (Appendix C) the interaction of the electric field with the ionospheric media that generates the optical emissions. Each essential part of the coupled model is described below.

4.1. APPENDIX A: GENERATION OF THE ELECTRIC FIELDS

A current pulse propagates with speed $\beta = v/c$ along a fractal structure. At the nth line element with orientation $\mathbf{L}_n$ and length L_n, which is parameterized by $l \epsilon [0, L_n]$, the current is given by $J_n(s_n, l, t) = I_n I(t - s_n/v + l/v)$, where s_n is the path length along the fractal (or if you prefer, a phase shift). The radiation field is the superposition, with the respective phases, of the small line current elements that form the fractal. For a set $\{\mathbf{r}_n, \mathbf{L}_n, I_n, s_n \mid n = 0, \ldots, N\}$ of line elements, the Fourier transformed hertz vector is given by

$$\Pi(\mathbf{x}, \omega) = \sum_{\{n\}} \frac{i\widehat{\mathbf{L}}_n}{\omega} \int_0^{L_n} \frac{J_n(s_n, l, \omega) \, e^{ik\|\mathbf{x} - \mathbf{r}_n - l\widehat{\mathbf{L}}_n\|}}{\| \mathbf{x} - \mathbf{r}_n - l\widehat{\mathbf{L}}_n \|} dl, \tag{5}$$

as a solution to Maxwell's equations. Here I(t) is the normalized temporal current profile, I_n is the strength of the current at the nth line element, s_n is the path length along the fractal, $\mathbf{r_n}$ is the position of the element, $J_n(s_n, l, \omega) = I_n I(\omega)e^{i\frac{\omega}{v}(s_n + l)}$ is the Fourier-transformed current strength at the nth line element, $\mathbf{x}$ is the position at which we measure the fields, ω is the frequency, and $k = \omega/c$. Values with the circumflex, e.g. $\hat{\mathbf{x}}$, indicate unit vectors, and $d_n = \| \mathbf{x} - \mathbf{r}_n \|$ means the standard distance from the line element to the field position.

To simplify the notation, we will denote $\mathbf{d}_n = \mathbf{x} - \mathbf{r}_n$. Note that, in general, the size of a single element $L_n \approx 100$ m is much smaller than the distance to the ionosphere $d_n \approx 80$ km, i.e., $L_n \ll d_n$. Therefore we can use the far-field approximation of the small elements to carry the above integral. Of course, the nonuniformity in the radiation pattern will be due to the phase coherence, or interference pattern, between the different elements that form the fractal. The electric field is then constructed from $\mathbf{E}(\mathbf{x}, \omega) = \nabla \times \nabla \times \Pi(\mathbf{x}, \omega)$. We then invert the Fourier transform of the field to real time and obtain the spatio-temporal radiation pattern, equivalent to Equation (1), due to the fractal discharge structure, which is given by

$$\mathbf{E}(\mathbf{x}, t) = \sum_n \frac{\beta I_n}{c d_n (1 - \beta(\widehat{\mathbf{L}}_n \cdot \widehat{\mathbf{d}}_n))}$$

$$\left(\widehat{\mathbf{L}}_n \left[I + \frac{c}{d_n} I_1 + \frac{c^2}{d_n^2} I_2 \right]_{t-\tau_2}^{t-\tau_1} - \widehat{\mathbf{d}}_n (\widehat{\mathbf{L}}_n \cdot \widehat{\mathbf{d}}_n) \left[I + \frac{3c}{d_n} I_1 + \frac{3c^2}{d_n^2} I_2 \right]_{t-\tau_2}^{t-\tau_1} \right)$$

where

$$I_1(t) = \int_{-\infty}^t d\tau \, I(\tau) \qquad I_2(t) = \int_{-\infty}^t d\tau \int_{-\infty}^\tau d\tau' \, I(\tau')$$

can be calculated exactly for the current described above, with

$$\tau_1 = \frac{d_n}{c} + \frac{s_n}{v} \qquad \tau_2 = \tau_1 + \frac{(\widehat{\mathbf{L}}_n \cdot \widehat{\mathbf{d}}_n)L_n}{c} + \frac{L_n}{v}$$

corresponding to the causal time delays from the two endpoints of the line element. The electric field in the far field approximation ($kL_n < 1$ and $kL_n < kd_n$) is given by

$$\mathbf{E}(\mathbf{x}, t) = \sum_n \frac{\beta I_n I(\tau) \mid_{t-\tau_2}^{t-\tau_1}}{cd_n(1 - \beta(\widehat{\mathbf{L}}_n \cdot \widehat{\mathbf{d}}_n))} \{\widehat{\mathbf{L}}_n - \widehat{\mathbf{d}}_n(\widehat{\mathbf{L}}_n \cdot \widehat{\mathbf{d}}_n)\} \qquad (6)$$

where the second term is small if we consider the radiation upward.

4.2. APPENDIX B: NONLINEAR PROPAGATION

As the lightning-induced fields propagate in the upper atmosphere, the field changes the properties of the medium by heating the electrons while experiencing absorption. The solution to Maxwell's equations for the propagation of the electric field is a nonlinear wave equation

$$\nabla^2 \mathbf{E} - \nabla(\nabla \cdot \mathbf{E}) - \frac{4\pi}{c^2}\frac{\partial}{\partial t}(\widehat{\sigma}\mathbf{E}) - \frac{1}{c^2}\frac{\partial^2}{\partial t^2}(\widehat{\varepsilon}\mathbf{E}) = 0, \qquad (7)$$

where the medium is incorporated in the conductivity $\widehat{\sigma}$ and in the dielectric $\widehat{\varepsilon}$ tensors (Gurevich, 1978). For the heights and frequencies of interest, $\widehat{\sigma}$ and $\widehat{\varepsilon}$ reach a steady state much faster than the timescale of the field variation, i.e., $1/\omega$. The solution in the ray approximation is therefore

$$E^2(\widehat{r}s, t) \simeq \frac{E^2(0, t - \frac{s}{c})}{s^2} H(t - \frac{s}{c})e^{-\csc(\chi)\int_0^z \kappa(z, E^2)dz}, \qquad (8)$$

where $\sin(\chi) = z/\sqrt{x^2 + y^2 + z^2}$ is the elevation angle of the point $\mathbf{r} = \widehat{r}s = \{x, y, z\}$, $H(t)$ is the step function, $\kappa(z, E^2) = \omega_e^2 v_e/c(\Omega^2 + v_e^2)$ is the absorption coefficient of the wave, $\omega_e^2 = 4\pi n_e e^2/m$ is the plasma frequency, and $\Omega = eB/mc$ is the electron gyrofrequency. The nonlinearity is incorporated self-consistently through $v_e = v_e(z, |E|)$, the electron-neutral collision frequency, due to the non-Maxwellian nature of the electron distribution function that the fields generate.

4.3. APPENDIX C: INTERACTION WITH THE IONOSPHERIC MEDIA

The electron distribution function in the presence of an electric field is strongly non-Maxwellian, requiring a kinetic treatment. The kinetic treatment will provide $v_e = v_e(z, |E|)$ and the excitation rates of the different electronic levels. We use an existing Fokker–Planck code, which has been developed for the description of ionospheric RF breakdown (Tsang et al., 1991; Papadopoulos et al., 1993). For given values of E and the neutral density n, the electron distribution function $f(\mathbf{v})$ is found by solving numerically the Fokker–Planck equation

$$\frac{\partial f}{\partial t} - \frac{1}{3m\mathrm{v}^2}\frac{\partial}{\partial \mathrm{v}}(\mathrm{v}^2 \nu(\mathrm{v})\widetilde{\epsilon}(E, \nu(\mathrm{v}))\frac{\partial f}{\partial \mathrm{v}}) = \$(f),$$ (9)

where

$$\widetilde{\epsilon}(E, \nu) = \frac{e^2 E_o^2}{m[\Omega_B^2 + \nu^2]}\{1 + (\frac{\Omega_B}{\nu})^2 \cos^2 \theta_o\}$$ (10)

is the quiver energy, which depends nonlinearly on the steady state averaged collisional frequency ν_{col}, $\$$ is the operator which describes the effect of the inelastic collisions (Tsang *et al.*, 1991), θ_o is the angle between the electric and magnetic fields, and $\nu(v)$ is the velocity dependent electron collisional frequency. We have assumed that the frequency of the electromagnetic fields satisfies $\omega \ll \Omega_B, \nu_e$. We have also assumed that $(\Omega_B/\nu)^2 \cos^2 \theta_o < 1$, which is the less energetically beneficial case. It is satisfied at any height of the lower ionosphere in the equatorial region or at the height below 85 km at middle latitude.

The relevant transport coefficients can be computed from

$$\nu = 4\pi N_{N_2} \int f(\mathrm{v})\mathrm{v}^3 \sigma(\mathrm{v})d\mathrm{v},$$ (11)

with σ as the relevant cross-section. In this manner we can compute (a) the electron neutral collisional frequency ν_{col}, (b) the electron ionization rate ν_i including attachment effects, (c) and the excitation rate of the different neutral and ion levels. We will concentrate on the 1st excited state of N_2, i.e., ν_{ex}, which seems to dominate the spectrum of red Sprites (Mende *et al.*, 1995; Hampton *et al.*, 1996). We assume a standard neutral atmosphere (Figure 4a), a night time electron density profile n_e (Figure 4b) and we plot the ionization electric field E_i as a function of height, such that $\nu_i(h, E_i) = 0$, in Figure 4c. The averaged steady state collisional rate ν_{col} is shown in Figure 4d. The ν_i and ν_{ex} at height $h = 80$ km are displayed in Figure 4e and 4f respectively, as a function of the normalized electric field E/E_i from the Fokker–Planck.

The excitations are then followed by optical emissions, and the number of photons emitted per second per cubic centimeter is given by $\nu_{ex}n_e$ in a first order estimation. In order to compare with observations, it is convenient to average in time the number of photons over the duration of the discharge T (approximately milliseconds), i.e., $< \nu_{ex}n_e > = \int_0^T \nu_{ex}n_e dt/T$. The intensity of the radiative transition in Rayleighs is given by

$$I = \frac{10^{-6}}{4\pi} \int < \nu_{ex}n_e > dl,$$ (12)

where the integral is carried along the visual path of the detector (column integrated). For $\widetilde{\epsilon} > 0.1$ eV, ionization is initiated, but for simplicity we are going to consider the case in which the electric field is below the ionization threshold, otherwise a self-consistent equation for the electron density in space must be included in the analysis.

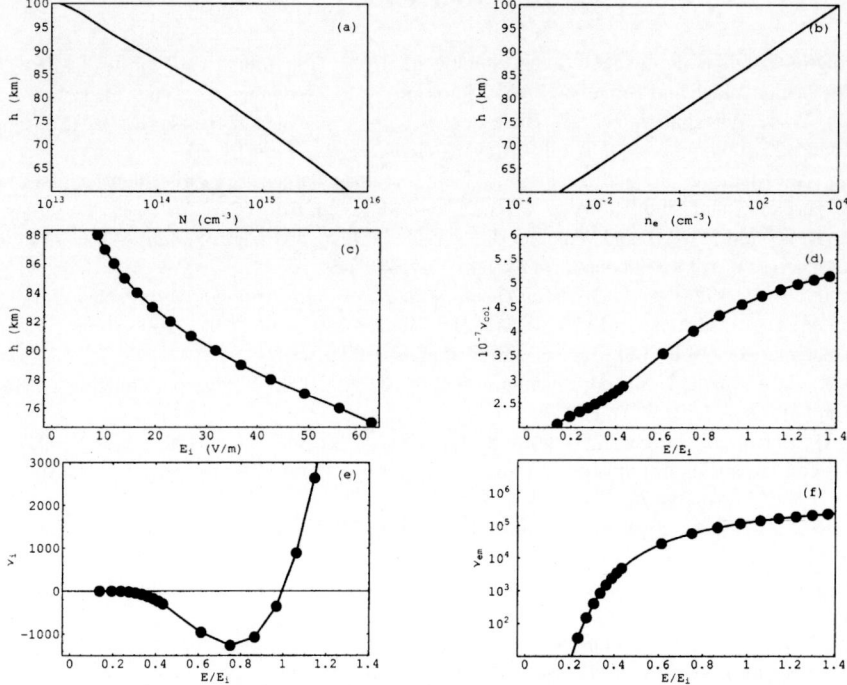

Figure 4. (a) N_{N_2} (top left). (b) n_e (top right). (c) E_i (middle left). (d) v_{col} (middle right). (e) v_i (bottom left). (f) v_{ex} (bottom right).

A model of the Sprite spectrum generated by this procedure, but including other excitation levels and absorption, is discussed in Milikh *et al.* (1997). It is also important to note that the transport coefficients can also be estimated from a semi-empirical approach as described in Pasko *et al.* (1997) (see Valdivia *et al.* (2002) for a comparison).

Acknowledgements

We acknowledge useful discussions with K. Papadopoulos, G. Milikh, and A. Gurevich. We acknowledge support by FONDECYT project N° 1030727, FONDECYT project N° 1020152, and AFOSR. We thank the hospitality of NITP/ CSSM at the University of Adelaide.

References

Barrington-Leigh, C. P., Inan, U. S. and Stanley, M.: 2001, 'Identification of Sprites and Elves with Intensified Video and Broadband Array Photometry', *J. Geophys. Res.* **106**(A2), 1741.

Boeck, W. L., Vaughan Jr., O. H., Blakeslee, R., Vonnegut B. and Brook M.: 1992, 'Lightning Induced Brightening in the Airglow Layer', *Geophys. Res. Lett.* **10**, 99.

Femia, N., Niemeyer, L. and Tucci, V.: 1993, 'Fractal Characteristics of Electrical Discharges: Experiments and Simulations', *J. Phys D. Appl. Phys.* **26**, 619–627.

Franz, R. C., Nemzek R. J. and Winckler, J. R.: 1990, 'Television Image of a Large Upward Electrical Discharge Above a Thunderstorm System', *Science* **249**, 48–51.

Gurevich, A. V.: 1978, *Nonlinear Phenomena in the Ionosphere,* Springer-Verlag, New York.

Hampton, D. L., Heavner, M. J., Wescott, E. M. and Sentman, D. D.: 1996, 'Optical Spectral Characteristics of Red Sprites', *Geophys. Res. Lett.* **23**(1), 89.

Gerken, E. A., Inan, U.S. and Barrington–Leigh, C. P.: 2000, 'Telescopic Imaging of Sprites', *Geophys. Res. Lett.* **27**(17), 2637.

Inan, U., Barrington–Leigh, C., Hansen, S., Glukhov, V. S., Bell, T. and Rairden, R.: 1997, 'Rapid Lateral Expansion of Optical Luminosity in Lightning Induced Ionospheric Flashes Referred to as 'Elves',' *Geophys. Res. Lett.* **24**, 583–586.

Kerr, R. A.: 1994, 'Atmospheric Scientists Puzzle over High Altitude Flashes, *Science* **264**, 1250–1251.

Lyons, W. A.: 1994, 'Characteristics of Luminous Structures in the Stratosphere above Thunderstorms as Imaged by Low-light Video, *Geophys. Res. Lett.* **21**, 875–878.

Lyons, W. A.: 1996, 'Sprite Observations above the U.S. High Plains in Relation to their Parent Thunderstorm Systems', *J. Geophys. Res.* **101**, 29,641–29,652.

Le Vine, D. M. and Meneghini, R.: 1978, 'Simulation of Radiation from Lightning Return Strokes: The Effects of Tortuosity', *Radio Sci.* **13**, 801–809.

Mazur, V., Shao, X. and Krehbiel, P. R.: 1998, ''Spider' Lightning in Intracloud and Positive Cloud-to-ground Flashes', *J. Geophys. Res.* **103**, 19811.

Mende, S. B., Rairden, R. L., Swenson, G. R. and Lyons, W. A.: 1995, 'Sprite Spectra; N2 1 PG Band Identification', *Geophys. Res. Lett.* **22**, 2633–2636.

Milikh, G. M., Papadopoulos, K. and Chang, C. L.: 1995, 'On the Physics of High Altitude Lightning', *Geophys. Res. Lett.* **22**, 85–88.

Milikh G. M., Valdivia, J. A. and Papadopoulos, K.: 1997, 'Model of Red Sprite Optical Spectra', *Geophys. Res. Lett.* **24**, 833–836.

Moudry, D., Stenbaek-Nielsen, H., Sentman, D. and Wescott, E.: 2002, 'Imaging of Elves, Halos and Sprite Initiation at 1 ms Time Resolution', J. Atmos. Sol. Terr. Phys., (submitted).

Niemeyer, L., Pietronero, L. and Wiesmann, H. J.: 1984, 'Fractal Dimension of Dielectric Breakdown', *Phys. Rev. Lett.* **52**, 1033–1036.

Papadopoulos, K., Milikh, G., Gurevich, A. V., Drobot, A. and Shanny, R.: 1993, 'Ionization Rates for Atmospheric and Ionospheric Breakdown', *J. Geophys. Res.* **98**, 17,593–17,596.

Pasko, V. P., Inan, U. S., Taranenko, Y. N. and Bell, T.: 1995, 'Heating, Ionization and Upward Discharges in the Mesosphere due to Intense Quasi-electrostatic Thundercloud Fields', *Geophys. Res. Lett.* **22**, 365–368.

Pasko, V. P., Inan, U. S., Bell, T. F. and Taranenko, Y. N.: 1997, 'Sprites Produced by Quasi-electrostatic Heating and Ionization in the Lower Ionosphere', *J. Geophys. Res.* **102**, 4529–4561.

Pasko, V. P., Inan, U. S. and Bell, T. F.: 1998, 'Spatial Structure of Sprites', *Geophys. Res. Lett.* **25**(12), 2123.

Pasko, V. P., Inan, U. S. and Bell, T. F; : 2000, 'Fractal Structure of Sprites', *Geophys. Res. Lett.* **27**, 497.

Pasko, V. P., Stanley, M. A., Mathews, J. D., Inan, J. S. and Wood, T. G.: 2002, 'Electrical Discharge from a Thundercloud Top to the Lower Ionosphere *Nature* **416**, 152.

Raizer, Y. P., Milikh, G. M., Shneider, M. N. and Novakovski, S. V.: 1998, 'Long Streamers in the Upper Atmosphere Above Thundercloud', *J. Phys. D. Appl. Phys.* **21**, 3255.

Rowland, H. L., Fernsler, R. F., Huba, J. D. and Bernhardt, P. A.: 1995, 'Lightning Driven EMP in the Upper Atmosphere', *Geophys. Res. Lett.* **22**, 361–364.

Rowland, H. L.: 1998, 'Theories and Simulations of Elves, Sprites and Blue Jets', *J. Atmos. Solar Terr. Phys.* **60**, 831.

Sentman, D. D. and Wescott, E. M.: 1993, 'Observations of Upper Atmospheric Optical Flashes Recorded from an Aircraft', *Geophys. Res. Lett.* **20**, 2857–2860.

Stanley, M., Krehbiel, P., Brook, M., Moore, C., Rison, W. and Abrahams, B.: 1999, 'High Speed Video of Initial Sprite Development', *Geophys. Res. Lett.* **26**(20), 3201.

Sentman, D. D., Wescott, E. M., Osborne, D. L., Hampton, D. L. and Heavner, M. J;: 1995, 'Preliminary Results from the Sprites94 Aircraft Campaign, 1, Red Sprites', *Geophys. Res. Lett* **22**, 1205–1208.

Stanley, M., Krehbiel, P., Brook, M., Moore, C., Rison, W. and Abrahams, B.: 1999, 'High Speed Video of Initial Sprite Development', *Geophys. Res. Lett.* **26**, 3201–32–4.

Tsang, K., Papadopoulos, K., Drobot, A., Vitello, P., Wallace, T. and Shanny, S.: 1991, 'RF Ionization of the Lower Ionosphere', *Radio Sci.* **20**(5), 1345–1360.

Uman, M. A.: 1984, *Lightning*, Dover, New York.

Vaughan, O. H. Jr., Blakeslee, R., Boeck, W. L., Vonnegut, B., Brook, M. and McKune Jr., J.: 1992, 'A Cloud-to-space Lightning as Recorded by the Space Shuttle Payload-bay TV Cameras', *Mon. Weather Rev.* **120**, 1459–1461.

Valdivia, J. A., Milikh, G. and Papadopoulos, K.: 1997, 'Red Sprites: Lightning as a Fractal Antenna', *Geophys. Res. Lett.* **24**, 3169–3172.

Valdivia, J. A., Milikh, G. M. and Papadopoulos, K.: 1998, 'Model of Red sprites Due to Intracloud Fractal Lightning Discharges', *Radio Sci.* **33**, 1655–1668.

Valdivia, J. A., Gomberoff, L. and Chian, A. C.: 2002, 'Can Density Fluctuations be Responsible for the Spatially Structured Optical Emissions?' *Geophys. Res. Lett.* (submitted).

Vecchi, G., Labate, D. and Canavero, F.: 1994, 'Fractal Approach to Lightning Radiation on a Tortuous Channel', *Radio Sci.* **29**, 691–704.

Veronis, G., Pasko, V. P. and Inan, U. S.: 1999, 'Characteristics of Mesospheric Optical Emissions Produced by Lightning Discharges', *J. Geophys. Res.* **104**(A6), 12645.

Villanueva, Y, Rakov, V. A. and Uman, M. A.: 1994, 'Microsecond-scale Electric Field Pulses in Cloud Lightning Discharges', *J. Geophys. Res.* **99**, 14353.

Werner, D. H. and Werner, P. L.: 1995, 'On the Synthesis of Fractal Radiation Patterns', *Radio Sci.* **30**, 29–45.

Wescott, E. M., Stenbaek-Nielsen, H. C., Sentman, D. D., Heavner, M. J., Moudry, D. R. and Sao Sabbas, F. T.: 2001, 'Triangulation of Sprites, Associated Halos and their Possible Relation to Causative Lightning and Micro-meteors', *J. Geophys. Res.* **106**(A6), 10467.

Williams, E. R.: 1988, 'The Electrification of Thunderstorms', *Sci. Am.*, 88–89, **259**.

Winckler, J. R., Lyons, W. A., Nelson, T. E. and Nemzek, R. J.: 1996, 'New High-resolution Ground-based Studies of Sprites', *J. Geophys. Res.* **101**, 6997–7004.

Yamada, Y., Fukunishi, H., Nakamura, T. and Tsude, T.: 2001, 'Breaking of Small Scale Gravity Wave and Transition to Turbulence Observed in OH Airglow', *Geophys. Res. Lett.* **28**, 2153–2156.

Zabotin, N. A. and Wright, J. W.: 2001, 'Role of Meteoric Dust in Sprite Formation', *Geophys. Res. Lett.* **28**(13), 2593.

IV: SPACE WEATHER/SPACE CLIMATE

SPACE WEATHER: ITS EFFECTS AND PREDICTABILITY

DAVID G. COLE

IPS Radio and Space Services, P.O. Box 1386 Haymarket, Sydney NSW 1240, Australia

Abstract. Terrestrial technology is now, and increasingly, sensitive to space weather. Most space weather is caused by solar storms and the resulting changes to the Earth's radiation environment and the magnetosphere. The Sun as the driver of space weather is under intense observation but remains to be adequately modelled. Recent spacecraft measurements are greatly improving models of solar activity, the interaction of the solar wind with the magnetosphere, and models of the radiation belts. In-situ data updates the basic magnetospheric model to provide specific details of high-energy electron flux at satellite orbits. Shock wave effects at the magnetopause can also be coarsely predicted. However, the specific geomagnetic effects at ground level depend on the calculation of magnetic and electric fields and further improvements are needed. New work on physical models is showing promise of raising geomagnetic and ionospheric predictability above the synoptic climatological level.

Key words: geomagnetic storms, ionospheric prediction, space weather

1. Space Weather

Many terrestrial systems include technology that is designed to operate under 'normal' space conditions and are vulnerable to extremes of space weather. Reliance of our technical society on space systems, coupled with periods of solar activity whose intensity is not yet predictable, makes space weather a prime factor in today's society.

Space weather refers to conditions on the sun and in the solar wind, magnetosphere, ionosphere and thermosphere that can influence the performance and reliability of space-borne and ground-based technological systems and can endanger human life and health. Adverse conditions can cause disruption of satellite operations, communications, navigation, and electric power grids, leading to a variety of socioeconomic losses. Space weather environment includes the Sun, interplanetary space between the Sun and Earth, the magnetosphere and the thermosphere down to ground level. It also includes space debris such as meteors and asteroids that enter the Earth's neighbourhood.

1.1. SOLAR ACTIVITY

The geo-effectiveness of a space disturbance is dependent on factors such as the location of solar activity, the complexity of the interplanetary environment between

Space Science Reviews **107**: 295–302, 2003.
© 2003 *Kluwer Academic Publishers.*

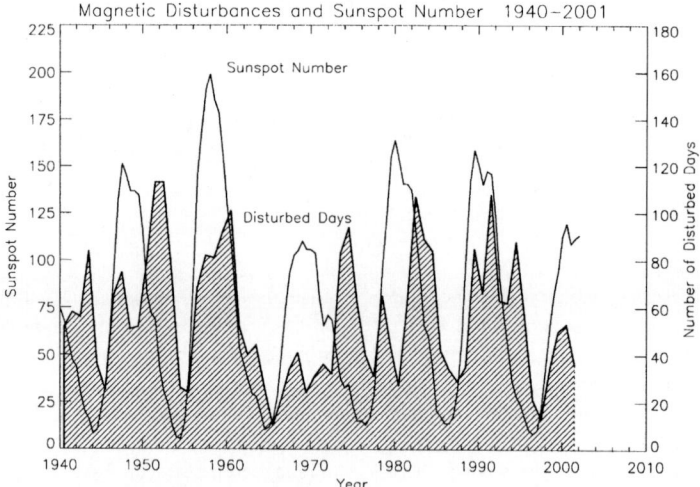

Figure 1. Occurrence of geomagnetic activity (Ap ≥ 25) with solar sunspot cycle behaviour (after R.Thompson, Proc. STPW96, CRL Tokyo, 1997).

the Sun and the Earth, and the level of radiation resistance of the technology affected. The Sun is the primary driver of space weather, and its cycles of activity give us some satisfaction as to its predictability. Although the most severe solar activity occurs around the cycle maximum, its influence on the magnetosphere and ionosphere continues through solar minimum (Figure 1).

Major events occur through coronal mass ejection (CME) from the sun's atmosphere or through high-speed solar wind streams associated with coronal holes. Great strides in our understanding of the Sun's convection zone have been made since the advent of helioseismology and the results of the SOHO and GONG experiments (Leibacher, 1998). We now have data on the underlying magnetic field structures and the possibility of 'seeing' active regions on the far side of the sun (Lindsey and Braun, 2000). Fine resolution images from satellites such as Yokkoh and TRACE have given clues to the processes of solar field eruption and reconnection of magnetic fields in the lower corona. However, the forecasting of solar events remains inadequate. Neither the timing nor the intensity of solar disturbances are modelled accurately enough to give more than a few hours of qualitative warning.

The planning of space projects requires predictions of long-term solar cycle activity. Of the methods predicting solar cycle behaviour, the geomagnetic-precursor technique (Thompson, 1993) has proven to be reliable. Future data on the strength of the polar magnetic fields near solar minimum may give better indication of the incoming cycle of activity (Schatten *et al.*, 1996).

Predictions of transient events such as flares and CMEs rely on the appearance of the underlying active region growth and remain probabilistic in nature. High-resolution space-based imagery of magnetic flux emerging from the photosphere is likely to increase warning times.

1.2. INTERPLANETARY SPACE

The solar wind (SW) is a constant outward plasma flow from the solar corona, among which there are occasional high speed streams (8–900 km sec^{-1}) caused by CMEs or coronal holes. The plasma cloud ejected from the corona carries magnetic field structure through the SW. The complexity, and in particular the overall direction of the magnetic field, determines its interaction with the Earth's magnetosphere. When the interplanetary magnetic field (IMF) turns southward, the IMF connects with the magnetospheric field allowing charged particles to enter the magnetosphere, stream down magnetospheric field lines and enter the inner plasmasphere through the magnetotail. The interaction of the space plasma on the magnetosphere, the shock of the interaction and the subsequent electric fields and currents drive many of the space weather effects detected on Earth.

1.3. MAGNETOSPHERE

During geomagnetic storms, intense changes in magnetospheric electric currents and the entry of energetic particles along magnetospheric field lines transfer energy from the solar wind to the radiation belts and to polar regions down to ionospheric heights.

The two radiation belts trapped within the magnetosphere react to variations in the solar flux. In quiescent times, the outer radiation belt containing energetic electrons (≥ 2 MeV) is located at about 2.5 to 6.6 Earth radii (10 000–36 000 km from the Earth's surface). The inner belt is made up of energetic protons (above 10 MeV) and lies at 1.2 to 1.8 Earth radii, a height of about 4000 km from the surface.

Ring currents of electrons and lower energy ions (≤ 0.05 MeV) reside within the outer belt and their variation with space weather modulates the magnetic field at the Earth's surface. If the IMF behind the shock is southward pointing, the CME plasma can enter the magnetosphere directly, its energy being partitioned between the auroral ionosphere, magnetospheric currents, and magnetotail injection of plasma into the polar regions. The trapped electron intensity and density in the outer radiation belt increases during the main phase of a geomagnetic storm and during substorms. This creates partial ring currents that contribute to ionospheric disturbances at high latitudes.

Global MHD models (Scholer, 2001) can describe the plasma environment from about three Earth radii out, but do not have the resolution required for the inner magnetosphere. Particle dynamical models are needed for this region. As a predictive tool, a magnetospheric model should be able to calculate the time-dependent electric and magnetic fields, the electric currents and particle distributions that cause geomagnetic and ionospheric storms (Toffoletto *et al.*, 2003).

Data from satellites monitoring real-time radiation belt flux, spectral intensity, and magnetospheric fields at specific sites can be used to adjust the basic model of radiation belt electron intensity and spectra to give values at other locations (Moorer and Baker, 2001).

Strong correlation between solar wind velocity and relativistic electron fluxes remains one of the best predictors of relativistic electron events although such events can be associated with CMEs and impulsive interplanetary shock waves as well as high solar wind streams and coronal holes (Reeves *et al.*, 2001).

One vital factor in predicting the onset of magnetospheric storms is the direction of the IMF after a solar eruption. While the structure of CME clouds at the Earth appear to remain much as it was when it was ejected from the Sun's surface, the magnetic field in the leading magnetic flux cannot be predicted reliably (Chao and Chen, 2001).

1.4. MAGNETOSPHERE – IONOSPHERE

Once the interplanetary energy has coupled to the magnetosphere, magnetospheric plasma convection is enhanced and field aligned currents accelerate plasma from the magnetotail into the polar ionosphere. The precipitation of particles into the polar atmosphere excites visual auroras, increases the conductivity of the lower ionosphere (D and E regions) in the auroral oval region and allows magnetospheric electric fields to drive an intense auroral electrojet.

The entry of magnetospheric energy into the polar regions heats the thermosphere, changing the distribution of atmospheric constituents and hence ionospheric electron density profiles. This is generally observed as reduced ionospheric critical frequencies and increased heights of maximum density.

1.5. SPACE WEATHER FORECASTING

Many countries operate space environment warning centres that share data and predictions through the auspices of the International Space Environment Service (ISES, 2003). Current space weather prediction services tend to present real-time data, whether it be solar, IMF, magnetospheric, geomagnetic or ionospheric data, or indicators of space disturbances. Thus, the Australian Space Forecast Centre provides hourly updates of the status of space weather conditions (solar, ionospheric, geomagnetic) (Wilkinson *et al.*, 2000). Space weather forecasts are prepared from analysis of real-time data.

While spacecraft give upstream information of CME characteristics before they reach Earth, ground-based monitors allow large-scale equipment, such as solar radio spectrographs, to measure the Sun-Earth event. Ground-based equipment also verifies that a space disturbance is in progress at ground level. A mix of space-based and ground-based data provides coverage of the total space weather event as it approaches and interacts with the magnetosphere and ionosphere.

Real-time maps of polar currents and electric potential can be derived in near-real-time from ground-based magnetometers (Kamide *et al.*, 2003) while many of the observed features of recent geomagnetic storms have been reproduced by coupling a magnetosphere-ionosphere model with a thermosphere-ionosphere model (Raeder *et al.*, 2001).

The forecasting of high-latitude electrojet current intensity is used as a first step in estimating the ground-induced currents and their coupling into power systems (Kappenman, 2001) as a prediction model.

2. Space Weather Effects

Electromagnetic emission from the sun can degrade systems in space and radio systems on Earth. Changes in the spectrum and intensity of the radiation belts caused by high-speed solar streams and CMEs have affected the operation of spacecraft through vehicle damage, deterioration of solar cells, semiconductor damage, or through electric charging of the spacecraft. Changing fields in the magnetosphere can induce currents in the ionosphere and, at ground level, in terrestrial power systems and long pipe lines that may cause damage that is costly to industry. Variations in ionospheric conditions, subsequent to magnetospheric changes, influence the operation of radio systems such as short-wave communications and radar.

2.1. SPACE RADIATION ENVIRONMENT

Geostationary satellites operate near the top of the outer radiation belt, low Earth orbiting satellites operate within the inner radiation belt. Both environments are populated by charged particles whose energy and number depend on space weather conditions.

Satellite systems can be impaired through direct penetration of the electronics by high velocity solar protons, by deep or surface-charging or by surface damage. Surface damage, such as degradation of solar cells, is caused by low energy particles and radiation. Effects on satellite hardware have been reviewed by Baker (2001) and described by the Geological Survey of Canada (Canada, 2002).

Those working in space or undertaking extra-vehicular activity outside a space station or spacecraft beyond the protection of the Earth's atmosphere can experience harmful radiation doses during major solar events unless protected. Radiation hazards are less for most aircraft but as general aviation develops into the stratosphere, and to greater altitudes, the dosage experienced by passengers during a major space event can become significant, particularly over polar routes, where the protection of the Earth's magnetic field is lowest.

2.2. IONOSPHERIC EFFECTS

2.2.1. *Radio Communication*

Solar flare events produce x-ray emission that can cause short-term (minutes) increases in electron density at low ionospheric heights. These shortwave fade-outs can be severe enough to destroy radio communications through attenuation at HF and cause phase shifts at VLF frequencies.

As a result of magnetospheric precipitation, the ionisation levels of the lower levels of the polar ionosphere greatly increase and cause severe absorption of HF and VHF radio signals. These polar cap absorption (PCA) events can last from days to weeks and precede the long-term disturbance of the mid-latitude ionosphere.

Ionospheric storms generally manifest as reduced electron density at F-region heights during the main phase of the disturbance, gradually recovering to normal over a few days.

2.2.2. *Satellite and Navigation*

Navigation systems, consisting of constellations of Earth-orbiting satellites, use the propagation delay from satellite-to-receiver to measure the range from several satellites to determine the position of the receiver. Unexpected changes in the ionospheric section of the propagation path cause errors in range, and hence in position. Such changes in electron density can be caused by a solar disturbance.

2.3. GEOMAGNETIC EFFECTS

A geomagnetic storm will induce electric currents in conductors at ground level. If power lines, railway lines, steel pipelines or telecommunication cables, are long in terms of east-west extent, the currents can be large enough to cause costly damage (Canada, 2002).

Induced currents flowing through power transformers can trip relays and take out power lines or burn out transformers. In heavily loaded systems it is possible to have a failure of the whole system, as happened in the Canadian Hydro-Quebec system during a magnetic storm in March 1989.

Long pipeline corrosion occurs through chemical reactions that take place through current leakage between pipe and earth. To prevent the electro-chemical corrosion, a voltage between the pipe and ground is maintained. If the back emf is not altered to counter the current induced by the geomagnetic storm, pipe corrosion is increased.

3. Summary

The effects of major space weather events have demonstrated the cost in terms of lost productivity and damaged equipment but the costs run deeper than that. When technological systems are impaired or fail, then the socio-economic activity dependent on them is in jeopardy. The Quebec power blackout in 1989 had a net cost of about $13.2 million; damaged equipment accounting for about $6.5 million of that estimated cost (Canada, 2002). Telecommunication failure through the loss of a satellite may also cause safety-of-life risks. These risks are now reflected in the insurance premiums associated with space technology.

The warning of a potential threat from space weather on terrestrial and space systems can be provided a few hours ahead on the basis of observations of CMEs and verified by in-situ data from spacecraft, but the specific severity of the effect requires further development of models of the solar wind and magnetosphere. Collaborative research through international cooperation, such as in the Community Coordinated Modeling Center (CCMC) (Hesse *et al.*, 2003), is helping to improve and create the next generation of space weather models. It is the goal of the World Space Environment Forum to develop our ability to predict the behaviour of the Sun-Earth environment through contributions from many countries.

Recent data from satellites have increased progress on space weather prediction. There are now models specifying the magnetospheric environment with local accuracy when used in conjunction with in-situ data; polar radars deliver real-time maps of magnetospheric and ionospheric currents and magnetic fields; and networks of ionospheric and geomagnetic monitors give constant information on ionospheric conditions. Much has been achieved in the last few years but the ability to forecast the total Sun-Earth event days, or even hours, ahead remains as a challenge.

References

Baker, D. N.: 2001, 'Satellite Anomalies due to Space Storms', in I. A. Daglis (ed.), *Space Storms and Space Weather Hazards*, Kluwer Academic Publishers, NATO Science Series, Dordrecht.

Chao, J. K. and Chen, H. H.: 2001, 'Prediction of Southward IMF Bz', in P. Song, H. J. Singer, and G. L. Siscoe (eds), *Space Weather*, AGU Geophysical Monograph Series, Volume 125, Washington.

Geological Survey of Canada: 2002, 'Space Weather Effects on Technology', *http://www.spaceweather.gc.ca/effects_e.shtml*.

Hesse, M., Kuznetsova, M., Rastaetter, L., Keller, K. and Falasca, A.: 2003, 'Space Weather Model Testing and Validation at the Community Coordinated Modeling Center', this volume.

ISES: 2003, *http://www.ises-spaceweather.org/about_ises/index.html*.

Kamide, Y., Kihn, E. A. and Cliver, E. W.: 2003, 'Real-time Specification of the Geospace Environment', this volume.

Kappenman, J. G.: 2001, 'An Introduction to Power Grid Impacts and Vulnerabilities from Space Weather', in I. A. Daglis (ed.), *Space Storms and Space Weather Hazards*, Kluwer Academic Publishers, NATO Science Series, Dordrecht.

Leibacher, J. W. and the GONG Project Team: 1998, 'The Global Oscillation Network Group (GONG) Project', in A. Wilson (ed.), *Structure and Dynamics of the Interior of the Sun and Sun-Like Stars*, SOHO 6/GONG 98, SP-418.

Lindsey, C. and Braun, D. C.: 2000, 'Basic Principles of Solar Acoustic Holography', *Solar Phys.* **192**/1–2, p261.

Moorer, D. F. Jr. and Baker, D. N.: 2001, 'Specification of Energetic Magnetospheric Electrons', in P. Song, H. J. Singer, and G. L. Siscoe (eds), *Space Weather*, AGU Geophysical Monograph Series, Volume 125, Washington.

Raeder, J., Wang, Y. and Fuller-Rowell, T. J.: 2001, 'Geomagnetic Storm Simulation with a Coupled Magnetosphere-ionosphere-thermosphere Model', in P. Song, H. J. Singer, and G. L. Siscoe (eds), *Space Weather*, AGU Geophysical Monograph Series, Volume 125, Washington.

Reeves, G. D., McAdams, K. L., Friedel, R. H. W. and Cayton, T. E.: 2001, 'The Search for Predictable Features of Relativistic Electron Events: Results from the GEM Storms Campaign', in P. Song, H. J. Singer, and G. L. Siscoe (eds), *Space Weather*, AGU Geophysical Monograph Series, Volume 125, Washington.

Schatten, K. H., Myers, D. J. and Sofia, S.: 1996, 'Solar Activity Forecast for Solar Cycle 23', *Geophys. Res. Lett.* **23**(6), p605–608.

Scholer, M.: 2001, 'Global Magnetospheric Modelling: Methods, Results and Open Questions', in I. A. Daglis (ed.), *Space Storms and Space Weather Hazards*, Kluwer Academic Publishers, NATO Science Series, Dordrecht.

Thompson, R. J.: 1993, 'A Technique for Predicting the Amplitude of the Solar Cycle', *Solar Phys.* **148**, p383.

Toffoletto, F., Sazykin, S., Spiro, R. and Wolf, R.: 2003, 'Modelling the Inner Magnetosphere', this volume.

Wilkinson P., Patterson, G., Cole, D., Yuile, C., Wang, Y -J., Tripathi, Y., Marshall, R., Thompson, R. and Phelan, P.: 2000, 'Australian Space Weather Services', *Adv. Space Res.* **26** (1), p233.

THE FEDSAT MICROSATELLITE MISSION

B. J. FRASER

Cooperative Research Centre for Satellite Systems, School of Mathematical and Physical Sciences, University of Newcastle, Callaghan, NSW 2308, Australia

Abstract. An Australian research microsatellite, FedSat with a complement of four payloads was launched from Tanegashima, Japan on 14 December 2002 into a near-circular sun synchronous 10:30 LT polar orbit at an inclination of 98.7° and altitude 800 km. Scientific experiments include a triaxial fluxgate magnetometer with a frequency response up to 100 Hz and a GPS receiver to monitor total electron content (TEC) and provide a precise orbit determination. Communications experiments include a Ka-band transponder and a UHF packet data service. A high performance computer payload will test reconfigurable computing technology.

Key words: microsatellite, magnetometer, magnetosphere

1. Introduction

Australia has recently reactivated its national space research program with the establishment of the Cooperative Research Centre for Satellite Systems (CRCSS). The specific objective of the Centre is to deliver a sustainable space industry and integrated education and training opportunities in space science, engineering and technology. The primary project of the CRCSS is to launch an Australian science and technology microsatellite mission. This has been realized through the development of FedSat, a 0.6 m cube microsatellite weighing 60 kg. Its purposes are to provide a research platform for space science, satellite communications and GPS studies. Currently the CRCSS combines the resources and skills of 12 organisations. The FedSat mission is not dedicated to a specific scientific or engineering goal but instead reflects the research interests of the CRCSS partners. This allows for experience to be gained in all areas where technology and applications are covered by four programs; Space Science (University of Newcastle; La Trobe University, Melbourne), Communications (University of South Australia, Adelaide; University of Technology, Sydney; Defence Scientific and Technical Organisation, Adelaide; CSIRO-TIP, Sydney), Satellite Systems (Queensland University of Technology, Brisbane), and Satellite Engineering (CSIRO-CRCSS, Canberra; Auspace Ltd, Canberra; Vipac Engineers and Scientists Ltd., Adelaide).

Space Science Reviews **107**: 303–306, 2003.
© 2003 *Kluwer Academic Publishers.*

Figure 1. FedSat showing the magnetometer head, folded 2.55 m boom and star camera.

2. The Fedsat Platform and Experiments

The satellite platform, shown in Figure 1, is based on a design originating from Space Innovations Limited (U.K.) and built by the CRCSS partners at Auspace in Canberra. The bus is a two tiered shelf structure with subsystems on the bottom shelf and payloads on the top shelf. The payloads are housed in three individual box enclosures on the top shelf. These include the UHF and Ka band communications payloads and the associated base-band processor in one enclosure. The second contains the GPS receiver, provided by NASA-JPL and the third contains the high performance computing equipment and the NewMag fluxgate magnetometer electronics. FedSat is flying three-axis stabilized with its base facing Earthward. The attitude control system (ACS) employs three rod magnetorquers, a three axis fluxgate magnetometer, three reaction wheels, three digital sun sensors and a star camera. The star camera is the cone-shaped instrument in Figure 1. The ACS will maintain pointing accuracy within ±2° of the target direction. Pointing knowledge is currently estimated as better than 2° but post-analysis should increase this to a few tens of arc-seconds using star camera data. The star camera is an improved version of the camera used on the SunSat microsatellite built by Stellenbosch University, South Africa. More details on FedSat are included in earlier papers by Barrington Brown *et al.* (1998) and Graham *et al.* (1999).

The remainder of this paper will concentrate on the Space Science Program and the Newcastle University Magnetometer Experiment (NewMag). NewMag was built for the Newcastle Space Physics Group (SPG) at the Institute of Geophysics and Planetary Physics UCLA in collaboration with C. T. Russell and his engineering team. It is based on the heritage of the POLAR (Russell *et al.*, 1995) fluxgate system and adapted for operation in a microsatellite. The dynamic range is ±65,000 nT at a sample rate of 10 vectors s^{-1} and a burst mode of 100 vectors s^{-1}, with a digitization error ± 0.2 nT. The fluxgate head is mounted at the end of a 2.55 m boom (Figure 1), similar to the Stellenbosch University SunSat magnetometer boom. Following FedSat launch the boom was successfully deployed on 13 January 2003. Further details of the NewMag instrumentation, which is now commencing to gather data, are included in Fraser *et al.* (2000, 2001).

3. The FedSat Space Science Program

The FedSat Space Science Program will undertake research on the structures and dynamics of the ionosphere and magnetosphere and their coupling, in relation to solar energy input, using magnetic field and GPS observations. NewMag data will study currents and wave fields and contribute to understanding space weather and its effects on technological systems. It will also map the geomagnetic field over the Australian region. GPS data will image the total electron content (TEC) and the electron density distributions in the Australian and Antarctic regions.

The behaviour of the near-Earth plasma system can be described in terms of the configuration of the geomagnetic field or the currents that are flowing. With respect to the coupling of the ionosphere to the magnetosphere field-aligned current (FAC) systems are important. The global FAC system is highly variable and it is difficult to obtain an instantaneous pattern of the currents on a short time scale. NewMag will monitor the high latitude FAC system and its fine structure in both hemispheres. It also will be used to calibrate the Iridium satellite constellation magnetometer data on a regular basis for global FAC pattern calibration (Waters *et al.*, 2001).

One of the strengths of the Newcastle SPG is its continuing studies of ULF and ion cyclotron waves in the magnetosphere and ionosphere using satellite data (e.g. Fraser *et al.*, 1996) and ground network and diagnostic data (e.g. Waters *et al.*, 1991; Menk *et al.*, 2000). With FedSat orbiting above the ionosphere this is a unique opportunity to study ULF and ion cyclotron wave transmission through the ionosphere. Of particular interest are the Pc3-5 ULF waves which are a consequence of solar wind energy establishing wave propagation and resonances in the magnetosphere. These may be Alfven mode field line resonances, compressional cavity mode or waveguide resonances, or simply compressional wave propagation. In order to be observed on the ground these waves must pass through the ionosphere. Combining FedSat NewMag ULF wave data in the 1–100 mHz band with Australian and Antarctic ground station magnetometer data will provide new information on identifying wave modes and measuring ionospheric transmission characteristics. Similar studies are planned for substorm associated Pi2 waves. Comparisons will be made with theoretical models following the work of Sciffer and Waters (2002). Similar studies will be undertaken on electromagnetic ion cyclotron (EMIC) wave packets in the 0.1–5 Hz band. These will include EMIC wave which propagate in the horizontal waveguide centered on the F2 region of the ionosphere (Fraser, 1975).

Acknowledgements

F. W. Menk, C. L. Waters, C. T. Russell, J. Means and A. Bish are thanked for their important contributions to this project. This research was carried out with financial support from the Commonwealth of Australia through the Cooperative Research Centres Program.

References

Barrington Brown, A. J., Wicks, A. N., Boland, L., Gardner, S. J. and Graham, E. C.: 1998, 'FedSat – An Advanced Microsatellite Based on a Microsil Bus', in *Proc. 12th Ann AIAA/USU Conf. on Small Satellites: Smaller than Small – the next generation, Utah State Univ., Logan.*

Fraser, B. J.: 1975, 'Ionospheric Duct Propagation and Pc1 Source Regions', *J. Geophys. Res.* **80**, 2790–2796.

Fraser, B. J., Singer, H. J., Hughes, W. J., Wygant, J. R., Anderson, R. R. and Hu, Y. D.: 1996, 'CRRES Poynting Vector Observations of Electromagnetic Ion-Cyclotron Waves Near the Plasmapause', *J. Geophys. Res.* **101**, 15331.

Fraser, B. J., Russell, C. T., Means, J. D., Menk, F. W. and Waters, C. L.: 2000, 'FedSat, an Australian Research Microsatellite', *Adv. Space Res.* **25**, 1325.

Fraser, B. J., Russell, C. T., Means, J. D., Menk, F. W. and Waters, C. L.: 2001, 'The FedSat NEWMAG Magnetic Field Experiment', *ANARE Reports 146*, 405–420.

Graham, E. C.: 1999, 'FedSat: an Australian Research Microsatellite Mission', *IAF Congress, pp. IAA-98-IAAA.11.1.01, Melbourne.*

Menk, F. W., Waters, C. L. and Fraser, B. J.: 2000, 'Field Line Resonances and Waveguide Modes at Low Latitudes; 1. Observations', *J. Geophys. Res.* **105**, 7747–7761.

Russell, C. T., Snare, R. C., Means, J. D., Pierce, D., Dearborn, D., Larson, M., Barr, G. and Le, G.: 1995, 'The GGS/POLAR Magnetic Fields Investigation', *Space Sci. Rev.* **71**, 563.

Sciffer, M. D. and Waters, C. L.: 2002, 'Propagation of ULF Waves Through the Ionosphere: Analytic Solutions for Oblique Magnetic Fields', *J. Geophys. Res.* **107**, A10, 1297, doi 1029/2001JA000184, 2002.

Waters, C. L., Menk, F. W. and Fraser, B. J.: 1991, 'The Resonance Structure of Low Latitude Field Line Resonances', *Geophys. Res. Lett.* **18**, 2293–2296.

Waters, C. L., Anderson, B. J. and Liou, K.: 2001, 'Estimation of Global Field Aligned Currents Using the Iridium System Magnetometer Data', *Geophys. Res. Lett.* **28**(11), 2165.

REAL-TIME SPECIFICATIONS OF THE GEOSPACE ENVIRONMENT

Y. KAMIDE[1], E. A. KIHN[2], A. J. RIDLEY[3], E. W. CLIVER[4] and Y. KADOWAKI[1]

[1] Solar-Terrestrial Environment Laboratory, Nagoya University, Toyokawa 442-8507, Japan
[2] NOAA National Geophysical Data Center, Boulder, Colorado 80305-3328, U.S.A.
[3] Space Physics Research Laboratory, The University of Michigan, Ann Arbor Michigan
48109-2143, U.S.A.
[4] AFRL/Space Vehicles Directorate, Hanscom AFB, Massachusetts 01731-3010, U.S.A.

Abstract. We report the recent progress in our joint program of real-time mapping of ionospheric electric fields and currents and field-aligned currents through the Geospace Environment Data Analysis System (GEDAS) at the Solar-Terrestrial Environment Laboratory and similar computer systems in the world. Data from individual ground magnetometers as well as from the solar wind are collected by these systems and are used as input for the KRM and AMIE magnetogram-inversion algorithms, which calculate the two-dimensional distribution of the ionospheric parameters. One of the goals of this program is to specify the solar-terrestrial environment in terms of ionospheric processes, providing the scientific community with more than what geomagnetic activity indices and statistical models provide.

Key words: geomagnetic storm, ionosphere, magnetosphere, solar wind, space weather, substorm

1. Introduction

The integrated understanding of complex interactions that couple the solar wind and the earth's magnetosphere and ionosphere is essential for space weather predictions. It is often said, however, that space weather research is behind surface (or troposphere) weather research by some fifty years in terms of their ability of quantitatively forecasting, for example, tomorrow's weather. To monitor the degree of how much the geospace is disturbed, the scientific community is still relying heavily on geomagnetic indices, such as *Kp, Dst*, and *AE*, which are scalar quantities expressing only global magnetic activity, although we are all aware that magnetospheric and ionospheric disturbances are highly localized.

Space weather has recently been becoming increasingly important in many respects, including the needs from the modern high-technology society that depends on satellite communications. This indicates that even though we are still far from a complete understanding of energy flow/transformation processes in the entire solar-terrestrial system, we are being expected by society to make predictions, at least, of major geomagnetic storms. It is thus very important to realize that the terminology 'space weather studies' be used to mean both basic research of solar-terrestrial relationships and its applications for space weather predictions including the ad-

Space Science Reviews **107**: 307–316, 2003.
© 2003 *Kluwer Academic Publishers.*

vancement of numerical modeling techniques and the construction of algorithms for predicting space 'events'.

At the Solar-Terrestrial Environment Laboratory (STEL), space weather studies are being conducted in basically two ways: understanding basic multi-scale processes occurring over the boundaries of various plasma regions in the solar-terrestrial system, and developing algorithms to predict geomagnetic storms/substorms. To promote these projects, STEL has installed a high-technology computer system, called Geospace Environment Data Analysis System (GEDAS). It represents a way to promote integrated studies by combining ground-based and satellite-based observations as well as modeling and simulation research. GEDAS intends to connect similar systems throughout the world on a near real-time basis. It is important to note that what is needed in space weather predictions is not only specifying global disturbances but also locating a particular phenomenon observed at one site within a global perspective.

The purpose of this paper is to report on-going research projects, in particular, of ionospheric electric fields and currents, using the GEDAS in collaboration with other institutions around the world.

2. What is GEDAS?

While observations from the earth's surface are considered to be a type of 'remote' sensing for solar-terrestrial processes and thus indirect, they nevertheless generate high spatial/temporal resolutions. On the other hand, satellite observations, being in-situ and 'direct,' provide only 'point' measurements along satellite orbits. It is thus necessary to evaluate crucially merits/demerits of each observation. What GEDAS does in practice can be summarized in the following way:

1. Local and global viewpoints: Without identifying the location of what a particular satellite is measuring within the entire solar-terrestrial system, one may not be able to discuss the physics of these measurements self-consistently. It is expected that by relying on GEDAS in which all available data are compared, one is able to pinpoint the location of localized, explosive phenomena, such as substorm expansions, within large-scale energy flows and transformation.

2. Tests for real-time data in simulation models: To understand nonlinear interactions among various plasma regions in the system, numerical modeling using basic equations is required. Under the GEDAS system, real-time data can be used as initial and boundary conditions for computer simulations, permitting researchers to forecast space weather events that may take place in the near future. Any researcher can be a project leader, using the real-time resources available through GEDAS to drive models or applications in their own research projects. Fresh ideas

can also be tested almost instantly against real data so that they can be improved quantitatively.

3. Calculations of Ionospheric Parameters

For one of the active projects currently underway using GEDAS, we have begun to collect ground magnetometer data on a near real-time basis in an attempt to compute the worldwide distribution of ionospheric parameters, such as ionospheric electric fields and currents. Figure 1 outlines how this project operates. Beginning with ground-based observations, our interest lies, as indicated by the main vertical line, in estimating ionospheric electric fields and currents at high latitudes. We also obtain field-aligned currents which connect the high-latitude ionosphere and the outer magnetosphere. This is a typical example of the so-called inversion problems.

We intend to use real-time, or near real-time, ground magnetometer data eventually from 50–70 observatories, which are combined with data from more direct observations by satellites and radars. This joint effort of STEL, the NOAA National Geophysical Data Center (NGDC), the NOAA Space Environment Center, the National Center for Atmospheric Research, and the

University of Michigan, will use operationally updated versions of the KRM and AMIE programs (Kamide *et al.*, 1981; Richmond and Kamide, 1988; Ridley *et al.*, 1998) to compute the instantaneous, two-dimensional distribution of ionospheric electrodynamic parameters at high latitudes.

The uneven distribution of ground magnetometers on the earth's surface is one of the inevitable problems we are facing. The AMIE code, along with solar wind observations (Zwickl *et al.*, 1998), is first used to calculate the overall distribution of the electric potential, which represents more or less a statistical pattern of the ionospheric potential, commensurate with the solar wind condition. For this process, an empirical model and ground magnetometer data only from a selected set of observatories are utilized. Once the global pattern is calculated, we use the result as the boundary condition to calculate more detailed structures of ionospheric parameters in a limited region on the basis of the KRM method (e.g., Sato *et al.*, 1995).

In our practical scheme, three algorithms are currently being referred to: KRM, rt-AMIE, and local-KRM. The KRM algorithm was originally designed to calculate ionospheric parameters, such as electric fields and currents and field-aligned currents, as well as Joule heating, on the basis of ground magnetometer data. It first computes the magnetic potential that is a best fit to the distribution of ground magnetic perturbations and then estimates electric potential patterns in the ionosphere using an ionospheric conductance model. The real-time AMIE algorithm, rt-AMIE in short, is a simplified version of AMIE which is a technique of calculating similar ionospheric parameters from all available information, such as ground magnetometer data and satellite observations of field-aligned currents, and even

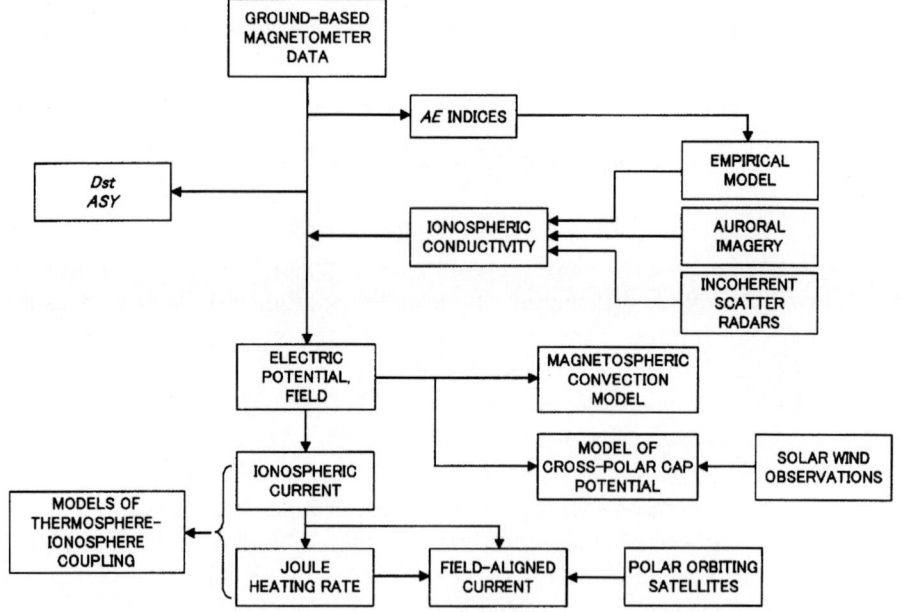

Figure 1. One of the GEDAS projects using the magnetogram-inversion technique, in which several ionospheric parameters are to be estimated on a real-time basis. The output will become important input for simulation/modeling studies at other institutions.

from empirical models of electrostatic potentials. The rt-AMIE algorithm that we are currently using depends on statistical models of the electric potential developed by Papitashvili *et al.* (1994) and Weimer (1995). These potential patterns in the models are given as a function of solar wind conditions, measured by the ACE spacecraft. The KRM and rt-AMIE algorithms have their own advantages and certain disadvantages. Relying on the strengths of each algorithm, a new algorithm, which is called the local-KRM algorithm, has been installed at GEDAS (Shirai *et al.*, 2002). Local-KRM is in a sense an effective combination of KRM and rt-AMIE. These two are used for separate estimates of ionospheric parameters in regions or local time sectors with good station coverage (by KRM) and poor coverage (by rt-AMIE). Care is taken to remove discontinuities, across the boundaries between these different regions.

At present, the number of stations providing real time, or near real-time, data is between 20–50, depending on the availability of data. Figure 2 presents a schematic diagram showing data flow in our scheme. First, ground magnetometer data and solar wind data from ACE are assembled. These data are being used for running the rt-AMIE program at NOAA/NGDC. The output from rt-AMIE, along with the original magnetometer data, is forwarded to GEDAS and is used as the boundary condition for the local-KRM calculation.

The local-KRM output includes equivalent currents, electric potential patterns, ionospheric currents, and field-aligned currents. At present we are calculating these

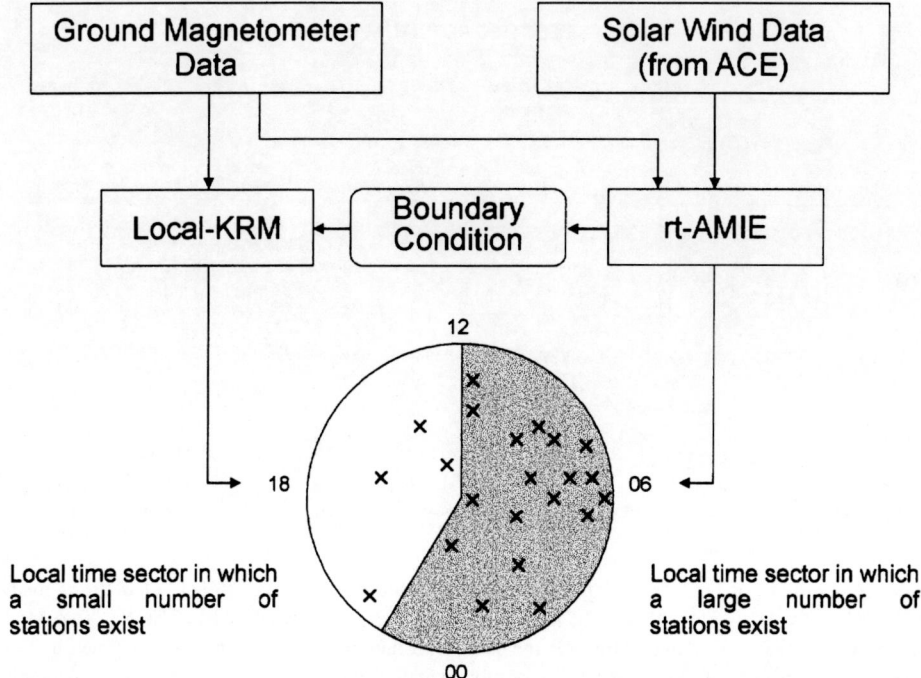

Figure 2. Schematic illustration showing data flow of the present project. Calculations of rt-AMIE are conducted at NOAA/NGDC whereas those of local-KRM are at STEL.

ionospheric parameters every 10 minutes with a grid size of 1 degree in latitude and 1 hr in MLT, using the ionospheric conductivity model of Ahn *et al.* (1998): see http://gedas22.stelab.nagoya-u.ac.jp/index.html. The whole procedure, from ground-based observations to the local-KRM output, takes at present about 20–30 minutes, depending on how quickly the data transport and actual calculations can be made.

Figures 3a and b present examples of the electric potential calculated for 1410 UT of July 17, 2002, just before the WISER workshop in Adelaide, from rt-AMIE (based on the Weimer (1995) model) and local-KRM, respectively. It is seen that the two patterns are quite similar to each other in that typical twin-vortex potential patterns can be identified in both potential distributions. The dusk-side pattern is nearly a duplication of that of rt-AMIE of the Weimer model. A potential minimum can be seen in 17–19 MLT hours at latitudes near 68° in both distributions. One can notice that in the morning sector, where data from a number of stations were available, details of the potential patterns in the two plots differ from each other, including the MLT extension of the peak potentials. The difference in the total potential difference between the two methods is attributable to an underestimate of the maximum potential on the dawn side in the rt-AMIE calculation resulting from a statistical model in rt-AMIE.

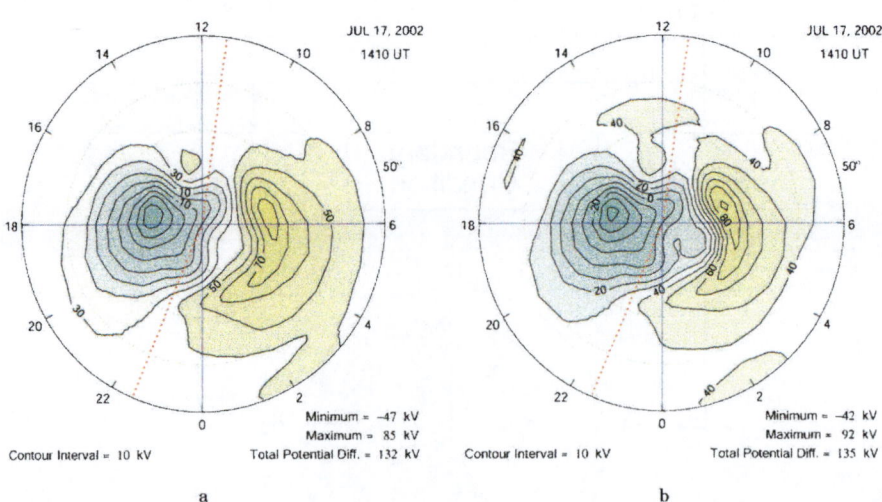

Figure 3. Example of real-time calculations of the electric potential: (a) global calculations made at NGDC on the basis of rt-AMIE; (b) local-KRM calculations made at STEL/GEDAS. Two MLT sectors are divided by a red dotted line. On the dawn side, data from a large number of magnetometers were available at this UT, whereas data from only a small number of magnetometers were available on the dusk side. The outer circle is for 50° in geomagnetic latitude.

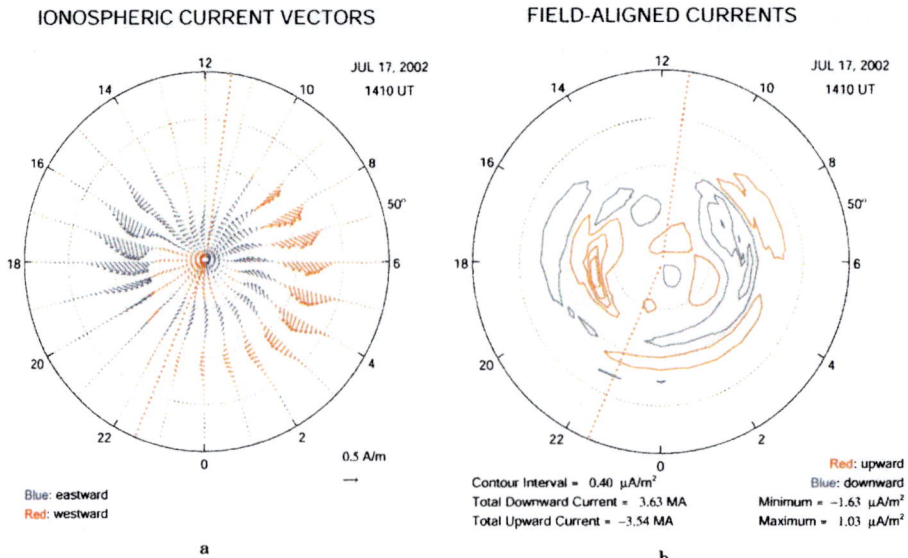

Figure 4. Local-KRM output, corresponding to the electric potential distribution shown in Figure 3: (a) Ionospheric current vectors; (b) field-aligned currents. The outher circle is for 50° in geomagnetic latitude.

The distributions of ionospheric current vectors and field-aligned currents displayed in Figures 4a and 4b, respectively, show a pair of the eastward and westward auroral electrojets and the so-called Region 1/2 field-aligned currents. Since an intense substorm expansion was not in progress at this time, the overall distribution of the ionospheric currents and their divergence is not characterized by an intrusion of an intense westward electrojet into the midnight sector, such that the eastward and westward electrojets are rather 'nicely' separated. The corresponding field-aligned currents, in which Region 1 current dominates, were also centered at dawn and dusk hours.

As Figure 1 indicates, the ionospheric conductivities can be normalized in the future using real-time observations by radars and global auroral distribution by polar-orbiting satellites. Once the ionospheric parameters are computed in this way, the output can be sent to institutions around the world where our output will become input for other modeling studies. For example, our electric field distribution can be mapped to the magnetosphere and is therefore useful for tracing particles in the magnetosphere: see Kamide *et al.* (2000) for some preliminary studies. Joule heating from our calculations can be used as input for calculating neutral winds in the thermosphere, which will modify the original electric field. There is no doubt that our output is also valuable for understanding the 'present' status of the auroral electrojets, which are critical to forecasting the strength of induced currents (Boteler *et al.*, 1998; Pirjola *et al.*, 2000).

4. Related GEDAS Projects

Other ongoing projects in the GEDAS framework include the following:

Solar Wind-Magnetosphere Coupling: An MHD simulation model of solar wind/magnetosphere interactions (Ogino *et al.*, 1994) is in operation in real time using the solar wind data from ACE. The validity of the potential distribution calculated from this MHD model can be tested immediately by comparing with the AMIE/KRM calculations. Any inconsistency or disagreement between the results from the two approaches must be accounted for in terms of the assumptions employed in the modeling techniques. In the present case, the inconsistency or disagreement results primarily from the 'one way' MHD simulation in which no relevant ionospheric boundary condition is included. Relying on the realistic distribution of the ionospheric potential from the KRM/AMIE calculations, the simulations can be upgraded.

Prediction of Solar Wind Speed: To better understand the propagation of CMEs, corotating interaction regions, and other structures, observations of interplanetary scintillation (IPS) of natural radio sources are conducted on a routine basis at STEL (Kojima *et al.*, 1998). Since IPS measurements are biased by line-of-sight integration, however, a computer assisted tomography technique is employed to obtain the longitudinal and latitudinal distributions of the speed and electron density fluctu-ations (Jackson *et al.*, 1998). Solar wind predictions are being performed using this program, in which IPS data are transferred to the University of California at San Diego in near real-time. Predictions of the solar wind speed near the Earth for the following days are being derived from this tomography technique. The predicted solar wind speed is displayed on the world-wide-web (http://www.sec.noaa.gov/ace/MAG_SWEPAM_24h.html) along with the ACE-observed solar wind speed.

5. Outlook

This paper has presented the recent progress we have made in our joint GEDAS program, with a special emphasis on the near real-time specification of ionospheric electrodynamics. The scheme we have developed can provide the scientific com-munity with two-dimensional mapping of electric fields and currents in the iono-sphere, as shown in Figures 3 and 4. Our final goal is to predict accurately space weather events when all details are coded properly in the computer system and when the 'present' condition of the sun is given. There should be no doubt that in order to achieve this degree of accuracy, a super computer of extremely high speed and capacity would be necessary.

Toward this goal, the project we have described in this paper is quite promising in, at least, two respects. First, GEDAS provides the scientific community with specifications of the geospace environment well beyond what are available from the popular geomagnetic activity indices. In addition, since the data products the GEDAS programs provide are based on real-time recordings of magnetic perturb-ations from a number of stations, the output should be more realistic than average potential patterns using a large number of 'point' measurements. By now it is well understood that the sun does not decide everything that will occur in the near-earth environment, and that the solar wind gives only the boundary condition for the magnetosphere in which various nonlinear plasma processes take place.

Second, it is important to note that recent space weather modeling efforts in-cluding global MHD and other modular models have also been improved, begin-ning to contribute to understanding the effects of solar wind disturbances on the magnetosphere and ionosphere (e.g., Raeder *et al.*, 1998; Papadoupoulos *et al.*, 1999; Gombosi *et al.*, 2000; Tanaka, 2000). It is important to integrate these sim-ulation models and real-time observations to increase our fidelity of space weather specifications and predictions. We are confident that computer systems, such as GEDAS, will contribute considerably to this integration.

Finally, two areas are noted which are of crucial importance for the success of space weather studies. They are the need for effective combinations (or interplays) of: (1) Observations and Modeling, and (2) Nowcasting and Forecasting. These will hopefully be accomplished within the GEDAS concept.

(1) *Observations and Modeling*: Spacecraft observations alone are inadequate to cover every plasma region in the solar-terrestrial system. Under appropriate initial/boundary conditions using real-time observations in the upstream solar wind, a numerical solution can be achieved, which would then be used to predict the next magnetospheric and ionospheric event, the validity of which would be readily checked against real-time data.

(2) *Nowcasting and Forecasting*: As demonstrated in the present paper, it is now possible to nowcast the global distribution of the electric potential in the ionosphere primarily from ground-based magnetometer observations. The time history of this potential distribution, which is very similar to the air pressure weather maps, can be extrapolated to the next moment using some efficient statistical methods. We must realize that the solar wind alone do not determine every process in the magnetosphere and ionosphere, but the history of the inner magnetosphere and the polar ionosphere determines where and how a substorm of what magnitude will take place.

Acknowledgements

We are grateful to Dr. H. Shirai who developed an early scheme of the local-KRM program, and Drs. V. O. Papitashvili and D. R. Weimer who kindly provided their models for our project. We also thank Drs. B.-H. Ahn, H. W. Kroehl, and A. S. Sharma for their illuminating discussions during the preparation of this manuscript.

References

Ahn, B. -H., Richmond, A. D., Kamide, Y., Kroehl, H. W., Emery, B. A., de la Beaujardire, O. and Akasofu, S. -I.: 1998, 'An Ionospheric Conductance Model Based on Ground Magnetic Disturbance Data', *J. Geophys. Res.* **103**, 14,769.

Boteler, D. .H., Pirjola, R. J. and Nevanlinna, H.: 1998, 'The Effects of Geomagnetic Disturbances on Electrical Systems at the Earth's Surface', *Adv. Space Res.* **22**, 17.

Gombosi, T. I., De Zeeuw, D. L., Groth, C. P. T. and Powell, K. G.: 2000, 'Magnetospheric Configuration for Parker-spiral IMF Conditions: Results of a 3D AMR MHD Simulation', *Adv. Space Res.* **26**, 139.

Jackson, B. V., Jick, P. L., Kojima, M. and Yokobe, A.: 1998, 'Heliospheric Tomography using Interplanetary Scintillation Observation, 1. Combined Nagoya and Cambridge data', *J. Geophys. Res.* **103**, 12,049.

Kamide, Y., Richmond, A. D. and Matsushita, S.: 1981, 'Estimation of Ionospheric Electric Fields, Ionospheric Currents, and Field-aligned Currents from Ground Magnetic Records', *J. Geophys. Res.* **86**, 801.

Kamide, Y., Shue, J. -H., Hausman, B. A. and Freeman, J. W.: 2000, 'Toward Real-time Mapping of Ionospheric Electric Fields and Currents', *Adv. Space Res.* **26**, 213.

Kojima, M., Tokumaru, M., Watanabe, H., Yokobe, A., Asai, K., Jackson, B. V. and Hick, P. L.: 1998, 'Heliospheric Tomography using Interplanetary Scintillation Observation, 2. Latitude and Heliospheric Dependence of Solar Wind Structure at 0.1–1 AU', *J. Geophys. Res.* **103**, 1981.

Ogino, T., Walker, R. J. and Ashour-Abdalla, M.: 1994, 'A Global Magnetohydrodynamic Simulation of the Response of the Magnetosphere to a Northward Turning of Interplanetary Magnetic Field', *J. Geophys. Res.* **99**, 11,027.

Papitashvili, V. O., Belov, B. A., Faermark, D. S., Feldstein, Y. I., Golyshev, S. A., Gromova, L. I. and Levitin, A. E.: 1994, 'Electric Potential Patterns in the Northern and Southern Polar Regions Parameterized by the Interplanetary Magnetic Field', *J. Geophys. Res.* **99**, 13,251.

Papadoupoulos, K., Goodrich, C. C., Wiltberger, M., Lopez, R. E. and Lyon, J. G.: 1999, 'The Physics of Substorms as Revealed by the ISTP', *Phys. Chem. Earth* **24**, 189.

Pirjola, R., Viljanen, A., Pulkkinen, A. and Amm, O.: 2000, 'Space Weather Risk in Power Transmission Systems', *Phys. Chem. Earth* **25**, 333.

Raeder, J., Berchem, J. and Ashour-Abdalla, M.: 1998, 'The Geospace Environment Modeling Grand Challenge: Results from a Global Geospace Circulation Model', *J. Geophys. Res.* **103**, 14,787.

Richmond, A. D. and Kamide, Y.: 1988, 'Mapping of Electrodynamic Features of the High-latitude Ionosphere from Localized Observations: Technique', *J. Geophys. Res.* **93**, 5741.

Ridley, A. J., Moretto, T., Ernstroem, P. and Clauer, C. R.: 1998, 'Global Analysis of Three Traveling Vortex Events during the November 1993 Storm using the Assimilative Mapping of Ionospheric Electrodynamic Technique', *J. Geophys. Res.* **103**, 26,349.

Sato, M., Kamide, Y., Richmond, A. D., Brekke, A. and Nozawa, S.: 1995, 'Regional Estimation of Electric Fields and Currents in the Polar Ionosphere', *Geophys. Res. Lett.* **22**, 283.

Shirai, H., Kamide, Y., Kihn, E. H., Hausman, B., Shinohara, M., Nakata, H., Isowa, M., Takada, T. K., Watanabe, Y. and Masuda, S.: 2002, 'Near Real-time Calculation of Ionospheric Electric Fields and Currents using GEDAS', *Proc. of COSPAR Coll.* (in press).

Tanaka, T.: 2000, 'The State Transition Model of the Substorm Onset', *J. Geophys. Res.* **105**, 21,081.

Weimer, D. R.: 1995, 'Models of High-latitude Electric Potentials Derived with a Least Error Fit of Spherical Harmonic Coefficients', *J. Geophys. Res.* **100**, 19,595.

Zwickl, R. D., Doggett, K., Sahm, S., Barrett, B., Grubb, R., Detman, B. R., Raben, V. J., Smith, C. A., Riley, P., Gold, R. E., Mewaldt, R. A. and Maruyama, T.: 1998, 'The NOAA Real-Time Solar-Wind (RTSW) system using ACE data', *Space Sci. Rev.* **86**, 633.

SOLAR INFLUENCE ON EARTH'S CLIMATE

NIGEL MARSH and HENRIK SVENSMARK

Danish Space Research Institute, Juliane Maries Vej 30, DK-2100 Copenhagen Ø, Denmark

Abstract. An increasing number of studies indicate that variations in solar activity have had a significant influence on Earth's climate. However, the mechanisms responsible for a solar influence are still not known. One possibility is that atmospheric transparency is influenced by changing cloud properties via cosmic ray ionisation (the latter being modulated by solar activity). Support for this idea is found from satellite observations of cloud cover. Such data have revealed a striking correlation between the intensity of galactic cosmic rays (GCR) and low liquid clouds (<3.2 km). GCR are responsible for nearly all ionisation in the atmosphere below 35 km. One mechanism could involve ion-induced formation of aerosol particles (diameter range, $0.001-1.0$ μm) that can act as cloud condensation nuclei (CCN). A systematic variation in the properties of CCN will affect the cloud droplet distribution and thereby influence the radiative properties of clouds. If the GCR-Cloud link is confirmed variations in galactic cosmic ray flux, caused by changes in solar activity and the space environment, could influence Earth's radiation budget.

Key words: climate, GCR, sun

1. Introduction

In this paper we discuss the possibility that solar activity and the space environment can influence terrestrial climate. Outlined in the flow diagram of Figure 1 are the two major routes by which the sun can influence this region of the atmosphere. This involves both solar irradiance providing the energy driving terrestrial climate, and solar wind shielding of Galactic Cosmic Rays (GCR) which are responsible for ionisation in the lower 35 km of the atmosphere. Changes in Total Solar Irradiance (TSI) over the 11 year solar cycle is about 0.24 W m^2 at the top of Earth's atmosphere, which is currently believed to be too small to have had a significant influence on climate. However, indirect effects may be much larger, and there are two mechanisms which have recently gained considerable attention. The first involves the UV part of the solar spectrum which varies by up to $10-100\%$ over a solar cycle. The role of UV in stratospheric ozone production is thought to influence the temperature and circulation of the stratosphere. Through dynamic coupling with the troposphere these changes may influence surface climate, but the extent of the coupling is currently uncertain (Haigh, 1996; Shindell *et al.*, 1999). The second indirect effect, and the one that will be considered here, is the possible role of atmospheric ionisation, resulting from GCR, in the production of new aerosol. Aerosols acting as Cloud Condensation Nuclei (CCN) play an important role in

Space Science Reviews **107**: 317–325, 2003.
© 2003 *Kluwer Academic Publishers.*

Marsh and Svensmark

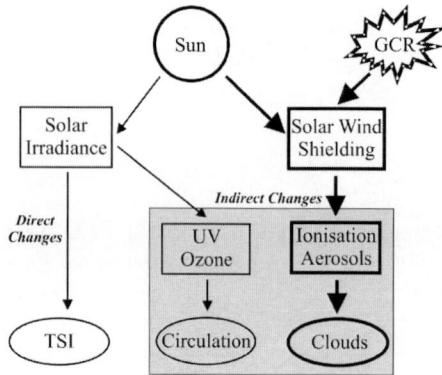

Figure 1. Possible routes for a solar influence on climate.

determining cloud properties. Clouds cover between 60–70% of the globe at any time, and influence the radiation budget by both reflecting incoming short wave radiation and trapping outgoing long wave radiation. A systematic change in the number of CCN will affect cloud properties, and thus would have an impact on Earth's climate.

2. GCR – Cloud Correlation

Using various satellite observations of cloud amount, a correlation has been found between total cloud cover averaged over the oceans and GCR (Svensmark and Friiss-Christensen, 1997; Svensmark, 1998). However, this result gave no indication as to the cloud type(s) affected, which is necessary for understanding the nature of any GCR-cloud mechanism and for estimating its potential radiative impact on climate. Marsh and Svensmark (2000b), have shown that the GCR-cloud correlation is limited to low cloud when using IR (at wavelength 10 μm) cloud observations provided by the International Satellite Cloud Climate Project (ISCCP version D2) between July 1983–September 1994 (Figure 2a). Recently, an extension to this data has been made available up to September 2001. From Figure 2a it can be seen that the correlation between low cloud and GCR breaks down after 1994. In addition a number of other ISCCP-D2 cloud parameters display a step function at around the same time. This is particularly apparent in the global average of high cloud amount shown in Figure 2b. Interestingly, between September 1994 and January 1995 ISCCP experienced a gap in available polar orbiting reference satellites required to inter-calibrate the five satellites used to provide global coverage of cloud observations (ISCCP homepage, 2002). An estimate of the effect of this calibration gap on ISCCP low cloud amount can be made by comparing it with independently observed cloud amounts obtained from the

Special Sounder Microwave Instrument (SSMI) (Ferraro *et al.*, 1996). SSMI data is available over oceans from July 1987–present with an 18 month gap between June 1990–December 1991. Since the SSMI cloud data is at a much lower resolution than ISCCP data, it is important to determine regions where the two data sets are observing similar cloud properties. These regions are found where the correlation coefficient at each grid point is >0.5. Limiting the period over which the correlation coefficients are found to July 1987–June 1990 allows for an independent test between the ISCCP and SSMI long term trends after January 1992 (Marsh and Svensmark, 2003). Figure 2c displays the regional averages over the full period of available data. While initially there is a good agreement between the two curves, after 1994 the long term trends diverge while the month to month fluctuations are still correlated. A correction factor can be estimated from the difference between the long term trends. When adding this correction factor to the globally averaged ISCCP-D2 low cloud cover (Figure 2a, dashed curve), the correlation with GCR is then found to exist for all available data (Marsh and Svensmark, 2003).

3. Ionisation – CCN – Cloud Properties

Svensmark (1998) suggested that a physical mechanism to explain the GCR-cloud link could involve the effects of atmospheric ionisation on aerosol chemistry or the phase transition of water vapour. Previously, Ney (1959) had observed that atmospheric ionisation was 'the meteorological variable subject to the largest solar cycle modulation in denser layers of the atmosphere'. This lead Dickinson (1975) to speculate that ionisation might modulate sulphate aerosol formation which when activated as CCN would influence cloud radiative properties.

   Low clouds generally consist of liquid water droplets which have formed where water vapour has condensed onto atmospheric aerosols acting as CCN. Artificial inputs of aerosol, e.g. ship exhaust or emissions from chimneys, into regions of low cloud formation tend to result in large increases of CCN locally, and it is possible to observe their influence on cloud radiative properties (King *et al.*, 1993). However, observing the impact of natural variations in CCN is not so straight forward. The fundamental physics of new particle formation is not understood and there are many complex processes involved in the growth from newly formed ultra fine aerosol $\sim 0.001~\mu m$ up to CCN $\sim 1.0~\mu m$. Atmospheric ionisation has been suggested as potentially important for influencing any one of the stages in the production and growth phases of aerosol. Numerical simulations focusing on the role of ionisation in the production of new aerosol have been successful at reproducing observations of new particle formation over the Pacific where classical nucleation theories have failed (Clarke *et al.*, 1998; Yu and Turco, 2000). Further, these model results suggest that the production of new aerosol is sensitive to small changes in ionisation in regions of low cloud formation <3.2 km, while relatively insensitive at higher altitudes (Yu, 2002). Although it is currently uncertain whether

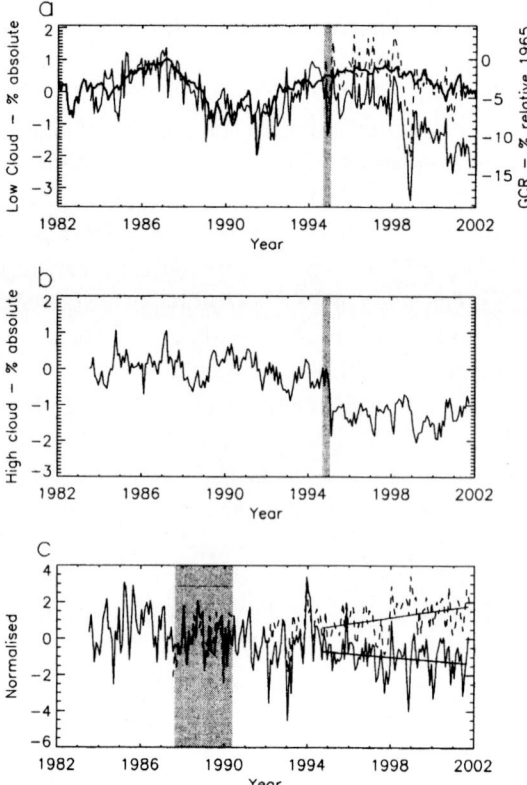

Figure 2. Monthly averages of cosmic rays (thick solid) and globally averaged anomalies (solid) of (a) ISCCP-D2 low cloud amounts, and (b) ISCCP-D2 high cloud amounts over the period of available ISCCP-D2 data. The dashed portion of the curve in (a) includes a drift term calculated by adding the difference between the linear trends seen in (c) to the ISCCP-D2 low cloud amount for each month. The anomalies are found by subtracting the climatic annual cycle for each month, averaged over the available period of ISCCP observations (July 1983–September 2001), before averaging over the globe. The shaded region in (a) and (b) denotes the gap in available ISCCP calibration satellites. (c) Spatial average of ISCCP low cloud amount (solid) and SSMI cloud amount (dashed) over regions of significant correlation ($r > 0.5$, $p > 99\%$). Both ISCCP low cloud amount and SSMI cloud amount have been normalised to their respective mean and variance over the highlighted period July 1987–June 1990. The thick lines represent the linear trends used in (a) for the respective curves starting July 1994.

this dependency on ionisation is maintained through the growth phase up to CCN sizes, it is consistent with the strong positive correlation found between low clouds and GCR.

The flow diagram in Figure 3 shows the link between ionisation and cloud radiative properties, assuming that a relationship between ionisation and CCN exists. An increase in CCN will lead to an increase in the number of cloud droplets and hence changes in liquid water content affecting the long wave properties (Han *et al.*, 2002). Increasing the droplet number will also lead to a decrease in droplet

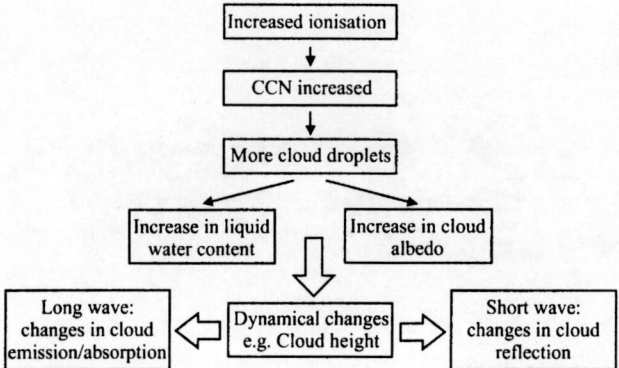

Figure 3. Possible route for an influence from atmospheric Ionisation on cloud radiative properties.

size which will affect a clouds albedo, and hence its short wave properties. Thus if there is a systematic change in atmospheric ionisation, such as over a solar cycle, a response is expected in both long wave and short wave cloud radiative properties.

Figure 4 shows a global map of the correlation coefficient found at each grid point between GCR and ISCCP-D2 low cloud top temperature (Marsh and Svensmark, 2000b). A region centered around the equator displays strong positive correlation coefficients. Figure 5 shows the global averages of ISCCP-D2 cloud optical depth for low and mid liquid stratus clouds. The optical depth observations are found using visible wavelengths (VIS) which are less certain then the IR observations and so the results are less robust. Despite these uncertainties Figure 5 indicates that a significant correlation is found between GCR and cloud optical depth. These results suggest that both long (IR) and short wave (VIS) properties of low clouds are responding to changes in atmospheric ionisation due to variations in solar activity.

4. Past Cloud Radiative Impact on Climate

Tropospheric temperatures observed with radiosondes over the period 1958–2001 display significant variability at a number of different time-scales. From monthly data the effects of El Niño and volcanic eruptions are particularly evident. These features are largely removed when filtering with a three year running mean, and the low pass Tropospheric temperatures show a remarkably good agreement with changes in solar activity (Figure 6) (Svensmark *et al.*, 2003). From Figure 6 an increase in reconstructed TSI of $\triangle F_s = 1$ W m^2, is seen to coincide with a 0.4 K increase in tropospheric temperature. To estimate whether the increase in TSI can explain the temperature change a simple sensitivity analysis can be performed where, $\triangle T = \lambda * \triangle F$. Here, $\triangle T$ is the response to a change in radiative forcing $\triangle F$ at top of the atmosphere. The climate sensitivity parameter $\lambda = 0.6$, is found from the average response of various climate models to a doubling of CO_2. Accounting

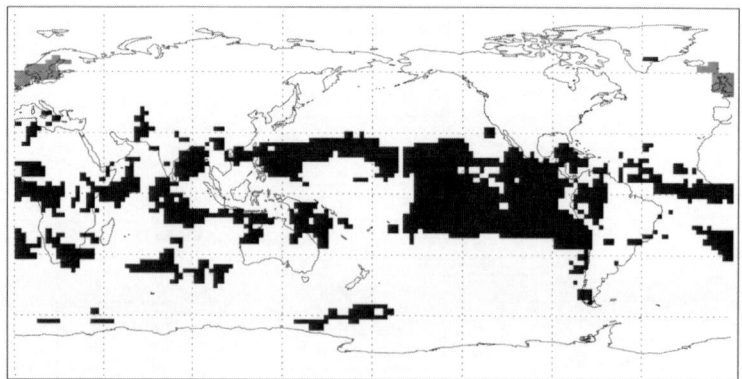

Figure 4. Point correlation maps for the period July 1983–August 1994 between GCR and ISCCP-D2 low cloud amounts. The correlation coefficient, r, is calculated from the 12 month running means at each grid box. The shading corresponds to regions where r > 0.6 (dark grey) and r < −0.6 (light grey) with p > 95% significance. White pixels indicate regions with either no data or an incomplete monthly time series.

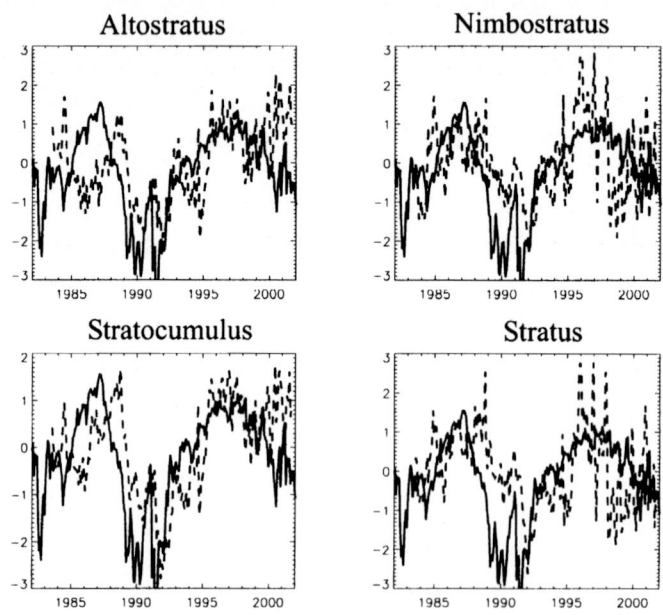

Figure 5. Monthly averages of cosmic rays (solid) and globally averaged anomalies (dashed) of ISCCP-D2 liquid stratus cloud optical depth. Anomalies are found as in Figure 1.

for Earth's average albedo, $\alpha = 0.3$, and the geometric effect, the TSI forcing at the top of the atmosphere is found to be $\triangle F = (1-\alpha)/4 * \triangle F_s = 0.2$ W m^2, which results in only a 0.1 K increase in temperature in contrast to the 0.4 K observed. Clearly, changes in TSI alone are too small to explain tropospheric temperatures and an amplification factor is required.

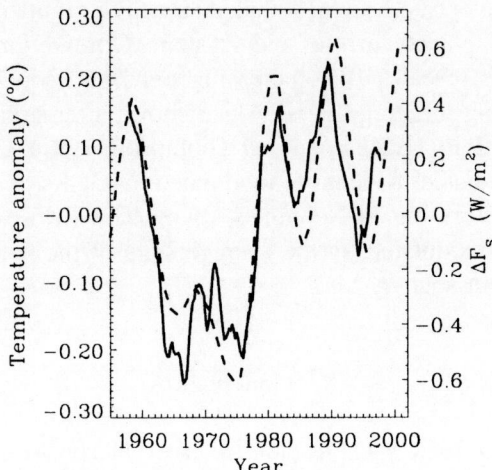

Figure 6. Tropospheric temperatures (solid) obtained from radiosondes (Parker *et al.*, 1997), together with reconstructed TSI (dashed), $\triangle F_s$, using re-scaled sunspot numbers as a proxy (Lean *et al.*, 1995). Both data sets have been low pass filtered with a three year running mean.

Using GCR as a proxy for low cloud changes, Marsh and Svensmark (2000a) estimated the changes in cloud radiative forcing over this period to be $\sim$0.5 W m^2. If this forcing factor is included in the above calculation then $\triangle T$ is estimated to be $\sim$0.4 K as observed. Although this is rather speculative, it does indicate that the contribution to climate change from an indirect response to changes in solar activity are potentially important.

An estimate of cloud radiative forcing for the past 100 years was found to be 1.5 W m^2 (Marsh and Svensmark, 2000a) by observing that the solar magnetic flux has more than doubled over the past century (Lockwood *et al.*, 1999). Over the same period global surface temperature has been observed to increase by $\sim$0.6 K. Other indications suggesting that the sun has played a role in past terrestrial climate variability is reflected in the close agreement between cosmogenic isotopes and various paleoclimate proxies. Bond *et al.* (2001) showed that Ice Rafted Debris (IRD) from north atlantic marine cores closely followed changes in solar activity for the past 12 000 yrs. IRD deposits are the result of melting icebergs which have originated from the ice margins and traveled under the influence of surface winds and currents out over the north Atlantic. Fluctuations in IRD reflect changes in north Atlantic climate and the close agreement with cosmogenic isotopes ^{14}C and ^{10}Be suggests a solar influence. Another example can be found from studies of δ^{18}O obtained from stalagmites by Neff *et al.* (2001) in the caves of Northern Oman. Here δ^{18}O reflects changes in monsoon precipitation and the position of the Inter-Tropical Convergence Zone (ITCZ). δ^{18}O measurements between 6000 and 9500 BP show a close agreement with ^{14}C again implicating the sun in long term climate variability.

Finally, evidence exists to suggest that solar activity influences the atmospheric properties of other planets in the solar system. Observations in the brightness changes of Neptune reveal a 10% increasing trend between 1970–2000. The residual after removing this trend is ~2% and shows a remarkable agreement with changes in solar activity (Lockwood and Thompson, 1991). One possible explanation is that ion mediated nucleation, modulated by GCRs, is affecting the albedo of methane clouds forming in Neptune's atmosphere (Moses *et al.*, 1989). This suggests that IMN modulated by the average state of the heliosphere could be a general feature of our solar system.

5. Conclusions

Inter-annual trends in ISCCP-D2 IR cloud data reveal a positive correlation between cosmic ray intensity and two parameters: (1) low cloud amount and, (2) low cloud top temperature. ISCCP-D2 VIS cloud data suggests that cloud optical thickness in liquid stratus clouds is also positively correlated with GCR. If confirmed, this would provide considerable support for a GCR-cloud mechanism. Numerical simulations of new aerosol production, suggest that production is most sensitive to changes in ionisation from GCR in the lower troposphere. Assuming that this is reflected in the final CCN distributions, this is consistent with the observations of a GCR-low cloud correlation.

Estimates of changes in cloud radiative forcing indicate that if a GCR-cloud mechanism exists it could have a significant impact on the global radiation budget. Past observations of climate change appear to follow changes in the cosmic ray flux lending support to the suggestion that solar activity and the space environment have influenced terrestrial climate.

References

Bond, G. *et al.*: 2001, 'Persistent Solar Influence on North Atlantic Climate during the Holocene'. *Science* **294**, 2130.

Clarke, A. D. *et al.*: 1998, 'Particle Nucleation in the Tropical Boundary Layer and its Coupling to Marine Sulfur Sources.', *Science* **282**, 89–92.

Dickinson, R.: 1975, 'Solar Variability and the Lower Atmosphere', *Bull. Am. Met. Soc.* **56**, 1240.

Ferraro, R., Weng, F., Grody, N. and Basist, A.: 1996, 'An Eight-year (1987–1994) Time Series of Rainfall, Snow Cover, and Sea Ice derived from SSM/I Measurements', *BAMS* **77**, 891.

Haigh, J. D.: 1996, 'The Impact of Solar Variability on Climate', *Science* **272**, 981.

Han, Q., Rossow, W., Zeng, J. and Welch, R.: 2002, 'Three Different Behaviors of Liquid Water Path of Water Clouds in Aerosol-cloud Interactions', *J. Atmos. Sci.* **59**, 726.

ISCCP homepage: 2002, 'ISCCP Calibration Coefficients', http://isccp.giss.nasa.gov/docs/calib.html.

King, M., Radke, L. and Hobbs, P.: 1993, 'Optical Properties of Marine Stratocumulus Clouds Modified by Ships', *J. Geophys. Res.* **98**, 2729.

Lean, J., Beer, J. and Bradley, R.: 1992, 'Reconstruction of Solar Irradiance Since 1610: Implications for Climate Change', *Geophys. Res. Lett.* **22**, 3195.

Lockwood, G. W. and Thompson, D. T.: 1991, 'Solar Cycle Relationship Clouded by Neptune's Sustained Brightness Maximum', *Nature* **349**, 593.

Lockwood, M. R., Stamper, R. and Wild, M. N: 1999, 'A doubling of the Sun's Coronal Magnetic Field during the Past 100 years', *Nature* **399**, 437.

Marsh, N. D. and Svensmark, H.: 2000a, 'Cosmic Rays, Clouds and Climate', *Space Sci. Rev.* **94**, 215.

Marsh, N. D. and Svensmark, H.: 2000b, 'Low Cloud Properties Influenced by Cosmic Rays', *Phys. Rev. Lett.* **85**(23), 5004.

Marsh, N. D. and Svensmark, H.: 2003, 'GCR and ENSO Trends in ISCCP-D2 Low Cloud Properties', *J. Geophys. Res.* **108**(D6), 4195 DOI 10.1029/2001JD001264.

Moses, J. I., Allen, M. and Yung, Y. L.: 1989, 'Neptune's Visual Albedo Variations over a Solar Cycle: A Pre-voyager Look at Ion-induced Nucleation and Cloud Formation in Neptune's Troposphere.', *Geophys. Res. Lett.* **16**, 1489.

Neff, U., Burns, S. J. Mangini, A., Mudelsee, M., Fleitmann, D. and Matter, A.: 2001, 'Strong Coherence between Solar Variability and the Monsoon in Oman between 9 and 6 kyr Ago', *Nature* **411**, 290.

Ney, E. R.: 1959, 'Cosmic Radiation and the Weather', *Nature* **183**, 451.

Parker, D., Gordon, M., Cullum, D., Sexton, D., Folland, C. and Rayner, N.: 1997, 'A New Gridded Radiosonde Temperature Data Base and Recent Temperature Trends', *Geophys. Res. Lett.* **24**, 1499–1502.

Shindell, D., Rind, D., Balabhandran, N., Lean, J. and Lonergan, P.: 1999, 'Solar Cycle Variability, Ozone, and Climate', *Science* **284**, 305.

Svensmark, H.: 1998, 'Influence of Cosmic Rays on Climate', *Phys. Rev. Lett.* **81**, 5027.

Svensmark, H., Marsh, N. D. and Sjölander, B.: 2003, 'Solar Influence on Tropospheric Temperatures' (in preparation); also http://www.dsri.dk/~hsv.

Svensmark, H. and Friis-Christensen, E.: 1997, 'Variation of Cosmic Ray Flux and Global Cloud Coverage – A Missing Link in Solar-Climate Relationships', *J. Atm. Sol. Terr. Phys.* **59**, 1225.

Yu, F.: 2002, 'Altitude Variations of Cosmic Ray Induced Production of Aerosols: Implications for Global Cloudiness and Climate', *J. Geophys. Res.* (in press).

Yu, F. and Turco, R. P.: 2000, 'Ultrafine Aerosol Formation via Ion-mediated Nucleation', *Geophys. Res. Lett.* **27**(6), 883.

SPACE WEATHER RESEARCH IN CHINA

FENGSI WEI[1], XUESHANG FENG[1], JIAN-SHAN GUO[1], QUANLIN FAN[1] and
JIAN WU[2]

[1]*Laboratory for Space Weather, Center for Space Science and Applied Research,
Chinese Academy of Sciences, P.O.Box 8701, Beijing 100080 China*
[2]*Ministry of Information industry, Beijing, China*

Abstract. Recent progress in space weather research are briefly presented here from three aspects: establishment or improvement in observation systems, such as extra-soft X-ray detector and γ-ray detector onboard the spacecraft 'Shen Zhou 2', new solar radio broad-band spectrometer, magnetometer–chain, ionosonde and digisonde–chain, laser-lidar system and VHF radar; partial topic progresses included in CMEs, multi-streamer structures, evolution of interplanetary magnetic field $\mathbf{B}_z$ component, regional properties of traveling ionospheric disturbances, a fully-nonlinear global dynamical model for the middle and upper atmosphere, and a combined prediction method for geomagnetic disturbances; and space weather activity, such as 'Meridian Project' – a national major scientific project, 'International Space Weather Meridian Circle Program' – a suggestion of internationalization of 'Meridian Project', 'Space Weather Research Plan' – a major research plan from National Natural Science Foundation of China (NNSFC) and other space weather activities.

Key words: space/ground-based observation, space weather activities, space weather research

1. Introduction

In the past decade, one of the most important achievements in solar–terrestrial physics research is the realization that there exists adverse space weather above 20–30 km or up to thousands kilometers above the earth. Such adverse conditions can do damage to our high technological systems such as satellites, communication, navigation, tracking, electric network and so on. As we are facing the adverse space weather, solar–terrestrial physics is entering into a new epoch of safeguarding human activity, which generates a new discipline 'space weather'. Today, space weather is rapidly becoming a major scientific activity worldwide, and thus forming national program of various countries (Robinson and Behnke, 2001; Withbore, 2001; Daly and Hilgers, 2001; Kamide, 2001; Panasyuk, 2001; Rostoker, 1997). Many scientists (Baker, 1998; Strangeway, 1998; Shea and Smart, 1998; Hoskinen, 1999) have clarified their academic viewpoints of space weather. In the early 1990s, Chinese scientists had realized that 'the day of space weather as a new discipline is approaching' (NNSF Report, 1992); 'Space weather will become a new systematic science discipline as we are entering the next century' (Wei, 1994). The study of global structure of solar plasma output and magnetic field was suggested

Space Science Reviews **107**: 327–334, 2003.
© 2003 *Kluwer Academic Publishers.*

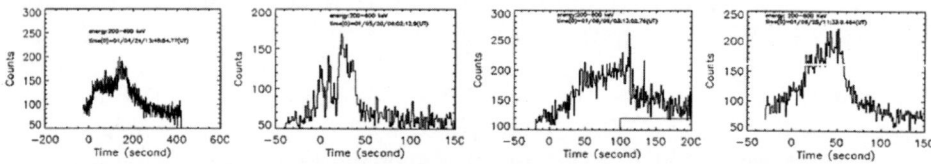

Figure 1. Some examples of the γ-ray bursts recorded by the spacecraft 'Shen Zhou 2'.

as a rebirth of space weather (Wei *et al.*, 1994) 'This project will contribute to the establishment as well as development of space weather in the next century' (Meridian Project (1994)). As a developing country, Chinese scientists pay more attention to the feedback of solar-terrestrial physics to our society, especially in initial step. Space weather, due to its challenge to science in understanding adverse space weather events and its importance to human activities, is becoming a common enterprise. Space weather in China is always supported by Chinese government since the beginning of the 1990s. Some major projects were proposed and are being conducted in succession. The observational equipments, research progress about NNSF's topics and the basic points for some major projects are briefly presented in this paper.

2. Observations

Some observation instruments have been established or improved in recent years, which are presented below.

a) An extra–soft X-ray detector and a γ-ray detector constructed by the Purple Mountain Observatory and onboard the spacecraft 'Shen Zhou 2' have been successfully operated since January 14, 2001. For the first time in China solar X-ray and γ-ray bursts have been observed from space; Figure 1 gives some examples;

b) An ionosonde and digisonde-Chain consisting of Beijing, Xinxiang, Wuhan, Hainan and Antarctic Zhongshan stations is being updated or constructed;

c) A Laser-lidar Chain, which consists of Beijing, Wuhan, Hainan and Antarctic Zhongshan stations, is under construction, in which trial observations of Laser-lidar are made in Wuhan Institute of Physics and Mathematics and Wuhan University. Figure 2a gives an example;

d) VHF radar is being constructed in Wuhan University. Figure 2b shows the plan of VHF Radar System of Wuhan University.

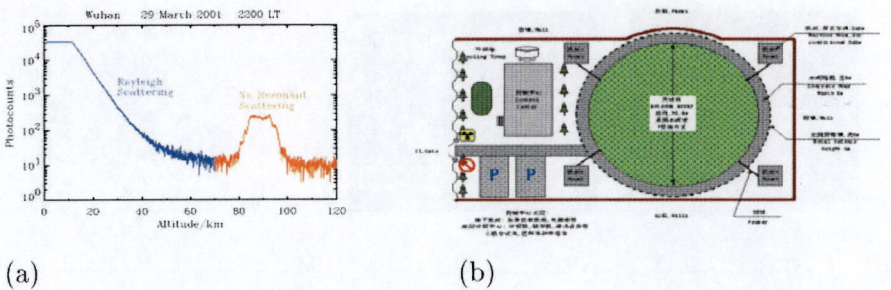

(a)   (b)

Figure 2. (a) Lidar photo count profile obtained on the night of 29 March 2001 at Wuhan University Campus; (b) The plan of VHF (MST) Radar System of Wuhan University.

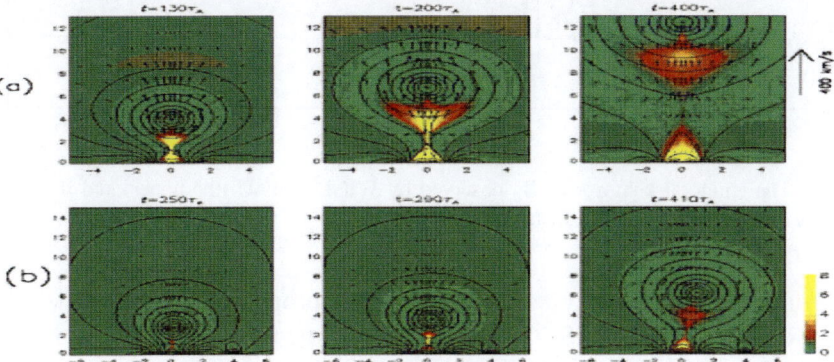

Figure 3. The simulated evolution of CMEs. (a) When the emerging flux appears below the rope. (b) When the emerging flux appears on the edges of the rope. In the figure color indicates temperature, arrows show the velocity, while the solid lines represent magnetic field lines.

3. Research

a) A new model of CMEs triggered by emerging flux has been proposed (Chen and Shibata, 2000). 2D numerical simulations are performed to study the magnetic reconnection during magnetic flux emerging from below the photosphere. It is shown that when the reconnection-favored emerging flux appears the flux rope would lose its equilibrium and be ejected, by which a current sheet is formed below the flux rope. When the current density exceeds a critical value, fast reconnection ensues, leading to the fast eruption of the flux rope, i.e., CMEs, and flaring arcades near the solar surface. The numerical results (Figure 3) can explain recent observations existing a strong correlation between CMEs and reconnection-favored emerging flux.

b) A new numerical procedure of asymmetric corona with multi-streamer structures called the magnetic field fitting-modification method, is proposed (Li *et al.*, 2001). It is used for generating a planar asymmetric corona with multi-streamer structures, as shown in SOHO/LASCO observations, and for providing an approx-

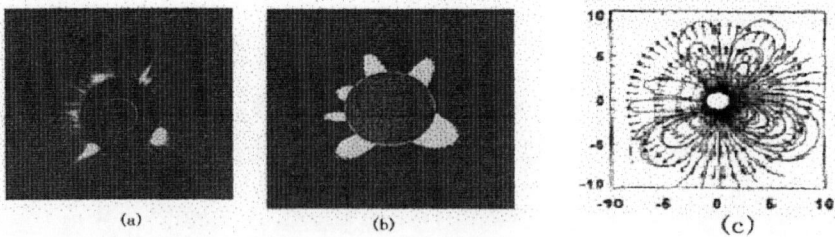

Figure 4. SOHO/LASCO observation of August 1999 CME event (a), computed density t = 50000 s (b) and simulated magnetic field and velocity at t = 50000 s (c).

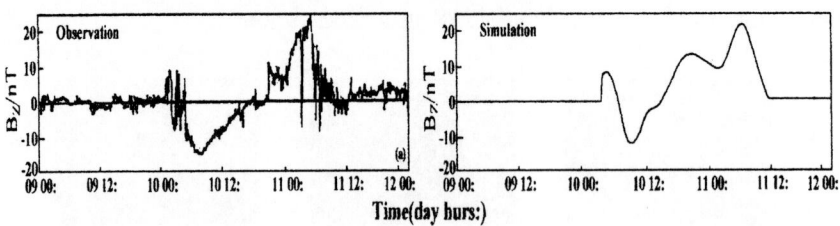

Figure 5. The temporal behavior of **B**$_z$ observed by WIND at 1AU (the upper) and its simulated result (the lower) in the period of January 1997 event.

imately real initial-boundary conditions of CME propagating through the corona. As an example, SOHO/LASCO observation of August 1999 CME event and the simulation results are given in Figure 4.

c) Numerical study of the southern component **B**$_z$ of IMF is made by using 3D-time dependent MHD equations with initial-boundary conditions constructed based on the source surface magnetic field observation. The simulated result for January 1997 event is given in Figure 5 (Shi *et al.*, 2001). The simulated result is basically consistent with the WIND's observation for the **B**$_z$ of IMF.

d) Regional properties of travelling ionospheric disturbances (TID) were studied by adopting a new analysis technique (Wan *et al.*, 1998). It is interesting to point out that the regions of the most TID sources are located in the lee sides of the bulging terrain of Qinghai-Tibet Plateau, where some violent meteorology events such as vortexes probably occurred. Figure 6 shows the vortex occurrence rate (the right) and the TID source in the troposphere (the left). It is obvious that around the east part of Qinghai-Tibet Plateau the distribution of the vortex occurrence shapes very like that of the source occurrence.

e) A fully-nonlinear global dynamical model for the middle and upper atmosphere has been proposed by using a full-implicit-continuous-Eulerian (FICE) scheme and taking the atmospheric motion equations in spherical coordinates as governing equations (Zhang and Yi, 1999). Numerical experiments show that the model is capable of simulating the long time (30 hours) evolution of atmospheric waves (Figure 7).

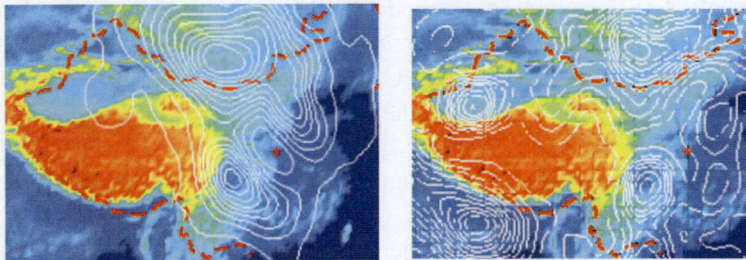

Figure 6. Contour maps showing the geographical distribution of the occurrence of TID sources in the troposphere (the left), and tropospheric vortexes. The background indicates the topography of Qinghai-Tibet Plateau. The asterisk is the location of the observation point (reflection point of the HF observation).

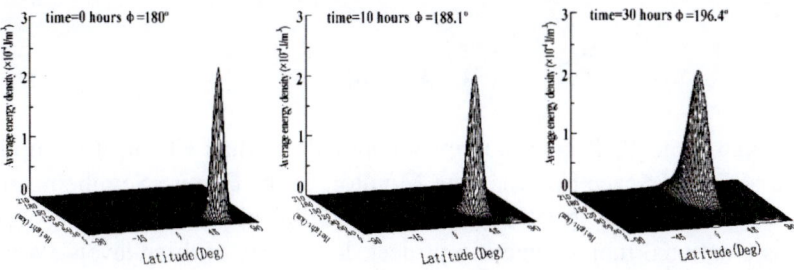

Figure 7. Contour maps showing the geographical distribution of the occurrence of TID sources in the troposphere (the left), and tropospheric vortexes. The background indicates the topography of Qinghai-Tibet Plateau. The asterisk is the location of the observation point (reflection point of the HF observation).

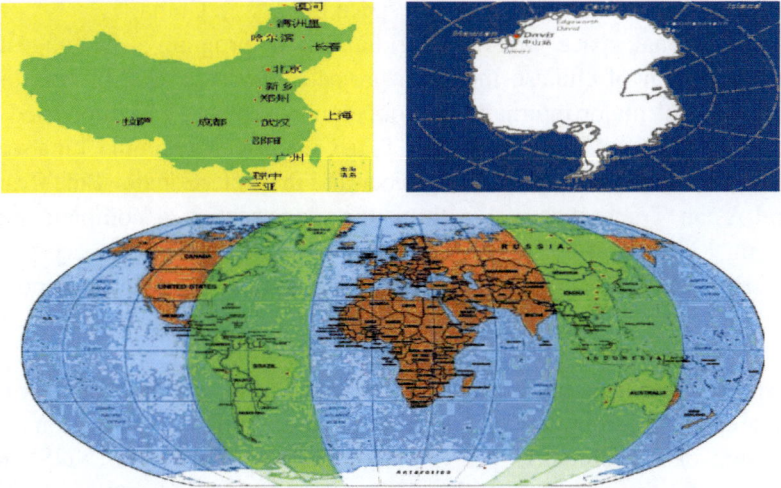

Figure 8. The distributions for 'Meridian Project' station chain of the Estern hemisphere and the relative Western hemisphere station chain.

f) A so-called ISF prediction method for geomagnetic disturbance has been tested for 24 major disastrous weather events (1980–1999). It combines three aspects: I – Interplanetary Scintillation Observation (IPS), S – Solar wind storm propagation dynamics and F – fuzzy mathematics. The test results indicate that: Initial time T – Its relative errors $\Delta T_{pred}/T_{obs} \leqslant 30\%$ for 78.3% of events; Amplitude of the geomagnetic disturbances, ΣK_p – taking account of $\mathbf{B}_z$ of IMF, $\Sigma K_{p,pred}/\Sigma K_{p,obs} \leqslant 10\%$ for 41.6% of all events.

g) Effects of space weather event to the radio-transmission also occur frequently in China. For June 2000 space weather event, radio-transmission in China had suffered serious disturbance and was broken off for long 17 hours in some places. There was almost no communication frequency band available from Xinxiang to Manzhouli and had also suffered serious disturbance both in High Frequency (HF) radio wave and in satellite-communication.

4. Activities

A. 'Meridian project'. It is an abbreviation for 'Meridian Chain of Comprehensive Ground-Based Space Environment Monitors in the Eastern Semihemisphere in China'. It was approved in 1997 by the National Science and Technology Leading Group as a national major scientific project. It consists of three levels (Wei et al., 2001): A Chinese multi-station chain along 120° E to monitor space environment, as shown in Figure 8; A comprehensive ground-based system to monitor space environment variations based on the multi instruments and facilities; and an integrated scientific project to gather Space environment monitor system, Data & Communication system and Research & Forecast System. This project is conducted by uniting the existing research facilities in space physics in China.

B. 'International space weather meridian circle program' – a suggestion of internationalization of chinese meridian project (ISWMCP, 2002). It is listed as one of the several major international cooperation project to be supported by the Ministry of Science and Technology (MST) in China, such as south China sea Monsoon Experiment, Chinese Meridian project and the Pole-Equator-Pole Transects-2-Austral-Asian Transect, etc. We think that constructing a complete meridian circle along 120° E and 60° W is a common cherished desire of space physicists in the world . Main objectives of ISWMCP are: establish a standing cooperative organization; enact collaborative research plans; and develop a working mode of joint research and so on (Wei et al., 2002).

C. 'Space weather research program'. It has been listed one of the six priority research areas to be supported by the national natural science foundation of China in the period of the tenth five-year plan (2001–2005) of China (NNSF Report, 2002). Scientific frontier, new technique and influence on human activities are included in this program.

D. Other space weather activities. A major grant named 'Adverse Disturbance Process in the Solar – Terrestrial Space and its Influence on Human Activities', supported by Chinese NNSF, is under implementation (1999–2003); about 50 scientists from 14 institutes and universities are involved in this major project (Wei and Fang, 1995); Four Chinese workshops on space weather were held in May 1997, August 1998, October 1999 and September 2001, respectively.

5. Conclusion

Space Weather is rapidly being developed as a new scientific field, which has connected the requirement of human activity with the developments of solar-terrestrial physics. Space weather has become a very active scientific frontier in space environment. Space weather research is faced with challenges in science, that is how to make the various space weather 'pictures' of the different space regions in the solar-terrestrial system become a complete space weather 'cinema'. Therefore, to understand the cause-effect chain processes is our main target in the future; Space weather is a human's common enterprise. Currently, it is in a significant developing stage, various international cooperation forms are all necessary for space weather study. 'International space weather Meridian circle program' will be a long operative international cooperation form based on the ground-based observations in the space weather area. It is useful to distribute the ground-based observations appropriately and to plan and use limited resource from various countries participated in the program reasonably.

Chinese space physicists hope to do their best together with the space physicists over the world, in order to promote the development of space weather science.

Finally, by making use of this opportunity, we would like to express our heartfelt thanks to you for your interests in space weather research of China.

Acknowledgements

The authors thank for Prof. Fang Chen, Prof. Wan Weixing and Prof. Yi Fan for their help in the paper's preparation. This report is supported by Chinese NNSF (Grant No. 49990450 and 49925412) and Chinese MBRP (Grant No. G200078405).

References

Baker, D. N.: 1998, *Adv. Space Res.* **22**, 7.
Baker, D. N. *et al.*: 1998, *EOS Trans. Am. Geophys. Union* **79**, 477.
Chen, P. F. and Shibata, K: 2000, *Astrophys. J.* **545**, 524.
Daly, E. J. and Hilgers, A.: 2001, in P. Song, H. J. Singer and G. L. Siscoe (eds.), *Space Weather*, Geophysical Monograph 125, AGU, Washington, D.C., p. 53.

Hoskinen, H. E. J.: 1999, *International SCOSTEP Newsletter* **2**(1), 19.

Kamide, Y.: 2001, in P. Song, H. J. Singer and G. L. Siscoe (eds.), *Space Weather*, Geophysical Monograph 125, AGU, Washington, D.C., p. 59.

Li, J., Wei, F. and Feng, X: 2001, *Geophys. Res. Lett.* **28**(7), 1359.

Meridian Project: 1994, Report to Department of Science and Technology.

NNSF Report: *Development Strategic Report of Natural Sciences-Space Physics*, 1996, suggested in 1992, edited by NNSF of China, Science Press, China.

NNSF Report: 2002, *21th Strategic Keys-Earth Sciences*, edited by NNSF of China, Chinese science and technology publisher, Beijing.

Panasyuk, M.: 2001, in P. Song, H. J. Singer and G. L. Siscoe (eds.), *Space Weather*, Geophysical Monograph 125, AGU, Washington, D.C., p. 65.

Robinson, R. M. and Behnke, R. A.: 2001, in P. Song, H. J. Singer and G. L. Siscoe (eds.), *Space Weather*, Geophysical Monograph 125, AGU, Washington, D.C., p. 125.

Rostoker, G.: 1997, *The Space Weather Initiative in Canada*.

Shea, M. A. and Smart, D. F: 1998, in X. Feng, F. Wei and M. Dryer (eds.), *Advances in solar connection with transient interplanetary phenomena*, International Academic Publishers, Beijing, p. 225.

Shi, Y., Wei, F. and Feng, X.: 2001, *Science in China (A)*, **30**, 61.

Strangeway, R. J.: 1998, *Adv. Space Res.* **22**, 3.

Wan, W. *et al.*: 1998, *Geophys. Res. Lett.* **25**, 3775.

Wei, F.: 1994, *Prog. Geophys.* **9**(3), 86.

Wei, F. and Fang, C.: 1995, *Process Study of Adverse Disturbances in Solar-terrestrial Space and its Effect on Human Activities*, Report to the National Natural Science Foundation of China.

Wei, F. *et al.*: 1994, *Blueprint for the Study of Space Physics of China*.

Wei, F. *et al.*: 2001, *The Meridian Project (ZIWU Project)-An Introduction*, Chinese Meridian Project Working Group.

Wei, F. *et al.*: 2002, International Space Weather Meridian Circle Program Proposed by Chinese Meridian Project Working Group, Beijing.

Withbore, G. L.: 2001, in P. Song, H. J. Singer and G. L. Siscoe (eds.), *Space Weather*, Geophysical Monograph 125, AGU, Washington, D.C. p. 45.

Zhang, S. and Yi, F.: 1999, *J. Geophys. Res.* **104**(D12), 14261.

V: SPACE PLASMA PHYSICS/ASTROPHYSICS

MHD NUMERICAL SIMULATIONS OF PROTO-STELLAR JETS

ADRIANO HOTH CERQUEIRA[1] and ELISABETE M. DE GOUVEIA DAL PINO[2]

[1] *Universidade Estadual de Santa Cruz, Departamento de Ciências Exatas e Tecnológicas, Rodovia Ilhéus-Itabuna, km 16, Ilhéus, Bahia, Brazil CEP 45650-000*
[2] *Universidade de São Paulo, Departamento de Astronomia, Geofísica e Ciências Atmosféricas, Rua do Matão 1226, Cidade Universitária, São Paulo, SP, Brazil CEP 05508-900*

Abstract. We will summarize in this paper the effects that the presence of the magnetic field can cause to proto-stellar jet dynamics, structure and emission line properties, and the differences between two- and three-dimensional numerical simulations will be emphasized.

Key words: Herbig-Haro objects, magnetohydrodynamics, numerical technique, stellar formation

1. Introduction

Young stellar objects produce highly collimated outflows (the *Herbig–Haro*, hereafter HH jets) that may extend up to parsec-length scales in the interstellar medium (hereafter ISM). These HH jets propagate supersonically into the ISM (with Mach numbers of the order of $M \sim 20$ that imply velocities of the order of 300–1000 km s^{-1}; see Reipurth and Bally (2001) for a recent review), and the interaction of them with their surrounding medium is responsible for creating concentrated, emission line structures (the HH objects). Loss of energy due to radiative cooling behind the shock fronts is responsible to create not only the emission in the HH jets, but is also believed to impose strong constraints to the jet dynamics, as we can see from numerical simulations (see below), and is also suggested by recent laboratory experiments (Lebedev *et al.*, 2002) of cooled jets.

The role played by the magnetic field in the production mechanism, as well as, in the structure and evolution of these systems in the ISM seems to be a key point. Recent studies show that even a poorly collimated magnetohydrodynamic (MHD) wind formed in an accretion/proto-star system may evolve in the large scales to a highly collimated, jet-like outflow (see Gardiner, Frank and Hartmann, 2002). On the other hand, numerical simulations in two- and three-dimensions of magnetized jets have shown the importance of MHD effects to the structure, evolution and detailed shock emission properties of HH jets (e.g., Cerqueira and de Gouveia Dal Pino, 2001a; Gardiner *et al.*, 2000; Stone and Hardee, 2000; O'Sullivan and Ray, 2000). In this short communication, we will try to show the most relevant aspects of the results from our 3-D simulations of radiative cooling, MHD jets and their comparison with previously published results.

Space Science Reviews **107**: 337–340, 2003.
© 2003 *Kluwer Academic Publishers.*

2. Numerical Technique, Results and Conclusions

Our numerical simulations were carried out using the so-called Smooth Particle Hydrodynamics Technique (Monaghan, 1999). The 3-D code is designed to simulate jets using a large range of different initial conditions. The jet is initially introduced in the computational domain either with constant velocity (steady state) or intermittently (in order to access the effects caused by velocity variability, such as, the formation of multiple internal working surfaces or *knots*; see Cerqueira and de Gouveia Dal Pino (2001b) and references therein). The system of equations solved are the usual, ideal MHD equations (see Cerqueira and de Gouveia Dal Pino (1999) for details[1]). We assume a fully ionized (Hydrogen) gas with an ideal equation of state, and a radiative cooling rate (due to collisional de-excitation and recombination behind the shocks) given by the coronal cooling function tabulated for the interstellar gas (e.g., Dalgarno and McCray, 1972). For magnetized runs, we can start the simulation with three different initial magnetic configurations: an initially constant longitudinal magnetic field parallel to the jet axis permeating both, the jet and the ambient medium; a force-free helical magnetic field which also extends to the ambient medium (see, e.g., Figure 1 of Cerqueira and de Gouveia Dal Pino, 1999); and a purely toroidal magnetic field permeating the jet only (see, e.g., Figure 1 of Stone and Hardee, 2000). In the toroidal configuration, the jet gas pressure has a radial profile with a maximum at the jet axis (Stone and Hardee, 2000). For the other models (including the pure hydrodynamic models, which are also simulated for comparison with the MHD models), a top-hat profile was assumed for the thermal and magnetic pressures. Both, steady-state and pulsed jets have been simulated.

The MHD SPH simulations of high Mach number ($M \sim 10-20$), heavy (with densities that are higher than that of the ambient medium by a factor of ~ 10), radiative cooling jets show that the effects of magnetic fields are dependent on both the field-geometry and intensity. Our results[2] show that the presence of a helical or a toroidal field tends to affect more the characteristics of the fluid, compared to the purely HD calculation, than a longitudinal field. However, the relative differences which are detected in 2-D simulations involving distinct magnetic field geometries (Stone and Hardee, 2000), seem to decrease in the 3-D calculations (Cerqueira and de Gouveia Dal Pino, 1999, 2001a, 2001b). A striking phenomenum that develops in some 2-D MHD simulations consists in the accumulation of high pressured gas between the two shocks that develop at the head of the jet. This enforces their separation by an extension of several jet radius. This structure, called nose cone,

---

[1] Due to the lack of space, we will not show here the employed MHD system of equations, the equations for the initial magnetic field configurations, or any figures. The reader could look into the papers for more details.

[2] We have, actually, done simulations using a plasma β parameter close to unity. It should be noted that we do not expect, from observations, magnetic filed intensities much higher than those provided by this value of β which implies B $\sim 10-100$ μG.

is absent in 3-D MHD models (Cerqueira and de Gouveia Dal Pino, 1999, 2001a, 2001b). We should emphasize that observations do not show direct evidence for nose cones at the head of protostellar jets, which is consistent with our 3D results.

The Kelvin–Helmholtz (K–H) and Rayleigh–Taylor (R–T) instabilities in our 3-D MHD SPH simulations behave slightly different than in pure hydrodynamical SPH jets, for typical HH jet parameters. Weak *pinches* due to the K–H instabilities develop along the flow, but they do not play an important role in the formation of the HH emission knots along the jets (e.g., Cerqueira and de Gouveia Dal Pino, 1999). Hydromagnetic pinches have also been detected in simulations involving toroidal magnetic fields. These can be particularly strong in the presence of intense fields, for example, close to the jet source (e.g., Frank *et al.*, 1998). R–T instability, which occurs close to the jet head (see Cerqueira, de Gouveia Dal Pino and Herant, 1997) in both pure hydrodynamic and MHD jets is inhibited by the presence of a perpendicular magnetic field. However, such effects are more pronounced in a steady-state jet. In a pulsed jet, the continuous impact of the multiple internal working surfaces (or knots) with the leading one promotes a complex morphological structuring in the jet head, and this effect has a major role to inprint morphological signatures. In 3-D calculations, magnetic fields that are initially nearly in equipartition with the gas tend to affect essentially the detailed structure behind the shocks at the head and internal knots, mainly for non-longitudinal magnetic field topologies. In such cases, the intensity of the emission lines behind the internal knots is increased relative to that in the pure hydrodynamical jet, and this could be eventually used as a diagnostic of the presence of magnetic fields in these jets (Cerqueira and de Gouveia Dal Pino, 2001b). Further 3-D MHD studies are still required as the detailed structure and emission properties of the jets seem to be sensitive to multidimensional effects when magnetic forces are present.

Acknowledgements

The authors would like to thank the project PRONEX (41.96.0908.00) for partial financial support. A.H.C. acknowledges FAPESB, PROPP-UESC and PRODOC/ UFBa for partial financial support.

References

Cerqueira, A. H. and de Gouveia Dal Pino, E. M.: 1999, 'Magnetic Field Effects on the Structure and Evolution of Overdense Radiatively Cooling Jets', *Astrophys. J.* **510**(2), 828–845.

Cerqueira, A. H. and de Gouveia Dal Pino, E. M.: 2001a, 'On the Influence of Magnetic Fields on the Structure of Protostellar Jets', *Astrophys. J. Lett.* **550**(1), L91–L94.

Cerqueira, A. H. and de Gouveia Dal Pino, E. M.: 2001b, 'Three-dimensional Magnetohydrodynamic Simulations of Radiatively Cooling, Pulsed Jets', *Astrophys. J.* **560**(2), 779–791.

Cerqueira, A. H., de Gouveia Dal Pino, E. M. and Herant, M.: 1997, 'Magnetic Field Effects on the Head Structure of Pro tostellar Jets', *Astrophys. J. Lett.* **489**, L185–L188.

Dalgarno, A. and McCray, R. A.: 1972, 'Heating and Ionization of HI Regions', *Ann. Rev. Astron. Astrophys.* **10**, 375.

Frank, A., Ryu, D., Jones, T. W. and Noriega-Crespo, A.: 1998, 'Effects of Cooling on the Propagation of Magnetized Jets', *Astrophys. J. Lett.* **494**, L79–L82.

Gardiner, T. A., Frank, A. and Hartmann, L.: 2002, 'Stellar Outflows Driven by Magnetized Wide-Angle Winds', *Astrophys. J.* (submitted: astro-ph/0202243).

Gardiner, T. A., Frank, A., Jones, T. W. and Ryu, D.: 2000, 'Influence of Magnetic Fields on Pulsed, Radiative Jets', *Astrophys. J.* **530**(2), 834–850.

Lebedev, S. V., Chittenden, J. P., Beg, F. N., Bland, S. N., Ciardi, A., Ampleford, D., Hughes, S., Haines, M. G., Frank, A., Blackman, E. G. and Gardiner, T.: 2002, 'Laboratory Astrophysics and Collimated Stellar Outflows: The Production of Radiatively Cooled Hypersonic Plasma Jets' *Astrophys. J.* **564**, 113–119.

Monaghan, J. J.: 1999, 'Smoothed Particle Hydrodynamics', in Sh. Miyama, K. Tomisaka and T. Hanawa (eds), *Numerical Astrophysics*, Vol. 240, Kluwer Academic Publishers, Astrophysics and Space Science Library, Boston, P. 357.

O'Sullivan, S. and Ray, T. P.: 2000, 'Numerical Simulations of Steady and Pulsed Non-adiabatic Magnetised Jets from Young Stars', *Astron. Astrophys.* **363**, 355–372.

Reipurth, B. and Bally J.: 2001, 'Herbig-Haro Flows: Probes of Early Stellar Evolution' *Ann. Rev. Astron. Astrophys.* **39**, 403–455.

Stone, J. M. and Hardee, P. E.: 2000, 'Magnetohydrodynamic Models of Axisymmetric Protostellar Jets', *Astrophys. J.* **540**(1), 192–210.

IN SEARCH OF THE ORIGIN OF THE HIGHEST ENERGY COSMIC RAYS

R. W. CLAY and B. R. DAWSON FOR THE PIERRE AUGER COLLABORATION

Physics Department, University of Adelaide, Adelaide 5005, Australia

Abstract. The origin of the highest energy cosmic rays is a long-standing problem in astrophysics. We know such particles exist, but their rarity has hampered the discovery of their sources. A new international collaboration is constructing the Pierre Auger Observatory in Argentina. With a collecting area of 3000 km^2, this detector will collect data at a rate ten-times larger than any previous experiment. An introduction outlining the astrophysical problem of the high energy cosmic rays will be followed by a description of the unique Auger Observatory.

Key words: Pierre Auger project, UHE cosmic rays

1. The Highest Energy Cosmic Rays

Cosmic rays with energies above 10^{19}eV arrive at Earth at a very low rate. Great efforts have been expended by several groups over thirty years to detect about 1000 particles above this energy. Only a handful of events have been seen above 10^{20} eV. The mystery of these particles is typified by the 3.2×10^{20} eV cosmic ray observed by the Fly's Eye detector in 1991 (Bird *et al.*, 1995). The origin of the particle is unknown. At such a high energy, and with its assumed charge (that of a proton or light nucleus), its path through the cosmos would have been relatively unaffected by galactic and intergalactic magnetic fields. Yet no likely astrophysical source is known along the arrival direction, within the maximum possible source distance imposed by collisions with photons of the microwave background.

Models for sources of the highest energy cosmic rays have extreme difficulty accommodating the observed energies. However, we have incontrovertible evidence that such particles exist. Where microscopic particles gain this macroscopic energy, and how they reach us, are the mysteries that have brought together a large number of physicists to form the Pierre Auger collaboration.

Cosmic ray arrival rates at these energies are roughly one per square kilometre per year above 10^{19} eV, falling to a rate of at most one per square kilometre per century above 10^{20} eV. Their low rate is clearly a problem, but their high energy is an advantage, since their paths through the cosmos are not likely to have been affected too much by galactic and intergalactic magnetic fields (Clay *et al.*, 1998). If we can detect enough particles, the arrival direction map will likely reflect the distribution of cosmic sources.

Space Science Reviews **107**: 341–344, 2003.
© 2003 *Kluwer Academic Publishers.*

A cosmic ray striking the Earth's atmosphere will initiate a huge cascade of subatomic particles that travels through the atmosphere at a speed close to the speed of light. For the highest energy cosmic rays this cascade, or 'extensive air shower', will contain more than 10^{11} particles and will cover several square kilometres at ground level.

It is generally accepted that the highest energy cosmic rays are extragalactic in origin – if they were accelerated within our galaxy a clear directional anisotropy would have been identified by earlier experiments. In the 1960's it was realised that if these particles were indeed of extragalactic origin, there would be a clear signature in the cosmic ray energy spectrum, provided that the sources uniformly populated extragalactic space. That signature would be an end to the spectrum at around 6×10^{19} eV (the Greisen-Zatsepin-Kuzmin or GZK cut-off) caused by interactions between the cosmic rays and photons of the universal 3 K microwave background. More energetic cosmic rays would rapidly lose energy through this mechanism, and reach Earth with depleted energies. This effectively imposes a distance limit on the sources of the highest energy particles, of order 50 Mpc. (Other attenuation processes affect cosmic ray nuclei and photons, imposing similar distance limits.) Thus the shape of the cosmic ray energy spectrum, together with information on particle arrival directions, can give important information on cosmic ray origin. Several excellent recent reviews of this problem and interpretations of the observations have been published (Yoshida and Dai, 1998; Cronin, 1999; Nagano and Watson, 2000).

The consensus (though not unanimous) view is that cosmic rays at the highest energies are likely to be protonic and extragalactic. No clear departure from isotropy has been seen above 10^{19} eV, and the presence of the GZK cut-off is unclear. Recent results from the HiRes and AGASA experiments (Jui et al., 2001; Sakaki et al., 2001) have not clarified the issue, and appear to be in some conflict. Thus, despite the extreme efforts of several groups over many years, we are still limited by event statistics at the highest energies.

Theoretical activity has been at a high level in recent years, and the suggested possibilities for the cosmic ray sources are numerous (Bhattacharjee and Sigl, 2000; Bertou, Boratav and Letessier-Selvon, 2000). 'Classical' ideas centre on large scale relativistic shocks in active galaxy jets or in colliding galaxies. These ideas have been joined by conjecture on the nature of gamma-ray burst objects and the possibility that these objects also accelerate cosmic rays. An alternative view is that the highest energy cosmic rays are not the products of traditional acceleration, but rather they come from the decay of massive particles released from topological defects in space-time. Most models, including the exotic ones, have predictions in terms of the shape of the cosmic ray energy spectrum, the nature of the particles (e.g. hadrons, gamma-rays) or the expected arrival direction distribution.

2. The Pierre Auger Observatory

Planning for the Auger Observatory began more than a decade ago. The project was named after the 1930's cosmic ray pioneer and co-founder of CERN, and has grown to include over 200 members from 15 countries. An observatory will be built in both the northern and southern hemispheres. The southern observatory in Mendoza Province, Argentina, is funded and is under construction. By the end of 2004 a 3 000 km^2 area will be instrumented with an array of 1600 surface detectors placed on a 1.5 km grid. The atmosphere above the array will be monitored on dark nights by air fluorescence detectors placed at four positions on the array's perimeter (Auger Collaboration, 1997).

Each surface detector is a tank of water (10 square metres in area, 1.2 m depth) instrumented with three photomultiplier tubes. When air shower particles traverse a tank, Cerenkov light is generated and detected. This relatively simple system is able to discriminate between the muonic and electromagnetic components of the air shower. A typical 10^{19} eV air shower will trigger five surface detectors, growing to 15 detectors at an energy of 10^{20} eV. This sampling of the shower allows estimates of the original cosmic ray's energy, arrival direction and mass. Each surface detector is solar powered and transmits its trigger information and data via high frequency radio.

The air fluorescence technique was developed in the 1980's by the Fly's Eye experiment. The passage of a cosmic ray shower causes atmospheric nitrogen to fluoresce, and the faint near-UV light is detected by arrays of large mirrors (each 3.4 m diameter) with fast photomultiplier cameras (440 pixels each). Four fluorescence detector sites will be arranged around the perimeter of the array area with each site housing 6 mirror units. The fluorescence detectors will independently measure the cosmic ray energy, arrival direction and mass.

In the first quarter of 2002 Auger began operating its 'Engineering Array', effectively one-fortieth of the total observatory (Auger Collaboration, 2001). Construction of the full system will begin in mid-2002 and continue until the end of 2004. Serious data collection will begin in early 2003 with approximately one-quarter of the final observatory in operation. In particular, two of the four fluorescence sites will be completed by that time.

The Auger project is innovative in many areas, with its size being the most obvious advance over previous experiments. The southern surface detector array will collect approximately 5000 events a year above 10^{19} eV, a ten-times greater rate than the High Resolution Fly's Eye, the previous largest observatory. In addition, Auger is the first true hybrid detector of the highest energy cosmic rays, with its combination of the surface detector array and four fluorescence detector sites. This will allow a number of important cross-checks, and for the first time a surface array's energy assignment method will be calibrated by the intrinsically calorimetric fluorescence technique (Sommers, 1995). The hybrid nature of the system also provides us with a new reconstruction method, where we use information from both the surface array and a fluorescence detector to determine the shower axis geometry. High quality axis reconstruction can be obtained with data from a single fluorescence eye (Dawson *et al.*, 1996).

Analysis is underway on data from the Engineering array. The careful design process is paying dividends with high quality data from the fluorescence and surface detectors. In the meantime, construction of the bulk of the observatory is continuing at a rapid pace.

References

Auger Collaboration: 1997, *Pierre Auger Project Design Report*, 2nd ed., Fermilab, Batavia Ill., http://www.auger.org/admin/DesignReport/index.html.

Auger Collaboration: 2001, 'Series of Contributions', in *Proc. 27th International Cosmic Ray Conference*, Hamburg, 2: 699–787.

Bertou, X., Boratav, M. and Letessier-Selvon, A.: 2000, 'Physics of Extremely High Energy Cosmic Rays', *Int. J. Mod. Phys.* **A15**, 2181.

Bhattacharjee, P. and Sigl, G.: 2000, 'Origin and Propagation of Extremely High Energy Cosmic Rays', *Phys. Rep.* **327**, 109–247.

Bird, D. J. *et al.*: 1995, *Astrophys. J.* **441**, 144.

Clay, R. W., Cook, S., Dawson, B. R., Smith, A. G. K. and Lampard, R.: 1998, *Astropart. Phys.*, 9221–225.

Cronin, J. W.: 1999, 'Cosmic Rays: The Most Energetic Particles in the Universe', *Rev. Mod. Phys.* **71**, S165.

Dawson, B. R., Dai, H. Y., Sommers, P. and Yoshida, S.: 1996, *Astropart. Phys.* **5**, 239.

Jui, C. H. (HiRes Collab): 2001, in *Proc. of 27th International Cosmic Ray*, Hamburg, **1**, 354.

Nagano, M. and Watson, A. A.: 2000, 'Observations and Implications of the Ultrahigh-energy Cosmic Rays', *Rev. Mod. Phys.* **72**, 689.

Sakaki, N. (AGASA Collab): 2001, in *Proc. of 27th International Cosmic Ray*, Hamburg, 1, 333.

Sommers, P.: 1995, *Astropart. Phys.* **3**, 349.

Yoshida, S. and Dai, H. Y.: 1998, *J.* Phys. **G24**, 905–938.

WAVE-PARTICLE-ELECTRIC FIELD SYNERGETIC AURORAL ELECTRON ACCELERATION

ALTAIR SOUZA DE ASSIS

Universidade Federal Fluminense, Caixa Postal 100294, 24001-970 – Niterói – RJ – Brazil

Abstract. We discuss afresh the problem of the auroral electron acceleration based on the controversy reports of Bryant, D. A. *et al.*: 1992, *Phys. Rev. Lett.* **68**, 37, and Borovsky, J.: 1992, *Phys. Rev. Lett.* **69**, 1054, related to which mechanism is more tenable to accelerate auroral electrons: dc electric field generated somehow in aurora or wave-particle interaction due to auroral wave turbulence? Here, we show that both mechanisms are important, and what is most likely to happen in aurora is that the turbulence and the dc electric field structure will assist each other so as to synergetically accelerate those electrons.

Key words: auroral electron acceleration, electric field, synergy, turbulence

1. Introduction

In earlier papers, space plasma scientists have shown the importance of static electric fields to the auroral acceleration process and reported that intense auroral electron fluxes were generated by quasistatic potential structures (Arnoldy *et al.*, 1974; Borovsky, 1992; McIlwain, 1960; Mozer *et al.*, 1980). On the other hand, Bryant and collaborators (Bryant *et al.*, 1992), have shown that auroral electron fluxes can also be formed by pure wave turbulence activities. These two theories have successfully explained a majority of ground and spacecraft measurements showing that the acceleration pattern is related to the structure of the observed background dc electric fields or wave turbulence. Extending this discussion further, Bryant *et al.* (1992), criticized the former acceleration model saying that double-layer could not accelerate auroral electrons to create auroras, and cited several references and physical conjectures to support their ideas. Then Borovsky (Borovsky, 1992) answered this criticism, and showed that double layers indeed do accelerate auroral electrons and can then form auroras. However, after all the discussions the former still maintains his position that auroral electron acceleration by double-layer is fundamentally untenable (Bryant, 1994; Bryant, 1999; McClements, 1999). Later new papers by the same authors, on the same subject, were published, leaving untouched the possibility of the synergy in the turbulence-double-layer auroral electron acceleration. However, examining observations of rockets and satellites, it is clear that the two structures coexist, and the flux enhancement is clearly seen (Karlsson, 2001; Ivchenko, 2002). Though the reports on such events are few in literature, their existence cannot readily be explained by the two current theories cited above, and therefore a further explanation is necessary. In this paper, we present a theoretical discussion that supports the conjecture of wave-particle interaction assisting the auroral electron acceleration due to a dc electric field working so as to enhance the electron flux. We access the field aligned electron acceleration using the Fokker–Planck equation, and the turbulence is modelled by the weak turbulence theory. The result is valid for the turbulence induced by lower hybrid waves-lhw, kinetic Alfven waves-KAW, and electromagnetic ion cyclotron waves-emicw.

II – Basic Equation (de Assis and Leubner, 1994):

The electron flux is given by the Maxwellian-bound equation:

$$\frac{\partial}{\partial t} f(v_{||}, t) = \frac{\partial}{\partial v_{||}} \left[D_{wave}(v_{||}, t) \frac{\partial}{\partial v_{||}} f(v_{||}, t) \right] + \frac{eE_{DL}}{m_e} \frac{\partial}{\partial v_{//}} f(v_{||}, t)$$
$$+ \frac{\partial}{\partial v_{||}} \left\{ \nu_A \left[v_{||} f(v_{||}, t) + (v_{the})^2 \frac{\partial}{\partial v_{||}} f(v_{||}, t) \right] \right\} \quad , \quad (1)$$

where $f(v_{||}, t)$ is the time averaged electron velocity distribution function, which evolves in the low time scale due to turbulence [Cherenkov damping], anomalous collisions induced by turbulence and the dc electric field. Here: $D_{wave}(v_{||}, t)$, E_{DL}, $\nu_A v_{||}$, $\nu_A(v_{the})^2$, v_{the}, e, m_e are the wave diffusion coefficient, the field aligned dc electric field, the anomalous friction term [driven by turbulence], the anomalous collisional diffusion[driven by turbulence], the electrons thermal speed, the electron's charge, and the electron's mass, respectively. The last term on the right hand side of Equation (1) is used to permit the electron distribution function to return to the original starting equilibrium state, it simulates, due to the turbulence, the effect of the Fokker–Planck's Coulomb collision operator. The Maxwellization of the electrons by turbulence is called the 'Langmuir paradox' (Kadontsev, 1965). Considering now the steady-state regime Equation (1) can be written as follows:

$$\frac{d}{dv_{||}} \left\{ \nu_A \left[v_{||} f(v_{||}) + v_{the}^2 \frac{d}{dv_{||}} f(v_{||}) \right] \right\} +$$
$$\frac{d}{dv_{||}} \left[D_{wave}(\frac{d}{dv_{||}} f(v_{||})) \right] + \frac{eE_{DL}}{m_e} \frac{d}{dv_{||}} f(v_{||}) + A\delta(v_{||}) = 0 \quad , \quad (2)$$

where A is a source term and $\delta(v_{//})$ is the Dirac's delta distribution. In the above equation, A should be interpreted as the electron flux induced by the dc electric field E_{DL} assisted by the wave turbulence $D_{wave}(v_{//})$. The auroral emission is related to A, and this is the important quantity to be known. After introducing the step function $H(x)[\frac{d}{dx} H(x) = \delta(x)]$ in the source term $A\delta(x) \Rightarrow A\frac{d}{dx} H(x)$, in Equation (2), all terms are total derivatives with respect to $v_{//}$, and can then be immediately integrated to yield,

$$\Pi(v_{||})f(v_{||}) + \Theta(v_{||})\frac{df}{dv_{||}} = C - AH(v_{||}), \quad (3)$$

where a convenient regrouping of terms has been carried out. In Equation (1): $\Pi(v_{||}) = v_{||} v_A + e E_{DL}/m_e$ and $\Theta(v_{||}) = D_{wave}(v_{||}) + D_C^A$. In order to determine the constant of integration C, we consider the limit $v_{||} \to +\infty$, where $H(v_{||})$ is equal to unity, $df(v_{||})/dv_{||}$ as well as v_A are equal to zero. Hence,

$$f(+\infty)\frac{e E_{DL}}{m_e} = C - A \tag{4}$$

The term on the left-hand side of (4) is the flux of the runaway electrons, which in the steady-state must be equal to the source constant A. Consequently, consistency requires that the integration constant C be zero. As a consequence, we can write the auroral electron flux as:

$$A = \frac{e E_D^A f(v_{//} = 0)}{\sqrt{2\pi} m_e} \left(\frac{E_{DL}}{E_D^A}\right)^{\frac{3}{2}} \exp[-[\frac{1 + \frac{v_{phase//}}{v_C^A}\frac{D_{wave}}{D_C^A}}{2\frac{E_{DL}}{E_C^A}(1 + \frac{D_{wave}}{D_C^A})}]], \tag{5}$$

where E_D^A is the anomalous Dreicer Field. The electron flux induced by the dc electric field structure is enhanced by the presence of the turbulence since the exponential, in Equation (5), A goes to zero smoother if $D_{wave} \neq 0$. Since A depends on $f(v_{//} = 0)$ [the local plasma density], the flux will be also enhanced for any enhancement in the local plasma density. Solving Equation (5), for the general case, numerically, we can show: (1) the electron fluxes can be enhanced by orders of magnitude over that without turbulence, (2) for any wave phase velocity condition and turbulence spectral width, the enhancement is weaker the weaker the double-layer electric field, (3) the enhancement is caused by the Cherenkov damping and it saturates when the plateau in the distribution function is reached, (4) the spectral width of the turbulence plays a more important role on the enhancement than the spectral shape does, and finally (5) the following synergetic effects can be present in this kind of acceleration process:

(a) Electric field-turbulence: $[A_{E_{DL} \neq 0}^{D \neq o} > A_{E_{DL}=0}^{D \neq 0} + A_{E_{DL} \neq 0}^{D=0}]$,

(b) Turbulence-turbulence $[A_{E_{DL} \neq 0}^{D_{wave1}^+ D_{wave2}^{\neq 0}} > A_{E_{DL} \neq 0}^{D_{wave1} \neq 0} + A_{E_{DL} \neq 0}^{D_{wave2} \neq 0}]$.

In conclusion, we have shown that auroral magnetic field aligned dc electric field structures and auroral Alfvenic and/or lower hybrid turbulence can work synergetically so as to enhance the observed auroral electron fluxes induced by the former. Therefore, it is not possible to explain all the auroral electron energy observations, by the satellites and rockets (Karlsson, 2001; Ivchenko, 2002), if one considers the acceleration by these two mechanisms separately, the synergetic effects must be considered as well, being the main case the enhancement observed at the edges of auroral arcs. The distribution $f(v_{||}, t)$ depends on the local plasma density, and so the auroral electron fluxes. In collisional ionospheres, the anomalous term should be replaced by the normal collisional one. Since we are dealing with field-aligned acceleration, in a collisionless auroral plasmas, 2D effects are not important and therefore we neglect them here.

Acknowledgements

The travel's financial support, to attend the WSEF2002, was provided by FAPERJ The Rio de Janeiro State Research Foundation. This work was done within the framework of the Associateship Scheme of the International Centre for Theoretical Physics – ICTP, Trieste, Italy. The author would like to thank ICTP and the Royal Institute of Technology, Stockholm, Sweden for the research support. I would like to thank also to Prof. Dr. A. C. L. Chian for the most kind hospitality and care during my stay in Adelaide, and G. Marklund, T. Karlsson and C-G Fälthammar, KTH-Alfven Laboratory, Stockholm – Sweden, for useful discussion on auroral electron dynamics.

References

Arnoldy, R. L., Lewis, P. B. and Isaackson, P. O.: 1974, *J. Geophys. Res.* **79**, 4208.
Borovsky, J. E.: 1992, *Phys. Rev. Lett.* **69**(7), 1054–1056.
Borovsky, J. E.: 1993, *J. Geophys. Res.* **98**, 6101–6120.
Bryant, D. A.: 1994, *Contemporary Phys.* **35**(3), 165–169.
Bryant, D. A., Birgham, R. and de Angeli, U.: 1992, *Phys. Rev. Lett.* **68**(1), 37–39.
Bryant, D. A.: 1999, *Electron Acceleration in the Aurora and Beyond.* Book, Institute Physics Publishing, Bristol.
de Assis, A. S. and Leubner, C.: 1994, *Astron. Astrophys.* **281**, 588–593.
Ivchenko, N.: 2002, *Alfven Waves and Spatio-Temporal Structuring in the Auroral Ionosphere.* PhD thesis, KTH, Stockholm, Sweden.
Kadomtsev, B. B.: 1965, *Plasma Turbulence, Book, Section 5b,* Academic Press, NY.
Karlsson, T.: 2001, Auroral Electric Fields from Satellite Observations and Numerical Modeling, PhD thesis, KTH, Stockholm, Sweden.
McClements, K. G.: 1999, *Nuclear Fusion* **39**(8), 1071–1073.
McIlwain, C. E.: 1960, *J. Geophys. Res.* **65**, 2727.
Mozer, F. S. , Cattell, C. A., Hudson, M. K., Lysak, R. L., Temerin, M. and Torbert, R. B.: 1980, *Space Sci. Rev.* **27**, 155.

SPECTRUM OF GLOBAL IDEAL-MAGNETOHYDRODYNAMIC THREE-DIMENSIONAL BALLOONING MODES

R. L. DEWAR

Department of Theoretical Physics & Plasma Research Laboratory, The Australian National University, Canberra ACT 0200, Australia

Abstract. The class of pressure-driven plasma instabilities known as ballooning modes may be responsible for such diverse phenomena as high-beta disruptions in tokamaks, solar flares and magnetospheric substorms. In this paper the theory of the spectrum of unstable eigenvalues of the linearized ideal magnetohydrodynamic (MHD) equations of motion in non-axisymmetric toroidal equilibria is sketched, comparing and contrasting systems with open field lines and systems with toroidally confined field lines. The need to regularize ideal MHD to keep the wavenumber finite, and the relevance of quantum chaos theory to understand the structure of the spectrum, is pointed out.

Key words: ballooning instability, quantum chaos, stellarator

1. Introduction

This paper reviews recent work on the novel geometric effects encountered when ideal magnetohydrodynamics (MHD) is used to calculate the eigenspectrum of linearized normal modes in finite-β nonaxisymmetric toroidal plasma containment devices (stellarators). If we consider systems that are interchange stable, then these instabilities are ballooning modes, which have also been put forward as a possible mechanism for disrupting the cross-tail current at magnetospheric substorm onset (Bhattacharjee, Ma and Wang, 1998).

It should be noted however that, although the magnetosphere is both toroidal and non-axisymmetric, it is less like a stellarator than another class of laboratory magnetic confinement device – the magnetic mirror – because, like the mirror, the magnetic field lines are of finite extent in at least one direction.

The problem of loss of energy along the field lines in mirrors has led to their virtual abandonment as candidates for fusion reactors, the bulk of magnetic confinement fusion research being focussed on the axisymmetric 'tokamak' class of experiment. In these devices the ignorability of the toroidal coordinate implies that the magnetic field line 'dynamics' (a Hamiltonian system) is integrable, so the field lines are of infinite extent and lie on nested invariant tori. This greatly improves the insulating properties of the magnetic field.

However in recent years there has been a resurgence of interest in the non-axisymmetric 'stellarator' class due to their relative immunity toward current-driven

Space Science Reviews **107**: 349–352, 2003.
© 2003 *Kluwer Academic Publishers.*

instabilities, which can lead to catastrophic disruption of tokamak discharges. The more complicated geometry of stellarators has required the development of more sophisticated methods for theoretical and computational analysis.

One theoretical problem inherent to nonaxisymmetric devices is the generic nonintegrability of the magnetic field (Hudson and Dewar, 1999). However the Kolmogorov–Arnol'd–Moser (KAM) theorem gives hope that careful design and construction can lead to the retention of a large measure of invariant tori, and this has in practice proved to be the case. In the following we shall be perturbing around a nonaxisymmetric MHD equilibrium in which the gradient of the pressure p is balanced by the plasma current $\mathbf{j}$ and magnetic field $\mathbf{B}$ (i.e. $\nabla p = \mathbf{j} \times \mathbf{B}$). We shall assume the magnetic field is integrable, so that all field lines lie on invariant tori (magnetic surfaces) labelled with a dimensionless parameter $s \in [0, 1]$. We use a general curvilinear angular coordinate system on each magnetic surface, with the toroidal angle ζ and poloidal angle θ being chosen in such a way that the magnetic field lines appear straight when graphed in θ-ζ space. The slope of the field line, $q(s) = d\zeta/d\theta$, defines the winding number for the torus, $1/q$ being known as the *rotational transform*. We shall assume $q(s)$ is a monotonic function, so that q may be used as an alternative to s as a magnetic surface label.

2. Ballooning Equation

The ideal MHD equations for a linearized displacement $\boldsymbol{\xi}$ with time dependence cos or sin ωt are the Euler–Lagrange equations that extremize the time-averaged Lagrangian $L = \omega^2 K - \delta W$, where $\omega^2 K$ is the kinetic energy and δW is the quadratic perturbation in the potential energy of the plasma. We order typical wavevectors $\mathbf{k}_\perp$ perpendicular to the field lines to be large compared with the typical inverse system scale length L^{-1}, i.e. $k_\perp L \gg 1$, but keep $k_\parallel L = O(1)$ so that the modes are 'flute like', $k_\parallel/k_\perp \ll 1$. Taking a simplified kinetic energy one may express $\boldsymbol{\xi}$ in terms of a scalar stream function, φ, as $\boldsymbol{\xi}_\perp = \mathbf{B} \times \nabla\varphi/B^2$ (Dewar, 1997). Then the eigenvalue problem reduces to a scalar wave equation

$$
\mathbf{B}\cdot\left\{ \frac{1}{B^2}\nabla\left[\frac{B^2}{\mu_0}\mathsf{P}_\perp\cdot\nabla\left(\frac{\mathbf{B}\cdot\nabla\varphi}{B^2} \right) \right] \right\} + \nabla\left(\frac{\nabla p \times \mathbf{B}, \boldsymbol{\kappa} \times \mathbf{B}\cdot\nabla\varphi}{B^4} \right)
$$
$$
+ \nabla\left(\frac{\boldsymbol{\kappa} \times \mathbf{B}\nabla p \times \mathbf{B}\cdot\nabla\varphi}{B^4} \right) + \omega^2\nabla\left(\frac{\rho}{B^2}\mathsf{P}_\perp\cdot\nabla\varphi \right) = 0 ,
\tag{1}
$$

where ρ is the mass density, $\boldsymbol{\kappa} \equiv \mathbf{e}_\parallel\cdot\nabla\,\mathbf{e}_\parallel$ is the local *curvature vector* of the magnetic field lines and $\mathsf{P}_\perp$ is the *perpendicular projection operator* $\mathsf{I} - \mathbf{e}_\parallel\mathbf{e}_\parallel$, $\mathbf{e}_\parallel$ being the unit vector in the direction of $\mathbf{B}$.

Because of the assumption of large $k_\perp L$ we can make further analytical progress by using a form of the Wentzel–Brillouin–Kramers (WKB) approximation in directions transverse to $\mathbf{B}$. Thus we set $\varphi = \widehat{\varphi}\exp(iS)$, where the phase or *eikonal* S is constant on a field line but varies rapidly from field line to field line. Thus the

$k_\parallel$ spectrum is absorbed into the amplitude factor $\widehat{\varphi}$, varying on the equilibrium scale. Defining $\mathbf{k} \equiv \mathbf{k}_\perp \equiv \nabla S$, the PDE Equation (1) is reduced at leading order to a dispersion equation by the replacement rule $\nabla \mapsto i\mathbf{k}$. However, unlike the case in more conventional WKB theory, the resulting leading order dispersion relation degenerates to the trivial result $\omega^2 = 0$. This is simply a statement of the fact that the frequencies of Alfvén and slow magnetosonic waves reach their minima of zero when $k_\parallel$ vanishes.

To find a nontrivial dispersion relation, we must go to next order, where, implicitly, $k_\parallel \neq 0$. Then we must keep the operator nature of $\nabla_\parallel \equiv d/dl$ (where l is the distance along a field line) in its action on $\widehat{\varphi}$ and equilibrium quantities. The resulting ODE along a field line, known as the *ballooning equation*, gives a local dispersion relation when we solve it as an eigenvalue equation for ω^2, thus effectively projecting the field line onto a point in a surface of section transverse to the field line, reducing the WKB problem to a two-dimensional one.

The boundary conditions under which the ballooning equation is to be solved depend on the geometry – in systems with open field lines like the magnetosphere or magnetic mirrors one can assume a line tying boundary condition at the end points, but in toroidal confinement the field lines extend to infinity so we take the eigenfunction to vanish at $l = \pm\infty$, recovering periodicity by an infinite superposition of these aperiodic solutions (Dewar and Glasser, 1983).

As in (Dewar and Glasser, 1983) we write the magnetic field as $\mathbf{B} = \nabla\alpha \times \nabla\psi$, where $\psi(s)$ is a flux function and $\alpha \equiv \zeta - q\theta$ is the *field-line label*.

The local ballooning eigenvalues depend on s, α, and the direction (but not the magnitude) of $\mathbf{k}_\perp$. This is expressed in terms of the parameter $\theta_k \equiv k_q/k_\alpha$, which has angle-like symmetry properties. The global eigenvalues are determined by ray tracing (Cooper, Singleton and Dewar, 1996) so we need to know the detailed dependence of the eigenvalue on all three parameters, ψ, α and θ_k.

A full understanding of this dependence requires the ballooning eigenvalue equation to be solved many times in a parameter scan over a three-dimensional mesh. Such studies have been carried out for the the Australian stellarator H-1NF (Cuthbert, 1999; Cuthbert and Dewar, 2000), Japanese stellarators LHD (Cuthbert *et al.*, 1998) and Heliotron J (Yamagishi, Nakamura and Kondo, 2001; Yamagishi *et al.*, 2002), and in a design study for the planned US stellarator NCSX (Redi *et al.*, 2002).

Understanding the results has required introducing concepts from other fields of physics that are not typically encountered in MHD theory, in particular the concepts of Anderson localization (Cuthbert and Dewar, 2000) and quantum chaos (cuthbert, Dewar and Ball, 2001). The latter refers to the theory of global eigenvalue spectra when the quasi-classical (WKB) approximation encounters chaotic ray trajectories.

It was shown in (Cuthbert, Dewar and Ball, 2001) that the ray dynamics was strongly chaotic for toroidally localized ballooning modes in stellarators, once MHD was regularized by the *ad hoc* insertion of a short-wavelength cutoff to keep

$|\mathbf{k}_\perp|$ bounded. If this regularization was not done, the rays escaped to infinity in the $\mathbf{k}_\perp$ sector of their 4-dimensional phase space.

Physically, such a regularization occurs through finite-Larmor-radius (FLR) effects not included in ideal MHD. The simplest way (Tang, Dewar and Manickam, 1982) to include some FLR regularization is to replace ω^2 by $\omega(\omega-\omega_*)$ in the local ballooning dispersion relation, where ω_* is the local drift frequency. A study of the effect of non-axisymmetry on ballooning modes in magnetic mirrors in the 1980s (Nevins and Pearlstein, 1988) using FLR regularization found quantum chaos to be relevant to the spectral problem in these devices also. It is a subject of ongoing research to determine the full implications of quantum chaos theory for laboratory and space plasma problems.

References

Bhattacharjee, A., Ma, Z. W. and Wang, X.: 1998, *Phys. Plasmas* **5**, 2001.
Cooper, W. A., Singleton, D. B. and Dewar, R. L.: 1996, *Phys. Plasmas* **3**, 275; Erratum: Phys. Plasmas **3**, 3520 (1996).
Cuthbert, P.: 1999, 'Ballooning Instabilities in Three-Dimensional Toroidal Plasmas', Ph.D. thesis, The Australian National University.
Cuthbert, P. and Dewar, R. L.: 2000, *Phys. Plasmas* **7**, 2302.
Cuthbert, P., Dewar, R. L. and Ball, R.: 2001, *Phys. Rev. Letters* **86**, 2321.
Cuthbert, P., Lewandowski, J. L. V., Gardner, H. J., Persson, M., Singleton, D. B., Dewar, R. L., Nakajima, N. and Cooper, W. A.: 1998, *Phys. Plasmas* **5**, 2921.
Dewar, R. L.: 1997, *J. Plasma Fusion Res.* **73**, 1123.
Dewar, R. L. and Glasser, A. H.: 1983, *Phys. Fluids* **26**, 3038.
Hudson, S. R. and Dewar, R. L.: 1999, *Phys. Plasmas* **6**, 1532.
Nevins, W. M. and Pearlstein, L. D.: 1988, *Phys. Fluids* **31**, 1988.
Redi, M. H., Johnson, J. L., Klasky, S., Canik, J., Dewar, R. L. and Cooper, W. A.: 2002, *Phys. Plasmas* **9**, 1990.
Tang, W. M., Dewar, R. L. and Manickam, J.: 1982, *Nucl. Fusion* **22**, 1079.
Yamagishi, O., Nakamura, Y. and Kondo, K.: 2001, *Phys. Plasmas* **8**, 2750.
Yamagishi, O., Nakamura, Y., Kondo, K. and Nakajima, N.: 2002, *Phys. Plasmas* **9**, 3429.

STABILITY AND WAVES OF TRANSONIC LABORATORY AND SPACE PLASMAS

J. P. GOEDBLOED

FOM-Institute for Plasma Physics, P.O. Box 1207, 3430BE Nieuwegein, the Netherlands
(e-mail: goedbloed@rijnh.nl)

Abstract. The properties of magnetohydrodynamic waves and instabilities of laboratory and space plasmas are determined by the overall magnetic confinement geometry and by the detailed distributions of the density, pressure, magnetic field, and background velocity of the plasma. Consequently, measurement of the spectrum of MHD waves (MHD spectroscopy) gives direct information on the internal state of the plasma, provided a theoretical model is available to solve the forward as well as the inverse spectral problems. This terminology entails a program, viz. to improve the accuracy of our knowledge of plasmas, both in the laboratory and in space. Here, helioseismology (which could be considered as one of the forms of MHD spectroscopy) may serve as a luminous example. The required study of magnetohydrodynamic waves and instabilities of both laboratory and space plasmas has been conducted for many years starting from the assumption of static equilibrium. Recently, there is a outburst of interest for plasma states where this assumption is violated. In fusion research, this interest is due to the importance of neutral beam heating and pumped divertor action for the extraction of heat and exhaust needed in future tokamak reactors. Both result in rotation of the plasma with speeds that do not permit the assumption of static equilibrium anymore. In astrophysics, observations in the full range of electromagnetic radiation has revealed the primary importance of plasma flows in such diverse situations as coronal flux tubes, stellar winds, rotating accretion disks, and jets emitted from radio galaxies. These flows have speeds which substantially influence the background stationary equilibrium state, if such a state exists at all. Consequently, it is important to study both the stationary states of magnetized plasmas with flow and the waves and instabilities they exhibit. We will present new results along these lines, extending from the discovery of gaps in the continuous spectrum and low-frequency Alfvén waves driven by rotation to the nonlinear flow patterns that occur when the background speed traverses the full range from sub-slow to super-fast.

Key words: MHD waves, transonic plasmas

1. MHD Waves: Two Paradigms

Magnetohydrodynamics (MHD) describes a vast territory of plasma dynamics in the universe. Due to scale-invariance of the MHD equations (Goedbloed, 2002), laboratory tokamak plasmas as well as space and astrophysical plasmas of all relevant sizes may be analyzed by the same analytical and numerical techniques. The present paper aims at demonstrating the fertility of this point of view.

Recall from the history of 20th century physics that significant insight and subsequent progress comes from the investigation of *generic physical problems*, like computing the quantum mechanical structure of the spectrum of the hydrogen

Space Science Reviews **107**: 353–360, 2003.
© 2003 *Kluwer Academic Publishers.*

atom. In solar physics, an analogous problem is the computation and measurement of the (classical) spectrum of sound waves of the sun making it possible to determine the internal structure of the sun ($\equiv$ helioseismology). Similarly, *MHD spectroscopy* (Goedbloed *et al.*, 1993) may be conducted for a wide variety of magnetized plasmas in tokamaks, sunspots, solar coronal loops, and astrophysical plasmas of all kinds, like accretions disks about compact objects. The common approach to all these problems is the computation of the spectrum of a linear operator and comparison with the observed frequencies of the system of interest. This, in turn, eventually leads to much improved knowledge of the internal structure and dynamics of that system.

The theoretical study of MHD spectroscopy of plasmas involves a split in a background equilibrium and the linear perturbations of it, resulting in the traditional paradigm of **static plasmas** ($\mathbf{v} = 0$):

– Equilibrium:

$$\nabla p - \rho \mathbf{g} = \mathbf{j} \times \mathbf{B}, \qquad \mathbf{j} = \nabla \times \mathbf{B}, \qquad \nabla \cdot \mathbf{B} = 0; \tag{1}$$

– Spectrum:

$$\mathbf{F}(\boldsymbol{\xi}) = -\rho \omega^2 \boldsymbol{\xi}. \tag{2}$$

Although extremely successful for tokamaks, this paradigm needs to be modified when plasma background flows are present, as in almost all astrophysical plasmas, to account for **stationary plasmas** ($\mathbf{v} \neq 0$):

– Equilibrium:

$$\nabla \cdot (\rho \mathbf{v}) = 0,$$
$$\rho \mathbf{v} \cdot \nabla \mathbf{v} + \nabla p - \rho \mathbf{g} = \mathbf{j} \times \mathbf{B}, \qquad \mathbf{j} = \nabla \times \mathbf{B},$$
$$\mathbf{v} \cdot \nabla p + \gamma p \nabla \cdot \mathbf{v} = 0, \tag{3}$$
$$\nabla \times (\mathbf{v} \times \mathbf{B}) = 0, \qquad \nabla \cdot \mathbf{B} = 0;$$

– Spectrum:

$$\mathbf{F}_{\text{static}}(\boldsymbol{\xi}) + \nabla \cdot \left[\rho (\mathbf{v} \cdot \nabla \mathbf{v}) \boldsymbol{\xi} - \rho \mathbf{v} \mathbf{v} \cdot \nabla \boldsymbol{\xi} \right]$$
$$+ 2i \rho \omega \mathbf{v} \cdot \nabla \boldsymbol{\xi} + \rho \omega^2 \boldsymbol{\xi} = 0. \tag{4}$$

Here, the linear force operator $\mathbf{F}(\boldsymbol{\xi})$ is the well-known expression derived by Bernstein *et al.*, (1958) leading to the linear eigenvalue problem (2) in terms of ω^2, and background plasma flow implies modification to the quadratic eigenvalue problem (4) derived by Frieman and Rotenberg (1960).

It is clear that the analysis of stationary plasmas is substantially more complicated than that of static plasmas, both as regards the generation of the required

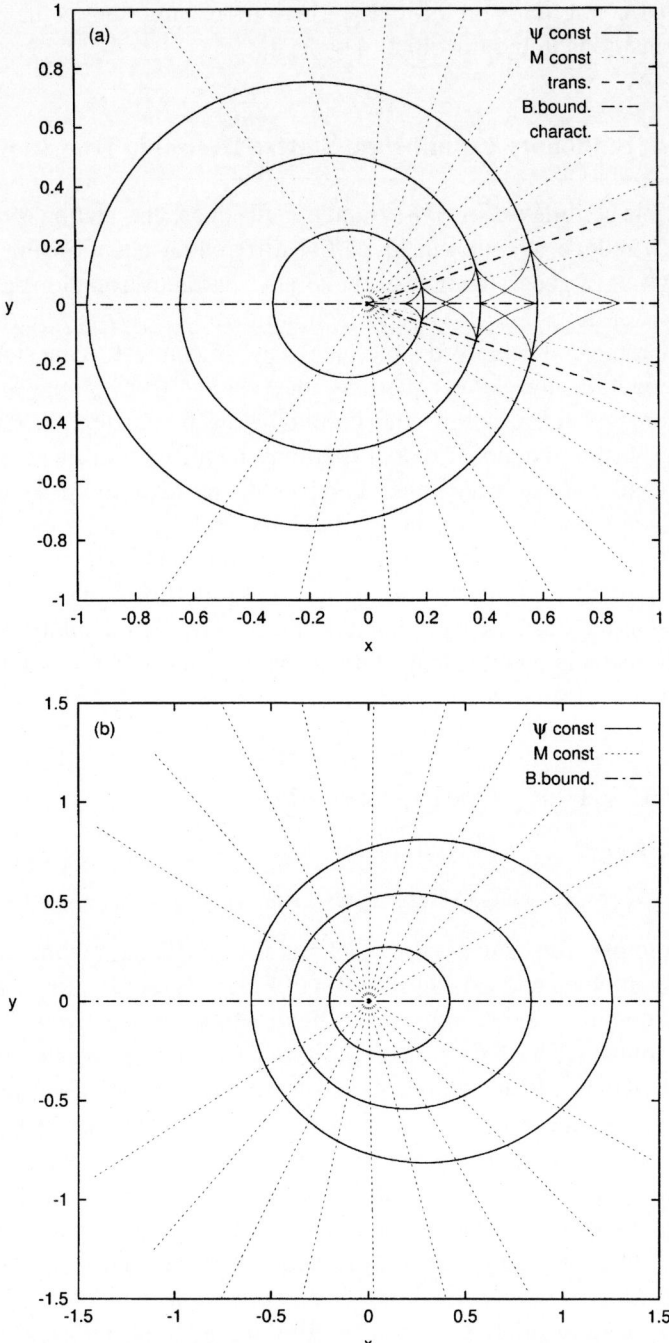

Figure 1. (a) Streamlines and characteristics in sub-slow (1st elliptic, 1st hyperbolic) flow regimes;
(b) Streamlines in slow (2nd elliptic) flow regime.

equilibria and the computation of the associated spectra of waves and instabilities. However, that is not the main difficulty. Much more important is the possibility of transonic transitions in the flow that may occur.

2. Stationary Equilibrium States: Transonic Transitions

For translation symmetric or axi-symmetric plasmas, the stationary equilibrium equations (3) reduce to a nonlinear partial differential equation for the poloidal magnetic flux ψ (a generalization of the Grad–Shafranov equation) and an algebraic equation for the poloidal Alfvén Mach number $M \equiv \sqrt{\rho} v_p / B_p$ (the Bernoulli equation), that have to be solved simultaneously. In contrast to the standard Grad–Shafranov equation, the latter equations may change from elliptic to hyperbolic depending on the magnitude of the poloidal flow parameter M. One difficulty of transonic plasma dynamics is that this may happen somewhere in the middle of the region of interest where the character of the flow suddenly changes dramatically. Even more dramatic, though fundamentally revealing the true nature of MHD transonic flows, are the slow and fast limiting line singularities M_{SL} and M_{FL} and the Alfvén singularity $M_A \equiv 1$, where the flow simply fails to remain continuous. On the basis of these singularities, four fundamentally different flow regimes may be distinguished, that again fall apart in seven regimes when ellipticity or hyperbolicity is taken into account:

$$
\begin{aligned}
&0 < M < M_{SL} \qquad &&\text{(1st ell. \& hyp. } sub\text{-}slow\text{)},\\
&M_{SL} < M < 1 \qquad &&\text{(2nd hyp. \& ell. } slow\text{)},\\
&1 < M < M_{FL} \qquad &&\text{(3rd ell. \& hyp. } fast\text{)},\\
&M_{FL} < M < \infty \qquad &&\text{(4th hyp. } super\text{-}fast\text{)},
\end{aligned}
\tag{5}
$$

As an example, consider the flow patterns for translation symmetric stationary equilibria (Goedbloed and Lifschitz, 1997) of Figure 1, representing periodic flows in the first (sub-slow) and second (slow) flow regimes, respectively. The two flow patterns are obtained by fixing a certain parameter controlling the magnitude of the poloidal Alfvén Mach number M. However, if one increases this parameter continuously from the value corresponding to Figure 1a to that of Figure 1b, the situation of Figure 2 is encountered: Suddenly, the periodic flow pattern is broken open and two possible solutions (sub-slow and slow) are obtained occupying the same wedge-shaped region on one side of the limiting line characteristics where $M = M_{SL}$. Trying to reflect them with respect to those lines and then joining them by means of the MHD shock conditions fails because the conditions simply cannot be satisfied: These singularities are really limiting in all meanings of the word.

The positive side of this is that we encounter in the stationary flow patterns an essential feature of the three MHD waves in disguise: The stationary equilibrium equations depend on the parallel flow operator $\mathbf{v} \cdot \nabla$, causing the Doppler shifts of

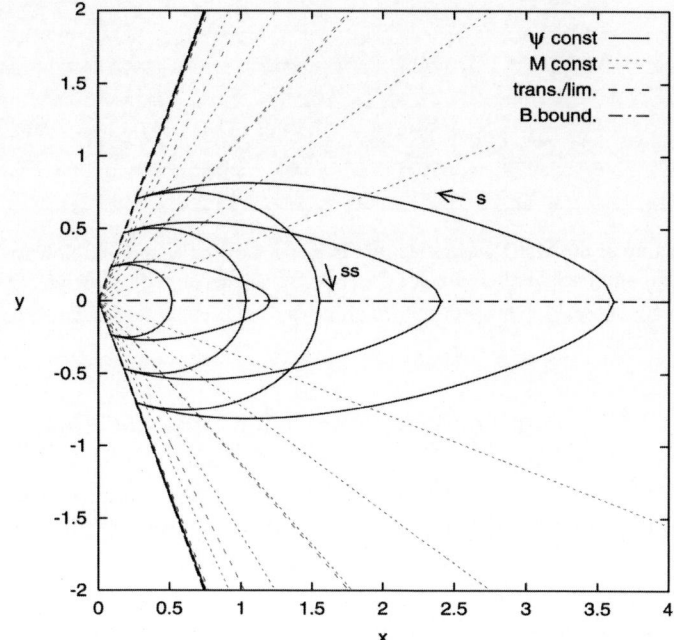

Figure 2. Two sets of streamlines in sub-slow (1st elliptic, 1st hyperbolic) and slow (2nd hyperbolic, 2nd elliptic) flow regimes bounded by a limiting line singularity.

the frequencies of the waves, in precisely the same fashion as the time-dependent spectral equations depend on the differential operator $\partial/\partial t$, producing the eigenvalues ω for the normal modes. The reason is that the original MHD equations depend on the Lagrangian time derivative $D/Dt \equiv \partial/\partial t + \nabla \cdot \mathbf{v}$. This operator produces the waves through the first part and the spatial characteristics through the second (Goedbloed, 2002).

The negative side of this interesting connection is that the computation of the waves and instabilities of transonic plasmas becomes a real tour the force, at least if one wishes to maintain the depth and precision obtained in tokamak spectral theory, where accurately calculating the shape of the magnetic surfaces and the anisotropy of the waves with respect to them are crucial to predict stability of a particular magnetic confinement experiment. It appears that the only way out at present is to restrict the equilibrium flows to be in the elliptic flow regimes. Fortunately, from Equation(5), three of those regimes are available, viz.: (1) a *sub-slow* regime, where $M < M_c$, (2) a *slow* regime, where $M_s < M < 1$, and a (3) *fast* regime, where $1 < M < M_f$. Here, the critical transition speeds M_c and $M_{s,f}$ depend on the local values of the physical variable, i.e. these transition values are not known beforehand but are to be determined together with the solutions. Hence, staying in the elliptic flow regimes is a delicate numerical problem.

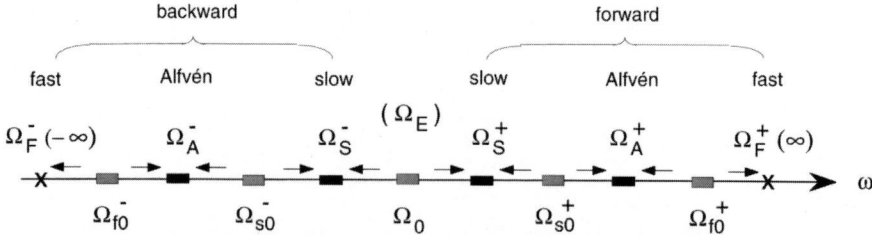

Figure 3. Structure of the MHD wave spectrum in plasmas with background flow for small inhomogeneity: forward and backward continua $\Omega_S^\pm$, $\Omega_A^\pm$, $\Omega_F^\pm$ are separated by regions Ω_0, $\Omega_{s0}^\pm$, $\Omega_{f0}^\pm$ of non-monotonicity of the discrete spectrum which, otherwise, is either Sturmian ($\rightarrow$) or anti-Sturmian ($\leftarrow$) along the real ω-axis.

3. Computing Wave Spectra for Transonic Flows

With the proviso of elliptic flows, the standard paradigm, expressed by Equations (3) and (4), for investigating the spectrum of stationary plasmas may be applied. For the static case of Equations (1) and (2), the structure of the spectrum centers about the slow, Alfvén, and fast continuous spectra ω_S^2, ω_A^2, and $\omega_F^2 \equiv \infty$, where the three MHD waves become purely polarized in the three orthogonal directions associated with the magnetic surfaces and the field lines. For stationary plasmas, the left-right symmetry of the spectrum is lifted by the Doppler shift, so that the continua become:

$$\Omega_S^\pm = \pm\omega_S + \mathbf{k} \cdot \mathbf{v}, \quad \Omega_A^\pm = \pm\omega_A + \mathbf{k} \cdot \mathbf{v}, \quad \Omega_F^\pm = \pm\infty. \tag{6}$$

Hence, we may now encounter a spectral structure as schematically illustrated in Figure 3.

This simple structure is only a local presentation. Of course, the global structure of the spectrum is determined by the curvatures of the magnetic field and flow lines and by the detailed distributions of all the equilibrium variables. Hence, one needs to compute the stationary equilibrium states with sufficient accuracy to justify a spectral analysis. This problem has been solved and incorporated in our new stationary equilibrium code FINESSE (Beliën *et al.*, 2002). The only restriction of this code is that the poloidal flows should be in the elliptic regimes.

Next, one needs to solve for the ideal and resistive waves and instabilities of this system, described by ψ and M^2, and five flux functions. The formidable analysis has been completed and implemented in a Galerkin scheme with finite elements for the normal and Fourier harmonics for the poloidal direction. This leads to a large non-symmetric eigenvalue problem which is solved by means of the Jacobi–Davidson algorithm (Sleijpen and Van der Vorst, 1996) which permits us to zoom in onto a target eigenvalue and then produce all eigenvalues in a wide neighborhood of it with unprecedented accuracy. This has been incorporated in our new spectral code PHOENIX (Van der Holst, 2002). First results on spectra with transonic poloidal flows are published elsewhere (Beliën *et al.*, 2001; Goedbloed, 2002).

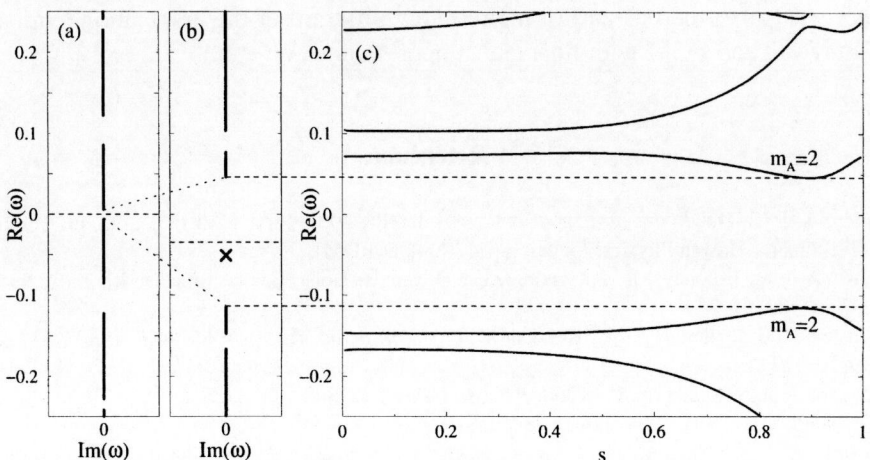

Figure 4. Gap in the continuous spectrum widened by toroidal flow, with a global Toroidal Flow-induced Alfvén Eigenmode (TFAE) appearing in the gap.

4. Perspective

Let us now return to our starting point and illustrate the potential of MHD spectroscopy by the example of a toroidally rotating tokamak equilibrium. This example does not require the full machinery of transonic poloidal flows ($M = 0$ here), but it does demonstrate the importance and peculiarities of the Doppler shifts in a highly inhomogeneous curved environment. Figure 4 shows the continuous spectra without and with toroidal flow. At low-frequency, due to the toroidal flow, a tiny gap in the continuum (caused by geodesic curvature (Goedbloed, 1975)) is Doppler shifted and significantly widened so that a new global mode, the Toroidal Flow-induced Alfvén Eigenmode (TFAE) (Van der Holst *et al.*, 2000), may appear there. Both the gap and the global mode inside are due to the interaction of the forward and backward $m = 2$ Alfvén modes. Such low-frequency modes are characteristic of the toroidal flow profile and, hence, they may be used to determine it, thus providing another example of MHD spectroscopy.

The tokamak magnetic topology does not differ significantly from that of a thick accretion disk. Hence, most of the modes, and certainly the methodology of computing MHD spectra of stationary axi-symmetric equilibria are relevant for a new chapter in plasma-astrophysics that we have called magneto-seismology of accretion disks (Keppens *et al.*, 2002).

Acknowledgements

The author wishes to thank Sander Beliën, Bart van der Holst, Rony Keppens, Sascha Lifschitz, and Stefaan Poedts for many years of fruitful collaboration. This work was performed as part of the research program of the association agreement of Euratom and FOM with financial support from NWO and Euratom.

References

Beliën, A. J. C., Goedbloed, J. P. and Van der Holst, B.: 2001, Proc. 28th Eur. Conf. on Controlled Fusion and Plasma Physics, Madeira, p. 1309 (CD-ROM).

Beliën, A. J. C., Botchev, M. A., Goedbloed, J. P., van der Holst, B. and Keppens, R.: 2002, *J. Comp. Phys.* **182**, 91.

Bernstein, I. B., Frieman, E. A., Kruskal, M. D. and Kulsrud, R. M.: 1958, *Proc. Roy. Soc. London* A244, 17.

Frieman, E. and Rotenberg, M.: 1960; *Rev. Mod. Phys.* **32**, 898.

Goedbloed, J. P.: 1975, *Phys. Fluids* **18**, 1258–1268.

Goedbloed, J. P.: 2002, in *New Plasma Horizons*, International Topical Conference on Plasma Physics, 3–7 September 2001, Faro, Ed. Lennart Stenflo, *Physica Scripta* **T98**, 43–47.

Goedbloed, J. P. and Lifschitz, A.: 1997, *Phys. Plasmas* **4**, 3544.

Goedbloed, J. P., Huysmans, G. T. A., Holties, H., Kerner, W. and Poedts, S.: 1993, *Plasma Phys. Contr. Fusion* **35**, B277–292.

Van der Holst, B., Beliën, A. J. C. and Goedbloed, J. P.: 2000, *Phys. Rev. Lett.* **84**, 2865; *Phys. Plasmas* **7**, 4208.

Van der Holst, B., Beliën, A. J. C. and Goedbloed, J. P.: 2002, to be published.

Keppens, R., Casse, F. and Goedbloed, J. P.: 2002, *Astrophys. J.* **569**, L121–L126.

Sleijpen, G. L. G. and Van der Vorst, H. A.: 1996, *SIAM J. Matrix Anal. Appl.* **17**, 401.

WAVE INDUCED ENERGETIC PARTICLE GENERATION AND SPACE PLASMA MODELING

MANFRED P. LEUBNER

Institute for Theoretical Physics, University of Innsbruck, Innsbruck, Austria and Space Research Institute, Austrian Academy of Sciences, Graz, Austria

Abstract. An increasing number of high-resolution spacecraft observations provide access to details of energetic electron and ion velocity-space distribution structures. Since resonant wave-particle interaction processes depend considerably on the distribution function details, space plasma modeling is of particular interest for studies of a variety of plasma environments as planetary magnetospheres, the interplanetary medium or solar flares. After summarizing the most popular particle acceleration processes we focus on wave-powered energization mechanisms induced by Landau interaction and demonstrate from a time-evolutionary scenario that power-law distributions, highly favored by observations in recent years, are generated resonantly by an Alfvén wave spectrum and possibly saturate. This process is further stimulated in non-uniform magnetic field configurations where multiple wave packets at different phase velocities provide the energy source for a continuous acceleration process. Moreover, in this conjunction we demonstrate that in particular κ-distributions are a consequence of a generalized entropy concept, favored by nonextensive statistics, which provides the missing link for power-law plasma models from fundamental physics. With regard to in situ space observations examples are provided illuminating that for non-thermal plasma characteristics the particular structure of the velocity-space distribution dominates as regulating mechanism for the wave-particle interaction process over effects related to changes in space plasma parameters.

Key words: astrophysical plasmas, nonextensive entropy, non-thermal distributions, wave-particle interaction

1. Introduction

A variety of in situ space observations provide clear evidence of the ubiquitous presence of non-thermal plasma properties, as suprathermal particle populations, loss-cone structures or non-gyrotropic conditions at plasma boundary layers. The associated generation mechanisms of non-Maxwellian particle components as sources of the free energy along with accurate analytical representations of observed structures in view of wave-particle interaction analyses are of crucial importance for further progress in fundamental space plasma physics.

Globally, we may classify particle distribution functions with regard to their symmetry properties into distributions subject to velocity dependence only and subject to both, velocity and space dependence. The symmetric, isotropic and gyrotropic Maxwellian $f_0(v)$ is modified in presence of an external magnetic field where symmetry is retained perpendicular to $\mathbf{B}_0$ and the velocity distribution $f_0(v_\parallel,$

Space Science Reviews **107**: 361–368, 2003.

$v_\perp$) becomes anisotropic but is still gyrotropic. Furthermore, this function can be generalized for space and astrophysical plasma applications in view of clear observational evidence, confirming the ubiquitous appearance of suprathermal tails (Leubner, 2000), into the anisotropic family of κ-distributions $f_0(v_\parallel, v_\perp, \kappa)$. This velocity space representation constitutes a power-law in particle speed and reduces to the two temperature Maxwellian for $\kappa \to \infty$. κ-distributions are most appropriate for space plasma modeling and found recently a justification from fundamental physics, since it was shown that they turn out as consequence of nonextensive entropy environments (Leubner, 2002a). The symmetry perpendicular to $\mathbf{B}_0$ is still retained after generalizing to a mixed suprathermal loss-cone distribution $f_0(v_\parallel, v_\perp, \kappa, l)$, where l measures the loss-cone strength, a model highly valuable in need of an accurate representation of magnetospheric conditions and recently introduced and applied to kinetic mirror mode analysis (Leubner and Schupfer, 2001). Finally, including also space dependence, no symmetry properties are left and the distribution function $f_0(v_\parallel, v_\perp, \phi + \Omega_0 t, x, \kappa)$ is anisotropic, non-gyrotropic as depending on the gyrophase angle ϕ and a spatial coordinate x and may be also time dependent if, for instance, rotating with the gyroperiode Ω_0. The most general situation provided by an analytical representation is found after including in addition a beam parallel to $\mathbf{B}_0$ and allowing a combination of ring structures, suprathermal tails and loss-cone properties as signature of magnetic field gradients.

Space plasmas are subject to three dominant deviations from multi-temperature Maxwellians: (a) suprathermal tails, today known as the 'normal' situation occurring anywhere in astrophysical environments (Mendis and Rosenberg, 1994) and best represented by the family of κ-distributions (Christon et al., 1991; Decker et al., 1995; Chaston et al., 1997; Janhunen and Olssen, 1998; Leubner, 2000; Leubner and Schupfer, 2000); (b) loss-cone distributions (Summers and Thorne, 1995; Leubner and Schupfer, 2001), which are inherent in any magnetic mirror structure as magnetospheric configurations and well modeled by Dory–Guest–Harris (DGH) type distributions (Dory et al., 1965) or substracted Maxwellians and (c) non-gyrotropic plasma conditions detected predominantly at thin plasma boundary layers (Motschmann et al., 1999; Gurgiolo, 2000) where an analytical form that accurately represents observations is introduced below.

Related examples of experimental verifications for suprathermal populations are (a) DE 2 observations of auroral particle distribution structures, Freja satellite electron spectra, high energy ring current particle detections (Lui and Rostoker, 1995), solar wind observations near 1 AU by WIND (Collier et al., 1996) and beyond by Ulysses as well as chromospheric/coronal plasma environments (Maksimovic et al., 1997). Loss-cone features (b) are found in any planetary magnetosphere, in solar (stellar) atmospheres and were observed recently in conjunction with the magnetic field of Jupiters moon Ganymed creating a large loss-cone in all particle distributions (Williams and Mauk, 1997). Finally, non-gyrotropic or gyrophase-bunched distributions (c) (Motschmann et al., 1999) are detected for

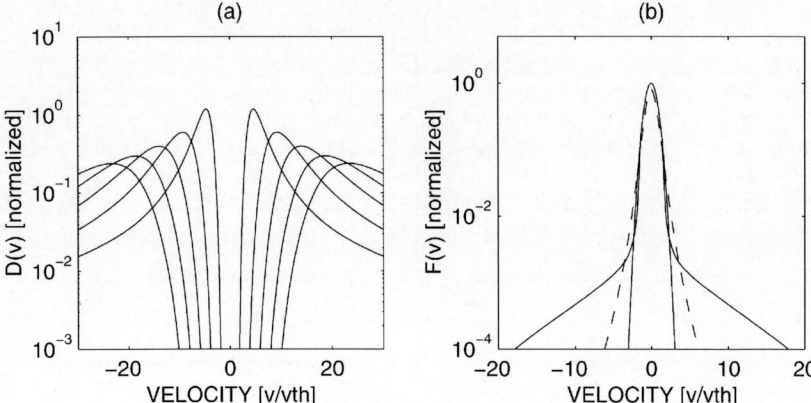

Figure 1. (a) Diffusion properties in increasing magnetic fields from 500 G (innermost curve) to 2500 G (outermost curve). (b) A Maxwellian, a κ-distribution and a saturated structure from synergetic acceleration subject to highly energetic electron populations.

ions at collisionless shocks, at the plasma sheath boundary and by GEOTAIL observations generally at reconnection regions (Frank *et al.*, 1994; Tu *et al.*, 1997; Gurgiolo, 2000) as well as at comets. Since electron and ion gyroradii are related as $r_{ge} \ll r_{gi}$ only a few observations of gyrophase-bunched electron structures are available today, as WIND detections at the Earth's bow shock.

2. Theoretical Outline and Results

Particle acceleration is a universal phenomenon in a large variety of astrophysical plasmas as in planetary magnetospheres, the interplanetary medium, in chromospheric and solar flare environments. Favored generation mechanisms of energetic particles include DC-electric fields, potential drops at magnetic field line reconnection structures, shock-drift acceleration and acceleration due to wave particle interaction. As potential particle acceleration mechanism kinetic Alfvén wave interaction (Leubner, 2000) is able to generate the entire energy scale ranging from suprathermal structures up to relativistic energies . Theoretically, high energy tail populations are created when simulating particle acceleration in response to a realistic broadband Alfvén wave spectrum within a Fokker–Planck approach. Furthermore, in the presence of magnetic field or density gradients multiple wave packets of different phase velocities are generated. This process results in a resonant particle acceleration out of the bulk of the distribution such that they become again resonant with wave packets of increasing phase speed, a mechanism of synergetic particle energization (Leubner, 1999).

Let the collective, time dependent development of the velocity space distribution $f(v, t)$ in response to a wave spectrum due to resonant Landau interaction and in response to particle collisions being regulated by the Fokker–Planck equation as

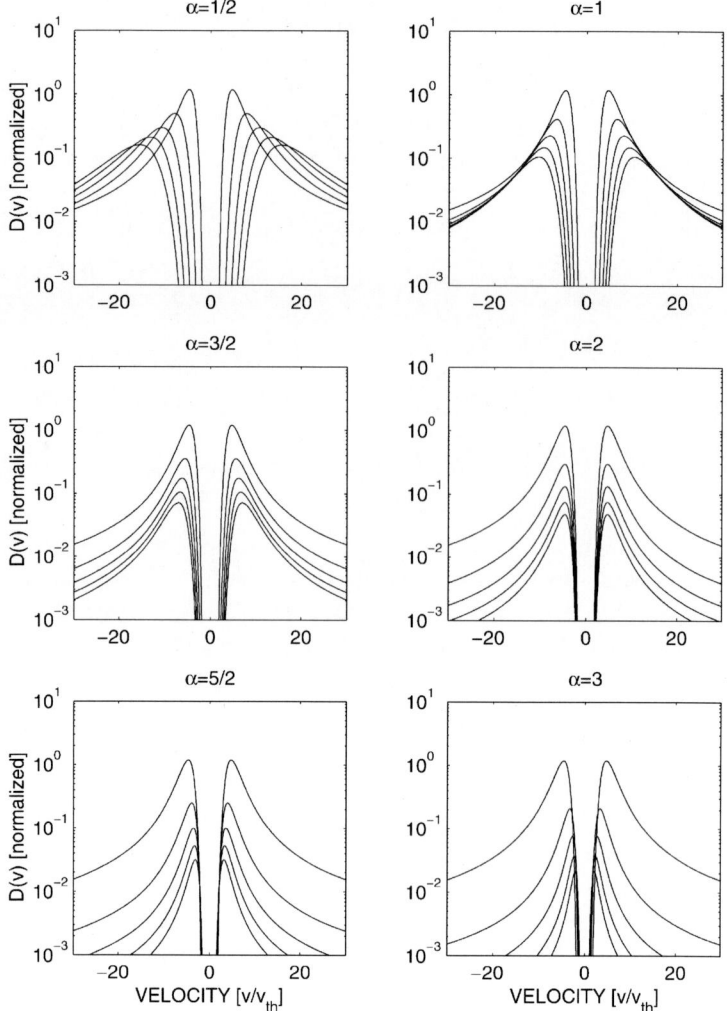

Figure 2. Snapshots of the changes of the diffusion properties in velocity space due to magnetic field gradients and density variations according to $B^{\alpha} = const. \times N$.

$$\frac{\partial f}{\partial t} = \frac{\partial}{\partial v} \nu(v) \left[vf + v_{th}^2 \frac{\partial f}{\partial v} \right] + \frac{\partial}{\partial v} D(v) \frac{\partial f}{\partial v} \tag{1}$$

Here $\nu(v)$ and $D(v)$ denote the velocity dependent collision and diffusion operators, respectively, v_{th} is the thermal speed and $f(v, t = 0)$ shall be represented by a starting Maxwellian equilibrium distribution.

The first term on the right hand side of Equation (1) takes care if a Maxwellian is to be restored. Furthermore, for appropriate approximations we note that for low collisionality changes in the pitch angle are negligible and that the main dynamics of Alfvén wave-particle energy exchange due to Landau interaction is regulated in the direction parallel to the ambient magnetic field $\mathbf{B}_0$. Introducing the quasi-linear,

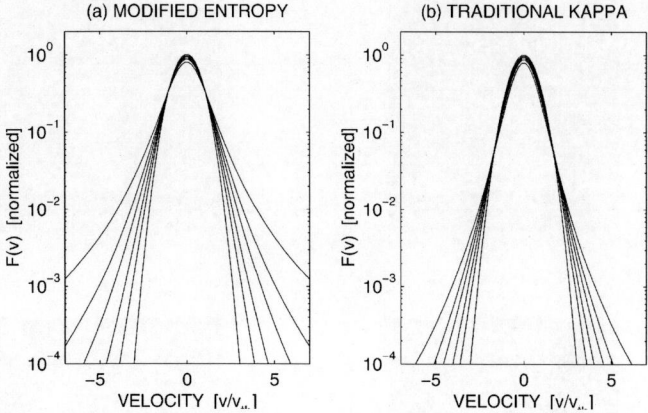

Figure 3. The family of κ-distributions. The panels (a) and (b) demonstrate the difference between the nonextensive entropy solution and the traditional κ-distributions. For both, $\kappa = 2$ corresponds to the outermost curve followed by values $\kappa = 3, 4, 6$ and 10. The innermost curve represents with $\kappa = \infty$ a Maxwellian.

one dimensional diffusion operator and adopting a broadband spectrum of Alfvén waves trapped inside an envelop of specific extension, Equation (1) provides the formalism for numerical simulations of wave-particle Landau energy exchange (Leubner, 2000). As result, in uniform space plasmas suprathermal particle populations can be generated by resonant energy transfer from a broadband Alfvén wave spectrum yielding power law (kappa-like) velocity distributions. In case of magnetic field or density gradients energetic particle populations are generated by synergetic effects in a multi-stage acceleration process. Wave packets of increasing phase velocity provide highly efficient particle acceleration up to relativistic energies since the gradients act as catalyst (Leubner, 1999), see Figure 1. It turns out that temperature variations have minor effects on the diffusion characteristics whereas magnetic field and density variations predominantly determine the resonance properties of the wave-particle interaction. Assuming $T = const.$ we can study a situation obeying the condition $B^\alpha = const. \times N$ as model of the field line convergence, see Figure 2, where for a constant plasma beta (equipartition) $\alpha = 2$.

Astrophysical environments are subject to long-range interactions requiring a generalization of the Boltzmann–Gibbs–Shannon entropy. A nonextensive entropy approach (Leubner, 2002a) yields an undisturbed distribution $f(v) = B_\kappa (1 + v^2/\kappa v_{th}^2)^{-\kappa}$ (B_κ is a normalization constant) that differs only in the exponent by a constant in comparison to the traditional form of κ-distributions and results in structures exhibiting more pronounced tails for the same κ values, see Figure 3. This provides naturally a justification of the basic analytical form of κ-distributions from fundamental physics and suggests to use the nonextensive entropy solution for future data fitting.

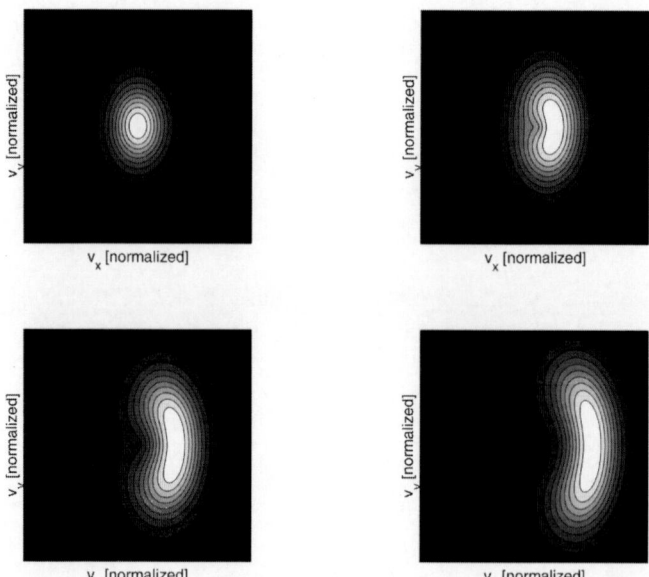

Figure 4. Contour plots of a non-gyrotropic distribution. The sequence of panels, from upper left to lower right, illuminate changes in the distribution structure due to decreasing magnetic field strength.

Magnetospheric observations require a generalization of Maxwellians into mixed suprathermal loss-cone distributions, provided e.g. by merging κ-distributions with the DGH-type loss-cone as

$$f_{\kappa j}(v_\parallel, v_\perp) = A_{\kappa j}(\frac{v_\perp}{v_{th\perp}})^{2j}\left[1 + \frac{v_\parallel^2}{\kappa v_{th\parallel}^2} + \frac{v_\perp^2}{\kappa v_{th\perp}^2}\right]^{-(\kappa+1)} \tag{2}$$

Here the parameter κ shapes predominantly the suprathermal tails of this distribution and j is a measure of the loss-cone strength. $v_{th\parallel,\perp}$ are the thermal speeds parallel and perpendicular to $\mathbf{B}_0$ and the normalization is performed with respect to the particle density N, where $A_{\kappa j}$ denotes the appropriate normalization constant. It was shown that the particular structure of the distribution modeled by the spectral index κ and loss-cone index j dominates as regulating mechanism for instability thresholds over changes in the plasma parameters β and temperature anisotropy (Leubner and Schupfer, 2001), thus demonstrating the importance of accurate space plasma modeling.

Finally, if plasma inhomogeneity scales are smaller than the charged particles gyroradius non-gyrotropic distributions are generated. This occurs predominantly at boundary layers and at reconnection structures and requires an appropriate model distribution in view of wave-particle interaction analysis. Introducing a ring distribution with a shift $v_{0\perp}$ and superimposing the dependence on the gyrophase angle Φ in a $x - y$ plane perpendicular to $\mathbf{B}_0$, where particle concentration maximizes at some gyrophase angle, yields in 'Maxwellian' notation

$$f_{gM} = A_{gM} \exp[-\frac{v_\parallel^2}{v_{th\parallel}^2} - \frac{(v_\perp - v_{0\perp})^2}{v_{th\perp}^2}] \exp[-\frac{(v_\perp \cos \Phi - v_{0\perp})^2}{v_{th\perp}^2}] \tag{3}$$

Equation (3) can be generalized easily into a κ-form (Leubner, 2002b) and models accurately structures observed at boundary layers, see Figure 4. Non-thermal features of the general representation f_{gM} (A_{gM} is a normalization constant) include temperature anisotropy, ring structures, stationary and, in case of time dependency, also non-stationary non-gyrotropy and can be extended for inclusion of suprathermal populations. This enables us to handle any combination with a minimum of free parameters providing excellent fitting of observed structures.

3. Summary

κ-like distributions fitting observed suprathermal particle populations can be generated by resonant interaction with Alfvén wave packets. In non-uniform plasmas relativistic particle energies are achieved via synergetic effects due to the interaction with wave packets of increasing phase velocity. A nonextensive entropy approach justifies naturally from long-range interactions the analytical form of the family of κ-distributions from fundamental physics. Furthermore, a mixed suprathermal loss-cone distribution is proposed modeling accurately mirror type environments and demonstrating that the particular structure of velocity distributions dominates as regulating mechanism for wave excitation over changes in plasma parameters. Finally, a highly general analytical representation of suprathermal gyrophase-bunched distributions is introduced, providing the basis for wave-particle interaction studies at space plasma boundary layers.

References

Chaston, C. C., Hu, Y. D. and Fraser, B. J.: 1997, *Geophys. Res. Lett.* **22**, 2913.

Christon, S. P., Williams, D. J., Mitchell, D. G., Huang, C. Y. and Frank, L. A.: 1991, *J. Geophys. Res.* **96**, 1.

Collier, M. R., Hamilton, D. C., Gloeckler, G., Bochsler, P. and Sheldon, R. B.: 1996, *Geophys. Res. Lett.* **23**, 1191.

Decker, D. T., Basu, B., Jasperse, J. R., Strickland, D. J., Sharber, J. R. and Winningham, J. D.: 1995, *J. Geophys. Res.* **100**, 21409.

Dory, R. A., Guest, G. E. and Harris, E. G.: 1965, *Phys. Rev. Lett.* **14**, 131.

Frank, L. A., Paterson, W. R. and Kivelson, M. G.: 1994, *J. Geophys. Res.* **99**, 14887.

Gurgiolo, C.: 2000, *Geophys. Res. Lett.* **27**, 3153.

Janhunen, P. and Olsson, A.: 1998, *Ann. Geophys.* **16**, 292.

Leubner, M. P.: 1999, in P. C. H. Martens, S. Tsuruta and M. A. Webber (eds.), *Highly Energetic Physical Processes and Mechanisms for Emission from Astrophysical Plasmas*, International Astronomical Union, IAU 195, 315.

Leubner, M. P., 2000, *Planet. Space Sci.* **48**, 133.

Leubner, M. P. and Schupfer, N.: 2000, *J. Geophys. Res.* **105**, 27387.

Leubner, M. P. and Schupfer, N.: 2001, *J. Geophys. Res.* **106**, 12993.

Leubner, M. P.: 2002a, astro-ph/0111444, *Astrophys. Space Sci.* (in press).

Leubner, M. P.: 2002b, *Planet. Space Sci.* (to appear).

Lui, W. W. and Rostoker, G.: 1995, *J. Geophys. Res.* **100**, 21897.

Maksimovic, M., Pierrard, V. and Lemaire, J. F.: 1997, *Astron. Astrophys.* **324**, 725.

Mendis, D. A. and Rosenberg, M.: 1994, *Ann. Rev. Astron. Astrophys.* **32**, 419.

Motschmann, U., Glassmeier, K. H. and Brinca, A. L.: 1999, *Ann. Geophys.* **17**, 613.

Summers, D. and Thorne, R. M.: 1995, *J. Plasma Phys.* **53**, 293.

Tu, J.-N., Mukai, T., Hoshino, M., Saito, T., Matusno, Y., Yamamoto, T. and Kokubun, S.: 1997, *Geophys. Res. Lett.* **24**, 2247.

Williams, D. J. and Mauk, B.: 1997, *J. Geophys. Res.* **102**, 24283.

SMALL-ANGLE SCATTERING AND DIFFUSION: APPLICATION TO RELATIVISTIC SHOCK ACCELERATION

R.J. PROTHEROE, A. MELI* and A.-C. DONEA

*Department of Physics & Mathematical Physics, The University of Adelaide,
Adelaide, SA 5005, Australia*

Abstract. We investigate ways of accurately simulating the propagation of energetic charged particles over small times where the standard Monte Carlo approximation to diffusive transport breaks down. We find that a small-angle scattering procedure with appropriately chosen step-lengths and scattering angles gives accurate results, and we apply this to the simulation of propagation upstream in relativistic shock acceleration.

Key words: cosmic rays, diffusion theory, relativistic shock acceleration

1. Introduction

Relativistic charged particle transport in magnetized astro-physical plasma is strongly affected by magnetic irregularities, and may be approximated by diffusion. Diffusive transport of particles having speed v can be simulated by a three-dimensional random walk with steps sampled from an exponential distribution with mean free path $\lambda = 3D/v$, where D (cm^2 s^{-1}) is the spatial diffusion coefficient, followed by large-angle (isotropic) scattering after each step (e.g., Chandrasekhar, 1943), and this gives good results for distances much larger than λ.

In diffusive shock acceleration at relativistic shocks problems arise when simulating particle motion upstream of the shock because the particle speeds, v, and the shock speed $v_{\text{shock}} = c(1 - 1/\gamma_{\text{shock}}^2)^{1/2}$ are both close to c, and so very small deflections are sufficient to cause a particle to re-cross the shock. Clearly, Monte Carlo simulation by a random walk with mean free path λ and large-angle scattering is inappropriate here, and in Monte Carlo simulations of relativistic shock acceleration at parallel shocks Achterberg *et al.* (2001) consider instead the diffusion of a particle's direction for a given angular diffusion coefficient D_θ (rad^2 s^{-1}). Similarly, for a given spatial diffusion coefficient D, Protheroe (2001) and Meli and Quenby (2001) adopt a random walk with a smaller mean free path, $\bar{\ell} \ll \lambda$, followed by scattering at each step by a small angle with mean deflection, $\bar{\theta} < 1/\gamma_{\text{shock}}$. See Bednarz and Ostrowski (2001) for a recent review of relativistic shock acceleration.

*Visiting from Imperial College, London

Space Science Reviews **107**: 369–372, 2003.
© 2003 *Kluwer Academic Publishers.*

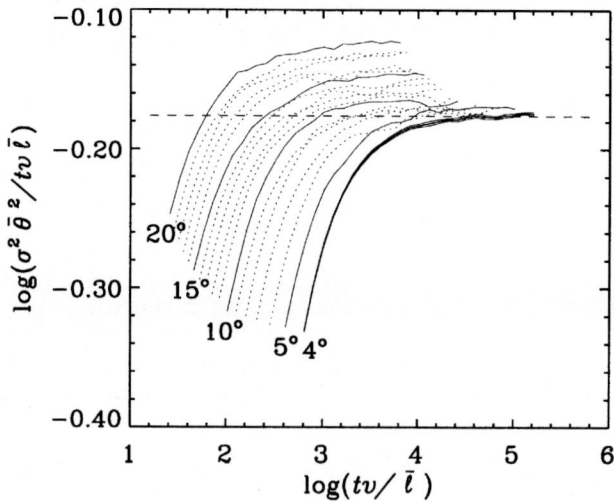

Figure 1. σ^2 vs. time for a 3D random walk with isotropic injection at the origin at $t = 0$. Step-lengths ℓ were sampled from an exponential distribution with mean $\bar{\ell}$ followed by small-angle scattering with scattering angle θ sampled from an exponential distribution with mean $\bar{\theta}$ (the numbers attached to the curves). Dashed line is $\sigma^2 = 2tv\bar{\ell}/3\bar{\theta}^2$. Curves for $5°$–$20°$ result from 10^4 simulations; $4°$ curve results from 8×10^4 simulations (width shows statistical error).

2. Small-angle Scattering and Diffusion

We consider propagation by small steps sampled from an exponential distribution with mean $\bar{\ell} \ll \lambda$, followed at each step by scattering by a small angle sampled from an exponential distribution with mean $\bar{\theta} \ll \pi$. The change in direction (θ_1, θ_2) may then be described as two-dimensional diffusion with angular diffusion coefficient $D_\theta = \bar{\theta}v_\theta/2$ (rad^2 s^{-1}) where $v_\theta = \bar{\theta}/\bar{t}$, and $\bar{t} = \bar{\ell}/v$ such that $D_\theta = \bar{\theta}^2v/(2\bar{\ell})$. The time t_{iso}, which gives rise to a deflection equivalent to a large-angle (isotropic) scattering, is determined by $(\sigma_{\theta_1}^2 + \sigma_{\theta_2}^2)^{1/2} = \sqrt{4D_\theta t_{\text{iso}}} \sim \pi/2$, giving $\lambda \sim vt_{\text{iso}} \propto \bar{\ell}/\bar{\theta}^2$ and a spatial diffusion coefficient $D \propto \bar{\ell}v/\bar{\theta}^2 \propto v^2/D_\theta$. By using a Monte Carlo method it is straightforward to test this, determine the constant of proportionality, and thereby make the connection between diffusion and small angle scattering.

The solution of the diffusion equation for a delta-function source in position and time $q(\vec{r}, t) = \delta(\vec{r})\delta(t)$ and an infinite diffusive medium is a three-dimensional gaussian with standard deviation $\sigma = \sqrt{2Dt}$ (Chandrasekhar, 1943). The results from several Monte Carlo random walk simulations are shown in Figure 1, from which we find that the expected dependence occurs for $\bar{\theta} < 5°$ at times $t > 10^5\bar{\ell}/v$. For this case, we see from Figure 1 that $\sigma^2 \to 2tv\bar{\ell}/3\bar{\theta}^2$, and so we obtain the connection between small-angle scattering and diffusion theory, namely $D \approx \bar{\ell}v/(3\bar{\theta}^2) \approx v^2/(6D_\theta)$.

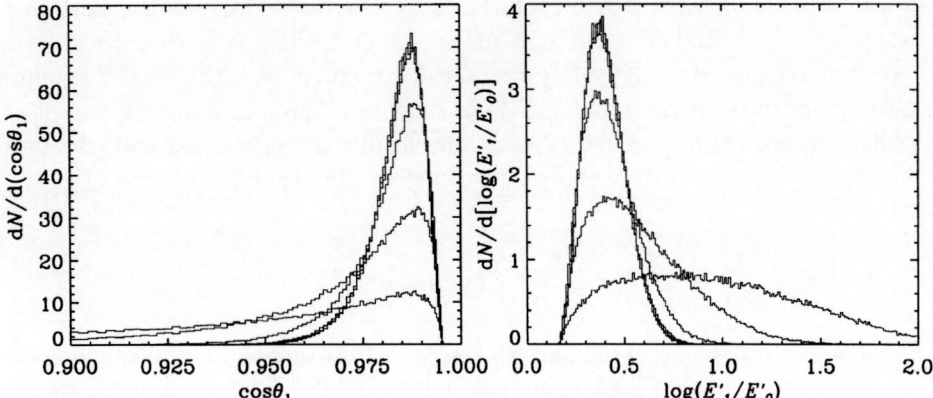

Figure 2. Small-angle scattering simulation of excursion upstream in diffusive shock acceleration at a parallel relativistic shock with $\gamma_{shock} = 10$. Results are shown for 10^5 injected particles and $\bar{\theta} = 10^{-2}/\gamma_{shock}$ (top histogram), $3 \times 10^{-2}/\gamma_{shock}$, $10^{-1}/\gamma_{shock}$, $0.3/\gamma_{shock}$, $1/\gamma_{shock}$ and $3/\gamma_{shock}$ (bottom histogram). Note that the top three histograms are almost indistinguishable.

3. Application to Relativistic Shock Acceleration

As viewed in the frame of reference of the upstream plasma, ultra-relativistic particles are only able to cross the shock from downstream to upstream if the angle θ between their direction and the shock normal pointing upstream is $\theta < \sin^{-1}(1/\gamma_{shock})$, where $\gamma_{shock} = (1 - \beta_{shock}^2)^{-1/2}$ and $\beta_{shock} = v_{shock}/c$. For highly relativistic shocks these particles cross the shock from downstream to upstream travelling almost parallel to the shock normal. Similarly, having crossed the shock, only a very slight angular deflection, by $\sim 1/\gamma_{shock}$ is sufficient to return them downstream of the shock. This change in particle direction gives rise to a change in particle energy E' and momentum p', measured in the downstream plasma frame (primed coordinates), of

$$\frac{E'_{n+1}}{E'_n} \approx \frac{p'_{n+1}}{p'_n} = \frac{1 - \beta_{12}\cos\theta_{n+1}}{1 - \beta_{12}\cos\theta_n} \qquad (1)$$

in an acceleration cycle (downstream $\rightarrow$ upstream $\rightarrow$ downstream), where β_{12} is the speed of the upstream plasma as viewed from the downstream frame.

In 'parallel shocks' the magnetic field is parallel to the shock normal, and so the pitch angle ψ is the angle to the shock normal and $v\cos\psi$ gives the component of velocity parallel to the shock. Thus the small-angle scattering method described in the previous section is used here to simulate particle motion upstream of a parallel relativistic shock, including the effects of pitch-angle scattering, for a given diffusion coefficient. We inject ultra-relativistic particles at the shock with downstream-frame energy E'_0 travelling upstream parallel to the shock normal, i.e. $\theta_0 = 0$. We follow a particle's trajectory until the shock catches up with it, and it crosses from upstream to downstream with an upstream-frame angle θ_1 to the shock normal and a downstream-frame energy E'_1. The simulation was performed for $\gamma_{shock} = 10$, and five different mean scattering angles $\bar{\theta}$, to determine the maximum $\bar{\theta}$-value that can safely be used for accurate simulation. The resulting distributions of $\cos\theta$ and $\log(E'_1/E'_0)$ are shown in Figure 2, and show that in this application one requires $\bar{\theta} < 0.1/\gamma_{shock}$. Our results are quite consistent with those of Achterberg *et al.* (2001), who used a diffusive angular step $\Delta\theta_{st} \leq 0.1/\gamma_{shock}$.

4. Conclusion

The standard Monte Carlo random walk approach to simulation of energetic charged particle propagation for a given spatial diffusion coefficient D can be extended to apply accurately to times much less than $\lambda/v = 3D/v^2$ by using a small-angle scattering procedure with steps sampled from an exponential distribution with mean free path $\bar{\ell} = \bar{\theta}^2\lambda$ followed at each step by scattering with angular steps sampled from an exponential distribution with mean scattering angle $\bar{\theta} < 0.09$ rad ($5°$). The spatial and angular diffusion coefficients are then $D \approx \bar{\ell}v/(3\bar{\theta}^2)$ and $D_\theta = \bar{\theta}^2 v/(2\bar{\ell})$, and are related by $D \approx v^2/(6D_\theta)$. In simulation of upstream propagation in relativistic shock acceleration one must use $\bar{\theta} < 0.1/\gamma_{shock}$ to obtain accurate results.

References

Achterberg, A., Gallant, Y.A., Kirk, J.G. and Guthmann, A.W.: 2001, *MNRAS* **328**, 393.
Bednarz, J. and Ostrowski, M.: 2001, *MNRAS* **310**, L13.
Chandrasekhar, S.: 1943, *Rev. Mod. Phys.* **15**, 1.
Meli, A. and Quenby, J.: 2001, in M. Simson *et al.* (eds), *27th Int. Cosmic Ray Conf.*, Vol. 7, p. 2742.
Protheroe, R.J.: 2001, in M. Simson *et al.* (eds), *27th Int. Cosmic Ray Conf.*, Vol. 6, p. 2006 and
    p. 2014.

ON STABILIZING ROLE OF PARALLEL INHOMOGENEOUS FLOW ON LOW-FREQUENCY SPACE FLUCTUATIONS

S. SEN

*Research School of Physical Sciences and Engineering, The Australian National University,
Canberra, ACT 0200, Australia
and
Department of Physics and Astrophysics, University of Delhi, Delhi 110 007, India*

Abstract. A fundamental reality throughout the space plasma is the existence of magnetic field-aligned flows. It is usually believed that the spatial transverse shear in the parallel flow destabilizes many low frequency oscillations and this may be the origin of low frequency oscillations in the ionosphere (V. V. Gavrishchaka *et al.*, 1998, *Phys. Rev. Lett.*, **80**, 728 and *Phys. Rev. Lett.*: 2000, **85**, 4285). Here we show that this notion of destabilizing influence of the shear in the parallel flow can be changed altogether if one takes the effect of the flow curvature (second spatial derivative) into account. The transverse curvature in the parallel flow can overcome the destabilizing influence of the shear and can render the low frequency modes stable. It is shown that unlike flow shear the effect of flow curvature is sign- and mode-dependent.

Key words: ballooning mode, flow profile, PVS instability, RT instability

Linear theory analysis and particle-in-cell simulation exhibits that a spatial transverse gradient in the ion drift parallel to the magnetic field can generate a broadband multimode spectrum (with frequencies varying from much lower of the ion cyclotron frequency to much higher of the ion cyclotron frequency). On the nonlinear terms, these unstable waves give rise to multiscale spatially coherent structures, substantial cross-field transport, ion energization, and phase-space diffusion (Gavrishchaka *et al.*, 1998, 2000). These signatures are usually compared with the Fast Auroral Snapshot satellite data in the upward current region. Although some similarities are observed in accordance with these theoretical models, there seems to be much more which actually does not conform to these theoretical predictions.

In this work we introduce a novel concept of the flow curvature (the second radial derivative) in addition to the usual flow shear (the first radial derivative) and find that it is the flow curvature which plays the crucial role in the linear mode stability. Unlike the flow shear the effect of flow curvature is found to be sign-dependent and also mode-dependent. For RT (Rayleigh–Taylor) type mode with positive curvature in the parallel flow the destabilizing influence of the parallel flow shear can be suppressed altogether and the mode can even become fully stable. On the other hand, for the negative curvature the unstable mode might be further destabilized than what is predicted by the shear alone. For the PVS (Parallel Velocity Shear) type mode it is just the opposite, positive curvature acts to destabilize the mode whereas negative curvature can render the mode completely stable. As

Space Science Reviews **107**: 373–381, 2003.
© 2003 *Kluwer Academic Publishers.*

it is only natural that all experimental flow profiles must in principle be curved rather than a pure straight line (only in this case flow shear will be relevant), the inclusion of curvature in the flow profile rather than the shear alone is therefore more realistic. This new observations may therefore be more appropriate and complete to compare with the observations of the Fast Auroral Snapshot satellite data. In the following, we choose two important space instabilities Rayleigh–Taylor and Parallel Velocity Shear instability to show how the new physics of flow curvature works.

1. Rayleigh–Taylor Instability

1.1. Introduction

Heavy fluid supported by a lighter fluid in the presence of the gravitational field gives rise to Raleigh–Taylor (RT) instability (Chandrashekhar, 1981). This instability commonly occurs both in collisionless and collisional domain in the magnetized plasmas, where the role of the light fluid is played by the magnetic field. The collisionless interchange type instability (ballooning mode) can exist in the earth's plasma sphere as well as in the laboratory plasma. These collisionless mode arises due to an unfavorable curvature in the magnetic field (simulating an effective gravity) in the presence of a pressure gradient. It is believed that the RT instability can play a major role in the onset of equatorial spread F (Ossakow, 1979). This instability is also known to be a major problem for a wide range of applications, from pulsed power technology to inertial confinement fusion.

Here, we develop a nonlocal theory of the RT mode in the presence of a radially varying parallel flow. Our full analytic analysis shows that the parallel flow curvature can stabilize the RT mode. The flow curvature also has a robust effect on the radial structure of the mode. The parallel flow shear, on the other hand, plays an insignificant role in this matter.

1.2. Stability Analysis

We carry out a nonlocal stability analysis and assume, for simplicity, a cartesian co-ordinate system and a two-fluid model for the Rayleigh–Taylor modes in the presence of a radially varying parallel flow. We assume a low-β collisionless plasma (hence neglecting any electromagnetic fluctuations) with both equilibrium density variation and magnetic shear in x, i.e., $n_o(x) = n_{oo} \exp(-x^2/2L_n^2)$, $\mathbf{B}(x) = B[\mathbf{e}_z + (x/L_s)\mathbf{e}_y]$. Here, L_s is the magnetic shear scale length and x is the distance from the mode rational surface defined by $\mathbf{k}.\mathbf{B} = 0$. We assume a uniform gravity force mg in the x direction for ions. We model the equilibrium parallel flow by a profile $V_{\parallel 0}(x) = V_{\parallel 00} + V_{\parallel 00}x/L_{v1} + V_{\parallel 00}x^2/2L_{v2}^2$, where the shear and the curvature contributions to the velocity profile are represented by the second and the third term in the Taylor series respectively; and $L_{v1} = (1/V_{\parallel oo}dV_{\parallel 0}(x)/dx)^{-1}$;

$L_{v2} = (1/V_{\|oo}d^2V_{\|o}(x)/dx^2)^{-1}$. Because of inhomogeneity in the x-direction, perturbations have the form $\phi(\mathbf{x}, t) = \phi(x)\exp i(k_y y + k_z z - \omega t)$. Ion inertia effects are retained to include the ion polarization drift, but electron inertia and ion and electron pressure are neglected for simplicity. We can write the linearized basic equation of continuity and momentum transfer for ions and electrons as follows:

$$\frac{\partial n_\alpha}{\partial t} + \nabla.(n_\alpha V_\alpha) = 0$$

$$E = -\frac{1}{c}(V_e \times B)$$

$$m_i(\frac{\partial}{\partial t} + V_i.\nabla)V_i = eE + \frac{eV_i \times B}{c} + m_i g$$

Here

$$\mathbf{V_e} = V_{\|0}(x) + V_E$$
$$\mathbf{V_i} = V_{\|0}(x) + V_E + V_p + V_g$$
$$\mathbf{V_E} = -c(\nabla_\perp\phi \times \mathbf{B})/B^2$$
$$\mathbf{V_p} = (\frac{ic\omega}{B\omega_{ci}})\nabla_\perp\phi$$
$$V_g = \frac{cm_i g}{eB}$$
$$\omega_{ci} = \frac{eB}{cm_i}$$
$$\nabla_\perp = ik_y\mathbf{e}_y + \mathbf{e}_x\frac{d}{dx}$$

Here α denotes species (e for electrons and i for ions), all other symbols are assumed to have the usual meaning unless otherwise stated explicitly. Parallel flow, $V_{\|0}$, has therefore two effects. First, it introduces a Doppler shift, $k_\| V_{\|0}(x)$, in all time derivatives and second, an extra term, $\mathbf{V}_E.\nabla V_{\|0}(x)$, representing radial convection of ion momentum. It is the second term which makes the effect of parallel flow shear completely different from that of the perpendicular flow shear. We eliminate the Doppler shift by performing a Galilean transformation in the $\mathbf{e}_\|$ direction. It is important to mention here that the spatial variation in the Doppler shift in the mode frequency due to *parallel* flow is negligible for flute-type modes ($k_\| << k_\perp$). It is probably obvious as $d^n/dx^n(k_\| V_o(x)) << d^n/dx^n(k_\perp V_o(x))$ due to the fact that $k_\| << k_\perp$. So, one can eliminate the Doppler shift performing a Galilean transformation in the $\mathbf{e}_\|$ direction. This has also been noted elsewhere. Now, using quasineutrality and the usual low frequency assumptions, we obtain radial eigenvalue equation:

$$\frac{d^2\phi}{dx^2} + p(x)\frac{d\phi}{dx} + q(x)\phi = 0 \tag{1}$$

where

$$p(x) = \frac{n_0'}{n_0}$$

$$q(x) = -k_y^2 + \frac{k_y^2 g}{\omega^2}\frac{n_0'}{n_0} + \frac{k_y^2 k_{\parallel} g}{\omega^3}\frac{dV_{\parallel}00}{dx}$$

In deriving Equation (1) we have retained terms up to the first order in $k_{\parallel}$ which is justified for the flute type mode ($k_{\parallel} \sim 0$). Now to remove the first derivative (with respect to x) term we make use of the transformation

$$\phi(x) = \psi(x)\exp(-\int^x \frac{p(\eta)}{2}d\eta)$$

when we get

$$\frac{d^2\psi}{dx^2} + Q(x)\psi = 0$$

where

$$Q(x) = q(x) - \frac{p'(x)}{2} - \frac{p^2(x)}{4}.$$

With the velocity and density profile described earlier, the radial eigenvalue equation now reduces to:

$$\frac{d^2\psi}{dx^2} + [T + Sx + Rx^2]\psi = 0 \qquad (2)$$

where

$$T = \frac{1}{L_n^2} - k_y^2$$

$$S = \frac{L_n k_y^2}{L_s L_{v1}} - \frac{k_y^2 g}{\omega^2 L_n^2}$$

$$R = \frac{L_n k_y^2}{L_s L_{v2}^2} - \frac{1}{4L_n^4}$$

In deriving Equation (2), we have assumed that $\omega \sim \sqrt{g/L_n} \sim V_{\parallel oo}$, which is usually true. Equation (2) is a simple Weber equation. Depending on the sign of R, we have two types of solutions. If $R < 0$, the solution satisfying the physical boundary condition, i.e., $\psi \to 0$ at $x = \pm\infty$ is given by

$$\psi(x) = \psi_o exp[-\sqrt{|R|}(x - x_o)^2]$$

where $x_o = S/2|R|$. So, in this case, the mode decays with x, i.e., it does not propagate and hence is intrinsically undamped. This solution therefore implies the existence of an unstable mode.

On the other hand, if $R > 0$, Equation (2) has the solution

$$\psi(x) = \psi_o exp[-i\sqrt{|R|}(x + x_o)^2]$$

Thus, in this case we have a non-localized mode with outgoing energy flux at $x = \pm\infty$. Because of the convective wave energy leakage the perturbation will decay in time in the absence of any energy source feeding the wave. The wave is therefore damped. So, positive parallel flow curvature has a stabilizing role on the RT instability. Parallel flow shear, on the other hand, has an insignificant role in this matter. It can shift the potential, but does not alter the quadratic structure. It also shifts the center of the mode away from the $x = 0$ rational surface. The main stabilizing effect comes from the quadratic term which forms an anti-well pushing the wave function away from $x = 0$. The corresponding dispersion relation is given by $T - \frac{S^2}{4R} = i\sqrt{R}$, which can be simplified for a weakly unstable situation ($\omega_i < \omega_r$) and without the velocity shear contribution as:

$$\left(\frac{k_y^4 g^2}{L_n^4}\right) \frac{-(\omega_r - \omega_i)^2 + 4\omega_r^2\omega_i^2 + 4i\omega_r\omega_i|(\omega_r^2 - \omega_i^2)|}{|[4\omega_r\omega_i(\omega_r^2 - \omega_i^2)]^2 - [(\omega_r - \omega_i)^2 - 4\omega_r^2\omega_i^2]^2|} = 4R(T - i\sqrt{R})$$

Here ω_r and ω_i are the real and the imaginary parts of the eigenfrequency, respectively. It is easy to see that without the velocity field the linear growth rate of the RT instability assumes the familiar form $\sqrt{g/L_n}$. From the dispersion relation it is clear that the parallel flow curvature can stabilize the RT instability if $\frac{L_n k_y^2}{L_s L_{v2}^2} > \frac{1}{4L_n^4}$. If we assume $L_{v2} \sim L_n$, $k_y\rho_i \sim 0.1$ (where ρ_i is the ion-gyroradius) and $\frac{L_n}{\rho_i} \sim 100$, then the condition of stability can be further simplified to $\frac{L_s}{L_n} < 400$. We emphasize here that, these assumptions, although are usually true, are made only to facilitate comparison with the experimental data and no generality whatsoever is lost thereby. We have therefore obtained a condition of stability for the RT mode in the presence of axial flow curvature, which is likely to be satisfied.

2. Parallel Velocity Shear Instability

2.1. INTRODUCTION

As strong parallel flows are rather ubiquitous in nature, occurring in the flanks of the earth's magnetosphere and along the auroral magnetic field lines the parallel velocity shear (PVS) instability is a major instability in the space physics. Here, we present a simple theory of this instability in the presence of parallel flow curvature. The destabilizing influence of the density gradient (drift-type mode) has been added so as to consider the most unstable situation.

2.2. STABILITY ANALYSIS

We will study the short-wavelength perturbation with a parallel velocity shear. By considering the flute-like perturbations ($k_\parallel \ll k_\perp$) with a parallel velocity shear, we intend to address an additional key issue whether the stabilization/destabilization mechanism is a general feature of all modes with structure parallel to the magnetic field. We adopt a two-fluid theory in a sheared slab geometry, $\mathbf{B} = B_o[\mathbf{z} + (x/L_s)\mathbf{y}]$, where L_s is the scale length of magnetic shear. We assume a background plasma with all inhomogeneities only in the radial direction, where perturbations have the form $\phi(\mathbf{x}, t) = \phi(x)\exp[i(k_y y + k_z z - \omega t)]$. For simplicity, we take the ions to be cold and omit the electron temperature gradient. We ignore finite gyroradius effects by limiting consideration to the wavelength domain $k_\perp \rho_i \ll 1$, where ρ_i is the ion gyroradius. We then write down the linearized equations of continuity and parallel motion for the ions as (Sen, Rusbridge and hastie, 1994; Sen and Weiland, 1995; Sen, 1995):

$$\frac{\partial n_i}{\partial t} + \nabla_\perp.[N(x)\mathbf{V}_{\perp i}] + \nabla_\parallel[(N + n)(V_{\parallel 0} + V_{\parallel i})] = 0$$

$$m_i n_i[\frac{\partial V_{\parallel i}}{\partial t} + (\mathbf{V}_E + \mathbf{V}_{\parallel i}).\nabla V_{\parallel o}] = -en_i \nabla_\parallel \phi$$

Here,

$$\nabla_\perp = ik_y \hat{e}_y + \hat{e}_x \frac{d}{dx},$$
$$\mathbf{V}_{\perp i} = \mathbf{V}_E + \mathbf{V}_{pi},$$
$$\mathbf{V}_E = -c(\nabla_\perp \phi \times \mathbf{B}_o)/B_o^2$$
$$\mathbf{V}_{pi} = i[c(\omega - k_\parallel V_{\parallel 0}(x))/B_o \omega_{ci}]\nabla_\perp \phi,$$
$$T_i = 0$$

Here x is the distance from the mode rational surface defined by $\mathbf{k}.\mathbf{B_o} = 0$, and $V_{\parallel 0}$ is the equilibrium parallel velocity. All other symbols are assumed to have the usual meaning unless otherwise stated explicitly. As noted earlier, the parallel flow, $V_{\parallel 0}$, has therefore two effects. First, it introduces a Doppler shift, $k_\parallel V_{\parallel o}(x)$, in all time derivatives and second, an extra term, $\mathbf{V}_E.\nabla V_{\parallel o}(x)$, representing radial convection of ion momentum. It is the second term which makes the effect of parallel flow shear completely different from that of the perpendicular flow shear (Sen, Rusbridge and Hastie, 1994). We eliminate the Doppler shift by performing Galilean transformations in the $\hat{e}_\parallel$ direction. Now, using quasineutrality and the usual low frequency and long wavelength assumptions we obtain the radial eigenvalue equation:

$$\rho_s^2(\frac{d^2}{dx^2} - k_y^2)\phi - (1 - \frac{\omega_e^* + i\gamma}{\omega} + \frac{dV_{\parallel 0}}{dx}\frac{k_y \rho_s}{\omega}\frac{x}{x_s} - \frac{x^2}{x_s^2})\phi = 0 \qquad (3)$$

where

$$\rho_s^2 = \frac{C_s^2}{\omega_{ci}^2}, \quad C_s^2 = \frac{T_e}{m_i}, \quad \omega_e^*(x) = -k_y\rho_s C_s/L_n(x), \quad \gamma = \omega_e^*\delta, \quad x_s^2 = \frac{\omega^2}{k_\parallel'^2 C_s^2},$$

$$k_\parallel' = k_y/L_s, \quad k_\parallel = k_\parallel'x, \quad L_n(x)^{-1} = |\, d\, lnN(x)/dx \,|.$$

Here, δ takes account of the dissipative effects of the electron Landau resonance.

To model the equilibrium parallel velocity we assume a simple general case of the variation of $V_\parallel(x)$ with the radial distance x:

$$V_{\parallel o}(x) = V_{\parallel oo} + V_{\parallel o}'x + \frac{1}{2}V_{\parallel o}''x^2$$

where

$$V_{\parallel o}' = \frac{V_{\parallel oo}}{L_{v1}}, \quad \frac{1}{2}V_{\parallel o}'' = \frac{V_{\parallel oo}}{L_{v2}^2}$$

where $V_{\parallel oo}$ is the velocity characterizing flow. In considering the problem with a spatial variation of $\omega_e^*(x)$, we treat the simple case in which $\omega_e^*(x)$ is to be peaked at the mode rational surface defined by $x = 0$ and has a parabolic profile: $\omega_e^*(x) \equiv \omega_o^*(1 - x^2/L_*^2)$, where L_* is the density gradient scale length and will be taken typically $\sim L_n$. This is to ensure that the mode we are investigating is located at the minimum of $\frac{1}{L_n(x)}$ or at the maximum of $\frac{dn}{dx}$ which is the driving term of drift-type modes and hence we are considering the most unstable situation. With the velocity profiles just described Equation (3) reduces to

$$\rho_s^2\frac{d^2\phi}{dx^2} + (\Lambda + Px^2 - Qx)\phi = 0 \tag{4}$$

where

$$\Lambda = (\frac{\omega_o^*+i\gamma}{\omega} - k_y^2\rho_s^2 - 1), \quad P = (\frac{L_n^2}{\rho_s^2 L_s^2} - 2\frac{V_{\parallel oo}}{C_s}\frac{L_n^2}{L_{v2}^2}\frac{1}{L_s\rho_s} - \frac{1}{L_n^2}), \quad Q = (\frac{V_{\parallel oo}}{C_s}\frac{L_n^2}{L_{v1}}\frac{1}{L_s\rho_s})$$

In deriving Equation (4), we have assumed the usual drift approximation, i.e., $\omega \sim \omega_o^* = k_y V_o^*$, where $V_o^* = |\rho_s C_s/L_n|$. We emphasize that these assumptions are made only to facilitate comparison with the experimental data and no generality whatsoever is lost thereby.

Equation (4) is a simple Weber equation. Depending on the sign of P, we have two types of solution. If $P < 0$, i.e.,

$$\frac{L_n^2}{\rho_s^2 L_s^2} < 2\frac{V_{\parallel o}}{C_s}\frac{L_n^2}{L_{v2}^2}\frac{1}{L_s\rho_s} + \frac{1}{L_n^2}, \tag{5}$$

the solution which satisfies the physical boundary condition, i.e., $\phi \to 0$ at $x = \pm\infty$ is given by

$$\phi(x) = \phi_o exp[-\frac{\sqrt{|P|}}{2\rho_s}(x - x_o)^2] \tag{6}$$

where, $x_o = \frac{|Q|}{2|P|}$. The wave therefore does not propagate and is intrinsically undamped.

On the other hand, if $P > 0$, Equation (4) has the solution

$$\phi(x) = \phi_o exp[-i\frac{\sqrt{|P|}}{2\rho_s}(x + x_o)^2]$$

(7)

Thus, we have now a non-localized mode carrying energy outward. Because of the convective wave energy leakage the perturbation will decay in time in the absence of any energy source feeding the wave. The wave is therefore damped.

The overall stability of the mode will be determined by the strength of the driving term modeled by the $i\delta$ term and is obtained from the dispersion relation

$$\gamma < \frac{\rho_s(\frac{L_n^2}{\rho_s^2 L_s^2} - 2\frac{V_{\parallel oo}}{C_s}\frac{L_n^2}{L_{v2}^2}\frac{1}{L_s\rho_s} - \frac{1}{L_n^2})^{1/2}\omega_o^*}{(1 + k_y^2\rho_s^2 + \frac{Q^2}{4P})}$$

(8)

A few interesting points emerge from relation (8). First, the sign of the flow shear has no effect on stability as it occurs through the Q^2 term in (8). It can also be concluded by noting the invariance of Equation (4) under the combined operation of reflection $x \rightarrow -x$ and change in sign of $Q \rightarrow -Q$. Second, it is the parallel flow curvature which actually plays the key role in the stability of the mode. Velocity shear, on the other hand, shifts the potential but does not affect the quadratic structure. However, the most important observation emerging from these studies (for example, see relation (5)) is that the positive parallel flow curvature acts to destabilize the mode. Now, relation (5) also allows us to make another additional important observation. We notice that for negative parallel flow curvature, the curvature acts to stabilize the mode. Flow curvature now forms an additional antiwell which pushes the wave function away from the mode rational surface, thereby enhancing stabilization.

3. Conclusion

In summary, in this work we introduce a novel concept of the flow curvature (the second radial derivative) and find that it is the flow curvature which plays the crucial role in the linear mode stability. Unlike the flow shear the effect of flow curvature is found to be sign-dependent and also mode-dependent. For RT type mode with the positive curvature in the parallel flow the destabilizing influence of the parallel flow shear can be suppressed altogether and the mode can even become fully stable. On the other hand, for the negative curvature the unstable mode might be further destabilized than what is predicted by the shear alone. For the PVS type mode it is just the opposite, positive curvature acts to destabilize the mode whereas negative curvature can render the mode completely stable. As it is only natural that all experimental flow profiles must in principle be curved rather than a pure straight line, the inclusion of curvature in the flow profile rather than the shear alone is only realistic. This new observations may therefore be more appropriate and complete to compare with the observations of the Fast Auroral Snapshot satellite data.

In this work we choose two important space instabilities Rayleigh–Taylor (RT) and Parallel Velocity Shear (PVS) instability to show how the new physics of flow curvature works and come to these important conclusions. On the experimental front, there seems to be enough indication on the STOR-M tokamak experiments that parallel flow curvature indeed plays a leading role in the mode stability (Sen *et al.*, 2002). However, to put this proposition on a firmer footing one needs work on other relevant low frequency modes and should carry out full nonlinear analysis which are the topics of future investigations. However, the basic stabilizing/destabilizing mechanism from the flow curvature is, however, quite unambiguously shown by our simple model and is not expected to be seriously modified.

References

Gavrishchaka, V. V. *et al.*: 1998, *Phys. Rev. Lett.* **80**, 728.
Gavrishchaka, V. V. *et al.*: 2000, *Phys. Rev. Lett.* **85**, 4285.
Chandrashekhar, S.: 1981, *Hydrodynamic and Hydromagnetic Stability*, Dover, New York.
Ossakow, S. L.: 1979, *Rev. Geophys. Space Phys.* **17**, 521.
Sen, S., Rusbridge, M. G. and Hastie, R. J.: 1994, *Nuc. Fusion* **34**, 87.
Sen, S. and Weiland, J.: 1995, *Phys. Plasma* **2**, 777.
Sen, S.: 1995, *Phys. Plasma* **2**, 2701.
Sen, S. *et al.*: 2002, *Phys. Rev. Lett.* **2**, 185001.

ALFVÉN WAVES IN THE CONTEXT OF SOLAR-LIKE STAR FORMATION: ACCRETION COLUMNS AND DISKS

MARIA JAQUELINE VASCONCELOS[1], VERA JATENCO-PEREIRA[2] and
REUVEN OPHER[2]

[1]*Universidade Estadual de Santa Cruz, Ilhéus, BRASIL (e-mail: mjvasc@uesc.br)*
[2]*IAG Universidade de São Paulo, São Paulo, Brasil*

Abstract. In this work we examine the damping of Alfvén waves as a source of plasma heating in disks and magnetic funnels of young solar like stars, the T Tauri stars. We apply four different damping mechanisms in this study: viscous-resistive, collisional, nonlinear and turbulent, exploring a wide range of wave frequencies, from $10^{-5}\Omega_i$ to $10^{-1}\Omega_i$ (where Ω_i is the ion-cyclotron frequency). The results show that Alfvénic heating can increase the ionization rate of accretion disks and elevate the temperature of magnetic funnels of T Tauri stars opening possibilities to explain some observational features of these objects.

Key words: accretion disks, stars: pre-main-sequence, turbulence, waves

1. Introduction

The application of Alfvén waves far from the solar context is not very common, although the use of magnetic field in astrophysics theories has becoming more and more frequent. The theory of protostellar accretion disks is not an exception to this fact. Although magnetic field has gaining importance, very few consider the possibility of the existence of Alfvén modes.

Protostellar accretion disks are objects thought to be formed around young, solar-like stars. Many observational signatures of Classical T Tauri stars are explained assuming the disk is described by the standard model (Lynden-Bell and Pringle, 1974; Shakura and Sunyaev, 1973) although it requires some modifications as, for example, the incorporation of the magnetospheric accretion model (Muzerolle *et al.*, 1998). Despite the model one uses to describe an accretion disk, a central role is played by viscosity, which is the responsible for accretion. Although many years have passed since the initial proposal of the existence of accretion disks, the mechanism behind viscosity remains unknown. The application of the most promising mechanism of angular momentum transport, the Balbus–Hawley instability (Balbus and Hawley, 1998), to protostellar accretion disks is not straightforward. This is because the temperatures obtained with the standard model, even if we consider irradiation by the central star (D'Alessio *et al.*, 1998), are not sufficient to make Balbus–Hawley instability to occur. In our work, we propose the damping

Space Science Reviews **107**: 383–386, 2003.
© 2003 *Kluwer Academic Publishers.*

of Alfvén waves can be another source of energy. We consider that the Alfvénic mode can be damped by two mechanisms: nonlinear and turbulent (Vasconcelos *et al.*, 2000).

Also, temperatures of the magnetospheric accretion funnels obtained from models are very low to explain the observations (Martin, 1996; Muzerolle *et al.*, 1998). We propose the damping of Alfvén waves could account for an increase of temperature of the funnels that can explain the observations (Vasconcelos *et al.*, 2002). We consider four damping mechanisms: nonlinear, turbulent, collisional and viscous-resistive.

2. Results

2.1. ALFVÉNIC HEATING OF PROTOSTELLAR ACCRETION DISKS

The heating generated by the damping of Alfvén waves is considered as an extra source of energy. Thus, the total heating rate, D_{tot}, is a sum of viscous and Alfvénic heating: $D_{tot} = D_{vis} + D_A$. The Alfvénic heating is written in terms of the nonlinear and turbulent Alfvénic heating rates, H_{nl} and H_{tur}, respectively. Moreover, $H_A = \Phi_\omega / L_A$, where Φ_ω is the Alfvén wave flux and L_A is the nonlinear or turbulent damping lengths. Since the Alfvén wave flux is known only for the Sun, we write the equations in terms of disk variables and use the standard model and the layered model (Gammie, 1996) in order to obtain radial profiles. The disk is divided in three different regions (regions 2, 3 and 4 – Vasconcelos *et al.* (2000)), according to Bell and Lin (1994)'s opacities. Moreover, we assume that $\varpi = F\Omega_i$ and $< \delta v^2 >= (f v_A)^2$, where F and f are free parameters and Ω_i is the ion-cyclotron frequency. We also assume the disk radiates as a blackbody. In the standard model, we assume three different vertical profiles, always with constant magnetic field: i) constant density and temperature, ii) exponentially varying density and constant temperature and, iii) exponentially varying density and temperature. We also take an $0.5 M_\odot$ young star, accreting at a rate of $10^{-8} M_\odot year^{-1}$. The parameters F and f are fixed in 0.1 and 0.002, respectively. Figure 1 shows our results. In the top row, the temperatures obtained using nonlinear damping are drawn for regions 2, 3 and 4. The bottom row shows temperatures obtained using nonlinear damping for the same regions. Temperatures obtained using standard model with different vertical profiles with Alfvénic heating are T_{const}, $T_{\rho var}$ and T_{var}. Temperature with Alfvénic heating in the layered model is T_{lay}. Effective temperature in standard model is T_{vis} and in the irradiated model is T_{irr}.

The turbulent damping of Alfvén waves produces a huge increase in disk temperatures, compared to temperatures that would be obtained if only viscous dissipation was taken into account. The increase in temperature caused by the nonlinear damping of Alfvén waves is not so great, but both lead to ionization fractions that permit Balbus–Hawley instability to occur (Vasconcelos, 2001)

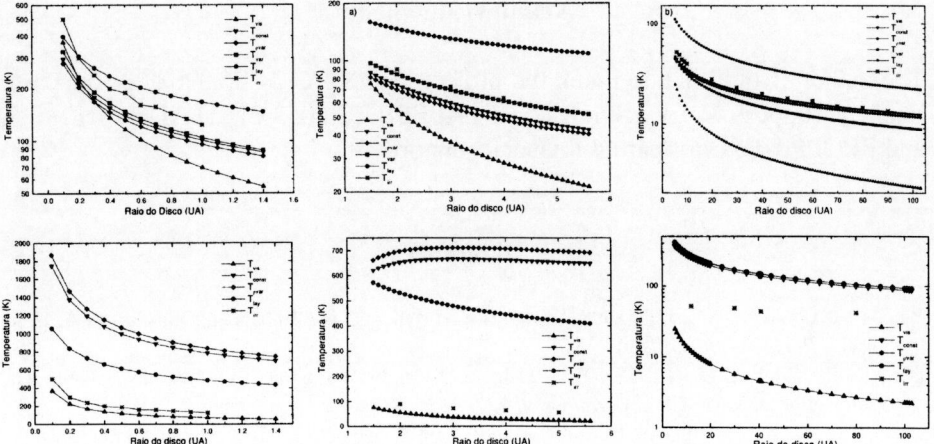

Figure 1. Temperatures for protostellar accretion disks.

2.2. Alfvénic heating on magnetospheric accretion models

In order to examine the heating of funnels of gas in magnetospheric accretion models, we consider four damping mechanisms: nonlinear, turbulent, collisional and viscous-resistive. We adopt the magnetospheric accretion model by Hartmann *et al.* (1994) with $\dot{M} = 10^{-7} M_\odot \, year^{-1}$ and $M = 0.8 M_\odot$. We use a relaxation procedure in order to solve the equations and a temperature profile which varies from ~7600 K to ~8200 K. We also solve Saha equation to obtain the ratio of protons to neutral atoms. Instead of a mean frequency for the waves, used previously (Vasconcelos *et al.*, 2000), we adopt a wave frequency spectrum, ranging from $10^{-1}\Omega_i$ to $10^{-5}\Omega_i$. Due to the interactive method used, f is no more a free parameter.

Our results have shown that the waves must be produced locally in order to Alfvénic heating be effective. In this case, temperatures obtained are able to reproduce the observations. Possible origins for the waves are Kelvin-Helmholtz instability and turbulence generated by, for example, intermittent accretion (Vasconcelos *et al.*, 2002).

3. Conclusions

We examine the damping of Alfvén waves as possible sources of heating in both protostellar accretion disks and magnetic funnels of Classical T Tauri stars. Our results show that Alfvénic heating contributes for the increase of the ionization fraction of the disks and also it is able to reproduce the temperature profile required to reproduce observations. Thus, we conclude that Alfvén waves are a viable heating mechanism for the environments considered in this work.

Acknowledgements

The authors would like to thank the project PRONEX (41.96.0908.00) for partial financial support. M.J.V acknowledges FAPESB, 62.0053/01-1-PADCT III/Milenio and PROPP/UESC for partial financial support.

References

Balbus, S. A. and Hawley, J. F.: 1998, 'Instability, Turbulence, and Enhanced Transport in Accretion Disks', *Rev. Mod. Phys.* **70**, 1–53.

Bell, K. R. and Lin, D. N. .C.: 1994, 'Using FU Orionis Outbursts to Constrain Self-regulated Protostellar Disk Models', *Astrophys. J.* **427**, 987–1004.

D'Alessio, P., Cantó, J., Calvet, N. and Lizano, S.: 1998, 'Accretion Disks around Young Objects. I. The Detailed Vertical Structure', *Astrophys. J.* **500**, 411–427.

Gammie, C. F.: 1996, 'Layered Accretion in T Tauri Disks', *Astrophys. J.* **457**, 355–362.

Hartmann, L., Hewett, R. and Calvet, N.: 1994, 'Magnetospheric Accretion Models for T Tauri stars. 1: Balmer Line Profiles Without Rotation', *Astrophys. J.* **426**, 669–687.

Lynden-Bell, D. and Pringle, J. E.: 1974, 'The Evolution of Viscous Discs and the Origin of Nebular Variables', *Mon. Not. R, Astron. Soc.* **168**, 603–637.

Martin, S. C.: 1996, 'The Thermal Structure of Magnetic Accretion Funnels in Young Stellar Objects', *Astrophys. J.* **470**, 537–550.

Muzerolle, J., Hartmann, L. and Calvet, N.: 1998, 'Magnetospheric Accretion Models for the Hydrogen Emission Lines of T Tauri Stars', *Astrophys. J.* **492**, 743–753.

Shakura, N. I. and Sunyaev, R. A.: 1973, 'Black Holes in Binary Systems. Observational Appearance', *Astron. Astrophys.* **24**, 337–355.

Vasconcelos, M. J., Jatenco-Pereira, V. and Opher, R.: 2000, 'Alfvénic Heating of Protostellar Accretion Disks', *Astrophys. J.* **534**, 967–975.

Vasconcelos, M. J.: 2001, PhD Thesis. Universidade de São Paulo.

Vasconcelos, M. J., Jatenco-Pereira, V. and Opher, R.: 2002, 'The Role of Damped Alfvén Waves on Magnetospheric Accretion Models of Young Stars', *Astrophys. J.* **574**, 847–860.

KINETIC EXCITATION MECHANISMS FOR ION-CYCLOTRON KINETIC ALFVÉN WAVES IN SUN-EARTH CONNECTION

YURIY VOITENKO[1,2] and MARCEL GOOSSENS[1]

[1]*Centre for Plasma Astrophysics, K.U.Leuven, Celestijnenlaan 200B, 3001 Heverlee, Belgium*
(e-mail: Yuriy.Voitenko@wis.kuleuven.ac.be)
[2]*Main Astronomical Observatory of the National Academy of Sciences of Ukraine,*
27 Zabolotnoho St., 03680 Kyiv, Ukraine

Abstract. We study kinetic excitation mechanisms for high-frequency dispersive Alfvén waves in the solar corona, solar wind, and Earth's magnetosphere. The ion-cyclotron and Cherenkov kinetic effects are important for these waves which we call the ion-cyclotron kinetic Alfvén waves (ICKAWs). Ion beams, anisotropic particles distributions and currents provide free energy for the excitation of ICKAWs in space plasmas. As particular examples we consider ICKAW instabilities in the coronal magnetic reconnection events, in the fast solar wind, and in the Earth's magnetopause. Energy conversion and transport initiated by ICKAW instabilities is significant for the whole dynamics of Sun-Earth connection chain, and observations of ICKAW activity could provide a diagnostic/predictive tool in the space environment research.

Key words: ion-cyclotron effects, kinetic Alfvén waves, kinetic instabilities, space plasmas

1. Introduction

Recent investigations suggest that high-frequency Alfvén waves (AWs) play a distinctive role in the heating of the solar corona and acceleration of the solar wind (Tu *et al.*, 1998; Li *et al.*, 1999; Cranmer *et al.*, 1999; Hollweg, 1999a). Attempts have been made to explain the presence of these high-frequency AWs under the assumption that they are excited at the coronal base by the numerous reconnection events (McKenzie *et al.*, 1995). Indeed, the energy released during magnetic reconnection can be more than sufficient for the required wave flux but the mechanism responsible for exciting high-frequency AWs remains unclear. A possible route for generating AWs with frequencies $\omega = 1 - 10^2 \text{ s}^{-1}$ involves MHD disturbances which accompany this magnetic reconnection activity. The cyclotron absorption of high-frequency parallel-propagating (slab) AWs, excited by reconnection events in the solar corona or by turbulent cascade (McKenzie *et al.*, 1995; Marsch and Tu, 1997), has been discussed as a heating mechanism. However, highly oblique AWs are more natural since high perpendicular wavenumbers are generated either by phase mixing (Goossens, 1991) or by turbulent cascade (Leamon *et al.*, 2000). The related kinetic effects of finite ion Larmor radius arising for oblique Alfvén wave propagation have to be taken into account in the heating theories.

Space Science Reviews **107**: 387–401, 2003.
© 2003 *Kluwer Academic Publishers.*

Kinetic and dispersive properties of (oblique) kinetic Alfvén waves (KAWs) are of interest from both a theoretical and a practical point of view, and are now under intensive investigation (Voitenko, 1998a–c; Hollweg, 1999b; Voitenko and Goossens, 2000b, and references therein). Contrary to ideal MHD Alfvén waves, KAWs can effectively interact with plasma particles via linear Cherenkov resonance (Hasegawa and Chen, 1976), and among themselves via nonlinear three-wave resonant interaction (Voitenko, 1998a, b; Voitenko and Goossens, 2000b). KAW instabilities have been shown to provide an efficient engine for the conversion of energy of beams and currents in the solar corona, solar wind, and Earth's magnetosphere (Voitenko et al., 1989; Hasegawa and Chen, 1992; Voitenko, 1995; Voitenko, 1998a–c; Voitenko and Goossens, 2000a; Stasiewicz et al., 2001). The presence of KAWs can lead to many various processes in space plasmas, like coronal loop heating (de Assis and Tsui, 1991), enhanced electron runaway (de Assis and Leubner, 1994), or current drive (Elfimov, de Azevedo and de Assis, 1996).

The common approach so far was to study the KAWs in the limit of weak parallel dispersion, $\omega_k / \Omega_p \ll 1$, and often in the limit of weak perpendicular dispersion, $k_\perp \rho_p \ll 1$, also (Ω_p is the ion (proton) cyclotron frequency and ρ_p is the ion gyroradius). However, KAWs with $k_\perp \rho_p \sim 1$ and $\omega_k / \Omega_p \lesssim 1$ are preferably excited by kinetic plasma instabilities (see Voitenko, 1995; Voitenko, 1998a, b), and the results previously reported in the literature for KAWs are not applicable in this case. Moreover, observational evidences for the competition of Landau and ion-cyclotron damping mechanisms for the Alfvén waves in solar wind have been reported recently (Leamon et al., 2000). It is evident that kinetic effects of the cyclotron wave-particle interaction may be responsible for the excitation and dissipation of the KAWs, and cannot be self-consistently included in a low-frequency model of KAWs, when $\omega_k / \Omega_p \ll 1$. Again, at frequencies $\omega_k / \Omega_p \sim 1$ the wave properties can be essentially different from those of low-frequency KAWs.

In the present paper we extend the kinetic theory of KAWs in the ion-cyclotron frequency range and describe the dispersive and dissipative properties of ion-cyclotron kinetic Alfvén waves (ICKAWs). In order to fully include the kinetic excitation/dissipation mechanisms for the ICKAWs and the consequent plasma heating in our analysis, we use the kinetic plasma theory – Maxwell–Vlasov set of equations. The interplay of ion Larmor radius and finite ion cyclotron frequency effects is taken into account allowing for a wide range of the 'kinetic variables', $\mu = k_\perp \rho_p$ and $\nu = k_\parallel \delta_p$. ρ_p is the ion (proton) gyroradius, and δ_p is the ion inertial length, $k_\parallel$ and $k_\perp$ are the wave vector components along and across the background magnetic field. The current-driven and ion-cyclotron beam-driven ICKAW instabilities are attended for the first time. The particular excitation mechanisms of ICKAWs by currents and ion beams are discussed in relation to coronal magnetic reconnection, fast solar wind, and Earth's magnetosphere.

The linear theory that we use here can be applied to the stability analysis of current and beam systems, and allows us to calculate the growth rate and source position of initially excited ICKAWs in the wavenumber space. The resulting ICKAW power spectrum should be calculated by use of the quasilinear and/or nonlinar theory, as has been done for KAWs by Voitenko (1998a–c) and by Voitenko and Goossens (2000a).

The present paper is restricted to a uniform and collisionless plasma. As the plasma is nonuniform, the parallel (λ_z) and perpendicular (λ_x) wavelengths should be much shorter than the length scales of the parallel (L_z) and perpendicular (L_x) plasma inhomogeneities, respectively. Because of the very small wavelengths of ICKAWs, $\lambda_x \sim \rho_p$ and $\lambda_z \sim \delta_p$, this condition usually implies that $L_x \gg \rho_p$ and $L_z \gg \delta_p$. In such plasmas ICKAWs can exist and can be excited locally. Again, ICKAWs are relatively high-frequency, $\omega \lesssim \Omega_p$, and do not experience any collisional effects in most space plasmas, where the collisional frequency is well below the ion-cyclotron frequency, $\nu_e \ll \Omega_p$. The effects of collisions at low wave frequencies (i.e., for KAWs) have been studied by Voitenko and Goossens (2000c).

2. ICKAW Theory: Basic Equations

Let us consider a fully ionized and uniform plasma in a background magnetic field B_0. Strictly speaking, all space plasmas are nonuniform, and the approximation of a uniform plasma means here that the parallel and perpendicular length scales of the plasma inhomogeneities are much longer than the respective wavelengths which appear in the problem: $L_x \gg \lambda_x$; $L_z \gg \lambda_z$. We focus on a magnetic plasma with a plasma/magnetic pressure ratio β in the range $m_e/m_p \ll \beta \ll 1$ as this is typical situation for many laboratory and space plasmas (m_e/m_p is the electron/ion mass ratio). To study magnetically non-compressible ($\tilde{B}_z = 0$) Alfvén waves, we use two fluctuating electromagnetic wave potentials, a scalar potential $\tilde{\phi}$ and (z-component of) a vector potential $\tilde{A}_z$ (Hasegawa and Chen, 1976). These potentials induce charge and current density perturbations $\tilde{q}_s$ and $\tilde{j}_s$ in the s-th plasma component, obeying the quasi-neutrality equation and Ampere's law in the z direction. In order to take into account all important kinetic properties of the ICKAWs, we calculate $\tilde{q}_s$ and $\tilde{j}_s$ by the use of the perturbed part, $\tilde{f}_s$, of the corresponding velocity distribution function ($F_s = F_s^0 + \tilde{f}_s$). For the gyroangle-averaged Fourier amplitudes we have (Voitenko, 1998a)

$$
\int_0^{2\pi} d\vartheta f_k^L = 2\pi \frac{e}{m} \sum_n \frac{J_n^2(\mu\zeta)}{-\omega_k + k_z V_z + n\Omega} \left[k_z(\phi_k - \frac{\omega_k}{k_z c} A_k)\frac{\partial}{\partial V_z} + \right.
$$
$$
\left. n\Omega \left((\phi_k - \frac{V_z}{c} A_k)\frac{\partial}{V_\perp \partial V_\perp} + \frac{1}{c} A_k \frac{\partial}{\partial V_z} \right) \right] F^0.
$$
(1)

We use a (factorized) drifting bi-Maxwellian distribution function for a plasma species s,

$$F_s^0 = \frac{n_s}{(2\pi)^{3/2} V_{Ts\parallel} V_{Ts\perp}^2} \exp\left(-\frac{V_\perp^2}{2V_{Ts\perp}^2}\right) \exp\left(-\frac{(V_z - V_{s0})^2}{2V_{Ts\parallel}^2}\right), \tag{2}$$

where V_{s0} is the bulk velocity and $V_{Ts\parallel}$ ($V_{Ts\perp}$) is the parallel (perpendicular) thermal velocity. The z axis is along the background magnetic field B_0. After performing the integration for the Fourier harmonic of Equation (1) with the distribution function (2), we obtain the charge perturbation in the s-th species as (Voitenko and Goossens, 2002)

$$q_{sk} = -\frac{\omega_s^2}{4\pi V_{Ts\parallel}^2} \sum_n \Lambda_n^{(s)} \times$$

$$\left[\left(1 + \frac{V_k}{\sqrt{2}V_{Ts\parallel}}\left(1 - \frac{V_{s0}}{V_k} - \Delta_s \frac{n\Omega_s}{\omega_k}\right) Z_n^{(s)}\right)\Phi + \right. \tag{3}$$

$$\left. \frac{n\Omega_s}{\omega_k}\left(\Delta_s + \frac{V_k}{\sqrt{2}V_{T\parallel}}\left(1 - \frac{V_{s0}}{V_k} - \Delta_s \frac{n\Omega_s}{\omega_k}\right) Z_n^{(s)}\right)\Psi\right].$$

and the perturbation of the parallel current density as

$$j_{sk} = e_s \int_{-\infty}^{+\infty} dV_z V_z \int_0^{+\infty} dV_\perp V_\perp \int_0^{2\pi} d\vartheta f_k^L = -V_k \frac{\omega_s^2}{4\pi V_{Ts\parallel}^2} \times$$

$$\sum_n \Lambda_n^{(s)}\left(1 - \frac{n\Omega_s}{\omega_k}\right)\left[\left(1 + \frac{V_k}{\sqrt{2}V_{Ts\parallel}}\left(1 - \frac{V_{s0}}{V_k} - \Delta_s \frac{n\Omega_s}{\omega_k}\right) Z_n^{(s)}\right)\Phi + \right. \tag{4}$$

$$\left. \frac{n\Omega_s}{\omega_k}\left(\Delta_s + \frac{V_k}{\sqrt{2}V_{Ts\parallel}}\left(1 - \frac{V_{s0}}{V_k} - \Delta_s \frac{n\Omega_s}{\omega_k}\right) Z_n^{(s)}\right)\Psi\right],$$

where $\Delta_s = 1 - T_{s\parallel}/T_{s\perp}$ is the thermal anisotropy, and $\Lambda_n^{(s)} \equiv \Lambda_n\left(\mu_s^2\right) = I_n\left(\mu_s^2\right)\exp\left(-\mu_s^2\right)$. $I_n\left(\mu_s^2\right)$ is the modified Bessel function, and $V_k = \omega_k/k_z$ is the wave phase velocity. We have introduced here the modified potentials $\Phi = \phi_k - A_k V_k/c$ and $\Psi = A_k V_k/c$. The plasma dispersion function is defined as

$$Z_n^{(s)} \equiv Z\left(\sigma_n^{(s)}\right) = \frac{1}{\sqrt{\pi}} \int_{-\infty}^\infty dy \frac{\exp\left(-y^2\right)}{y - \sigma_n^{(s)}},$$

where

$$\sigma_n^{(s)} = \frac{\omega_k - k_z V_{s0} - n\Omega_s}{\sqrt{2}k_z V_{Ts\parallel}}.$$

To find the dispersion relation for the IC KAWs we eliminate the wave potentials Ψ and Φ by means of two independent relations – quasineutrality condition, $\sum_s q_{sk} = 0$, and Ampére law in z direction, $4\pi V_k \sum_s j_{sk} = c^2 k^2 \Psi$. Combining these equations, and splitting the plasma dispersion function Z_n into real and imaginary parts, $Z_n = Z_{n\mathrm{Re}} + i Z_{n\mathrm{Im}}$, we get the dispersion equation for the ICKAW (Voitenko and Goossens, 2002):

$$\frac{V_k^2}{V_A^2} \left[\left(D_e + D_{\Phi j} \right) D_{\Psi q} - D_{\Psi j} \left(D_e + D_{\Phi q} \right) \right]$$

$$= \frac{T_e}{T_\perp} \mu_T^2 \left(D_e + D_{\Phi q} \right) + i \frac{T_e}{T_\perp} \mu_T^2 \left(\delta_e + \delta_{\Phi q} \right) \tag{5}$$

$$-i \frac{V_k^2}{V_A^2} \left[\left(1 + D_{\Phi j} \right) \delta_{\Psi q} + D_{\Psi q} \left(\delta_e + \delta_{\Phi j} \right) - D_{\Psi j} \left(\delta_e + \delta_{\Phi q} \right) - \delta_{\Psi j} \left(1 + D_{\Phi q} \right) \right].$$

The dispersive electron term is

$$D_e = 1 + \sigma_{e0} Z_{Re} \left(\sigma_{e0} \right) . \tag{6}$$

The dispersive ion terms are

$$D_{\Phi q} = \sum_s \frac{T_e}{T_{s\parallel}} \frac{n_s e_s^2}{n_0 e^2} \sum_n \Lambda_n^{(s)} \left[1 + \left(1 - \frac{n\Omega_s}{\omega_k - k_z V_{s0}} \Delta_s \right) \sigma_0 Z_{n\mathrm{Re}}^{(s)} \right]; \tag{7}$$

$$D_{\Psi q} = \sum_s \frac{T_e}{T_{s\parallel}} \frac{n_s e_s^2}{n_0 e^2} \sum_n \Lambda_n^{(s)} \frac{n\Omega_s}{\omega_k} \left(1 - \frac{n\Omega_s}{\omega_k - k_z V_{s0}} \Delta_s \right) \sigma_0 Z_{n\mathrm{Re}}^{(s)}; \tag{8}$$

$$D_{\Phi j} = D_{\Phi q} - D_{\Psi q}; \tag{9}$$

$$D_{\Psi j} = \sum_s \frac{T_e}{T_{s\parallel}} \frac{n_s e_s^2}{n_0 e^2} \sum_n \Lambda_n^{(s)} \left[-n^2 \frac{\Omega_s^2}{\omega_k^2} \Delta_s + \right.$$

$$\left. \left(1 - \frac{n\Omega_s}{\omega_k} \right) \frac{n\Omega_s}{\omega_k} \left(1 - \frac{n\Omega_s}{\omega_k - k_z V_{s0}} \Delta_s \right) \sigma_0 Z_{n\mathrm{Re}}^{(s)} \right]; \tag{10}$$

The dissipative (imaginary) terms are

$$\delta_e = \sqrt{\frac{\pi}{2}} \frac{V_k - V_{e0}}{V_{Te}}; \tag{11}$$

$$\delta_{\Phi q} = \sum_s \frac{T_e}{T_{s\parallel}} \frac{n_s e_s^2}{n_0 e^2} \sum_n \Lambda_n^{(s)} \left(1 - \frac{n\Omega_s}{\omega_k - k_z V_{s0}} \Delta_s \right) \sigma_0 Z_{n\mathrm{Im}}^{(s)}; \tag{12}$$

$$\delta_{\Psi q} = \sum_s \frac{T_e}{T_{s\parallel}} \frac{n_s e_s^2}{n_0 e^2} \sum_n \Lambda_n^{(s)} \frac{n\Omega_s}{\omega_k} \left(1 - \frac{n\Omega_s}{\omega_k - k_z V_{s0}} \Delta_s \right) \sigma_0 Z_{n\mathrm{Im}}^{(s)}; \tag{13}$$

$$\delta_{\Phi j} = \delta_{\Phi q} - \delta_{\Psi q}; \tag{14}$$

$$\delta_{\Psi j} = \sum_s \frac{T_e}{T_{s\parallel}} \frac{n_s e_s^2}{n_0 e^2} \sum_n \Lambda_n^{(s)} \left(1 - \frac{n\Omega_s}{\omega_k}\right) \frac{n\Omega_s}{\omega_k} \left(1 - \frac{n\Omega_s}{\omega_k - k_z V_{s0}} \Delta_s\right) \sigma_0 Z_{n\mathrm{Im}}^{(s)}; \quad (15)$$

The dispersion Equation (5) is quite general and may be used to study IC KAWs in a multi-component plasma with anisotropic temperatures of the plasma species moving with different velocities along $\mathbf{B}_0$. The Cherenkov beam-driven instability of ICKAWs and the influence of beams on the dispersion of unstable waves have been studied by Voitenko et al. (2001) and Voitenko and Goossens (2002). Here we neglect the beam dispersive effects which are weak for unequal beams. Instead, we introduce, in addition to the ($n = 0$) Cherenkov resonance, the $n = -1$ ion-cyclotron resonance with beam particles, and the Cherenkov resonance with current electrons.

Let us consider a simple plasma model consisting of a background current-carrying electron-proton (hydrogen) plasma with an additional beam-like proton component, and $\Delta_s = 0$ for all components. We model the electrons by a drifting Maxwellian distribution function with drift velocity $V_{e0}^2 \ll V_{Te}^2$, and then we expand all ionic $Z_{n\mathrm{Re}}^{(s)}$ in powers of $\sigma_n^2 \gg 1$, but electronic $Z_{0\mathrm{Re}}^{(e)}$ in powers of $\left(\sigma_0^{(e)}\right)^2 \ll 1$ (therefore $D_e \approx 1$). The condition $\sigma_n^2 \gg 1$ means that we consider waves with a phase velocity faster than the ion thermal velocity, and with a frequency not too close to the ion-cyclotron frequency.

Under these assumptions, Equation (5) can be greatly simplified (Voitenko and Goossens, 2002). Allowing for an imaginary part of the wave frequency, $\omega_k \to \omega_k + i\gamma_k$, $\gamma_k \ll \omega_k$, its solution can be found as

$$v_k = V_{(-)} \tag{16}$$

for the normalized wave phase velocity $v_k = V_k/V_A = \omega_k/(k_z V_A)$, and

$$\frac{\gamma_k}{\omega_k} = -\frac{1}{2} \frac{1}{V_{(-)}^2 \left(V_{(-)}^2 - V_{(+)}^2\right)} \frac{\left(v^2 V_{(-)}^2 - 1\right)}{v^2 \left(1 - \Lambda_0 - 2\Lambda_1\right)} \delta. \tag{17}$$

for the wave growth/damping rate. The functions $V_{(\pm)}$ are given by

$$V_{(\pm)}^2 = \frac{1 - t v^2 \mu^2 D_1 + v^2 K_0^2}{2v^2 (1 - D_1)} \pm \sqrt{\left(\frac{1 - t v^2 \mu^2 D_1 + v^2 K_0^2}{2v^2 (1 - D_1)}\right)^2 - \frac{K_0^2}{v^2 (1 - D_1)}}, \tag{18}$$

where $t = T_e/T_{p\perp}$, the variable $v^2 = k_z^2 \delta_p^2$ is introduced by the parallel ion dispersion, δ_p is the ion (proton) inertial length, $D_1 = 2\Lambda_1/(1 - \Lambda_0)$, and K_0 is the classic KAW dispersion function,

$$K_0^2 = \frac{\mu^2}{1 - \Lambda_0} + t\mu^2.$$

It is our aim to investigate the proton beam-driven and current-driven Cherenkov instabilities, and proton beam-driven ion-cyclotron instability. Hence, we keep in the dissipative part the Cherenkov (Landau) resonance terms for all plasma components, the $n = 1$ cyclotron resonance term for the background protons, and the $n = -1$ cyclotron resonance term for the beam: $\delta = \delta_e^L + \delta_p^L + \delta_{p+}^{IC} + \delta_{b-}^{IC}$, where

$$\delta_e^L = \mu^2 D_p \sqrt{\frac{\pi}{2}} \frac{V_k - V_{e0}}{V_{Te}};$$

$$\delta_p^L = \mu^2 D_p \sqrt{\frac{\pi}{2}} \frac{T_e}{T_{p\perp}} \Lambda_0 \frac{V_k}{V_{Tp}} \exp\left(-\frac{V_k^2}{2V_{Tp}^2}\right);$$

$$\delta_{p+}^{IC} = \sqrt{\frac{\pi}{2}} \Lambda_1 \left(\frac{1 + D_p}{v_z^2} - 2\frac{D_p}{v_z}\frac{V_k}{V_{Ap}} + \frac{n_p}{n_0}\mu^2 D_p\right) \times$$

$$\times \frac{V_k}{V_{Tp}} \exp\left(-\left(\frac{V_k - V_A/v_z}{\sqrt{2}V_{Tp}}\right)^2\right),$$

$$\delta_{b-}^{IC} = \sqrt{\frac{\pi}{2}} \Lambda_1 \left(\frac{1 + D_p}{v_z^2} + 2\frac{D_p}{v_z}\frac{V_k}{V_{Ap}} + \frac{n_p}{n_0}\mu^2\frac{D_p}{1 + D_b}\right) \times$$

$$\times \frac{V_k - V_{b0}}{V_{Tp}} \exp\left(-\left(\frac{V_k - V_{b0} + V_A/v_z}{\sqrt{2}V_{Tp}}\right)^2\right),$$

where $t_b = T_p/T_b$, and D_p is given by

$$D_p = \frac{n_p}{n_0}\frac{T_e}{T_{p\perp}}\left(1 - \Lambda_0 - 2\Lambda_1\frac{\omega_k^2}{\omega_k^2 - \Omega_p^2}\right). \tag{19}$$

Since the coefficient in front of δ in (17) is negative, the condition for ICKAW instability is $\delta < 0$. The electron current velocity V_{e0} and the proton beam velocity V_{b0} are destabilizing factors.

The low-frequency dispersive Alfvén wave has been known as the kinetic Alfvén wave in the domain $m_e/m_p < \beta < 1$, where it can be super-Alfvénic due to the effects of perpendicular thermal dispersion. As can be seen from Figure 1, the wave phase velocity, Equation (16), can vary from sub- to super-Alfvénic. Therefore, contrary to KAWs, the Landau resonance is accessible for ICKAWs not only with super-Alfvénic beams in a $m_e/m_p < \beta < 1$ plasma.

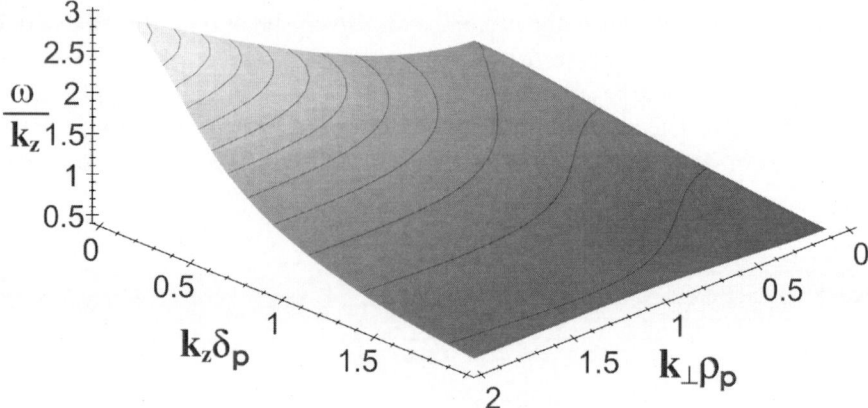

Figure 1. The parallel phase velocity of ICKAWs ω/k_z in units of the Alfven velocity V_A as a function of the normalized parallel ($k_z\rho_p$) and perpendicular ($k_\perp\rho_p$) wavenumbers. The electron/proton temperature ratio $T_e/T_p = 1$ and $\sqrt{2}V_{Tp}/V_A = 0.2$.

3. Current-driven ICKAW Instability

The growth rate of the current-driven KAW instability, given by Voitenko *et al.* (1989) and by Voitenko (1995), become inapplicable in the proton cyclotron frequency range where the instability is strongest. Here we analyze, for the first time, the current-driven ICKAW instability taking into account the effects of the finite proton cyclotron frequency. For a model of current-carrying plasma with Maxwellian protons and drifting Maxwellian electrons, the general expression (17) for the growth/damping rate becomes

$$\frac{\gamma_k}{\omega_k} = \sqrt{\frac{\pi}{8}} \frac{\mu^2 D_p \left(v^2 V_{(-)}^2 - 1\right)}{V_{(-)}^2 \left(V_{(-)}^2 - V_{(+)}^2\right) v^2 (1 - \Lambda_0 - 2\Lambda_1)} \frac{V_A}{V_{Te}} \left(\frac{V_{e0}}{V_A} - \frac{V_c}{V_A}\right). \tag{20}$$

Here the critical (not threshold!) velocity, V_c, for given ICKAW wavenumbers is given by

$$\frac{V_c}{V_A} = \frac{V_k}{V_A} \left\{ 1 + \Lambda_0 \frac{V_{Te}}{V_{Tp}} \exp\left(-\frac{V_k^2}{2V_{Tp}^2}\right) + \right.$$

$$\left. + \frac{V_{Te}}{V_{Tp}} \Lambda_1 \left(\frac{1 + D_p}{v_z^2 \mu^2 D_p} - \frac{2V_k}{v_z \mu^2 V_A} + 1\right) \exp\left(-\left(\frac{V_k - V_A/v_z}{\sqrt{2}V_{Tp}}\right)^2\right) \right\}. \tag{21}$$

When we minimize (21) with respect to wavenumbers, we find the turn-on condition for instability $V_{e0} > V_{thr}$, where $V_{thr} \gtrsim 0.5V_A$ in a low-β coronal plasma with a small fraction of resonant protons. The linear perpendicular wavenumber spectrum (20) of the unstable ICKAWs excited by the field-aligned current with $V_{e0} = 2V_A$ is shown in Figure 2 for the parallel wavenumber $k_z\delta_p = 0.35$. At

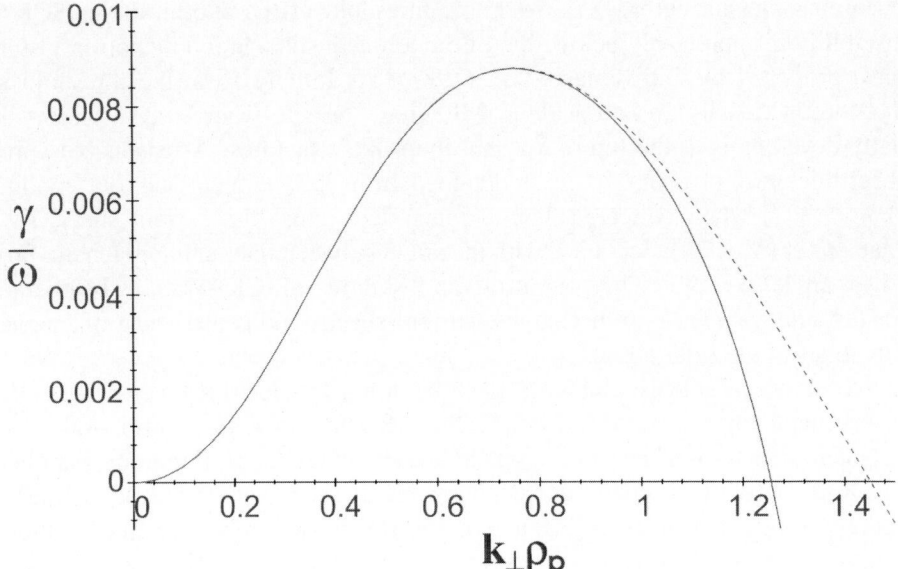

Figure 2. Normalized growth rate of current-driven ICKAW instability, γ/ω, with (solid curve) and without (dash curve) ion-cyclotron effects. Electron current velocity $V_{e0} = 2V_A \approx 0.1V_{Te}$, $T_e/T_p = 1$.

$k_z\delta_p \gtrsim 0.5$ the waves are stabilized by strong ion-cyclotron damping described by the last term in (21). As is seen from Figure 2, the ion-cyclotron damping reduces an upper cutoff for excited waves in perpendicular wavenumber space. With decreasing of $k_z\delta_p$ in the range $k_z\delta_p < 0.3$, ICKAWs smoothly transform into KAWs, and their instability range extends in μ well above unity. The corresponding maximum growth rate of the current-driven ICKAW instability $\gamma/\Omega_p \sim 0.01$ is attained at $k_z\delta_p \approx 0.45$ and $k_\perp\rho_p \approx 0.9$. The ICKAWs excited by the $V_{e0} = 2V_A$ current-driven instability have frequencies $\omega \gtrsim 0.6\Omega_p$. Such ICKAWs are not much different from KAWs in the sense that their perpendicular dispersion still dominates (they are super-Alfvénic), but they undergo ion-cyclotron interactions and transform the parallel energy of current electrons into perpendicular ion heating.

The problem of anomalous transport is of particular importance for space physics, because it is anomalous transport that makes a rapid energy release and transport possible in certain magnetic configurations. It is believed that the onset of fast release of energy occurs as soon as the current density exceeds a threshold value in regard to emerging or interacting magnetic flux. This is determined by the turn-on condition for the favorable instability. We suggest a current-driven kinetic Alfvén instability as the mechanism for producing anomalous dissipation of concentrated field-aligned currents in the solar corona and Earth's magnetopause.

For numerical estimate we take values typical for coronal active regions: $V_A = 10^8$ cm s^{-1}, $\Omega_p = 10^6$ s^{-1}, $\omega_{pe} = 10^{10}$ s^{-1}, $V_{Tp} = 4 \times 10^7$ cm s^{-1}. In the low-

β coronal plasma with $T_e/T_p \sim 1$, the threshold of the current-driven ICKAW instability is appreciably below that of the ion-acoustic and ion-cyclotron instabilities, proposed by Duijveman et al. (1981). As follows from the threshold and the current density variation along MF lines, the site where the instability initially develops is at the top of coronal loop-like structures. A reasonable current sheet thickness of about 10 ρ_p is obtained from the marginal stability condition ($V_{e0} \approx V_A$) when Ampere's law is integrated across the current sheet, $L_x \approx 2 \tan(\phi/2) (V_A/\Omega_p) (V_A/V_{e0})$, with the angle between interacting magnetic fluxes (shear angle) $\phi \sim 90°$. The current-driven instability of ICKAWs may be operative for fast energy release in discharges, current sheets, and coalescence of magnetic flux tubes in the solar corona.

Observations suggest that KAWs are the dominant electromagnetic fluctuations in the magnetopause (Johnson et al., 2001; Stasiewicz et al., 2001). Analysis of magnetic data shows that the magnetic shear angle ϕ is an important parameter. The KAW activity switches on when the magnetic shear angle ϕ reaches a threshold value of about 50 degrees. Johnson et al. (2001) consider this to be an indication of KAWs' excitation by the mode conversion of magnetosheat MHD waves that reach resonances in the magnetopause. For typical magnetopause parameters (thickness $L = 500$ km, $V_A = 300$ km s^{-1}, $\rho_p = 50$ km), the threshold angle of 50 degrees corresponds to a current velocity of the electrons close to the local Alfvén velocity. This observation suggests an alternative excitation mechanism for KAWs in the magnetopause by the current-driven ICKAW instability. The threshold bulk velocity of the electrons for this instability is close to the local Alfvén velocity. It is lowest among the known current instabilites for magnetopause parameters. Moreover, in a self-consistent approach, the thickness of the magnetopause itself should be determined by the back reaction of the magnetopause on the growing magnetic shear pumped by the solar wind. Again, the condition of marginal stability fits well with other observed magnetopause parameters.

It is quite possible that the turbulence of KAWs is developed in the magnetopause and produces spectra which differ substantially from linear/quasilinear spectra (Voitenko, 1998a, 1998b). Indeed, the frequency spectra of these fluctuations, measured by satellites, exhibit power laws with indeces $\lesssim -2.5$ (see Rezeau and Belmont, 2001, and references therein). Such spectra suggest an interpretation in terms of the inertial-range turbulence formed by the nonlinear wave-wave interactions, similar to Kolmogorov fluid turbulence. However, the interpretation of the spectral data is complicated by the Doppler shifts of wave frequencies due to large perpendicular wavenumbers. The angular wave frequencies ω_S, as seen in the satellite frame, are wave frequencies modified by the Doppler shifts introduced by the relative motion of the satellite with respect to the medium: $\omega_S = \omega - \mathbf{k} \cdot \mathbf{V}_S$ ($\mathbf{V}_S$ is the satellite velocity with respect to the medium). If the satellite velocity is not strictly parallel to $\mathbf{B}_0$, the small perpendicular wavelengths of KAWs produce large Doppler shifts which can dominate the measured $\omega_S \approx \mathbf{k}_\perp \cdot \mathbf{V}_{S\perp}$. A broadband KAW spectrum is excited by the current-driven instability around $k_{0\perp} \sim \rho_p^{-1}$ by super-Alfvenic currents, $V_{0e} \gtrsim 2V_A$. In this situation, the spectrum $W_k \sim k_\perp^{-3.5}$ ($W_k \sim k_\perp^{-4}$) should be formed in the range $k_\perp > \rho_p^{-1}$ ($k_\perp < \rho_p^{-1}$) by the direct (inverse) energy cascade from the source $k_{0\perp} \gtrsim \rho_p^{-1}$ ($k_{0\perp} \lesssim \rho_p^{-1}$) (Voitenko, 1998b). It is easy to show that the azimuthally symmetric perpendicular wavenumber spectrum with the power law index $-p$ ($W_k \sim k_\perp^{-p}$) can be seen in the satellite frame as the frequency spectrum with the power law index $-p + 1$ ($W_\omega \sim \omega_S^{-p+1}$). Therefore, the theoretically predicted perpendicular wavenumber spectra of KAWs can explain the frequency spectra $W_\omega \sim \omega_S^{-2.5}$, $\sim \omega_S^{-3}$, observed by satellites.

For the self-consistency of the local approximation, the perpendicular wavelengths of the excited ICKAWs should be much shorter than the perpendicular inhomogeneity length scale. Let us take the current sheet's width L_x as a measure of the perpendicular inhomogeneity length scale in the case of smooth variations across the sheet, with $\lambda_x \sim \rho_i$ for the KAWs excited by $V_{e0} \gtrsim 2V_A$ currents. The condition $L_x/\lambda_x \gg 1$ can be written as $\tan \phi/2 \gg V_{Tp}/V_A$. For typical values $V_{Tp}/V_A \approx 0.5$ in the magnetopause ($V_{Tp}/V_A \approx 0.2$ in corona), this condition is marginally satisfied for the threshold shear angle $\phi \approx 50°$ ($\phi \approx 20°$ in corona). As the nonuniformities and magnetic shears can be stronger across current sheets, the cross-sheet extension of wave fields may become limited by the nonuniformity gradients (see, e.g., Figure 2 by Johnson and Cheng, 1997). The local analysis then may give only an order-of-magnitude estimate of the current-driven instability because the local growth rate has to be modified by the wave profile in the x direction. These problems need further investigation.

4. Proton Beam-driven ICKAW Instability: n = −1 Beam-wave Ion-cyclotron Resonance

The proton beam-driven KAW instability has been studied taking into account the Cherenkov resonance of waves with beam protons (Voitenko, 1996; Voitenko, 1998c). The theory has been improved by extending the KAWs in the ion-cyclotron frequency range, and the same (Cherenkov) instability of ICKAWs has been studied by Voitenko, Goossens and Marsch (2001) and by Voitenko and Goossens (2002).

Here we include the effects due to an $n = -1$ ion-cyclotron (anomalous Doppler) resonance of ICKAWs with beam protons in the analysis. Preliminary parametric study of the ICKAW growth/damping rate (17) (with $V_{e0} = 0$) has shown that, for Alfvénic beams $0.6V_A < V_{b0} < 1.4V_A$, the beam-wave ion-cyclotron resonance is not so important for the ICKAW instability. In this case, as has been shown in previous papers (Voitenko et al., 2001; Voitenko and Goossens, 2002), the ($n = 0$) Cherenkov resonance with beam particles determines the wave growth, which is $\gamma/\Omega_p \sim 0.01$ for the relative beam number density $n_b/n_0 = 0.2$. However, the situation becomes much different at super-Alfvén beam velocities, $V_{b0} > 1.4V_A$. We demonstrate this in Figure 3 by use of (17) with $V_{b0} = 1.5V_A$. Even if,

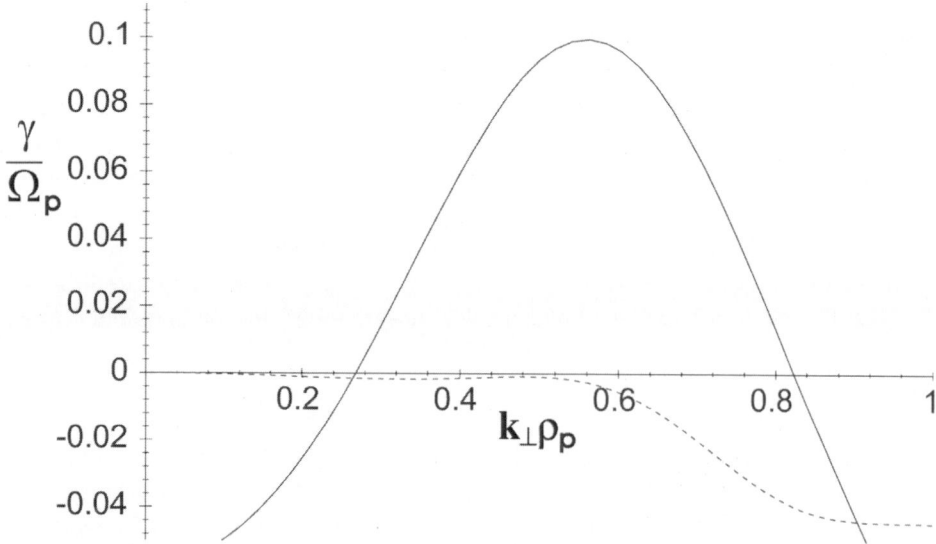

Figure 3. Normalized growth rate of the proton beam driven ICKAW instability γ / Ω_p vs. normalized perpendicular wavenumber $k_\perp \rho_p$ for the waves with parallel wavenumber $k_z \rho_p = 1.3$ (solid curve). The proton beam velocity is $V_b / V_A = 1.5$, the relative number density $n_b / n_0 = 0.2$, the temperature $T_{b\perp} / T_{p\perp} = T_e / T_{p\perp} = 2$, $V_{Tp} / V_A = 0.2$. There is no instability (dash curve) if we account only for Cherenkov resonances, ignoring the $n = -1$ wave-beam cyclotron resonance.

for the chosen plasma parameters, the Cherenkov resonance alone cannot provide instability, the ion-cyclotron wave-beam resonance give strong growth with $\gamma / \Omega_p \sim 0.1$.

Double-peaked anisotropic velocity distributions of protons and other ions have often been observed in the solar wind by *Helios* (Marsch, 1991). ICKAWs instabilities can provide the energy exchange mechanisms between the parallel energy of beam-like ion populations and the perpendicular energy of core protons in the solar wind (Voitenko *et al.*, 2001). There is also abundant observational evidence for ≥ 0.1 MeV (i.e. Alfvénic) proton beams in the solar corona (Simnett, 1995). These beams may be generated in the corona (upward) and in chromosphere (downward) by outflows from magnetic reconnection events at the base of the solar corona (see Voitenko, 1996, 1998 for the plasma model). Similar plasma configurations are often observed in-situ in the Earth's magnetosphere where the beams are accelerated by the magnetic reconnection in the tail.

Whatever the mechanism is that is responsible for injecting a fraction of the ions with Alfvénic velocities in the solar corona, solar wind, and Earth's magnetosphere, the beam-like ion populations can themselves become unstable and emit ICKAWs. The relaxation of beams and ICKAW instabilities can be complicated by the interplay between Cherenkov and ion-cyclotron wave-particle resonances. In particular, the initial ion-cyclotron ICKAW instability can relax the beam to a state which becomes unstable with respect to the Cherenkov ICKAWs instability.

The fate of excited ICKAWs is obvious. They have frequencies in the vicinity of the ion-cyclotron frequency where the cyclotron damping is also strong, and once they leave the unstable wavenumber-frequency domain due to the linear or nonlinear spectral transport, they immediately meet the conditions for very strong damping. These waves quickly damp heating the plasma. Since the beam-driven waves are in Landau resonance with the electrons and protons and in cyclotron resonance with the protons and the other ions, they are able to transform the kinetic energy of reconnection outflows into anisotropic heating. In a high-temperature plasma, $\sqrt{2}V_{Tp}/V_A \approx 0.5$, the excited waves have phase velocities that are not far from the ion thermal velocities, and the wave frequency is close to the proton cyclotron frequency. Hence, the ion heating in both the perpendicular and the parallel directions is expected to be stronger than the electron heating. For lower temperatures, $\sqrt{2}V_{Tp}/V_A \sim 0.1$, the direct electron heating at the Landau resonance becomes essential and competes with the (still strong) ion-cyclotron damping, resulting in perpendicular ion, and parallel electron heating.

The beam-driven ICKAW instability provides a very efficient engine for converting the energy of reconnection outflows from magnetic reconnection events into heat in the surrounding plasma (Voitenko and Goossens, 2002). Such impulsive plasma heating is observed in the low corona by Yohkoh and SOHO (blinkers, nano- and microflares) and attracts a growing interest (Shimizu et al., 1992; Berger et al., 1999; Roussev et al., 2001). At the same time, an extended coronal heating by high-frequency Alfvén waves excited during small but numerous reconnection events is expected for the double-beam distributions which become progressively more unstable during the upward propagation of the beams. Again, the ion-cyclotron $n = -1$ beam-ICKAW resonance amplifies the beam instability and makes it much less sensitive to details of beam particles distributions, facilitating energy transfer from reconnection outflows to plasma heating.

5. Conclusions

The combined action of finite ion cyclotron frequency and finite ion gyroradius effects has a profound influence on the kinetic Alfvén waves – KAWs. Parallel wave dispersion (finite frequency effects) in combination with perpendicular wave dispersion (finite gyroradius effects) make the KAW able to interact resonantly with sub-Alfvénic as well as super-Alfvénic beams. It is natural to call these waves ICKAWs – ion-cyclotron kinetic Alfvén waves. The phase velocity of the ICKAWs along the background magnetic field can be slower than the Alfvén velocity due to the parallel dispersion (finite ion cyclotron frequency effects, as in sub-Alfvénic ion-cyclotron AWs), but also faster than the Alfvén velocity due to the perpendicular dispersion (finite ion gyroradius effects, as in super-Alfvénic KAWs). Since these waves have a field-aligned component of electric field, they undergo Cherenkov resonant interaction and can be excited by particle beams and currents via the kinetic mechanism of inverse Landau damping. For the excitation of ICKAWs, the beam velocities should not be super-Alfvénic, as in the case of fast waves and usual low-frequency KAWs, but can extend in the sub-Alfvénic range.

We have developed new analytical expressions for ICKAWs taking into account both the parallel and the perpendicular wave dispersion, the Cherenkov resonant interaction with current electrons and beam protons, the $n = -1$ ion-cyclotron resonant interaction with beam particles, and the $n = 1$ ion-cyclotron resonant interaction with background plasma. The combined action of ion-cyclotron and finite gyroradius dispersive and dissipative effects leads to the reinforcement of the instability to $\gamma/\Omega_p \gtrsim 0.1$ at beam velocities $V_{b0} > 1.4V_A$. The decisive role is here played by the ion-cyclotron $n = -1$ resonant interaction with beam protons.

We analyzed a few particular kinetic excitation mechanisms for ICKAWs in the Sun-Earth connection. It is shown analytically, and observational evidences are provided, that ICKAWs are easily excited by field-aligned currents and ion beams in the solar corona, solar wind and Earth's magnetopause, affecting energy release, transport and transformation. As ICKAWs are sensitive to both the Cherenkov resonance and the ion-cyclotron resonance, they can serve as a mediator for the energy exchange between the parallel and perpendicular degrees of freedom in beam/current systems. The beam- and current-driven ICKAW instabilities attain their maximum growth rate at wave frequencies that are in the vicinity of the ion-cyclotron frequency, and the excited ICKAWs exert a strong influence upon the velocity distribution of the ions via ion-cyclotron interactions. Therefore, the ICKAWs that are excited by the parallel ion beams and currents can heat the ion components in the perpendicular direction, giving rise to an anisotropic ion heating, as is suggested by numerous observations.

The new properties under study make ICKAWs very important for the impulsive plasma heating in the energetic explosions that are observed in the low corona, transition region, and chromosphere, for the extended heating of ions in the solar corona and solar wind, and for the dynamics of the Earth's magnetopause.

Acknowledgement

Yu.V. acknowledges the financial support by the FWO-Vlaanderen, grants G.0344.98 and G.0178.3.

References

Berger, T. E., De Pontieu, B., Schrijver, C. J. and Title, A. M.: 1999, *Astrophys. J.* **519**, L97.
Cranmer, S. R., Field, G. B. and Kohl, J. L.: 1999, *Astrophys. J.* **518**, 937.
de Assis, A. S. and Tsui, K. H.: 1991, *Astrophys. J.* **366**, 324.
de Assis, A. S. and Leubner, C.: 1994, *Astron. Astrophys.* **281**, 588.
Duijveman, A., Hoyng, P. and Ionson, J. A.: 1981, *Astrophys.J.* **245**, 721.

Elfimov, A. G., de Azevedo, C. A. and de Assis, A. S.: 1996, *Solar Phys.* **167**, 203.

Goossens, M.: 1991, in: E. R. Priest and A. W. Wood (eds), *Advances in Solar System Magneto-hydrodynamics*, Cambridge University Press, Cambridge, England; New York, NY, p. 137.

Hasegawa, A. and Chen, L.: 1976, *Phys. Fluids.* **30**, 1924.

Hasegawa, A. and Chen, L.: 1992, *Ann. Geophysicae* **10**, 644.

Hollweg, J. V.: 1999a, *J. Geophys. Res.* **104**, 505.

Hollweg, J. V.: 1999b, *J. Geophys. Res.* **104**, 14,811.

Johnson, J. R. and Cheng, C. Z.: 1997, *Geophys. Res. Lett.* **24**, 1423.

Johnson, J. R., Cheng, C. Z. and Song, P.: 2001, *Geophys. Res. Lett.* **28**, 227.

Leamon, R. J., Matthaeus, W. H., Smith, C. W., Zank, G. P., Mullan, D. J. and Oughton, S.: 2000, *Astrophys. J.* **537**, 1054.

Li, X., Habbal, S. R., Hollweg, J. V. and Esser, R.: 1999, *J. Geophys. Res.* **104**, 2521.

Marsch, E.: 1991, in R. Schwenn and E. Marsch (eds), *Physics of the Inner Heliosphere*, Vol. II, Springer-Verlag, Heidelberg, Germany, p. 45.

Marsch, E. and Tu, C.: 1997, *Astron. Astrophys.* **319**, L17.

McKenzie, J. F., Banaszkiewicz, M. and Axford, W. I.: 1995, *Astron. Astrophys.* **303**, L45.

Priest, E. R.: 1990, in E. R. Priest and V. Krishan (eds), *Basic Plasma Processes on the Sun*, D. Reidel Publ. Co., Dordrecht, Holland, p. 271.

Rezeau, L. and Belmont, G.: 2001, *Space Sci. Rev.* **95**, 427.

Roussev, I., Galsgaard, K., Erdelyi, R. and Doyle, J. G.: 2001, *Astron. Astrophys.* **370**, 298.

Shimizu, T., Tsuneta, S., Acton, L. W., Lemen, J. R. and Uchida, Y.: 1992, *Publ. Astron. Soc. Japan* **44**, L147.

Simnett, G. M.: 1995, *Space Sci. Rev.* **73**, 387.

Stasiewicz, K., Seyler, C. E., Mozer, F. S., Gustafsson, G., Pickett, J. and Popielawska, B.: 2001, *J. Geophys. Res.* **106**, 29503.

Tu, C. Y., Marsch, E., Wilhelm, K. and Curdt, W.: 1998, *Astrophys. J.* **503**, 475.

Voitenko, Yu. M.: 1995, *Solar Phys.* **161**, 197.

Voitenko, Yu. M.: 1998a, *J. Plasma Phys.* **60**, 497.

Voitenko, Yu. M.: 1998b, *J. Plasma Phys.* **60**, 515.

Voitenko, Yu. M.: 1998c, *Solar Phys.* **182**, 411.

Voitenko, Iu. M., Krishtal', A. N. and Iukhimuk, A. K.: 1989, *Kosmicheskaia Nauka i Tekhnika (ISSN 0321-4508)*, no. 4, pp. 75–78. (in Russian).

Voitenko, Yu. M. and Goossens, M.: 1999, in: *Magnetic Fields and Solar Processes.* Proc. of Ninth European Meeting On Solar Physics, Florence, Italy, September 12–18, 1999, ESA SP-448, p. 735.

Voitenko, Yu. M. and Goossens, M.: 2000a, *Physica Scripta* **T84**, 194.

Voitenko, Yu. M. and Goossens, M.: 2000b, *Astron. Astrophys.* **357**, 1073.

Voitenko, Yu. M. and Goossens, M.: 2000c, *Astron. Astrophys.* **357**, 1086.

Voitenko, Yu. M. and Goossens, M.: 2002, *Solar Phys.* **206**, 285.

MKDVB AND CKB SHOCK WAVES

C. C. WU

Institute of Geophysics and Planetary Physics, University of California,
Los Angeles, CA 90095, U.S.A.

Abstract. This review presents results from MKDVB and CKB systems to illustrate the properties of magnetohydrodynamics intermediate shock waves. The intermediate shocks can have a family of shock structures; the shock structure and the evolution are related. Furthermore, there exists a class of time-dependent intermediate shocks.

Key words: MHD, Riemann problem, shock wave

1. Introduction to MHD Shock Waves

This review presents results from two simpler systems to illustrate the essential properties of magnetohydrodynamics (MHD) shock waves. We begin the paper with an introduction to the MHD shocks. Because of the presence of the magnetic field, the MHD waves are highly anisotropic. The three wave speeds are called the fast wave, intermediate (Alfvén) wave, and slow wave. The fast wave is the fastest among the three, while the slow wave is the slowest. The fast and slow waves have changes in velocities, transverse field magnitudes, density, and pressure, while the Alfvén waves have only a rotation of the transverse magnetic field and transverse velocity, but no changes in their magnitudes and no changes in the density and pressure. The MHD system can be written in conservative form and from which one derives Hugoniot jump conditions for discontinuities. The first systematic study of MHD waves and shocks from the viewpoint of quasilinear hyperbolic equations was made by Friedrichs (1954).

By the middle of the 1960's, many facts about the MHD shocks were known. The discontinuous solutions from the Hugoniot conditions must first satisfy the thermodynamic entropy increasing condition. This condition eliminates all the expansion shocks. The remaining admissible solutions can be classified into six families: the familiar fast and slow shocks, and four kinds of intermediate shocks. The intermediate shocks are distinguished by the fact that the flow speed is larger than the MHD intermediate speed upstream and is smaller than the intermediate speed downstream, whereas the flow speed remains entirely above or entirely below the intermediate speed across the fast and slow shocks. The direction of the shock plane magnetic field component rotates by 180 degrees across intermediate shocks. These solutions are all compressive and entropy increasing. In addition to these

Space Science Reviews **107**: 403–421, 2003.
© 2003 *Kluwer Academic Publishers.*

shock solutions, there are usual contact discontinuities, in which only the density can have an arbitrary jump, and rotational discontinuities, with transverse magnetic field and velocity rotation but no changes in thermodynamic state.

It was discovered that with these shocks and discontinuities, the solutions to piston problems, Riemann problems, etc. are nonunique (e.g., Jeffrey and Taniuti, 1964). There are too many shocks. To overcome this nonuniqueness, additional criteria were invoked. A common criterion is usually called evolutionarity (Akhiezer *et al.*, 1959; Polovin, 1961). It requires that a linearized perturbation of the shock possess a unique solution. This immediately translates into counting characteristics, that the number of waves emanating from the shock path should be one less than the number of shock jump conditions. Under this criterion, the intermediate shocks were rejected, and all the others remain. The other criterion, introduced by Germain (1960), is that with model shock structure equations, such as MHD Navier–Stokes equations, the discontinuities should have a structure. The intermediate shocks may or may not have structures depending on dissipation coefficients, and may have nonunique structures, and again were rejected. This theory of 1960's, which is based on analogy with fluid mechanics, will be referred as classical MHD shock theory in this review.

In the 1980's, Brio and Wu (1987, 1988) found the existence of intermediate shocks in numerical calculations of the MHD Riemann problems. They also noticed the important fact that the MHD equations are not genuinely nonlinear, since the flux lacks convexity. They suggested that the nonevolutionarity of the intermediate shocks and the nonconvexity of the equations are related. A series of studies based on MHD equations (Wu, 1987, 1988a, 1988b, 1990, 1991, 1995; Wu and Kennel, 1992c), small-amplitude MHD equations (Kennel *et al.*, 1990; Wu and Kennel, 1992a, 1992b, 1993), and related modified-Korteweg–de Vries–Burgers (MKDVB) equations (Wu, 1991) came to many new conclusions, which we summarize very briefly:

1) The MHD equations are not genuinely nonlinear, because of the nonconvexity. 2) Intermediate shocks are physical. All four kinds of intermediate shocks can be formed from steepening. Their structure is dissipation dependent. 3) The rotational discontinuity is unstable in the presence of dissipation, and breaks up into all kinds of waves. 4) Riemann solutions are unique in dissipative MHD. 5) There exist non-coplanar intermediate shocks: these are shocks where the upstream and downstream field and velocity in the structure have components out of the plane. 6) There exist time-dependent intermediate shocks, not satisfying Rankine–Hugoniot conditions, and for these shocks, even the upstream and downstream fields and velocities are not coplanar.

In this review, we discuss these properties based on the small amplitude MHD equations (Cohen–Kulsrud–Burgers, CKB, equations), and the related MKDVB equation. The historical account in this Introduction section follows a presentation by Chu and Wu (1991). The discussion on MKDVB shocks is from Wu (1991), and the results on CKB shocks are from works by Kennel *et al.* (1990) and Wu and

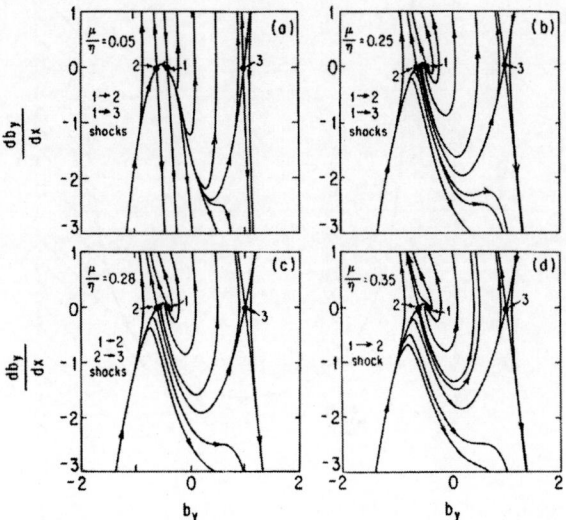

Figure 1. MKDVB shock trajectories using $\alpha = 1/4$, $\eta = 0.05$ and various ratios of μ/η for the case where the stationary state 1 is $b = -0.4$, state 2 is $b = -0.6$, and state 3 is $b = 1$.

Kennel (1992b, 1993). We note that Jacobs *et al.* (1995) has performed a thorough analysis of MKDVB shocks and the CKB shocks have also been studied by many researchers (e.g. Ruderman, 1989; Freistühler, 1992; Freistühler and Liu, 1993).

2. MKDVB Shock Waves

We consider the shock waves for the MKDVB equation, which is of the form

$$\frac{\partial b}{\partial t} + \alpha \frac{\partial}{\partial x}(b^3) = \eta \frac{\partial^2 b}{\partial x^2} + \mu \frac{\partial^3 b}{\partial x^3} \tag{1}$$

In this equation, there are both dissipative (η) and dispersive (μ) terms and α is a constant. Without the dissipative term, it is the usual MKDV equation. We will consider the case $\alpha > 0$; entirely analogous results may be obtained for $\alpha < 0$ (see below). This equation is not genuine nonlinear because the flux $f(b) = \alpha b^3$ is not a convex function for all b.

The stationary wave solution of (1) moving to the right with constant velocity V satisfies

$$\alpha \frac{\partial}{\partial x}(b^3) - V \frac{\partial b}{\partial x} = \eta \frac{\partial^2 b}{\partial x^2} + \mu \frac{\partial^3 b}{\partial x^3} \tag{2}$$

Integrating once with respect to x gives

$$\alpha b^3 - V b - F_0 = \eta \frac{\partial b}{\partial x} + \mu \frac{\partial^2 b}{\partial x^2} \tag{3}$$

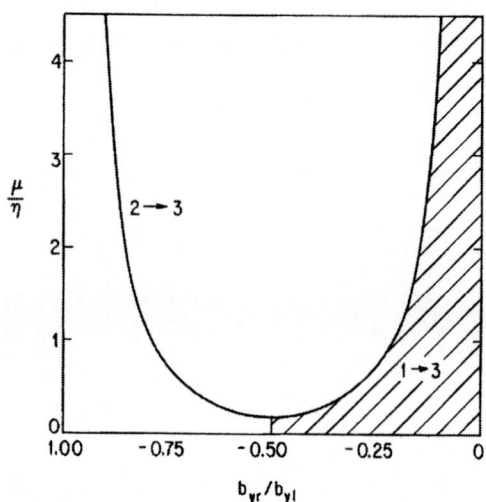

Figure 2. The dependence of the critical values of μ/η upon the ratio of the upstream to downstream b values. The $1 \rightarrow 3$ shocks can have shock structure solutions in the shaded area; the $2 \rightarrow 3$ shocks exist only for critical values. The $\mu = 0$ situation applies to cases with negative μ. The plot is for $\eta = 0.05$, $\alpha = 1/4$, and $b_l = 1$.

where F_0 is the constant of integration. Since both dissipation and dispersion are assumed important only within the shock layer, F_0 equals the flux $F = \alpha b^3 - Vb$ both upstream and downstream of the shock. This means that the upstream and downstream shock states are stationary points of (3), which satisfy the jump relation

$$\alpha b^3 - Vb = F_0 \tag{4}$$

Depending on α, V, and F_0, (4) may have one or three real solutions. When it has only one, there are no shocks. When it has three solutions, it is easy to see that their sum must be zero. Thus two of the solutions have the same sign and the third has the opposite sign. We call the third solution a stationary point (state) of type 3 and the other stationary points (states) of types 1 and 2. We define state 1 to have a smaller magnitude than state 2. Clearly the magnitude of state 3 is the sum of states 1 and 2.

The MKDVB equation has a characteristic wave with speed $\lambda = 3\alpha b^2$. For positive α, this wave corresponds to the fast MHD wave and the stationary states 1, 2, and 3 correspond to those in MHD. There are six possible jump relations among the three states. However, the $3 \rightarrow 1$, $3 \rightarrow 2$, and $2 \rightarrow 1$ transitions do not have shock structures. The $1 \rightarrow 2$ transitions always have structures, while the $1 \rightarrow 3$ and $2 \rightarrow 3$ transitions may have structures depending on η and μ. For both $1 \rightarrow 2$ and $1 \rightarrow 3$ shocks, the characteristics converge into them. For the $2 \rightarrow 3$ shock, the characteristics pass through it, since $\lambda > s$ (s denoting shock speed) for both upstream and downstream states. Thus the $2 \rightarrow 3$ transition is undercompressive.

For the $1 = 2 \rightarrow 3$ shock, where states 1 and 2 are the same with $b_r/b_l = -0.5$, the upstream $\lambda = s$.

Figure 1 shows the numerically computed trajectories of the shock structure equation (3) with condition (4) and $V = \alpha(b_l^2 + b_l b_r + b_r^2)$, where the subscripts l and r denote downstream and unstream states, respectively. It uses $\alpha = 1/4$, $\eta = 0.05$ and various ratios of μ/η (denoted by q) for a case where the stationary states 1, 2, and 3 are $b = -0.4$, $b = -0.6$, and $b = 1$, respectively. As shown in the figure, the $1 \rightarrow 2$ shock always exists, independent of q (including $q < 0$). On the other hand, the $1 \rightarrow 3$ shock can exist only when $q < 0.28$, and the $2 \rightarrow 3$ shock only exists at $q = 0.28$. When $q > 0.28$, neither $1 \rightarrow 3$ nor $2 \rightarrow 3$ can exist.

This dependence on the μ/η ratio is true for all MKDVB shocks. In Figure 2, the dependence of the critical value of q upon the ratio of the upstream to downstream b values is plotted. For a given q, the $2 \rightarrow 3$ shocks can have only one solution at one critical value of b_r^*/b_l, which is less than or equal to $-1/2$, and the $1 \rightarrow 3$ shocks can have shock structure solutions in the shaded area in $-(1 + b_r^*/b_l) \leq b_r/b_l \leq 0$. Note this plot depends on b_l, η, and α.

Now we show how the solutions of the MKDVB Riemann problem depend on the ratio of μ/η. We will solve numerically for the time evolution of an initial profile, $b(x)$, that varies between b_l at $x \rightarrow -\infty$ (downstream, to the left) and b_r at $x \rightarrow \infty$ (upstream, to the right). We choose $\alpha = 1/4$, $\eta = 0.05$, and $\mu = 0$ or 0.05, use the initial distribution $b(x) = b_t \cos(\theta(x))$, with $\theta(x) = (\theta_l + \theta_r)/2 - (\theta_l - \theta_r)\tanh(x)$ and $b_t(x) = (b_l + b_r)/2 - (b_l - b_r)\tanh(x)$, choose $b_l = 1$, $\theta_l = 0°$, $\theta_r = 180°$, $b_0 = 1$, and various values of b_r and integrate

$$\frac{\partial b}{\partial t} + \alpha \frac{\partial}{\partial x}(b(b^2 - b_0^2)) = \eta \frac{\partial^2 b}{\partial x^2} + \mu \frac{\partial^3 b}{\partial x^3} \tag{5}$$

over the range $-25 \leq x \leq 25$ or $-50 \leq x \leq 50$. Figure 3 summarizes the results of the calculations. Each row shows the results for different values of b_r. Let us begin with the situation where b_l and b_r are of the same sign. If $b_r > b_l$ a fast rarefaction wave will be created. If $b_r < b_l$ a steady fast shock will be generated. Since the $1 \rightarrow 2$ fast shocks always have shock structure solutions, the results are independent of the μ/η ratio. This is confirmed in the top two rows of Figure 3. (Because the stationary state 1 becomes a spiral point, an upstream wave exists in the right-hand panels of the second row.)

In the case that b_l and b_r are of opposite signs, $1 \rightarrow 3$ and $2 \rightarrow 3$ shocks are expected. From Figure 2, we see that when $\mu = 0$, $1 \rightarrow 3$ shocks in the range $-1/2 \leq b_r/b_l \leq 0$ can exist, but $2 \rightarrow 3$ shocks can not, except in the degenerate case $1 = 2 \rightarrow 3$, where states 1 and 2 are the same with $b_r/b_l = -0.5$. On the other hand, when $\mu = \eta = 0.05$, only $1 \rightarrow 3$ shocks in the range $-0.21 \leq b_r/b_l \leq 0$ and only one $2 \rightarrow 3$ shock with $b_r/b_l = -0.79$ can exist. These differences in the shock structure solutions account for the differences in the Riemam solutions. The third row of Figure 3 shows the results for $b_r/b_l = -0.2$. Since shock solutions exist for both $\mu = 0$ and $\mu = 0.05$ cases, we expect that the initial state to evolve into a single $1 \to 3$ shock, as the results confirm. The next row of the figure is for $b_r/b_l = -0.4$. Since when $\mu = 0$ and $\eta = 0.05$, the $1 \to 3$ shock structure with $b_r/b_l = -0.4$ exists, one expects the system to evolve into such a single $1 \to 3$ shock. However in the case where $\mu = \eta = 0.05$, the $1 \to 3$ shocks with $b_r/b_l = -0.4$ do not have structure solutions. Thus the system has to evolve into other waves. The possible solutions are either a fast shock or a fast rarefaction wave plus a $2 \to 3$ shock with $b_r/b_l = -0.79$. (Since fast waves converge into $1 \to 3$ shocks, it is not possible to have a fast shock or rarefaction wave plus a $1 \to 3$ shock.) In the present case, the fast wave is a fast shock since b changes from -0.4 to -0.79. The numerical results confirm these predictions.

Similarly, one can construct the solutions for the other two cases in the bottom two rows of Figure 3. For both $b_r = -0.6$ and $b_r = -1$, the system evolves into a fast rarefaction wave plus a $1 = 2 \to 3$ shock in the case $\mu = 0$. Since the upstream fast wave speed is the same as the $1 = 2 \to 3$ shock speed, the fast rarefaction wave is attached to the $1 = 2 \to 3$ shock. On the other hand, when $\mu = \eta = 0.05$, it evolves into a fast shock plus a $2 \to 3$ intermediate shock for the case $b_r = -0.6$, and into a fast rarefaction wave plus a $2 \to 3$ intermediate shock for the case $b_r = -1$. In sum, we show that, depending on the ratio of μ/η, the global solutions of the Riemann problem can be very different. We also show that the Riemann problems can be solved once the curve of critical value of μ/η of Figure 2 is known.

Since the $2 \to 3$ shock is undercompressive because the characteristics pass through it, its formation is thus different from the $1 \to 2$ and $1 \to 3$ formation. It can be formed by the shock interaction as in the Riemann problem. It can also form by wave steepening through the formation of $1 \to 3$ shock. For example, in a periodic system $0 \le x \le 1$, initial profile $b(x) = \sin(x)$ will first form $1 \to 2$ shocks, which then evolve into $1 \to 3$ shocks. They continue to evolve and the upstream to downstream ratio of b will reach $-1/2$ for the $\mu = 0$ case and form $1 = 2 \to 3$ shocks. For the case of $\eta = \mu = 0.05$, the upstream to downstream ratio of b of the $1 \to 3$ shock will reach its critical value, where it can be regarded as a combination of a $1 \to 2$ shock plus a $2 \to 3$ shock. After that, the system involves the two-shock combinations.

The above discussion applies to the situation where $\alpha > 0$. In the case $\alpha < 0$, the characteristic wave is related to the slow wave in MHD. The treatment is identical to the case for positive α. However, for $\alpha < 0$, $2 \to 3$ shocks exist only for negative μ, rather than positive μ in the positive α case.

In relation to the MKDVB equation, we mention the Korteweg–de Vries-Burgers (KDVB) equation, which models fluid mechanics. KDVB is related to (1) by replacing b^3 by b^2 and thus is genuinely nonlinear. It has two stationary points and a unique shock transition, which has shock solutions for all μ (Bona and Schonbek, 1985). Its Riemann solutions are unique as long as η is positive definite.

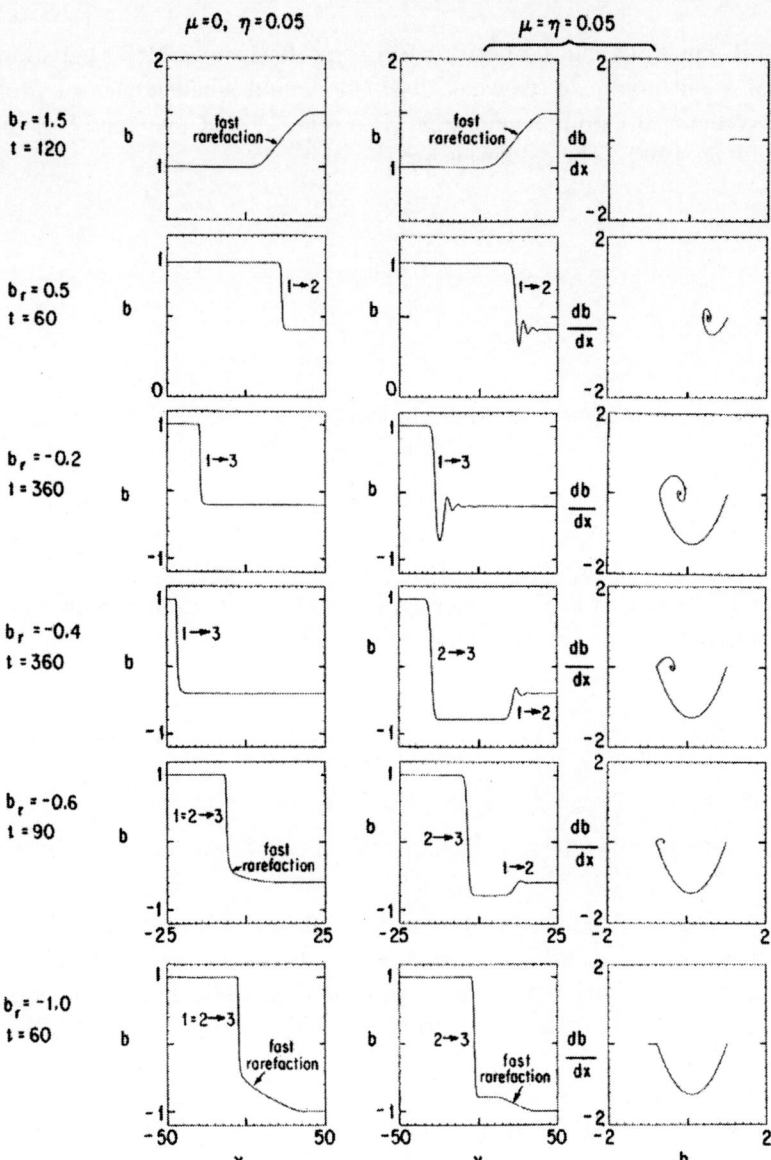

Figure 3. MKDVB Riemann problems. The left state is given by $b_l = 1$; the right states is given by several values of b_r. The results with $\mu = 0$ and $\eta = 0.05$ are on the left-hand panels, while the results with $\mu = \eta = 0.05$ are on the right-hand panels. This shows that the Riemann solutions depend on the values of μ and η.

3. CKB Shock Waves

3.1. CKB EQUATIONS

The CKB equations can be derived from the dissipative MHD equations in the limits of weak nonlinearity, weak dissipation, and small angles of propagation relative to the upstream magnetic field (Rogister, 1971; Cohen and Kulsrud, 1974; Kennel *et al.*, 1990). The equations are

$$\frac{\partial b_y}{\partial t} + \alpha \frac{\partial}{\partial \eta} \left(b_y (b^2 - b_0^2) \right) = \bar{R} \frac{\partial^2 b_y}{\partial \eta^2} \tag{6a}$$

$$\frac{\partial b_z}{\partial t} + \alpha \frac{\partial}{\partial \eta} \left(b_z (b^2 - b_0^2) \right) = \bar{R} \frac{\partial^2 b_z}{\partial \eta^2} \tag{6b}$$

The transverse magnetic field components are b_y and b_z; the magnitude is denoted by b or b_T, $b_T = \sqrt{b_y^2 + b_z^2}$; and b_0 is a constant. The coefficient α parametrizes the sign and strength of the nonlinear steepening. The dissipation length is denoted by $\bar{R}$. Note that some notations employed in this section are different from the previous section; for example, here η is a spatial variable.

The CKB equation has two underlying characteristic families. In a coordinate system such that the unperturbed magnetic field b_{z0} is equal to zero, the linearized Equations (6a) and (6b) are

$$\frac{\partial \delta b_y}{\partial t} + 3\alpha b_{y0}^2 \frac{\partial \delta b_y}{\partial \eta} = 0 \tag{7a}$$

$$\frac{\partial \delta b_z}{\partial t} + \alpha b_{y0}^2 \frac{\partial \delta b_z}{\partial \eta} = 0 \tag{7b}$$

where δb_y and δb_z are small perturbations and b_{y0} is the unperturbed field. This set describes small-amplitude waves of speed $3\alpha b_{y0}^2$ and αb_{y0}^2. The characteristic speeds are nearly equal when $b_{y0} \approx 0$. Thus, the characteristics underlying the CKB equation are non-strictly hyperbolic. In addition, like the MKDVB equation, they are not genuinely nonlinear.

Equation (7a) corresponds to MHD fast ($\alpha > 0$) or slow ($\alpha < 0$) waves and (7b) corresponds to MHD intermediate waves. As in MHD, the magnetic field varies its magnitude and does not rotate across a fast (slow) wave front; while the magnetic field rotates and does not change its magnitude across an intermediate wave front. Since both fast and slow wave speeds depend on the magnitude and not the direction of the magnetic field, a fast (slow) wave may steepen or rarefy; but an intermediate wave can do neither. Without dissipation, an intermediate wave will remain undistorted in time. However, fast (slow) waves will lead to compressive shocks and rarefaction waves.

3.2. CKB Shocks

A steady traveling wave moving to the right with constant velocity V obeys equations of the form (following the similar derivation in Section 2)

$$\alpha b_y b_T^2 - V b_y - F_{y0} = \bar{R} \frac{\partial b_y}{\partial \eta} \tag{8a}$$

$$\alpha b_z b_T^2 - V b_z - F_{z0} = \bar{R} \frac{\partial b_z}{\partial \eta} \tag{8b}$$

where F_{y0} and F_{z0} are constants of integration. Thus, the upstream and downstream shock states are stationary points of (8a, b), which satisfy the jump relations

$$\alpha b_y b_T^2 - V b_y = F_{y0} \tag{9a}$$

$$\alpha b_z b_T^2 - V b_z = F_{z0} \tag{9b}$$

One set of the stationary points is related by the so-called rotational discontinuities in MHD, across which b_T does not change, but the field direction may change, and $V = \alpha b_T^2$. However, these jump relations do not correspond to a steady traveling wave structure with finite width when dissipation is included (see below).

When b_T upstream and downstream have different magnitudes, they must be either parallel or anti-parallel to one another. Thus, CKB shocks are co-planar (like MHD shocks). It is then convenient to choose a reference frame such that $b_z = 0$ upstream and downstream. The jump relation (9a) for b_y becomes $\alpha b_y^3 - V b_y = F_{y0}$. Since this is the same equation as (4) for the MKDVB equation. The classification of shocks is the same as the MKDVB system. There are 3 stationary states. Consider the case $\alpha > 0$. The $1 \rightarrow 2$ shock corresponds to the MHD fast shock and the $1 \rightarrow 3$, $2 \rightarrow 3$ transitions correspond to two types of MHD intermediate shocks. Only fast waves converge into the $1 \rightarrow 2$ shock; only intermediate waves converge into the $2 \rightarrow 3$ shocks. But both fast and intermediate waves converge into a $1 \rightarrow 3$ shock, which is therefore over-compressive.

3.3. Structure of CKB Shocks

The solutions of the structure Equations (8a, b) may be ascertained from a local linear analysis in the vicinity of the stationary points. The stationary point of state 1 is an unstable node, where integral curves leave in all directions and the stationary point 3 is a stable node, where integral curves converge in all directions. Therefore point 1 can only serve as an upstream state and point 3 can only serve as a downstream state. The stationary point 2 is a saddle. The integral curve can enter the downstream point 2 along the b_y axis. On the other hand, only two solutions can leave the upstream point 2 saddle, and they must leave along the $\pm b_z$ axes.

Figure 4 show the numerically computed integral curves, which substantiate this linear analysis. It shows that (1) the $1 \rightarrow 2$ fast shock solution is unique; (2) there

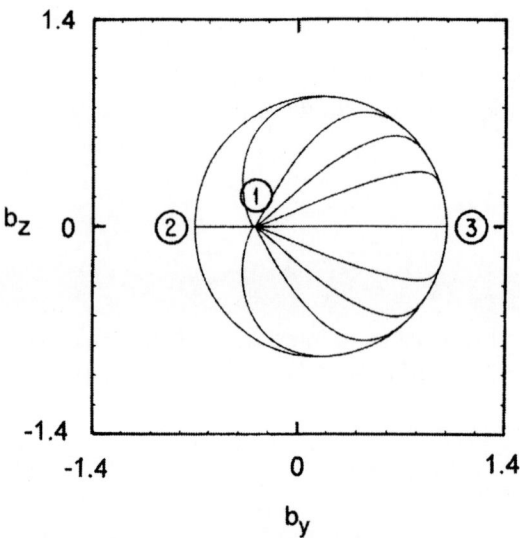

Figure 4. Shock structure solutions. Shown are shock structure solutions for $\bar{R} = 0.05$ for the stationary states $(b_y, b_z) = (-0.3, 0), (-0.7, 0)$, and $(1, 0)$ for state 1, 2, and 3, respectively. The stationary point 1 is an unstable node, the point 2 is a saddle, and the point 3 is a stable node.

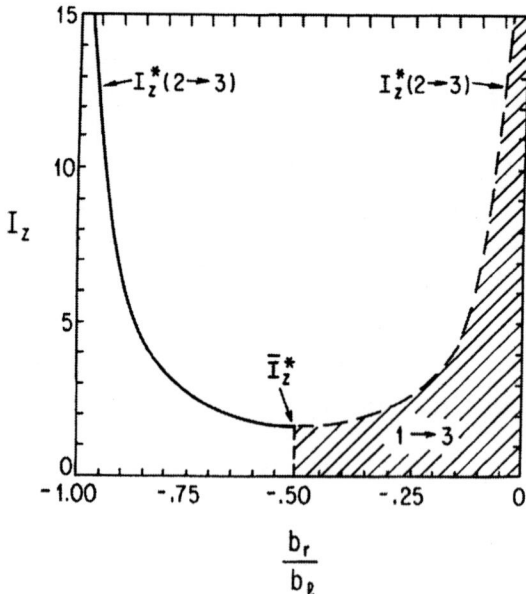

Figure 5. I_z across intermediate shocks. In this figure the dependence of allowed values of I_z upon the ratio of the upstream to downstream magnetic field, labeled as b_r/b_l, for $\bar{R} = 0.05$ is plotted. The $1 \rightarrow 3$ shocks may have any I_z in the shaded region in $-\frac{1}{2} \le b_r/b_l \le 0$; the $2 \rightarrow 3$ shocks can have two values of I_z (I_z^* and $-I_z^*$) in $-1 \le b_r/b_l \le -\frac{1}{2}$. (The plot is for positive I_z; a similar plot exists for negative I_z because of symmetry with respect to the change from b_z to $-b_z$.

is a one-parameter family of $1 \to 3$ structure solutions; (3) there is a pair of $2 \to 3$ intermediate shock solutions; (4) the other three transitions ($3 \to 1$, $3 \to 2$, and $2 \to 1$) do not have structure solutions.

3.4. THE GLOBAL INVARIANT I_z

The nonuniqueness of the intermediate shock solutions implies that simple specification of the upstream and downstream stationary points is not sufficient to specify a shock solution. The parameter labeling the $1 \to 3$ intermediate shock family is related to the out-of-plane magnetic flux. Since the $1 \to 3$ integral curves cannot intersect except at the stationary points, the curve C formed by the b_y axis and a given shock solution encloses a unique area A,

$$A = \frac{1}{2} \int_C (b_z db_y - b_y db_z) \tag{10}$$

Using (8a, b), we find that $A = \frac{1}{2R} F_{y0} I_z$, where $I_z = \int_{-\infty}^{\infty} b_z d\eta$ and the constant flux $F_{y0} = \alpha b_y^3 - s b_y$ upstream and downstream of the shock with s denoting shock speed. Thus the I_z parametrizes the family of $1 \to 3$ shock solutions. The quantity I_z can also be used to label the two $2 \to 3$ shock solutions. The two values of I_z for the $2 \to 3$ shock correspond to the upper and lower bounds of the $1 \to 3$ shocks. In the limit that the $2 \to 3$ shock approaches the $180°$ rotational discontinuity, where $s = \alpha b_{yu}^2$ and $b_{yd}^2 = b_{yu}^2$, I_z tends to ∞ as $\bar{R}/0$. Since b_z is bounded, this indicates an infinitely wide structure.

Figure 5 shows the allowed values of I_z obtained from numerical integration of the shock structure as a function of the ratio of the upstream to downstream magnetic field, $q = b_{yr}/b_{yl}$. The calculations are for $\bar{R} = 0.05$ and $b_{yl} = 1$. When $-1 \le q \le -\frac{1}{2}$, there is a pair of $2 \to 3$ intermediate shocks. In this case, $I_z = I_z^{+*}(q)$ and $I_z^{-*}(q)$, where $I_z^{\pm*}$ represent the upper and lower values of I_z. Because of symmetry with respect to the change from b_z to $-b_z$, $I_z^{-*}(q) = -I_z^{+*}(q)$. The upper branch corresponds to the $2 \to 3$ shocks in which the transverse magnetic field rotates in the left-hand sense, and the lower branch corresponds to the ones in which the magnetic field rotates in the right-hand sense. When $-\frac{1}{2} \le q \le 0$, there is a family of $1 \to 3$ intermediate shocks, whose I_z values are now bounded by $I_z = I_z^{+*}(q)$ and $I_z = I_z^{-*}(q)$.

In summary, the CKB theory shows that $1 \to 2$, $1 \to 3$ and $2 \to 3$ shocks posses structures, but not rotational discontinuities. This is a contrast from the classical MHD theory, where only $1 \to 2$ shocks and rotational discontinuities are allowed, while intermediate shocks are considered extraneous.

3.5. THE COPLANAR CKB RIEMANN PROBLEM

To discuss the interrelationship between the global evolution and structure of intermediate shocks, we consider the CKB Riemann problem. We solve numerically

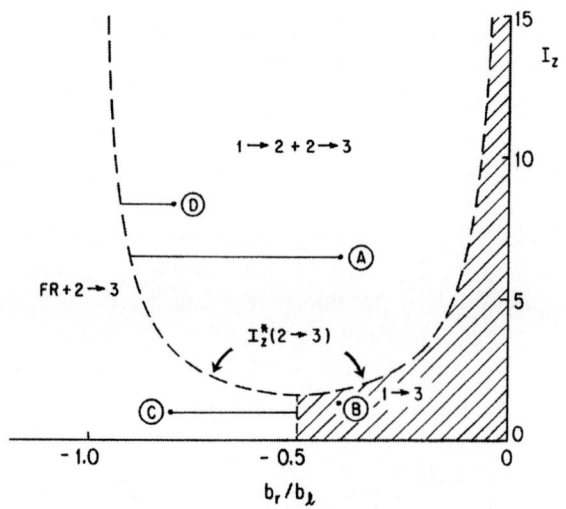

Figure 6. Summary of 180° Riemann problem. This figure shows the (I_z, b_r/b_l) parameter space in which the Riemann problem is specified, and characterizes the solutions in each region. The initial states of the four cases (A–D) presented in Figure 5 are marked by heavy dots. The solid lines indicate the evolution of the resulting 2 → 3 or 1 → 3 intermediate shocks.

for the time evolution of an initial field profile, $b_y(\eta)$ and $b_z(\eta)$, which are given by

$$b_y(\eta) = b(\eta) \cos \theta(\eta), \quad b_z(\eta) = b(\eta) \sin \theta(\eta) \tag{11a}$$

$$\theta(\eta) = \frac{1}{2}(\theta_l + \theta_r) - \frac{1}{2}(\theta_l - \theta_r) \tanh(\eta/\eta_w) \tag{11b}$$

$$b(\eta) = \frac{1}{2}(b_l + b_r) - \frac{1}{2}(b_l - b_r) \tanh(\eta/\eta_w). \tag{11c}$$

We choose $\alpha = \frac{1}{4}$, $\bar{R} = 0.05$, $b_{yl} = b_l = 1$, $b_{zl} = 0$, $\theta_l = 0$, and $b_0 = 1$, and various b_r, θ_r, and η_w. We integrate (6a, b) over a range L.

Let us define the transverse magnetic field moment, $\mathbf{I} = \int_L \mathbf{b} d\eta$. Integration of the CKB equation gives a relation

$$\frac{d\mathbf{I}}{dt} + \alpha[\mathbf{b}_r(b_r^2 - b_{T0}^2) - \mathbf{b}_l(b_l^2 - b_{T0}^2)] = 0 \tag{11}$$

which leads to an important distinction between the coplanar and noncoplanar Riemann problems. When the upstream and downstream magnetic field are co-planar, say in the $\pm y$ directions, then I_z is a global invariant of the coplanar Riemann problem. On the other hand, for the noncoplanar problem, I_z is not constant.

Here we summarize the results of the CKB coplanar Riemann problem. It was shown that the Riemann evolution of the coplanar case, where $b_{zl} = b_{zr} = 0$, can be predicted by specifying b_{yr}/b_{yl} together with I_z. Let b_r and b_l denote b_{yr} and

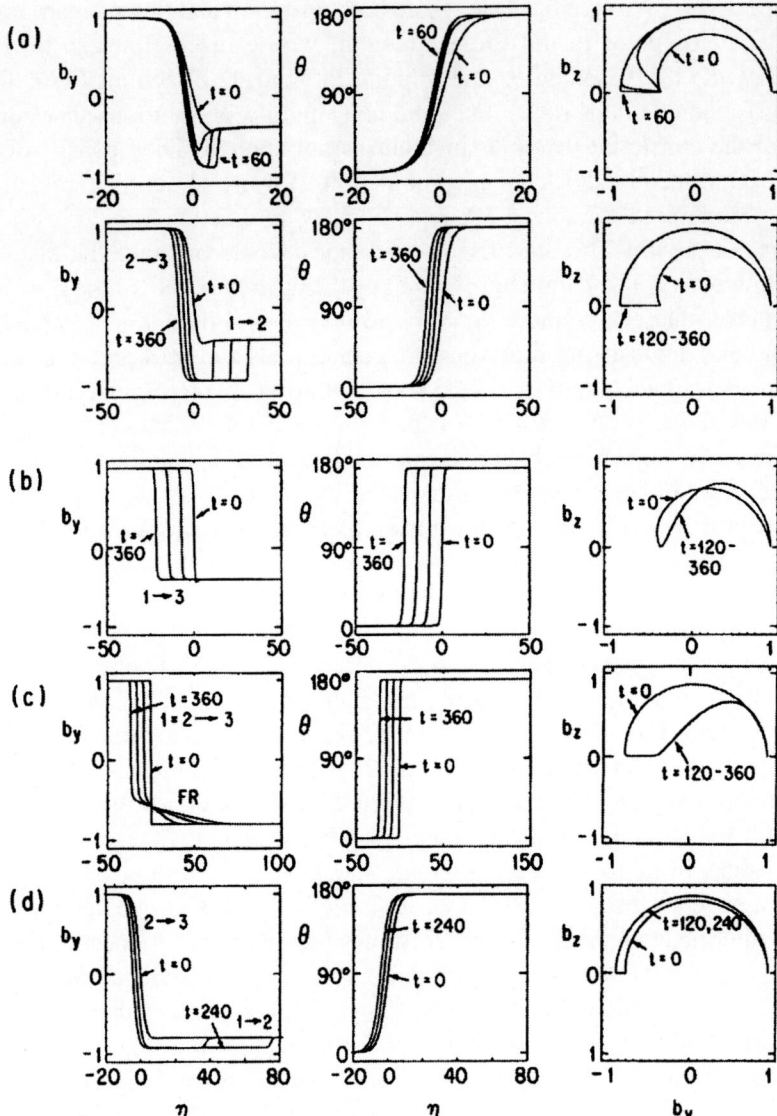

Figure 7. 180° Riemann problems. (a) shows the evolution of an initial state with $b_{yl} = 1$, $b_{yr} = -0.4$, and $I_z = 6.5$; the state evolves into a fast shock that separates from a $2 \rightarrow 3$ intermediate shock. (b) shows the evolution of an initial state with $b_{yl} = 1$, $b_{yr} = -0.4$, and $I_z = 1.3$; it evolves into a single $1 \rightarrow 3$ intermediate shock. (c) shows the evolution of an initial state with $b_{yl} = 1$, $b_{yr} = -0.8$, and $I_z = 1$; the state evolves into an attached fast rarefaction plus a $1 = 2 \rightarrow 3$ intermediate shock. (d) shows the evolution of an initial state with $b_{yl} = 1$, $b_{yr} = -0.8$, and $I_z = 5$; the state evolves into a fast shock that separates from a $2 \rightarrow 3$ intermediate shock. The top row of (a) shows the evolution at times $t = 0, 20, 40$, and 60. It shows that the final self-similar solution begin to emerge at about $t = 60$ when the separation between the $1 \rightarrow 2$ and $2 \rightarrow 3$ shocks is larger than their shock widths. The bottom row of (a), (b) and (c) show the evolutions for $t = 0$, 120, 240, and 360 and (d) shows the solution for $t = 0$, 120, and 240. They show that at large time when waves are well separated the solutions are self-similar as functions of η/t.

b_{yl}, respectively. We skip the case where both upstream and downstream transverse magnetic fields point in the same direction, whose discussion can be found in Kennel *et al.* (1990). When b_r and b_l point in opposite directions (180° Riemann problem), one can construct a one-parameter family of solutions that connect b_r and b_l if the constraint that I_z is invariant is not applied. They consist of either a $1 \rightarrow 3$ intermediate shock, or a combination of a fast shock or rarefaction with a $2 \rightarrow 3$ intermediate shock. (A fast shock or rarefaction wave plus a $1 \rightarrow 3$ shock is not possible, because the $1 \rightarrow 3$ shock would overtake the fast shock or rarefaction wave.) The parameter is given by the upstream state of the $2 \rightarrow 3$ shock. For instance, one can connect $b_l = 1$ and $b_r = -0.4$ by a $1 \rightarrow 3$ shock whose upstream and downstream transverse magnetic fields are -0.4 and 1, respectively. One can also connect them by a fast shock that brings the transverse magnetic field from -0.4 to b_m, $-0.5 \geq b_m > -1$, plus a $2 \rightarrow 3$ shock that changes b from b_m to 1. Since I_z is nonzero only across intermediate shocks, a unique solution is then obtained if the integral I_z across an isolated intermediate shock is the same as the initial I_z of the system. In the above example, a $1 \rightarrow 3$ shock is obtained if I_z is within the allowable range for its shock structure. A combination of a fast shock and a $2 \rightarrow 3$ is obtained if I_z is outside that range.

Figure 6 summarizes the results for a 180° Riemann problem. It shows the (I_z, b_r/b_l) parameter space together with the curve $I_z^*(2 \rightarrow 3)$. The types of solutions expected are indicated in the figure. In Figure 7 we give four examples for the evolution of the 180° Riemann problem. Their initial states are marked by $A-D$ in Figure 6 ; their evolutions of the resulting $2 \rightarrow 3$ or $1 \rightarrow 3$ intermediate shocks are also shown in Figure 6 by solid lines. Figure 7a shows the evolution of an initial state with $b_{yl} = 1$, $b_{yr} = -0.4$, and $I_z = 6.5$; the state evolves into a fast shock that separates from a $2 \rightarrow 3$ intermediate shock. Figure 7b shows the evolution of an initial state with $b_{yl} = 1$, $b_{yr} = -0.4$, and $I_z = 1.3$; it evolves into a single $1 \rightarrow 3$ intermediate shock. Figure 7c shows the evolution of an initial state with $b_{yl} = 1$, $b_{yr} = -0.8$, and $I_z = 1$; the state evolves into an attached fast rarefaction plus a $1 = 2 \rightarrow 3$ intermediate shock. Figure 7d shows the evolution of an initial state with $b_{yl} = 1$, $b_{yr} = -0.8$, and $I_z = 5$; the state evolves into a fast shock that separates from a $2 \rightarrow 3$ intermediate shock. The top row of Figure 7a shows the evolution at earlier times $t = 0, 20, 40$, and 60. It shows that the final self-similar solution begin to emerge at about $t = 60$ when the separation between the $1 \rightarrow 2$ and $2 \rightarrow 3$ shocks is larger than their shock widths. The bottom row of Figure 7a, Figure 7b, and 7c show the evolutions for $t = 0, 120, 240$, and 360 and Figure 7d shows the solution for $t = 0, 120$, and 240. They show that at large time when waves are well separated the solutions are self-similar as functions of η/t.

The above Riemann solutions differ from what one expects from the classical MHD theory. For an anti-parallel case (assuming $b_l > 0$ and $b_r < 0$), the solution according to the classical theory would include, if $b_l > |b_r|$, a fast shock that decreases b_y from b_{yr} to $-b_{yl}$, or, if $b_l < |b_r|$, a fast rarefaction wave that increases

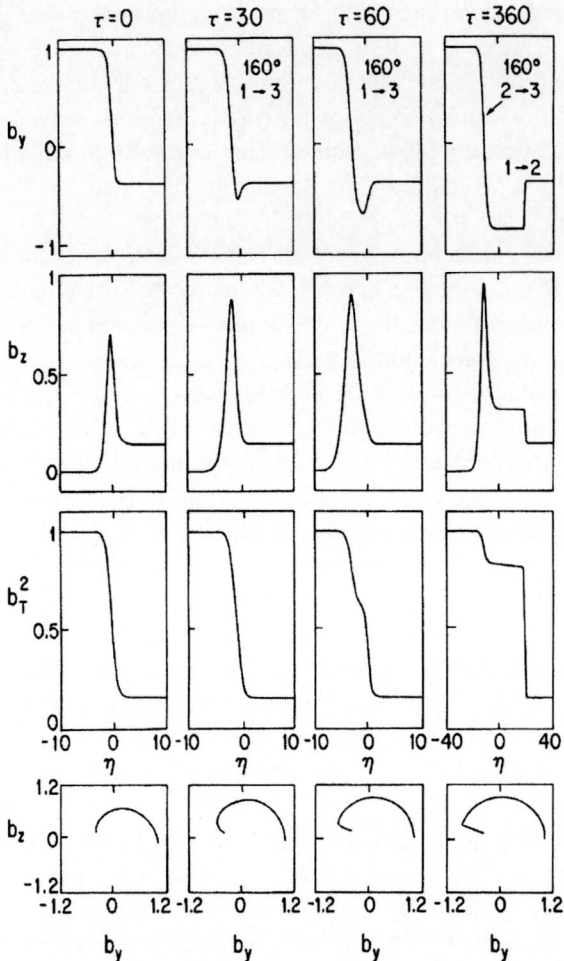

Figure 8. Noncoplanar Riemann problem: the evolution of a 160° field rotation. This system starts with a time-dependent $1 \rightarrow 3$ shock, then becomes a fast shock plus a time-dependent $2 \rightarrow 3$ intermediate shock.

b_y from b_{yr} to $-b_{yl}$, and a rotational discontinuity that changes b_y from $-b_{yl}$ to b_{yl}.

3.6. NONCOPLANAR RIEMANN PROBLEMS AND TIME-DEPENDENT INTERMEDIATE SHOCKS

In a coplanar 180° Riemann problem, b_y changes its sign across an intermediate shock in the final state. This is what one expects, because the intermediate shock is the only candidate that allows b_y to change sign and also possesses shock structure. What will happen when the system is not coplanar upstream and downstream? On one hand, we cannot use rotational discontinuities to rotate the field.

On the other hand, according to the Rankine–Hugoniot jump conditions, all the time-independent shocks, including intermediate shocks, are coplanar. Thus we expect a more complicated situation than in the coplanar case. To answer the above question, non-coplanar Riemann problems were solved numerically. The solutions of these non-coplanar problems are not self-similar in the variable η/t; new entities, which we called time-dependent intermediate shocks, are formed. These new shocks are time-dependent because they are non-coplanar, but they resemble shocks because both upstream and downstream intermediate character-istics converge into them; the magnetic field rotates by a fixed amount through these shocks. In the coplanar limit, these time-dependent shock-like structures go to time-independent intermediate shocks.

When the Riemann problem is not coplanar, I_z is not a global invariant, the solution will continue to evolve, and the b_z flux will continue to increase inside the time-dependent intermediate shock, which will then approach a broad rotational discontinuity as $t \to \infty$. As $t \to \infty$, the system will approach a fast shock plus a $2 \to 3$ time-dependent intermediate shock if $b_l > b_r$ and a fast rarefaction plus a fast shock plus $2 \to 3$ time-dependent intermediate shock if $b_l < b_r$. In the latter, the strength of the fast shock becomes very week at large t. Thus the flow structure at large t is similar to the result expected from the classical MHD theory, which predicts a solution that includes a fast shock (fast rarefaction) if $b_l > b_r$ ($b_l < b_r$) plus a rotational discontinuity. However, the rotational 'discontinuity' is infinitely wide in our dissipative system, whereas it was presumed to be thin in the classical theory of MHD shocks. The treatments also differ in the location of the discontinuity.

We now present a numerical example in detail. Figure 8 shows the evolution of a 160° rotation with $b_r = 0.4$ and $b_l = 1$, initialized with $\eta_w = 1$. The initial conditions are shown in the far-left column. We see that $b_{yl} = 1$, $b_{zl} = 0$ and $b_{yr} = -0.376$, $b_{zr} = 0.137$. The initial I_z is 1.15 across the transition layer, which is less than $I_z^* = 1.68$ for a $b_{yr}/b_{yl} = -0.4$ intermediate shock. The second, third, and fourth columns show $b_y(\eta)$, $b_z(\eta)$, $b_T^2(\eta)$, and $b_y - b_z$ hodograms, at $t = 30, 60$, and 360, respectively. If this system were coplanar, the initial state would evolve into a single $1 \to 3$ intermediate shock, because $I_z < I_z^*$. Indeed at $t = 30$ and 60, in the same time scale as the coplanar case, something like this occurs. However these structures are not intermediate shocks since the field rotation is 160°, not 180°. They do not satisfy the Rankine–Hugoniot jump condition, so they must be time-dependent. On the other hand, they are compressive because the fast and intermediate characteristics continue to converge into them. For these reasons, we call them time-dependent $1 \to 3$ intermediate shocks.

The total I_z of the system increases in time, see (11). This results in the increase of I_z across the 160° time-dependent $1 \to 3$ intermediate shock. As the I_z across the $1 \to 3$ shock increases above I_z^*, a fast shock followed by a $2 \to 3$ shock emerges, as the results at $t = 360$ show. The $2 \to 3$ shock at $t = 360$ is also time-dependent. It rotates the transverse magnetic field by 160°, and the intermediate characteristics still converge into it. The state between the $1 \rightarrow 2$ intermediate shock and the $2 \rightarrow 3$ intermediate shock is not constant in η and also evolves in time as we will show soon. (This differs from the coplanar case in which $I_z > I_z^*$, whose solution is self-similar in η/t and includes a $1 \rightarrow 2$ shock, followed by a constant state, and then by a $2 \rightarrow 3$ intermediate shock.) At large time, the system evolves into a fast shock plus a $2 \rightarrow 3$ time-dependent intermediate shock, which approaches the jump condition of a $160°$ rotational discontinuity (Wu and Kennel, 1992a, 1993) It has been shown that the time-dependent $2 \rightarrow 3$ shock has a well-defined unique structure. It has also been demonstrated that this time-dependent intermediate shock is an attractor of all the $160°$ Riemann problems, meaning that independent of the value of b_r in the Riemann problem, all field rotation occurs through the same $160°$ time-dependent shock (Wu and Kennel, 1993).

This $160°$ Riemann problem can be considered as a $180°$ $1 \rightarrow 3$ intermediate shock interaction with an intermediate wave that rotates the magnetic field by $20°$. Its evolution is, in a way, what Kantrowitz and Petschek (1996) have predicted based on the classical MHD theory. But there is a major difference. Since the shock frame fluid velocity is super-Alfvénic ahead and sub-Alfvénic behind the intermediate shock, the intermediate waves are convergent on the intermediate shock. Thus the intermediate wave cannot propagate upstream of the shock, neither can it stay behind the shock. According to the classical MHD theory, because shocks are considered as discontinuities and are planar, the out-of-plane magnetic field carried by the converging intermediate waves cannot exist within the intermediate shocks. Kantrowitz and Petschek thus concluded that once the intermediate wave (even with infinitesimal amplitude) interacts with the intermediate shock, the latter is expected to disintegrate instantly.

The numerical calculation shows that it takes a finite time for the $180°$ $1 \rightarrow 3$ intermediate shock to break up as a result of its interaction with an intermediate wave. This is because that the shock structure of the intermediate shock is included. Since I_z can be nonzero inside the shock structure, which means that intermediate waves can be trapped inside an intermediate shock. The interaction with an intermediate wave results in the change in the amount of I_z within the intermediate shock. The computation shows that the $1 \rightarrow 3$ shock begins to evolve only when the accumulated I_z is above I_z^*. In other words, an intermediate shock is stable as long as the perturbed amplitude is small, $I_z < I_z^*$.

3.7. IMPORTANCE OF TIME-DEPENDENT INTERMEDIATE SHOCKS

In a strict mathematical sense, there are no rotational discontinuities in the CKB system and there are no intermediate shocks when upstream and downstream fields are not coplanar. The time-dependent intermediate shock plays the role of connecting them together. It tends to an intermediate shock when the field rotation approaches $180°$ and it reaches a broad rotational wave when the flux within the shock layer is large compared to I_z. Since I_z is linearly related to $\bar{R}$, if the $\bar{R}$ scale length is small with respective to physical length of interest, the time-dependent intermediate shock may look thin in the physical system.

Acknowledgement

This work is supported in part by a NASA grant NAG 5-9111.

References

Akhiezer, A. I., Lubarski, G. J. and Polovin, R. V.: 1959, 'The Stability of Shock Waves in Magnetohydrodynamics', *Soviet Phys. – JETP* **8**, 507–511.

Brio, M. and Wu, C.C.: 1987, 'Characteristic Fields for the Equations of Magnetohydrodynamics', in B. Keyfitz and H. C. Kranzer (eds), *Nonstrictly Hyperbolic Conservation laws*, American Mathematical Society, Providence, R. I., pp. 19–23.

Brio, M. and Wu, C. C.: 1988, 'An Upwind Differencing Scheme for the Equations of Magneto-hydrodynamics', *J. Comp. Physics* **75**, 400–422.

Bona, J. L. and Schonbek, M. E.: 1985, 'Travelling-wave Solutions to the Korteweg–de Vries-Burgers Equation', *Proceedings of the Royal Society of Edinburgh, Section A (Mathematics)* **101**, 207–226.

Chu, C. K. and Wu, C. C.: 1991, 'Magnetohydrodynamic Shock Waves Revisited', in B. Engquist and B. Gustrafsson (eds), *The Third International Conference on Hyperbolic Problems, Theory, Numerical Methods and Application*, Chartwell-Bratt, p. 241.

Cohen, R. and Kulsrud, R.: 1974, 'Nonlinear Evolution of Parallel-propagating Hydromagnetic Waves', *Phys. Fluids* **17**, 2215–2225.

Friedrichs, K. O.: 1954, 'Non-linear Wave Motion in Magnetohydrodynamics', Los Alamos Rept. **2105**.

Freustühler, H.: 1992, 'Hyperbolic Systems of Conservation Laws with Rotationally Equivalent Flux Function', *Matematica Aplicada e Computacional.* **11**, 167–188.

Freistuhler, H. and Liu, T. P.: 1993, 'Nonlinear Stability of Overcompressive Shock Waves in a Rotationally Invariant System of Viscous Conservation Laws', *Comm. Math. Phys.* **153**, 147–158.

Germain, P.: 1960, 'Shock Waves and Shock-wave Structure in Magneto-fluid Dynamics', *Rev. Mod. Phys.* **32**, 951–958.

Jacobs, D., McKinney, B. and Shearer, M.: 1995, 'Travelling Wave Solutions of the Modified Korteweg–de Vries-Burgers Equation', *J. Diff. Equations* **116**, 448–467.

Jeffrey, A. and Taniuti, T.: 1964, *Nonlinear Wave Propagation*. Academic.

Kantrowitz, A. and Petschek, H.: 1966, 'MHD Characteristics and Shock Waves', in W. B. Kunkel (ed.), *Plasma Physics in Theory and Application*, McGraw-Hill, p. 148.

Kennel, C. F., Blandford, R. D. and Wu, C. C.: 1990, 'Structure and Evolution of Small Amplitude Intermediate Shock Waves', *Phys. Fluids* **B2**, 253–269.

Polovin, R. V.: 1961, 'Shock Waves in Magnetohydrodynamics', *Soviet Phys. Uspekhi* **3**, 677–588.

Rogister, A.: 1971, 'Parallel Propagation of Nonlinear Low-frequency Waves in High-beta Plasma', *Phys. Fluids* **14**, 2733–2739.

Ruderman, M. S.: 1989, 'Structure and Stability of Quasiparallel Small-amplitude Magnetohydro-dynamic Shocks', *Izv. Akad. Nauk SSSR, Mekh. Zhidk. Gaza* **4**, 153–160; *Fluid Dynamics* **24**, 618–628 (English translation).

Wu, C. C.: 1987, 'On MHD Intermediate Shocks', *Geophy. Res. Lett.* **14**, 668–671.

Wu, C. C.: 1988a, 'The MHD Intermediate Shock Interaction with an Intermediate Wave: Are Intermediate Shocks Physical?', *J. Geophys. Res.* **93**, 987–990.

Wu, C. C.: 1988b, 'Effects of Dissipation on Rotational Discontinuities' *J. Geophys. Res.* **93**, 3969–3982.

Wu, C. C.: 1990, 'Formation, Structure and Stability of MHD Intermediate Shocks', *J. Geophys. Res.* **95**, 8149–8175.

Wu, C. C.: 1991, 'New Theory of MHD Shock Waves', in M. Shearer (ed.), *Viscous Profiles and Numerical Methods for Shock Waves*, Chapter 17, SIAM, pp. 209–236.

Wu, C. C.: 1995, 'Magnetohydrodynamic Riemann Problem and the Structure of the Magnetic Reconnection Layer', *J. Geophys. Res.* **100**, 5579–5598.

Wu, C. C. and Kennel, C. F.: 1992a, 'Evolution of Small-amplitude Intermediate Shocks in a Dissipative and Dispersive System', *J. Plasma Phys.* **47**, 85–109.

Wu, C. C. and Kennel, C. F.: 1992b, 'Structure and Evolution of Time-dependent Intermediate Shocks', *Phys. Rev. Lett.* **68**, 56–59.

Wu, C. C. and Kennel, C. F.: 1992c, 'Structural Relations for Time-dependent Intermediate Shocks', *Geophy. Res. Lett.* **19**, 2087–2090.

Wu, C. C. and Kennel, C. F.: 1993, 'The Small Amplitude Magnetohydrodynamic Riemann Problem', *Phys. Fluids B (Plasma Physics)* **5**, 2877–2886.

VI: COMPLEX/INTELLIGENT SYSTEMS

COMPLEXITY, FORCED AND/OR SELF-ORGANIZED CRITICALITY, AND TOPOLOGICAL PHASE TRANSITIONS IN SPACE PLASMAS

TOM CHANG[1], SUNNY W. Y. TAM[1], CHENG-CHIN WU[2] and
GIUSEPPE CONSOLINI[3]

[1] Center for Space Research, Massachusetts Institute of Technology, Cambridge, MA, U.S.A.
[2] Department of Physics and Astronomy, University of California, Los Angeles, CA, U.S.A.
[3] Istituto di Fisica dello Spazio Interplanetario, Consiglio Nazionale delle Ricerche, Rome, Italy

Abstract. The first definitive observation that provided convincing evidence indicating certain turbulent space plasma processes are in states of 'complexity' was the discovery of the apparent power-law probability distribution of solar flare intensities. Recent statistical studies of complexity in space plasmas came from the AE index, UVI auroral imagery, and in-situ measurements related to the dynamics of the plasma sheet in the Earth's magnetotail and the auroral zone.

In this review, we describe a theory of dynamical 'complexity' for space plasma systems far from equilibrium. We demonstrate that the sporadic and localized interactions of magnetic coherent structures are the origin of 'complexity' in space plasmas. Such interactions generate the anomalous diffusion, transport, acceleration, and evolution of the macroscopic states of the overall dynamical systems.

Several illustrative examples are considered. These include: the dynamical multi- and cross-scale interactions of the macro-and kinetic coherent structures in a sheared magnetic field geometry, the preferential acceleration of the bursty bulk flows in the plasma sheet, and the onset of 'fluctuation induced nonlinear instabilities' that can lead to magnetic reconfigurations. The technique of dynamical renormalization group is introduced and applied to the study of two-dimensional intermittent MHD fluctuations and an analogous modified forest-fire model exhibiting forced and/or self-organized criticality [FSOC] and other types of topological phase transitions.

Key words: complexity, magnetotail, plasma sheet

1. Introduction

Research activity in space plasma physics is now arriving at an interesting juncture that becomes apparent when looking back at what has been accomplished in the past and looking forward to what will be required in the next decade. In this regard, it is noted that considerable observational and theoretical attention has been devoted towards the understanding of local, point observations of space plasma phenomena as they have been identified on US/European/Japanese research satellites. However, it is becoming clear that important questions that will be receiving attention in the coming years (particularly with the successful launches of the CLUSTER II and IMAGE satellites) are addressed toward global and multiscale issues: questions of energy and momentum transport within the magnetosphere,

Space Science Reviews **107**: 425–445, 2003.
© 2003 *Kluwer Academic Publishers.*

ionosphere and the Sun-Earth connection region, and of the nature of the particle populations, their source, entry, energization, diffusion, and ultimate loss from the systems. Thus, observations are becoming multi-spacecraft and/or multi-point in scope, and theoretical models are likewise being forced to confront issues of nonlocality and 'complexity'.

In this review, we discuss a number of such space plasma processes of complexity, with special emphasis on the dynamics of the magnetotail and its substorm behavior. We demonstrate that the sporadic and localized interactions of magnetic coherent structures arising from plasma resonances are the origin of 'complexity' in space plasmas. Such interactions, which generate the anomalous diffusion, transport and evolution of the macroscopic state variables of the overall dynamic system, may be modeled by a triggered localized chaotic growth equation of a set of relevant order parameters. The dynamics of such intermittent processes would generally pave the way for the global system to evolve into a 'complex' state of long-ranged interactions of fluctuations, displaying the phenomenon of forced and/or self-organized criticality (FSOC). The coarse-grained dissipation due to the intermittent fluctuations can also induce 'fluctuation-induced nonlinear instabilities' that can, in turn, reconfigure the topologies of the background magnetic fields.

The organization of this paper is as follows. In Section 2, we discuss the origin of the stochastically distributed coherent magnetic structures in the Earth's magnetotail. In Section 3, we consider the preferential acceleration of the coherent magnetic structures in the neutral sheet region and their relevance to the interpretation of the observed fast bursty bulk flows (BBF). In Section 4, we consider the localized merging processes of the coherent structures and their connection to the observed localized, sporadic reconnection signatures in plasmas (e.g., in the Earth's magnetotail). The concept and implications of the phenomenon of FSOC are then introduced. These ideas can lead naturally to the power-law scaling of the probability distributions and fractal spectra of the intermittent turbulence associated with the coherent structures. In Section 5, we provide some convincing arguments for intermittency and FSOC in the magnetospheric plasma sheet. In Section 6, the concept of invariant scaling, the dynamic renormalization group, and topological phase transitions are discussed. These are then related to the multifractal scaling across various physical scales and the fluctuation-induced nonlinear instabilities that may lead to the triggering of global magnetic field reconfigurations such as substorms. In Section 7, simple phenomenological and analogous intermittency models are described. The conclusion section then follows.

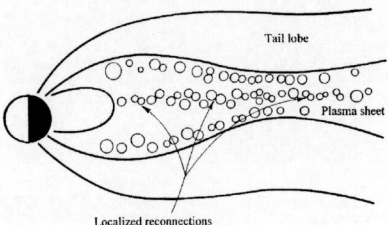

Figure 1. Cross-sectional view of flux tubes in the plasma sheet of the Earth's magnetotail.

2. Stochastically Distributed Coherent Structures in the Space Plasma Environment

In situ observations indicate that the dynamical processes in the plasma environment (e.g., the magnetotail region (Angelopoulos *et al.*, 1996; Lui, 1996; Lui, 1998; Nagai *et al.*, 1998)), generally entail localized intermittent processes and anomalous global transports. It was suggested by Chang (1998a, 1998b, 1998c; 1999) that instead of considering the turbulence as a mixture of interacting waves, such type of patchy intermittency could be more easily understood in terms of the development, interaction, merging, preferential acceleration and evolution of coherent magnetic structures.

Most field theoretical discussions begin with the concept of propagation of waves. For example, in the MHD formulation, one can combine the basic equations and express them in the following propagation forms:

$$\rho d\mathbf{V}/dt = \mathbf{B} \cdot \nabla \mathbf{B} + \cdots , \quad d\mathbf{B}/dt = \mathbf{B} \cdot \nabla \mathbf{V} + \cdots \tag{1}$$

where the ellipses represent the effects of the anisotropic pressure tensor, the compressible and dissipative effects, and all notations are standard. Equation (1) admits the well-known Alfvén waves. For such waves to propagate, the propagation vector $\mathbf{k}$ must contain a field-aligned component, i.e., $\mathbf{B} \cdot \nabla \rightarrow i\mathbf{k} \cdot \mathbf{B} \neq 0$. However, at sites where the parallel component of the propagation vector vanishes (the resonance sites), the fluctuations are localized. That is, around these resonance sites (usually in the form of curves), it may be shown that the fluctuations are held back by the background magnetic field, forming coherent magnetic structures in the form of flux tubes (Chang, 1998a, 1998b, 1998c; 1999). For the neutral sheet region of the magnetotail, these structures are essentially current filaments in the cross-tail direction (Figure 1). The results of a 2D MHD simulation showing such coherent structures are given in Figure 2.

Generally, there are various types of propagation modes (whistler modes, lower hybrid waves, etc.) in a space plasma. Thus, we envision a corresponding number of different types of plasma resonances and associated coherent magnetic structures that typically characterize the dynamics of a plasma medium under the influence of a background magnetic field.

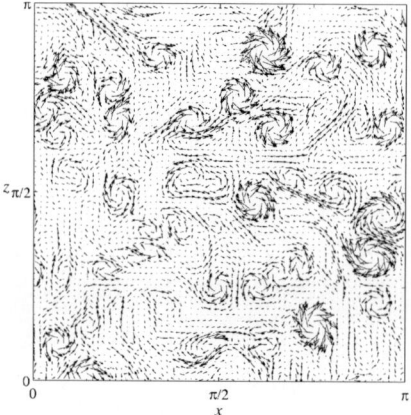

Figure 2. Two-dimensional MHD coherent structures generated by initially randomly distributed current filaments. Shown are the magnetic field vectors.

Such coherent structures may take on the forms of convective structures, propagating nonlinear solitary waves, pseudo-equilibrium configurations, and other varieties. Some of them may be more stable than others. These structures, however, generally are not purely laminar entities as they are composed of bundled fluctuations of all frequencies. Because of the very nature of the physics of complexity, it will be futile to attempt to evaluate and/or study each of these infinite varieties of structures; although some basic understanding of each type of these structures will generally be helpful in the comprehension of the full complexity of the nonlinear plasma dynamics.

These coherent structures will wiggle, migrate, deform and undergo different types of motions (including preferential acceleration), i.e., anisotropic stochastic randomization, under the influence of the local and global plasma and magnetic topology. It is this type of stochastic evolution and interaction of the coherent structures that characterize the dynamics of turbulent plasmas, not plane waves.

3. Preferential Acceleration of BBF

Consider the coherent magnetic structures discussed in the previous section, which in the neutral sheet region of the plasma sheet are essentially filaments of concentrated currents in the cross-tail direction. We expect that, in a sheared magnetic field that typically exists in the region under consideration, these multiscale fluctuations can supply the required coarse-graining dissipation that can produce nonlinear instabilities leading to 'X-point-like' structures of the averaged magnetic field lines. For example, for a sheared magnetic field $B_x(z)$, upon the onset of such fluctuation-induced nonlinear instabilities, the averaged magnetic field will generally acquire a z-component. Let us choose x as the Earth-magnetotail direction (positive toward the Earth) and y in the cross-tail current direction. Then the deformed magnetic

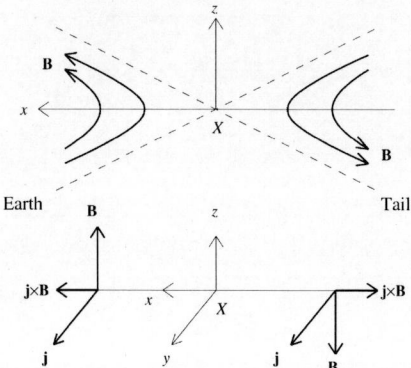

Figure 3. Schematic of a mean field X-point magnetic field geometry

field geometry after the development of a mean field 'X-point' magnetic structure will generally have a positive (negative) B_z component earthward (tailward) of the X-point. Near the neutral sheet region, the Lorentz force will therefore preferentially accelerate the coherent structures (which are essentially current filaments with polarities primarily pointing in the positive y-direction) earthward if they are situated earthward of the X-point and tailward if they are situated tailward of the X-point (Figure 3). These results would therefore match the general directions of motion of the observed BBF in the magnetotail (Nagai *et al.*, 1998).

We have performed preliminary two-dimensional numerical simulations to verify these conjectures. The simulations are based on a compressible MHD model that has been used by us in our previous studies of coherent magnetic structures (Wu and Chang, 2000a, 2000b; 2001; Chang and Wu, 2002; Chang *et al.*, 2002).

In the example, we have considered the motions of the coherent magnetic structures that developed in a sheared magnetic field upon the initial introduction of random magnetic fluctuations. These structures were generally aligned in the X-direction near the neutral line, $z = 0$, after some elapsed time. A positive B_z was then applied. It can be seen (from Figure 4; top panel: magnetic fields, bottom panel: flow vectors) that, after some additional elapsed time, the coherent structures (mostly oriented by currents in the positive y-direction), are generally accelerated in the positive x-direction; with one exception where a pair of magnetic structures with oppositely directed currents effectively canceled out the net effect of the Lorentz force on these structures. We have plotted the x-component of the flow velocities v_x due to the cumulative effect of the Lorentz (and pressure) forces acting on the flow (and in particular the coherent structures) after some duration of time has elapsed (Figure 5). We note that there are a number of peaks and valleys in the 3-dimensional display, with peak velocities nearly approaching that of the Alfvén speed mimicking the fast BBF that were observed in the magnetotail. It is to be noted that an individual BBF event may be composed of one, two, or several coherent magnetic structures.

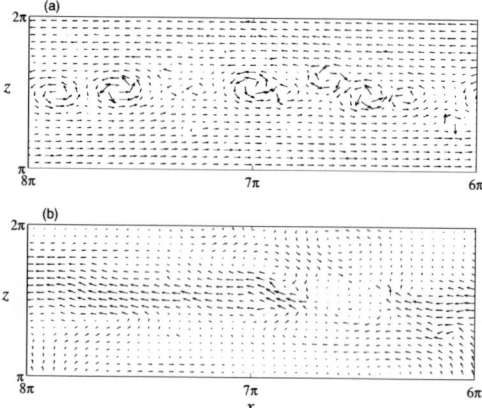

Figure 4. Effect of the Lorentz force on the motions of the coherent structures in a sheared magnetic field due to the application of a uniform field B_z. The magnetic fields (a) and flow velocities (b) are plotted in a domain $6\pi \leq x \leq 8\pi$ and $\pi \leq z \leq 2\pi$. The maximum velocity in (b) is 0.006.

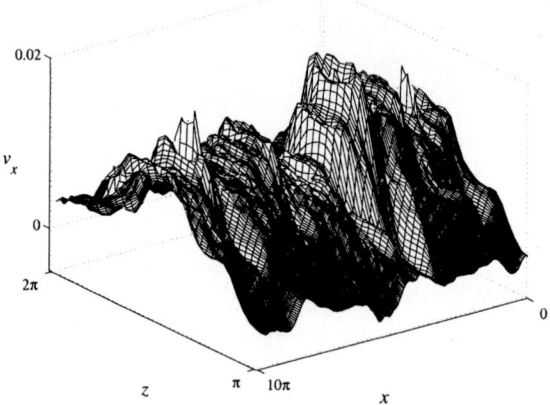

Figure 5. A 3D perspective plot of v_x. Its peak velocities are about 0.02, nearly approaching that of the Alfvén speed (about 0.025 based on $|B| = 0.025$). The maximum velocity is about 3 times higher than that at earlier time shown in Figure 4.

In Figure 6, we demonstrate in a self-consistent picture of what might occur near the neutral sheet region of the magnetotail. We injected randomly distributed magnetic fluctuations in a sheared magnetic field. The magnetic field geometry then underwent a fluctuation-induced nonlinear instability producing an X-point like mean field magnetic structure. The coherent structures (aligned in the neutral sheet region) are subsequently accelerated by the induced B_z away from the X-point in both the positive and negative x-directions. Contour plots of v_x after some elapsed time clearly indicate such effects.

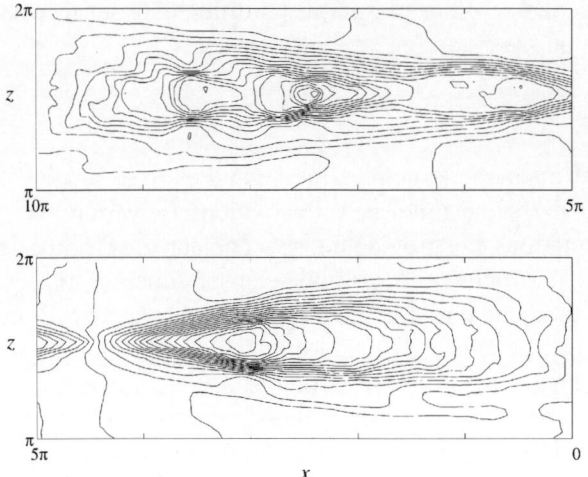

Figure 6. Contour plot of v_x in an X-point mean field magnetic geometry.

4. Localized Reconnections, FSOC and Power-Law Scaling

As the coherent magnetic structures approach each other, randomly or due to external forcing, they may merge or scatter. Near the neutral sheet region, most of these structures carry currents of the same polarity (in the cross-tail direction). As two current filaments of the same polarity migrate toward each other, a strong current sheet is generated. The instabilities and turbulence generated by this strong current sheet can then initiate the merging of these structures. The results of such merging processes might be the origins of the signatures of localized reconnection processes detected by ISEE, AMPTE, and other spacecraft. The observed localized reconnection signatures to date seem to take place mainly in domain sizes comparable to that of the ion gyroradius. Thus, very probably, most of these processes are influenced by microscopic kinetic effects. During these dynamic processes, the ions can approximately be assumed to be unmagnetized and the electrons strongly magnetized, and the plasma nearly collisionless. This can lead to electron-induced Hall currents. With a general magnetic geometry, whistler fluctuations may usually be generated.

Now, in analogy to the Alfvénic resonances, singularities of $k_\| = \mathbf{k} \cdot \mathbf{B} = 0$ can develop at which whistler fluctuations cannot propagate. These 'whistler resonances' can then provide the nuclear sites for the emergence of coherent whistler magnetic structures, which are the analog of the coherent Alfvénic magnetic structures but with much smaller scale sizes. The intermittent turbulence resulting from the intermixing and interactions of the coherent whistler structures in the intense current sheet region can then provide the coarse-grain averaged dissipation that allows the filamentary current structures to merge, interact, or breakup. In addition to the above scenarios, other plasma instabilities may set in when conditions are favorable to initiate the merging and interactions.

4.1. FSOC

As the coherent magnetic structures merge and evolve, larger coherent structures are formed. At the same time, new fluctuations of various sizes are generated. These new fluctuations can provide the new nuclear sites for the emergence of new coherent magnetic structures. After some elapsed time, we expect the distributions of the sizes of the coherent structures to encompass nearly all observable scales. It has been argued by Chang (1992; 1998a–1998c; 1999) and briefly reviewed in Section 6 that, when conditions are favorable, a state of dynamic criticality (FSOC) might be approached. At such a state, the structures take on all scale sizes with a power-law probability distribution of the scale sizes of the fluctuations, as well as power-law frequency (ω) and mode number (k) spectra of the correlations of the associated fluctuations. Analyses of existing observations in the intermittent turbulence region of the magnetotail and those conjectured from the AE index seem to confirm such predictions (Consolini, 1997; Lui, 1998; Angelopoulos et al., 1999; Lui et al., 2000; Uritsky et al., 2002). In addition, these multiscale coherent structures may render the coarse-grained dissipation that sometimes provides the seeds for the excitation of nonlinear instabilities leading to, for example, the onset of substorms in the Earth's magnetosphere.

5. Observational Evidences for FSOC and Intermittency in Space Plasmas

The first definitive observation that provided convincing evidence indicating certain turbulent space plasma processes are in states of forced and/or self-organized criticality (FSOC) was the discovery of the apparent power-law probability distribution of solar flare intensities (Lu, 1995). Recent statistical studies of complexity in space plasmas came from the AE index, UVI auroral imagery, and in-situ measurements related to the dynamics of the plasma sheet in the Earth's magnetotail. For example, the power-law spectra of AE burst occurrences as a function of AE burst strength has provided an important indication that the magnetosphere system is generally in a state of FSOC (Consolini, 1997, 2002). More recent evaluations of the local intermittency measure (LIM) (Farge et al., 1990) using the Morlett wavelet (Figure 7, from Consolini and Chang (2001)) clearly identified the burst contributions to the AE index, which appear as coherent time-frequency structures. These results strongly suggest that the Earth's plasma sheet, particularly during substorms times, is bursty and intermittent. Some of the salient features of the complexity studies using the UVI imagery and in-situ flow measurements in the plasma sheet region verifying these suggestions are briefly described below.

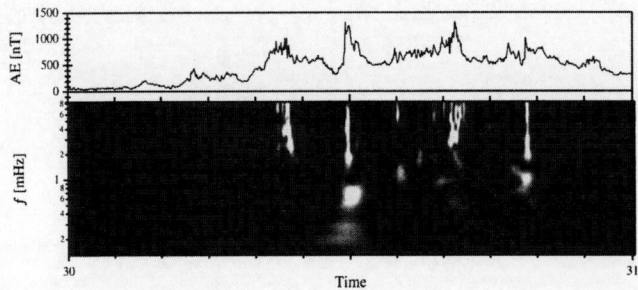

Figure 7. AE index behavior and the LIM measure for the substorm occurred on October 30, 1978.

5.1. AURORAL UVI IMAGES

UVI images provide detailed information on the dynamics of spatially distributed magnetotail activity covering extended observation periods. It has been shown that the positions of auroral active regions in the nighttime magnetosphere are correlated with the position of the plasma sheet instabilities (Fairfield *et al.*, 1999; Lyons *et al.*, 1999; Sergeev *et al.*, 1999; Ieda *et al.*, 2001; Nakamura *et al.*, 2001a, 2001b), whereas timing of the auroral disturbances provides good estimates for both small-scale isolated plasmoid releases (Ieda *et al.*, 2001) and for the global-scale substorm onset times (Germany *et al.*, 1998; Newell *et al.*, 2001).

Recently, Lui *et al.* (2000) used global auroral UVI imagery from the POLAR satellite to obtain the statistics of size and energy dissipated by the magnetospheric system as represented by the intensity of auroral emission on a world-wide scale. They found that the internal relaxations of the magnetosphere statistically follow power laws that have the same index independent of the overall level of activity. The analysis revealed two types of energy dissipation: those internal to the magnetosphere occurring at all activity levels with no intrinsic scale, and those associated with active times corresponding to global energy dissipation with a characteristic scale (Chapman *et al.*, 1998).

More recently, Uritsky *et al.* (2002) have performed an extensive analysis of the probability distributions of spatiotemporal magnetospheric disturbances as seen in POLAR UVI images of the nighttime ionosphere. This statistical study indicated stable power-law forms for both the probability distributions of the integrated size and energy of all 12300 auroral image events.

5.2. EVIDENCE OF INTERMITTENCY AND POWER-LAW BEHAVIOR OF BURSTY BULK FLOW (BBF) DURATIONS IN EARTH'S PLASMA SHEET

Statistical analyses based on in-situ data collected by the GEOTAIL satellite have yielded convincing proof of intermittent turbulence with power-law correlations (Angelopoulos *et al.*, 1999). It was shown that the magnetotail is generally in a bi-modal state: nearly stagnant, except when driven turbulent by transport efficient fast flows. Figure 8 displays the probability density functions (PDF) of the X-

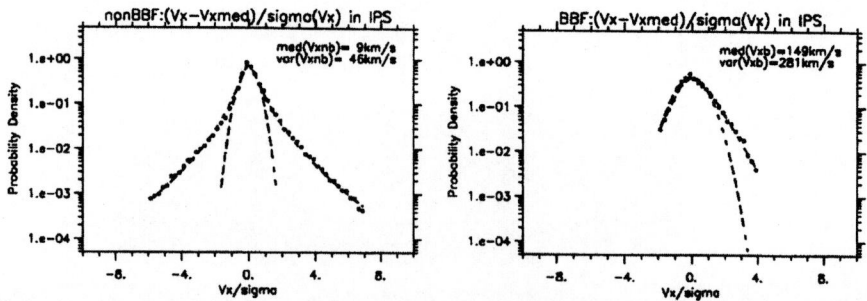

Figure 8. Probability distributions of the non-BBF and BBF X-component of the normalized flows along with best fits of Gaussian and Castaing functions (Angelopoulos *et al.*, 1999).

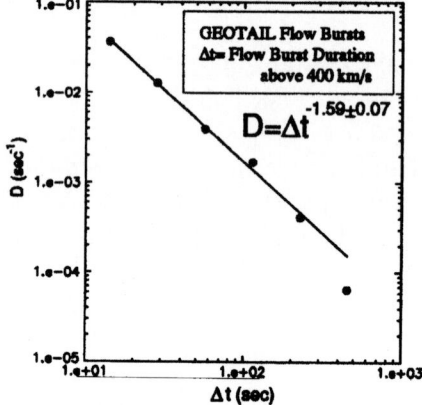

Figure 9. Probability density of BBF bursts in the plasma sheet (Angelopoulos *et al.*, 1999).

component of the inner plasma sheet flows for both the bursty bulk flow (BBF) and non-BBF populations. Both distributions are clearly non-Gaussian and both can be fitted nicely with the Castaing *et al.* (1990) distributions; thus, indicating intermittency for both components of flow in the plasma sheet.

Figure 9 shows the probability density function of the flow magnitude durations above 400 km s^{-1} in the inner plasma sheet (Angelopoulos *et al.*, 1999). The plasma sheet was selected on the basis of plasma pressure ($P_i > 0.01nPa$) but very similar results are obtained by confining the data in the near-neutral sheet region using plasma beta ($\beta_i > 0.5$). A power law is clearly indicated. The power-law behavior with spectral index ~ -1.6 remains when the velocity magnitude threshold is changed to a lower or higher value, and when different spatial regions of the magnetotail plasma sheet in the X-Y plane are considered, indicating that the results are quite robust. The results for linear binning in flow duration, Δt, are identical to the ones from logarithmic binning presented in the figure. Thus, there is strong indication that the energy dissipation in the magnetotail adheres to the behavior expected from FSOC.

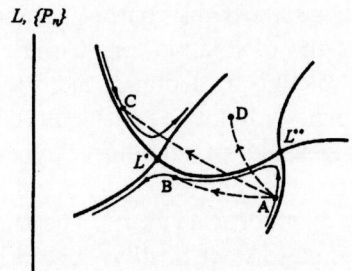

Figure 10. Renormalization-group trajectories and fixed points.

6. Invariant Scaling and Topological Phase Transitions

In the previous section, we mentioned the possibility of the existence of 'complex' topological states that can exhibit the characteristic phenomenon of dynamic criticality similar to that of equilibrium phase transitions. By 'complex' topological states we mean magnetic topologies that are not immediately deducible from the elemental (e.g., MHD and/or Vlasov) equations (Consolini and Chang, 2001). Below, we shall briefly address the salient features of the analogy between topological and equilibrium phase transitions. A thorough discussion of these ideas may be found in Chang (1992, 1999, 2001, and references contained therein).

For nonlinear stochastic systems near criticality, the correlations among the fluctuations of the random dynamical fields are extremely long-ranged and there exist many correlation scales. The dynamics of such systems are notoriously difficult to handle either analytically or numerically. On the other hand, since the correlations are extremely long-ranged, it is reasonable to expect that the system will exhibit some sort of invariance under scale transformations. A powerful technique that utilizes this invariance property is the technique of the dynamic renormalization group (Chang *et al.*, 1978, 1992, and references contained therein). As it is described in these references, based on the path integral formalism, the behavior of a nonlinear stochastic system far from equilibrium may be described in terms of a 'stochastic Lagrangian L'. Then, the renormalization-group (coarse-graining) transformation may be formally expressed as:

$$\partial L/\partial l = RL \tag{2}$$

where R is the renormalization-group (coarse-graining) transformation operator and l is the coarse-graining parameter for the continuous group of transformations. It will be convenient to consider the state of the stochastic Lagrangian in terms of its parameters $\{P_n\}$. Equation (2), then, specifies how the Lagrangian, L, flows (changes) with l in the affine space spanned by $\{P_n\}$, Figure 10.

Generally, there exist fixed points (singular points) in the flow field, where $dL/dl = 0$. At a fixed point (L^* or L^{**} in Figure 10), the correlation length should not be changing. However, the renormalization-group transformation requires that all length scales must change under the coarse-graining procedure. Therefore, to satisfy both requirements, the correlation length must be either infinite or zero. When it is at infinity, the system is by definition at criticality. The alternative trivial case of zero correlation length will not be considered here.

To study the stochastic behavior of a nonlinear dynamical system near a particular criticality (e.g., the one characterized by the fixed point L^*), we can linearize the renormalization-group operator R about L^*. The mathematical consequence of this approximation is that, close to criticality, certain linear combinations of the parameters that characterize the stochastic Lagrangian L will correlate with each other in the form of power laws. This includes, in particular, the (k, ω), i.e. mode number and frequency, spectra of the correlations of the various fluctuations of the dynamic field variables. In addition, it can be demonstrated from such a linearized analysis that generally only a small number of (relevant) parameters are needed to characterize the stochastic state of the system near criticality [i.e., low-dimensional behavior; see Chang (1992)].

6.1. SYMMETRY BREAKING AND TOPOLOGICAL PHASE TRANSITIONS

As the dynamical system evolves in time (autonomously or under external forcing), the state of the system (i.e., the values of the set of the parameters characterizing the stochastic Lagrangian, L) changes accordingly. A number of dynamical scenarios are possible. For example, the system may evolve from a critical state A (characterized by L^{**}) to another critical state B (characterized by L^*) as shown in Figure 10. In this case, the system may evolve continuously from one critical state to another. On the other hand, the evolution from the critical state A to critical state C as shown in Figure 10 would probably involve a dynamical instability characterized by a first-order-like topological phase transition because the dynamical path of evolution of the stochastic system would have to cross over a couple of topological (renormalization-group) separatrices. Alternatively, a dynamical system may evolve from a critical state A to a state D (as shown in Figure 10) which may not be situated in a regime dominated by any of the fixed points; in such a case, the final state of the system will no longer exhibit any of the characteristic properties that are associated with dynamic criticality. Alternatively, the dynamical system may deviate only moderately from the domain of a critical state characterized by a particular fixed point such that the system may still display low-dimensional scaling laws, but the scaling laws may now be deduced from straightforward dimensional arguments. The system is then in a so-called mean-field state. (For general references of symmetry breaking and nonlinear crossover, see Chang and Stanley (1973); Chang et al., (1973a, 1973b); Nicoll et al. (1974, 1976).)

7. Modeling of Dynamic Intermittency

7.1. TRIGGERED LOCAL ORDER-DISORDER TRANSITIONS

As noted in the previous section, the 'complex' behavior associated with the intermittent turbulence in magnetized plasmas may generally be traced directly to the sporadic and localized interactions of magnetic coherent structures. Coherent structures may merge and form more energetically favorable configurations. They may also become unstable (either linearly or nonlinearly) and bifurcate into smaller structures. The origins of these interactions may arise from the effects of the various MHD and kinetic (linear and nonlinear) instabilities and the resultant finer-scale turbulences.

Such localized sporadic interactions may be modeled (Chang and Wu, 2002) by the triggered (fast) localized chaotic growth of a set of relevant order parameters, $O_i (i = 1, 2, \cdots, N)$:

$$\partial O_i / \partial t = \psi_i(\mathbf{O}, \mathbf{P}; c_1, c_2 \cdots c_n; \tau_1, \tau_2 \cdots \tau_n), \tag{3}$$

where 'ψ_i' are functionals of $(\mathbf{O}, \mathbf{P})$, $\mathbf{P}(\mathbf{x}, t) = P_j(j = 1, 2, \cdots, M)$ are the state variables (or control parameters), $(\mathbf{x}, t)$ are the spatial and temporal variables, the c_n's are a set of triggering parameters and the τ's are the corresponding relaxation time scales characterizing the localized intermittencies, which, in general, are much smaller than those characterizing the evolutionary time scales for the state variables.

For the special case of a single one-dimensional order parameter O, let us assume that the driving terms may be expressed in terms of a real-valued state function (the local configurational free energy), $F(O, \mathbf{P}; c_1, c_2 \cdots c_n; \tau_1, \tau_2 \cdots \tau_n)$, and a noise term γ, such that

$$\partial O / \partial t = -\partial F(O)/\partial O + \gamma. \tag{4}$$

Generally, the topology of the value of the state function F will contain valleys and hills in the real space spanned by the order parameters O at each given $(\mathbf{x}, t)$. If F is real, continuous with continuous derivatives, then we may visualize F locally in terms of a real polynomial function of O. The topology of the state function $F(O)$ may take on forms that are (1) locally stable, Figure 11a, (2) locally nonlinearly unstable (bifurcation), Figure 11b, and (3) locally linearly unstable (bifurcation), Figure 11c. States (2) and (3) may be triggered by a critical parameter $c > 0$ (with a corresponding relaxation time scale τ with or without nonlinear fluctuations. For the triggering of a locally nonlinear instability, the triggering parameter would typically be some sort of a measure of the amplitude of the local nonlinear fluctuations. For the triggering of a local linear instability, the triggering parameter would generally be some sort of a measure of the amplitude of the local gradient of the state variable, P. (We note that a similar but symmetrical model based on the Landau–Ginsburg expansion has been considered by Gil and Sornette

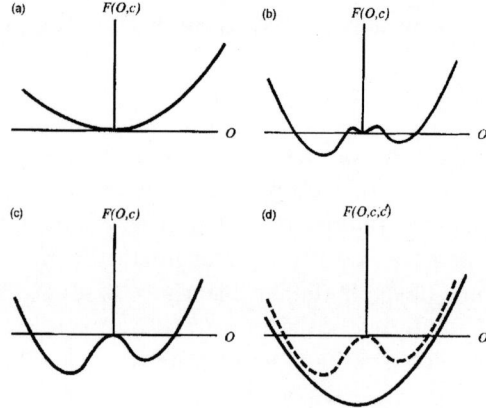

Figure 11. Topologies of state function $F(O)$.

(1996).) Alternatively, a locally bifurcated state such as state (3) may be influenced by another triggering parameter $c' > 0$ such that a more preferred local state similar to that of state (1) becomes energetically more favorable than that of the bifurcated state. See Figure 11d.

One may easily connect such local triggering behavior of order-disorder transitions to those of the localized, sporadic interactions of the magnetic coherent structures discussed in the previous section. For example, the situation depicted by Figure 11d can be interpreted as the merging of two coherent structures. Also, states (2) and (3) can be interpreted as the bifurcation of one coherent structure into two smaller coherent structures due to certain nonlinear or linear plasma instability. Because most of such processes are probably due to the result of certain kinetic effects, the interaction time-scales would generally be of the order of kinetic reaction-times and therefore much shorter than the system evolution-time to be described in the following section.

7.2. TRANSPORT EQUATIONS OF STATE VARIABLES

The (slow time) evolution equations of the state variables will generally consist of convective, forcing, random stirring, and transport terms. Typically, they may take on the following generic form (Chang and Wu, 2002):

$$\partial P_j/\partial t = \zeta_j(\mathbf{P}, \mathbf{O}; \nu_1, \nu_2 \cdots \nu_m) \tag{5}$$

where $\zeta_j (j = 1, 2, \cdots M)$ are functionals of $(\mathbf{P}, \mathbf{O})$ and the ν's are a set of time scales characterizing the long time system-wise evolution under the influence of the chaotic and anomalous growths of the triggered order parameters.

For a one-dimensional single real-valued state variable, one might envisage the following typical transport equation in differential form:

$$\partial P/\partial t = f + h(P, O)\,\partial P/\partial x + D\,\partial^2 P/\partial x^2 + g \tag{6}$$

where f is the forcing term, $h(P, O)$ is a convective function, g is a random noise, $D = D_0 H(O)$ is the anomalous transport coefficient with $H(O)$, a positive-definitive growing function of the magnitude of the absolute value of O, and D_0, a constant.

From this example (as well as the general expression, (5)), we note that the anomalous transport and convection can take on varying magnitudes (because of their dependence on the order parameter) and it is generally sporadic and localized throughout the system. Such behavior naturally leads to the evolution of fluctuations into all spatial and temporal scales, and would generally lead the global system to evolve into a 'complex' state of long-range interactions exhibiting the phenomenon of forced and/or self-organized criticality (FSOC) as has been discussed extensively in several of our previous publications (Chang, 1998a–1998c, 1999, 2001; Wu and Chang, 2000a, 2000b, 2001).

7.3. AN ILLUSTRATIVE EXAMPLE

We consider below a simple 2-D phenomenological model to mimic MHD turbulence. We introduce the flux function ψ for the transverse fluctuations such that $(-\partial \psi/\partial x, \partial \psi/\partial y) = (\delta B_y, \delta B_x)$ and $\nabla \cdot \delta \mathbf{B} = 0$. The coherent structures for such a system are generally flux tubes normal to the 2-D plane such as those simulated in Figure 2. Instead of invoking the standard 2-D MHD formalism, here we simply consider ψ as a dynamic order parameter. As the flux tubes merge and interact, they may correlate over long distances, which, in turn, will induce long relaxation times near FSOC. Assuming homogeneity, we model the dynamics of the flux tubes, in the crudest approximation, in terms of a classical Time-Dependent Landau–Ginsburg model as follows:

$$\partial \psi_k/\partial t = -\Gamma_k \, \partial F/\partial \psi_{-k} + f_k \tag{7}$$

where ψ_k are the Fourier components of the flux function, Γ_k an analytic function of k^2, and f_k a random noise which includes all the other effects that had been neglected in this crude model. We shall assume the state function to depend on the flux function ψ and the local 'pseudo-energy' measure ξ. For fluctuations, we shall assume that diffusion dominates over convection. Thus, in addition to the dynamic equation (7), we now include a diffusion equation for ξ. In Fourier space, we have

$$\partial \xi_k/\partial t = -D k^2 \, \partial F/\partial \xi_{-k} + h_k \tag{8}$$

where ξ_k are the Fourier components of ξ, $D(k)$ is the diffusion coefficient, $F(\psi_k, \xi_k, k)$ is the state function, and h_k is a random noise. By doing so, we separate the slow transport due to diffusion of the local 'pseudo-energy' measure ξ from the noise term f_k.

Under the dynamic renormalization group (DRG) transformation, the correlation function C for ψ_k should scale as:

$$e^{a_c l} C(k, \omega) = C(ke^l, \omega e^{a_\omega l}) \tag{9}$$

Figure 12. A snapshot of forest-fire model simulation. Black, gray and white pixels refer to empty sites, trees, and burning trees, respectively. The dimension of the square lattice is 100×100 sites with fixed boundaries.

where ω is the Fourier transform of the time t, l the renormalization parameter as defined in the previous section, and (a_c, a_ω) the correlation and dynamic exponents. Thus, $C/\omega^{a_c/a_\omega}$ is an absolute invariant under the DRG, or, $C \sim \omega^{-\lambda}$, where $\lambda = -a_c/a_\omega$. DRG analyses performed for Gaussian noises for several approximations (Chang, 2002) yield the value for λ to be approximately equal to $1.88-1.66$, and a value of 1.0 for the ω-exponent for the trace of the transverse magnetic correlation tensor. Interestingly, Matthaeus and Goldstein (1986) had suggested that such a magnetic correlation exponent might represent the dynamics of discrete structures in pseudo-2D MHD turbulence; thus, giving some credence to the model and the DRG analysis.

7.4. AN ANALOGOUS EXAMPLE – THE GLOBAL MODIFIED FOREST-FIRE MODEL

We demonstrate in this section an analogous model, which seems to encompass some of the basic characteristics of what are expected for the dynamical magnetotail. The model is the global generalization (Tam *et al.*, 2000) of a 'modified forest-fire model' originally introduced by Bak *et al.* (1990) and modified by Drossel and Schwabl (1992). Let us consider a rectangular grid of land, on which trees may grow at any given site (i, j) and any time step n with a probability p. At sites where there is a tree, there is a finite probability f that it might be hit by lightning and catch fire. If a tree catches fire at certain time step n, then its neighbor will catch fire at a subsequent time step $n + 1$. It is obvious that at any given time step n, there will be patches of green trees, patches of burning trees and patches of empty spaces (Figure 12). One may associate the growth of trees as the development of coherent structures, the burning trees as localized merging sites and empty spaces as quiescent states in the magnetotail, a picture simulating the sporadically growing of coherent magnetic structures with localized merging.

The incremental changes of the probabilities (p, f) and the densities of the green trees, burning trees and empty spaces (ρ_1, ρ_2, ρ_3), characterize the dynamics of the forest fire. In particular, (p, f) are the dynamic parameters, and the densities are the dynamic state variables. Following the discussions given in the previous section, we perform a renormalization (coarse-graining) transformation of the parameters (p, f) such that the phenomenon retains its essential basic characteristics. Let us symbolically denote this transformation as:

$$p' = F(p, f; \rho_1, \rho_2, \rho_3); \quad f' = G(p, f; \rho_1, \rho_2, \rho_3) \tag{10}$$

When the system reaches a steady state, the densities, (ρ_1, ρ_2, ρ_3), may be obtained from the mean field theory at any level of coarse-graining and are expressible in terms of the dynamic parameters (p, f).

In Figure 13, we show the affine space of transformations of (p, f) for a particular choice of renormalization-group procedure (Tam *et al.*, 2000) (a global generalization of the coarse-graining procedure suggested by Loreto *et al.* (1995)) in the physically meaningful region of (p, f) between (0, 1). We note that there are 4 distinct fixed points. Each fixed point has its own distinct dynamic critical behavior. In the figure, the renormalization trajectories are displayed as solid curves with arrows indicating the direction of coarse-graining. If the dynamic state of the forest-fire is at a state A near the fixed point (ii), then the system will exhibit the characteristic critical behavior prescribed by the invariance properties of this fixed point until sufficient coarse-graining has taken place such that the renormalized system approaches a point such as B in a region dominated by the fixed point (i). At this point, in the coarse-grained view, the dynamical system will now exhibit the critical behavior characterized by the fixed point (i). We have calculated this crossover (symmetry breaking) behavior for the power-law index of the probability distribution $P(s)$ of the scale sizes of the burnt trees (Figure 14). In terms of the magnetotail analogy, such type of crossover behavior might be associated with the change of physical behavior from the kinetic state to the MHD state.

We note that there is a distinct separatrix connecting the fixed points (iii) and (iv). Thus, transition from a dynamic state such as B to a dynamic state such as C generally cannot be smooth and is probably catastrophic. This would be the analog of the triggering of the onset of substorms due to coarse-grained dissipation in the magnetotail. Although not detailed here, there are many other properties of this analogous model that resemble the magnetotail dynamics.

8. Summary

We have presented a dynamical theory of 'complexity' for space plasmas far from equilibrium. The theory is based on the physical concepts of mutual interactions of coherent magnetic structures that emerge naturally from plasma resonances.

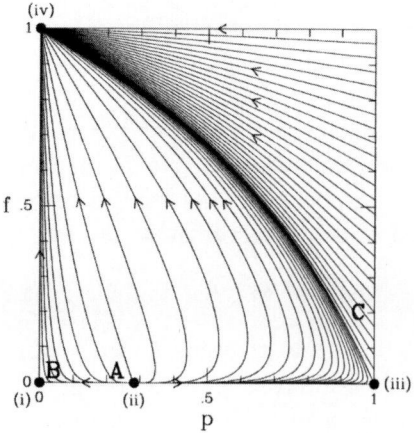

Figure 13. Renormalization-group trajectories for the modified forest-fire model. The four fixed points are indicated by the labels (i) to (iv).

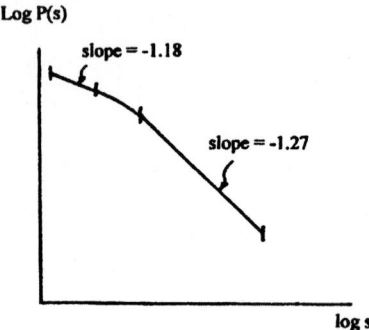

Figure 14. Crossover between the fixed points (ii) and (i), as indicated by the break in the probability distribution of the cluster size of burnt trees.

Models are constructed to represent the local intermittencies and global evolutional processes. Dynamical renormalization-group techniques are introduced to handle the stochastic nature of the model equations that exhibit 'complexity'. Examples are provided to illustrate the phenomena of anomalous transports, preferential accelerations, forced and/or self-organized criticality [FSOC], symmetry breaking and general topological phase transitions including the reconfigurations of mean magnetic field geometries due to coarse-grained dissipation.

Both the physical concepts and mathematical techniques discussed in this review are nontraditional. The readers are encouraged to consult the original references for further in-depth studies of these new emerging ideas.

Acknowledgements

The authors are indebted to V. Angelopoulos, S. C. Chapman, P. De Michelis, C. F. Kennel, A. Klimas, A. T. Y. Lui, D. Tetreault, N. Watkins, D. Vassiliadis, A. S. Sharma, M. I. Sitnov, and G. Ganguli for useful discussions. The work of TC and SWYT was partially supported by AFOSR, NSF and NASA. The research of CCW was supported by NASA. GC thanks the Italian National Research Council (CNR) and the Italian National Program for Antarctica Research (PNRA) for the financial support.

References

Angelopoulos, V., Coroniti, F. V., Kennel, C. F., Kivelson, M. G., Walker, R. J., Russell, C. T., McPherron, R. L., Sanchez, E., Meng, C. I., Baumjohann, W., Reeves, G. D., Belian, R. D., Sato, N., Fris-Christensen, E., Sutcliffe, P. R., Yumoto, K. and Harris, T.: 1996, 'Multi-point analysis of a BBF event on April 11, 1985', *J. Geophys. Res.* **101**, 4967.

Angelopoulos, V., Mukai, T. and Kokubun, S.: 1999, 'Evidence for Intermittency in Earth's Plasma Sheet and Implications for Self-organized Criticality', *Physics of Plasmas* **6**, 4161.

Bak, P., Chen, K. and Tang, C.: 1990, 'A Forest-fire Model and Some Thoughts on Turbulence', *Phys. Lett.* **A147**, 297.

Castaing, B., Gagne, Y. and Hopfinger, E.J.: 1990, 'Velocity Probability Density Functions of High Reynolds Number Turbulence', *Physica D* **46**, 177.

Chang, T. and Stanley, H. E.: 1973, 'Renormalization-group Verification of Crossover with Respect to Lattice Anisotropy Parameter', *Phys. Rev.* **B8**, 1178.

Chang, T., Hankey, A. and Stanley, H. E.: 1973a, 'Double-power Scaling Functions Near Tricritical Points', *Phys. Rev.* **B7**, 4263.

Chang, T., Hankey, A. and Stanley, H. E.: 1973b, 'Generalized Scaling Hypothesis in Multi-component Systems. I. Classification of Critical Points by Order and Scaling at Tricritical Points', *Phys. Rev.* **B8**, 346.

Chang, T., Nicoll, J. F. and Young, J. E.: 1978, 'A Closed-form Differential Renormalization-group Generator for Critical Dynamics', *Phys. Lett.* **67A**, 287.

Chang, T.: 1992, 'Low-dimensional Behavior and Symmetry Breaking of Stochastic Systems near Criticality – Can these Effects be Observed in Space and in the Laboratory?', *IEEE Trans. on Plasma Science* **20**, 691.

Chang, T., Vvedensky, D. D. and Nicoll, J. F.: 1992, 'Differential Renormalization-group Generators for Static and Dynamic Critical Phenomena', *Physics Reports* **217**, 279.

Chang, T.: 1998a, 'Sporadic, Localized Reconnections and Multiscale Intermittent Turbulence in the Magnetotail', in J. L. Horwitz, D. L. Gallagher and W. K. Peterson (eds), *Geospace Mass and Energy Flow*, American Geophysical, Union, Washington, D. C., AGU Geophysical Monograph **104**, p. 193.

Chang, T.: 1998b, 'Multiscale Intermittent Turbulence in the Magnetotail', in Y. Kamide *et al.* (eds), *Proc. 4th Intern. Conf. on Substorms*, Kluwer Academic Publishers, Dordrecht and Terra Scientific Publishing Company, Tokyo, p. 431.

Chang, T.: 1998c, 'Self-organized Criticality, Multi-fractal Spectra, and Intermittent Merging of Coherent Structures in the Magnetotail', in J. Büchner *et al.* (eds), *Astrophysics and Space Science*, Kluwer Academic Publishers, Dordrecht, the Netherlands, v. **264**, p. 303.

Chang, T.: 1999, 'Self-organized Criticality, Multi-fractal Spectra, Sporadic Localized Reconnections and Intermittent Turbulence in the Magnetotail', *Physics of Plasmas* **6**, 4137.

444   TOM CHANG ET AL.

Chang, T.: 2001, 'Colloid-like Behavior and Topological Phase Transitions in Space Plasmas:
    Intermittent Low Frequency Turbulence in the Auroral zone', *Physica Scripta* **T89**, 80.
Chang, T.: 2002, '"Complexity" Induced Plasma Turbulence in Coronal Holes and the Solar Wind',
    in *Solar Wind* **10** (in press).
Chang, T. and Wu, C. C.: 2002, '"Complexity" and Anomalous Transport in Space Plasmas', *Physics
    of Plasmas* **9**, 3679.
Chang, T., Wu, C. C. and Angelopoulos, V.: 2002, 'Preferential Acceleration of Coherent Magnetic
    Structures and Bursty Bulk Flows in Earth's Magnetotail', *Physica Scripta* **T98**, 48.
Chapman, S. C., Watkins, N.W., Dendy, R. G., Helander, P. and Rowlands, G.: 1998, 'A Simple
    Avalanche Model as an Analogue for Magnetospheric Activity', *Geophys. Res. Lett.* **25**, 2397.
Consolini, G.: 1997, 'Sandpile Cellular Automata and Magnetospheric Dynamics', in S. Aiello, N.
    Lucci, G. Sironi, A. Treves and U. Villante (eds), *Cosmic Physics in the Year 2000*, Soc. Ital. di
    Fis., Bologna, Italy, pp. 123–126.
Consolini, G. and Chang, T.: 2001, 'Magnetic Field Topology and Criticality in Geotail Dynamics:
    Relevance to Substorm Phenomena', *Space Sci. Rev.* **95**, 309.
Consolini, G.: 2002, 'Self-organized Criticality: A New Paradigm for the Magnetotail Dynamics',
    Fractals **10**, 275.
Drossel, B. and Schwabl, F.: 1992, 'Self-organized Critical Forest-fire Model', *Phys. Rev. Lett.* **69**,
    1629.
Fairfield, D. H., Mukai, T., Brittnacher, M., Reeves, G. D., Kokubun, S., Parks, G. K., Nagai, T.,
    Matsumoto, H., Hashimoto, K., Gurnett, D. A. and Yamamoto, T.: 1999, 'Earthward Flow Bursts
    in the Inner Magnetotail and Their Relation to Auroral Brightenings, AKR Intensifications,
    Geosynchronous Particle Injections and Magnetic Activity', *J. Geophys. Res.* **104**, 355–370.
Farge, M., Holschneider, M. and Colonna, J. F.: 1990, 'Wavelet Analysis of Coherent Two Di-
    mensional Turbulent Flows', in H. K. Moffat (ed.), *Topological Fluid Mechanics*, Cambridge
    University Press, Cambridge, p. 765.
Germany, G. A., Parks, G. K., Ranganath, H., Elsen, R., Richards, P. G., Swift, W., Spann, J. F.
    and Brittnacher, M.: 1998, 'Analysis of Auroral Morphology: Substorm Precursor and Onset on
    January 10, 1997', *Geophys. Res. Lett.* **25**, 3043–3046.
Gil, L. and Sornette, D.: 1996, 'Laudau-Ginzburg Theory of Self-organized Criticality', *Phys. Rev.
    Lett.* **76**, 3991.
Ieda, A., Fairfield, D. H., Mukai, T., Saito, Y., Kokubun, S., Liou, K., Meng, C. -I., Parks, G. K. and
    Brittnacher, M. J.: 2001, 'Plasmoid Ejection and Auroral Brightenings', *J. Geophys. Res.* **106**,
    3845–3857.
Loreto, V., Pietronero, L., Vespignani, A. and Zapperi, S.: 1995, 'Renormalization Group Approach
    to the Critical Behavior of the Forest-fire Model', *Phys. Rev. Lett.* **75**, 465.
Lu, E. T.: 1995, 'Avalanches in Continuum Driven Dissipative Systems', *Phys. Rev. Lett.* **74**, 2511–
    2514.
Lui, A. T. Y.: 1996, 'Current Disruptions in the Earth's Magnetosphere: Observations and Models',
    J. Geophys. Res. **101**, 4899.
Lui, A. T. Y.: 1998, 'Plasma Sheet Behavior Associated with Auroral Breakups', in Y. Kamide
    (ed.), *Proc. 4th Intern. Conf. on Substorms*, Kluwer Academic Publishers, Dordrecht and Terra
    Scientific Publishing Company, Tokyo, p. 183.
Lui, A. T. Y., Chapman, S. C., Liou, K., Newell, P. T., Meng, C. I., Brittnacher, M. and Parks, G. D.:
    2000, 'Is the Dynamic Magnetosphere an Avalanching System?', *Geophys. Res. Lett.* **27**, 911–
    914.
Lyons, L. R., Nagai, T., Blanchard, G. T., Samson, J. C. Yamamoto, T., Mukai, T., Nishida, A. and
    Kokubun, S.: 1999, 'Association Between Geotail Plasma Flows and Auroral Poleward Boundary
    Intensifications Observed by CANOPUS Photometers', *J. Geophys. Res.* **104**, 4485–4500.
Matthaeus, W. H. and Goldstein, M. L.: 1986, 'Low-frequency $1/f$ Noise in the Interplanetary
    Magnetic Field', *Phys. Rev. Lett.* **57**, 495.

Nagai, T., Fujimoto, M., Saito, Y., Machida, S. *et al.*: 1998, 'Structure and Dynamics of Magnetic Reconnection for Substorm Onsets with Geotail Observations', *J. Geophys. Res.* **103**, 4419.

Nakamura, R., Baumjohann, W., Brittnacher, M., Sergeev, V. A., Kubyshkina, M., Mukai, T. and Liou, K.: 2001a, 'Flow Bursts and Auroral Activations: Onset Timing and Foot Point Location', *J. Geophys. Res.* **106**, 10777–10789.

Nakamura, R., Baumjohann, W., Schodel, R., Brittnacher, M., Sergeev, V. A., Kubyshkina, M., Mukai, T. and Liou, K.: 2001b, 'Earthward Flow Bursts, Auroral Streamers, and Small Expansions', *J. Geophys. Res.* **106**, 10791–10802.

Newell, P. T., Liou, K., Sotirelis, T. and Meng, C. I.: 2001, 'Polar Ultraviolet Imager Observations of Global Auroral Power as a Function of Polar Cap Size and Magnetotail Stretching', *J. Geophys. Res.* **106**, 5895–5905.

Nicoll, J. F., Chang, T. and Stanley, H. E.: 1974, 'Nonlinear Solutions of Renormalization-group Equations', *Phys. Rev. Lett.* **32**, 1446.

Nicoll, J. F., Chang, T. and Stanley, H. E.: 1976, 'Nonlinear Crossover Between Critical and Tricritical Behavior', *Phys. Rev. Lett.* **36**, 113.

Sergeev, V. A., Liou, K., Meng, C. I., Newell, P. T., Brittnacher, M., Parks, G. and Reeves, G. D.: 1999, 'Development of Auroral Streamers in Association with Localized Impulsive Injections to the Inner Magnetotail', *Geophys. Res. Lett.* **26**, 417–420.

Tam, S. W. Y., Chang, T., Consolini, G. and de Michelis, P.: 2000, 'Renormalization-group Description and Comparison with Simulation Results for Forest-fire Models – Possible Near-criticality Phenomenon in the Dynamics of Space Plasmas', Trans. Amer. Geophys. Union, *EOS* **81**, SM62A-04.

Uritsky, V. M., Klimas, A. J., Vassiliadis, D., Chua, D. and Parks, G. D.: 2002, 'Scale-free Statistics of Spatiotemporal Auroral Emissions as Depicted by POLAR UVI Images: The Dynamic Magnetosphere is an Avalanching System', *J. Geophys. Res.* (in press).

Wu, C. C. and Chang, T.: 2000a, '2D MHD Simulation of the Emergence and Merging of Coherent Structures', *Geophys. Res. Lett.* **27**, 863.

Wu, C. C. and Chang, T.: 2000b, 'Dynamical Evolution of Coherent Structures in Intermittent Two-dimensional MHD Turbulence', *IEEE Trans. on Plasma Science* **28**, 1938.

Wu, C. C. and Chang, T.: 2001, 'Further Study of the Dynamics of Two-dimensional MHD Coherent Structures – A Large Scale Simulation', *J. Atmos. Sci. Terrest. Phys.* **63**, 1447.

DYNAMICAL SYSTEMS APPROACH TO SPACE ENVIRONMENT TURBULENCE

A. C.-L. CHIAN[1,2], F. A. BOROTTO[3], E. L. REMPEL[1,2], E. E. N. MACAU[2],
R. R. ROSA[2] and F. CHRISTIANSEN[4]

[1] *World Institute for Space Environment Research (WISER), NITP,*
University of Adelaide, SA 5005, Australia
[2] *National Institute for Space Research (INPE), P. O. Box 515,*
12221-970 São José dos Campos, São Paulo, Brazil
[3] *Universidad de Concepción, Departamento de Física, Concepción, Chile*
[4] *Solar-Terrestrial Physics Division, Danish Meteorological Institute,*
Lyngbyvej 100, DK-2100 Copenhagen, Denmark

Abstract. Space plasmas are dominated by waves, instabilities and turbulence. Dynamical systems approach offers powerful mathematical and computational techniques to probe the origin and nature of space environment turbulence. Using the nonlinear dynamics tools such as the bifurcation diagram and Poincaré maps, we study the transition from order to chaos, from weak to strong chaos, and the destruction of a chaotic attractor. The characterization of the complex system dynamics of the space environment, such as the Alfvén turbulence, can improve the capability of monitoring Sun-Earth connections and prediction of space weather.

Key words: Alfvén waves, chaos, solar wind, space plasmas, turbulence

1. Introduction

Space environment is a complex system whose dynamics depends on the interactions involving a large number of sub-systems. The Sun-Earth connections, which determine the space weather, are the result of a complex chain of interactions involving the solar interior, solar atmosphere, solar wind, magnetosphere, ionosphere and atmosphere coupling.

Spaces plasmas and atmospheres are dominated by waves, instabilities and turbulence. An accurate characterization of the dynamics of the space environment, from the interior of the Sun to the near-Earth space, is essential for improving the efficiency of monitoring of solar-terrestrial coupling and forecasting of space weather.

The nonlinear fluctuations of wave fields and plasma densities in the space environment can be modeled by nonlinear equations such as the derivative nonlinear Schrödinger equation and the Kuramoto–Sivashinsky equation. The dynamical systems concept provides a powerful tool to probe the onset of turbulence. In this paper, we review two approaches to analyze the space plasma dynamics in the

Space Science Reviews **107**: 447–461, 2003.
© 2003 *Kluwer Academic Publishers.*

presence of dissipation: low- and high-dimensional analysis, and discuss how these analyses can be applied to the study of Alfvén turbulence in the space environment. According to the inertial manifold theorem (Bohr *et al.*, 1998), in infinite-dimensional systems such as the Kuramoto–Sivashinsky equation, the number of 'relevant' degrees of freedom scales as some negative power of the dissipation parameter. In this paper we restrict our studies to relatively large values of dissipation in order to reduce the 'relevant' degrees of freedom. Hence, we treat the weak turbulence regime wherein the system is chaotic in time but remains coherent in space. For small values of dissipation, the system may evolve to well-developed turbulence involving complicated phenomena such as direct and inverse energy cascades, power-law spectra and inertial ranges.

2. Low-dimensional Analysis

2.1. THE DERIVATIVE NONLINEAR SCHRÖDINGER EQUATION

The nonlinear dynamics of a large-amplitude Alfvén wave traveling along an ambient magnetic field in the x-direction is described by the following driven-dissipative derivative nonlinear Schrödinger equation (DNLS) (Ghosh and Papadopoulos, 1987; Hada *et al.*, 1990; Chian *et al.*, 1998).

$$\partial_t b + \alpha \partial_x \left(|b|^2 b \right) - i \left(\mu + i\eta \right) \partial_x^2 b = S(b, x, t) , \qquad (1)$$

where η is the dissipative scale length, $b = b_y + ib_z$ is the complex transverse wave magnetic field normalized to the constant ambient magnetic field B_0, time t is normalized to the inverse of the ion cyclotron frequency $\omega_{ci} = eB_0/m_i$, space x is normalized to c_A/ω_{ci}, $c_A = B_0/(\mu_0\rho_0)^{1/2}$ is the Alfvén velocity, $\alpha = 1/[4(1-\beta)]$, $\beta = c_S^2/c_A^2$, $c_S = (P_0/\gamma\rho_0)^{1/2}$ is the acoustic velocity and μ is the dispersive parameter. We take the external driving force $S(b, x, t) = A \exp(ik\phi)$ to be a monochromatic left-hand circularly polarized wave with a wave phase $\phi = x - Vt$, where V is a constant wave velocity, A and k are arbitrary constants.

The first integral of (1) can be reduced to a system of ordinary differential equations by seeking stationary wave solutions with $b = b(\phi)$, yielding (Hada *et al.*, 1990)

$$\dot{b}_y - v\dot{b}_z = \partial H/\partial b_z + a\cos\theta , \qquad (2)$$

$$\dot{b}_z + v\dot{b}_y = -\partial H/\partial b_y + a\sin\theta , \qquad (3)$$

$$\dot{\theta} = \Omega , \qquad (4)$$

where $H = (\mathbf{b}^2 - 1)^2/4 - (\lambda/2)(\mathbf{b} - \hat{\mathbf{y}})^2$, the overdot denotes derivative with respect to the phase variable $\tau = \alpha b_0^2 \phi/\mu$, the normalized dissipation parameter

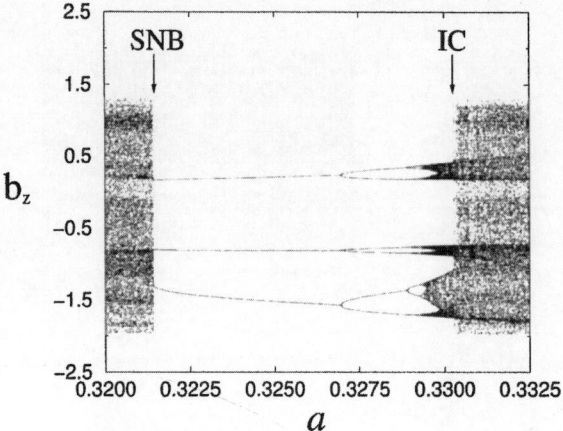

Figure 1. Bifurcation diagram $b_z(a)$ for $\Omega = -1$, $\nu = 0.02$, $\lambda = 1/4$ and $\alpha > 0$.

$\nu = \eta/\mu$, $b \to b/b_0$ (where b_0 is an integration constant), $\mathbf{b}=(b_y, b_z)$, $\theta = \Omega\phi$, $\Omega = \mu k/(\alpha b_0^2)$, $a = A/(\alpha b_0^2 k)$, $\lambda = -1 + V/(\alpha b_0^2)$. We assume $\beta < 1$, hence $\alpha > 0$.

We will use the following Poincaré section

$$\tau = \tau_p + nT, \qquad (n = 1, 2, \ldots) \tag{5}$$

where $T = 2\pi/\Omega$ is the period of the external force S and τ_p is the initial phase of $b(\tau)$. The associated Poincaré map is defined as

$$P : [b_y(\tau_p), b_z(\tau_p)] \to [b_y(\tau_p + nT), b_z(\tau_p + nT)]. \tag{6}$$

This Poincaré map represents the value of b at each period $2\pi/\Omega$.

2.2. INTERMITTENCY IN THE DNLS EQUATION

The chaos theory of nonlinear Alfvén waves can deepen our understanding of Alfvénic turbulence observed in the solar atmosphere and solar wind. In this section, we demonstrate how Alfvén intermittency can appear via chaotic transitions.

A bifurcation diagram for nonlinear Alfvén waves can be constructed from Equations (2)–(4) by varying the driver amplitude (a) while keeping other control parameters fixed ($\Omega = -1$, $\nu = 0.02$, and $\lambda = 1/4$). A left-hand driver is chosen. For every value of a, we drop the initial transient and plot the next 300 Poincaré points of the flow. Figure 1 illustrates a period-3 periodic window of the numerically computed bifurcation diagram (Chian *et al.*, 1998). A chaotic region terminates in a saddle-node bifurcation (SNB) at $a \sim 0.321382105$. At that point, the solutions become periodic with period-three. Beyond a certain driver amplitude ($a \sim 0.32692$), a sequence of period-doubling bifurcations, occurs. This cascade continues until chaos sets in, which eventually evolves into an interior

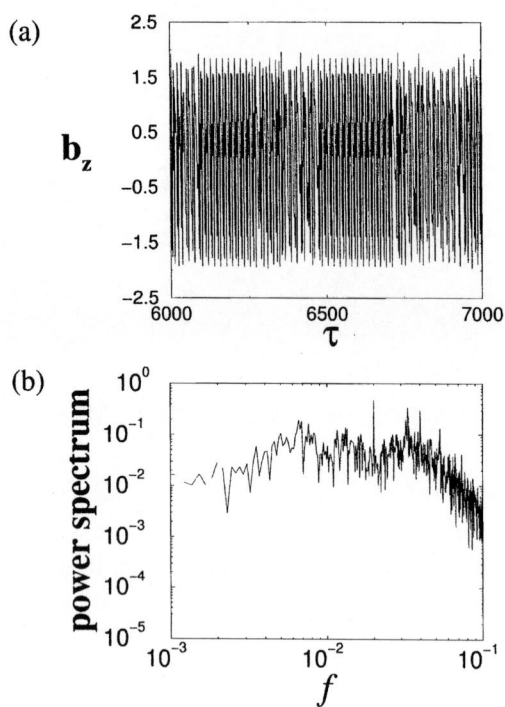

Figure 2. Example of the Pomeau-Manneville intermittent turbulence for $a = 0.3213795$ (a) $b_z(\tau)$, (b) $|b_z|^2$ as a function of f.

crisis (IC) via band merging, leading to the formation of a chaotic continuum at $a \sim 0.330249$.

Two types of temporal Alfvén intermittency can be identified in Figure 1: Pomeau–Manneville intermittency and crisis-induced intermittency (Chian *et al.*, 1998). The type-I Pomeau–Manneville intermittency is characterized by a time series containing nearly periodic laminar phases which are randomly interrupted by chaotic bursts, as exemplified by Figure 2. This intermittency occurs when a dynamical system is near a SNB. Figure 2a illustrates this type of intermittency at $a = 0.3213795$, just to the left of the SNB. For this value of a the system is chaotic. However, an approximately periodic behavior is seen in certain positions of the time series for $b_z(\tau)$ when the trajectory is close to the region of the phase space (b_y, b_z) occupied by the pre-SNB ($a > a_{SNB}$) periodic attractor. The power spectrum for the time series of Figure 2a is given in Figure 2b. The interior crisis-induced intermittency is characterized by a time series containing weakly chaotic laminar phases that are randomly interrupted by strongly chaotic bursts, as exemplified by Figure 3a. This intermittency occurs when a dynamical system is just after an IC. Figure 3a shows an intermittent time series for $b_z(\tau)$ corresponding to the strong chaotic attractor at $a = 0.33029$ in Figure 1, right after the interior crisis. For this parameter value, typical orbits spend most of the time near the banded pre-

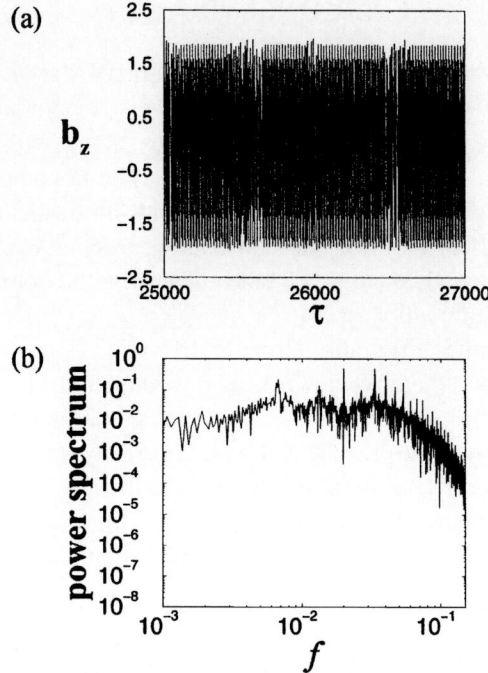

Figure 3. Example of the crisis-induced intermittent turbulence for $a = 0.33029$ (a) $b_z(\tau)$, (b) $|b_z|^2$ as a function of f.

crisis chaotic attractor in the phase space (b_y, b_z), with short bursts of strongly chaotic behavior. The power spectrum for the time series of Figure 3a is given in Figure 3b.

There is observational evidence of Alfvén chaos in the solar wind. A nonlinear time series analysis of Alfvén waves in the low-speed streams of the solar wind detected by the Helios spacecraft in the inner heliosphere shows that the Lyapunov exponent is positive, which indicates that the solar wind plasma is in a chaotic state (Macek and Redaelli, 2000). Alfvénic and MHD intermittent turbulence are easily encountered in the solar atmosphere and solar wind (Marsch and Liu, 1993; Walsh and Galtier, 2000; Patsourakos and Vial, 2002; Marsch and Tu, 1990). The observed solar Alfvénic intermittent turbulence shows power-law spectra. The power spectra, Figures 2b and 3b, of chaos-driven Alfvén intermittencies present power-law similar to the observations (Marsch and Tu, 1990). It is likely that the Alfvén Pomeau–Manneville intemittency and Alfvén interior-crisis intermittency play a relevant role in the intermittent heating and intermittent particle acceleration in the solar atmosphere.

2.3. BOUNDARY CRISIS IN THE DNLS EQUATION

In this section we show that the onset/destruction of Alfvén chaos can occur via a transition mechanism known as *boundary crisis*. A boundary crisis is a global bifurcation whereby a chaotic attractor vanishes along with its basin of attraction (Grebogi *et al.*, 1982; Chian *et al.*, 2002a). The boundary crises involve the tangency (or collision) of a chaotic attractor with an unstable fixed point or unstable periodic orbit (or its invariant stable manifolds) as some control parameter of the system is varied. The tangency takes places on the boundary of the basin of attraction of the chaotic attractor.

We construct another bifurcation diagram from Equations (2)–(4) by varying the dissipation control parameter ν and keeping the other control parameters fixed ($a = 0.1, \Omega = -1, \lambda = 1/4, \mu = 1/2$). We identified a complex plasma region, where five different attractors are present and a wealth of dynamical features are found (Chian *et al.*, 2002a). The bifurcation diagram of this region is given in Figure 4a. In this region the attractor A_1 remains a period-one limit cycle throughout, whereas the other four attractors appear only during certain intervals of ν. Only three out of five attractors (A_1, A_2 and A_3) are plotted in Figure 4a. Attractors A_4 and A_5 are confined within the region indicated by the bar in Figure 4a.

The bifurcation diagram Figure 4a shows that apart from a small range of ν wherein a single attractor (A_1) exists by itself, in most regions of the bifurcation diagram there is coexistence of two or more attractors.

Unstable periodic orbit (UPO) is the key for the characterization of nonlinear dynamical phenomena such as the boundary crisis. We determine the UPO from the numerical solution of (2)–(4) and analyze in detail the role played by the UPO in the onset of Alfvén boundary crisis. The complex plasma region we identified exhibits a large number of crises. In this section, we shall focus only on the characterization of the double boundary crises indicated in Figure 4a.

An examination of Figure 4a shows that a saddle-node bifurcation occurs for the attractor A_3 at $\nu \sim 0.0178162$ (attractor A_2 at $\nu \sim 0.02279$), giving rise to a pair of period-9 (period-3) stable and unstable periodic orbits, respectively. For each attractor, the resulting stable periodic orbit undergoes a cascade of period-doubling bifurcations which leads to the formation of a chaotic attractor. The chaotic attractor disappears abruptly at $\nu \sim 0.0174771$ for A_3 and at $\nu \sim 0.01514$ for A_2 due to double boundary crises. In both cases, the period-9 UPO of A_3, arising from the saddle-node bifurcation at $\nu \sim 0.0178162$, is involved.

The characterization of Alfvén double boundary crises can be performed using the Poincaré method. The Poincaré map of the middle branch of the strange attractor (SA) A_3 near the first crisis point $\nu \sim 0.0174771$ (marked BC in Figure 4b) is shown in Figure 5. The cross denotes one of the three Poincaré points associated with the middle branch of the period-9 UPO. The invariant stable manifolds (SM) associated with the period-9 UPO (saddle) are represented by the light lines. Figure 5 shows that at the crisis point the strange attractor collides head-on with the

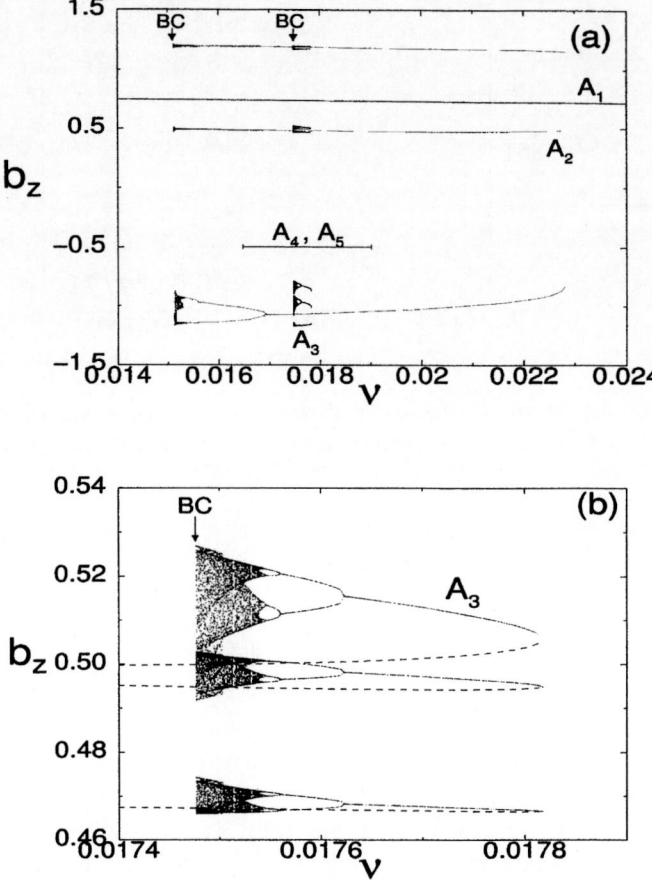

Figure 4. Bifurcation diagram $b_z(\nu)$ for $a = 0.1$, $\Omega = -1$, $\mu = 1/2$, $\lambda = 1/4$ and $\beta < 1$.
(a) attractors (A_1, A_2, A_3); (b) an enlargement of the middle branch of attractor A_3; BC denotes
boundary crisis. The dashed lines denote the unstable periodic orbit of period-9.

saddle and its stable manifolds. Before the first boundary crisis, the system has four
coexisting attractors (A_1, A_2, A_3, A_4). After the crisis, the attractor A_3 vanishes,
leaving the system with only three coexisting attractors (A_1, A_2, A_4).

The period-9 UPO survives the first boundary crisis that causes the destruction
of A_3 and continues to participate in the second boundary crisis that causes the
destruction of A_2. A Poincaré map of the middle branch of the strange attractor
A_2 just before the second crisis at $\nu = 0.01514$ is plotted in Figure 6. The three
crosses denote the three Poincaré points associated with the middle branch of the
period-9 UPO, and the light lines denote its stable manifolds. The second boundary
crisis occurs via a collision of the strange attractor with the p-9 UPO and its stable
manifolds. Before the second crisis, the system has two coexisting attractors (A_1,
A_2). After the crisis, the attractor A_2 vanishes, leaving the system with only the

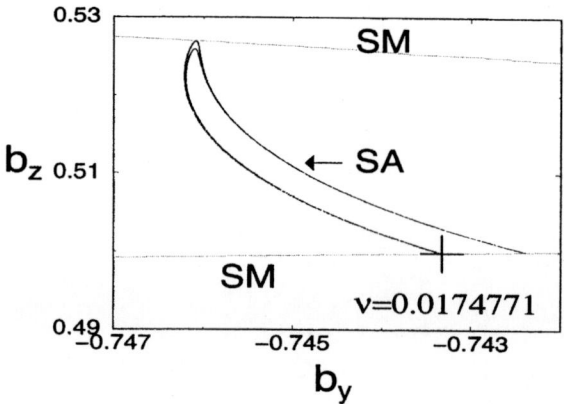

Figure 5. Poincaré map of strange attractor SA associated with the boundary crisis BC of Figure 4b. The cross denotes one of the Poincaré points of the middle branch of the unstable periodic orbit of period-9 and the light lines represent the stable manifolds (SM) of the period-9 saddle.

attractor A_1. Although we only showed the tangencies in the middle branch of the attractors A_2 and A_3, the same dynamics applies to the upper and lower branches.

Boundary crisis with its associated sudden appearance/disappearance of a chaotic attractor was seen in the numerical simulations of a theoretical heart model described by the Van der Pol oscillator (Abraham and Stewart, 1986). Recently, this phenomenon was observed in a dripping faucet laboratory experiment (Pinto and Sartorelli, 2000). These observations suggest that Alfvén boundary crisis is likely to take place in cosmic and laboratory plasmas. In fact, in a recent tokamak plasma experiment, a period-one unstable periodic orbit was identified in a weakly turbulent edge-localized-mode high-performance regime (ELM H mode) (Bak *et al.*, 1999). As shown in this section, the identification of UPO is essential for the characterization of chaotic system dynamics.

3. High-dimensional Analysis

In the previous sections, we studied Alfvén turbulence driven by a low-dimensional chaos in the space environment, using the derivative nonlinear Schrödinger equation. In this section, we investigate the nonlinear spatiotemporal dynamics of Alfvén waves governed by the Kuramoto-Sivashinsky equation, which under certain approximations, describes the phase evolution of the complex amplitude of the Ginzburg–Landau equation (Aranson *et al.*, 2002). It was demonstrated by Lefebvre and Hada (2000) that under the assumption of weak instability and wave-packet limit the derivative nonlinear Schrödinger equation reduces to a complex Ginzburg–Landau equation.

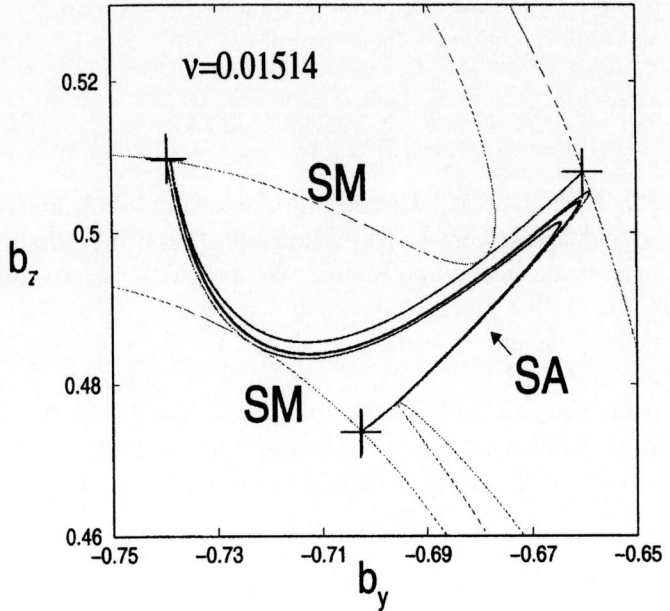

Figure 6. Dynamical structures of the boundary crisis (BC) of attractor A_2 at $\nu = 0.01514$. SA denotes the strange attractor, the crosses denote the Poincaré points of the middle branch of the unstable periodic orbit of period-9, the light lines denote the stable manifolds (SM) of the period-9 saddle.

3.1. THE KURAMOTO-SIVASHINSKY EQUATION

The one-dimensional Kuramoto–Sivashinsky equation can be written as (Chian *et al.*, 2002)

$$\partial_t u = -\partial_x^2 u - \nu \partial_x^4 u - \partial_x u^2, \tag{7}$$

where ν is a 'viscosity' damping parameter. We assume that $u(x,t)$ is subject to periodic boundary conditions $u(x,t) = u(x + 2\pi, t)$ and expand the solutions in a discrete spatial Fourier series

$$u(x,t) = \sum_{k=-\infty}^{\infty} b_k(t)e^{ikx}. \tag{8}$$

Substituting Equation (8) into Equation (7) yields an infinite set of ordinary differential equations for the complex Fourier coefficients $b_k(t)$

$$\dot{b}_k(t) = (k^2 - \nu k^4)b_k(t) - ik \sum_{m=-\infty}^{\infty} b_m(t)b_{k-m}(t), \tag{9}$$

where the dot denotes derivative with respect to t. Reality of $u(x,t)$ implies that $b_{-k} = b_k^*$. We restrict our investigation to the subspace of odd functions $u(x,t) =$

$-u(-x, t)$ assuming that $b_k(t)$ are purely imaginary by setting $b_k(t) = -ia_k(t)/2$, where $a_k(t)$ are real. Equation (9) then becomes

$$\dot{a}_k(t) = (k^2 - \nu k^4)a_k(t) - \frac{k}{2} \sum_{m=-\infty}^{\infty} a_m(t)a_{k-m}(t), \qquad (10)$$

where $a_0 = 0$, $1 \leq k \leq N$, N is the truncation order. We integrate the high-dimensional dynamical system given by Equation (10) using a fourth-order variable step Runge–Kutta integration routine. We choose $N = 16$, since numerical tests indicate that for the range of the control parameter ν used in this paper the solution dynamics remains essentially unaltered for $N > 16$. In all the computational results presented in this paper, higher order truncations yield the same numerical conclusions as the 16-mode truncation. In addition we test empirical truncated approximations against the decrease of the time step and the local truncation error accuracy. We adopt a Poincaré map with the $(N - 1)$ dimensional hyperplane defined by $a_1 = 0$, with $\dot{a}_1 > 0$.

3.2. HIGH-DIMENSIONAL INTERIOR CRISIS IN THE KS EQUATION

A bifurcation diagram can be obtained from the numerical solutions of the 16-mode truncation of Equation (10) by varying the control parameter ν and plotting the Poincaré points of one Fourier mode after discarding the initial transient. Figure 7a shows a period-3 (p-3) window where we plot the Poincaré points of the Fourier mode a_6 as a function of ν. The corresponding behavior of the maximum Lyapunov exponent is shown in Figure 7b. Evidently, the high-dimensional temporal dynamics of the K-S equation preserves the typical dynamical features of a low-dimensional dynamical system. The dotted lines in Figure 7a denote the Poincaré points of the p-3 unstable periodic orbit (UPO) which emerges via a saddle-node bifurcation at $\nu \sim 0.02992498$, marked SN in Figure 7a. In this section, we will analyze the role played by this p-3 UPO in the onset of interior crisis at $\nu_{IC} \sim 0.02992021$, marked IC in Figure 7.

The interior crisis at ν_{IC} occurs when the p-3 UPO collides head-on with the 3-band weak strange attractor evolved from a cascade of period-doubling bifurcations, as seen in Figure 7a. The interior crisis leads to a sudden expansion of a strange attractor, turning the weak strange attractor (WSA) into a strong strange attractor (SSA). The abrupt increase in the system's chaoticity after the interior crisis can be characterized by the value of the maximum Lyapunov exponent (λ_{max}), plotted in Figure 7b. At crisis ($\nu_{IC} \sim 0.02992021$), $\lambda_{max} \approx 0.35$, and after the crisis at $\nu = 0.02992006$, $\lambda_{max} \approx 0.62$. In Figure 7(c) we plotted the spatial correlation length ξ, which remains basically unaltered throughout the whole range of ν used in Figure 7. This means that there is little variance in the spatial disorder of the pattern. An inspection of the spatiotemporal pattern $u(x, t)$ after the crisis ($\nu = 0.02992006$), shown in Figure 8, reveals that for the chosen values of ν

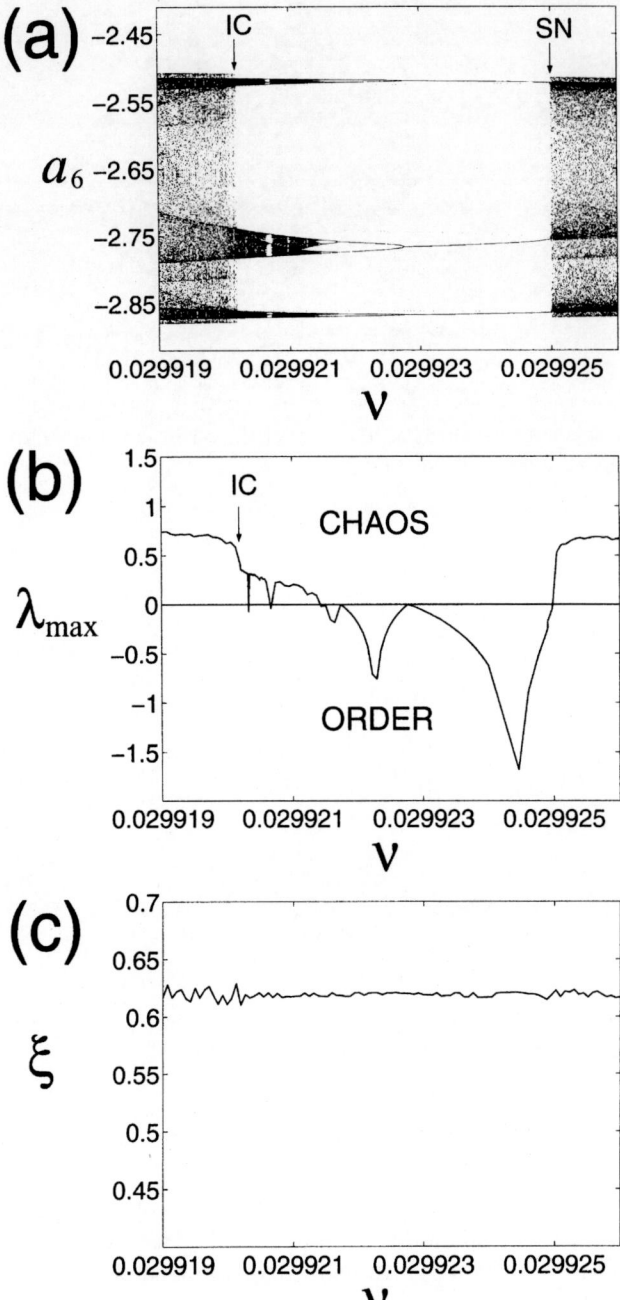

Figure 7. (a) Bifurcation diagram of a_6 as a function of ν. IC denotes interior crisis and SN denotes saddle-node bifurcation. The dotted lines represent a period-3 unstable periodic orbit. (b) Variation of the maximum Lyapunov exponent λ_{max} with ν. (c) Variation of the correlation length ξ with ν.

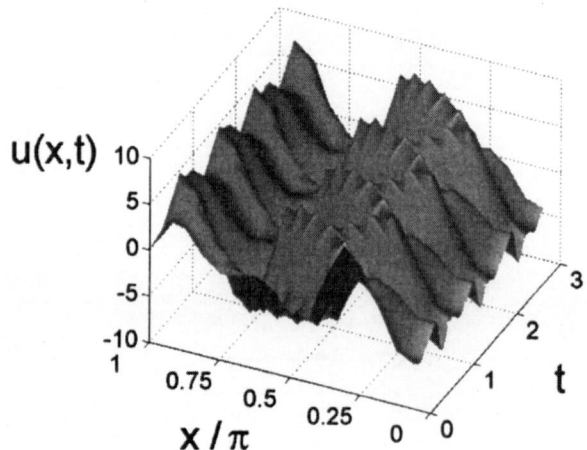

Figure 8. The spatiotemporal pattern of $u(x, t)$ after the crisis at $\nu = 0.02992006$. The system dynamics is chaotic in time but coherent in space.

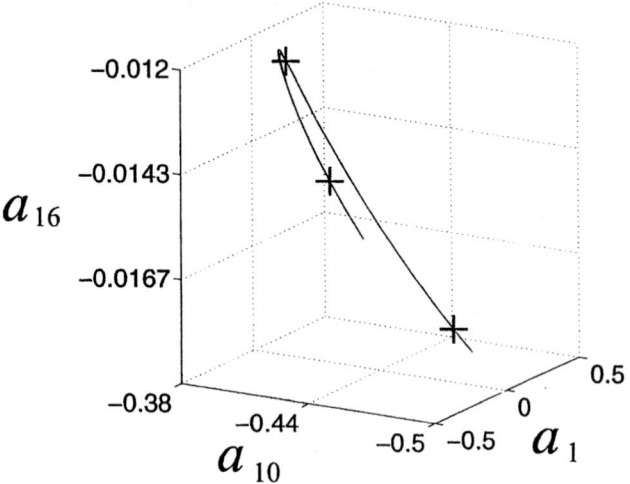

Figure 9. Three-dimensional projection (a_1, a_{10}, a_{16}) of the invariant unstable manifolds of the period-3 saddle (crosses) right after the crisis at $\nu = 0.02992020$.

and the spatial system size $L = 2\pi$, the dynamics of the Kuramoto–Sivashinsky equation is coherent in space.

At crisis, only one of the 16 stability eigenvalues of the p-3 UPO illustrated in Figure 7a has an absolute value greater than 1. This implies that its invariant unstable manifold is a one-dimensional curve embedded in the 15-dimensional Poincaré space. Of the remaining eigenvalues, one has absolute value equal to unity and all the other fourteen have absolute values less than one, implying that the invariant stable manifolds have dimension fourteen. Although we have adopted a 16-mode truncated system in our analysis, all the calculations performed can be

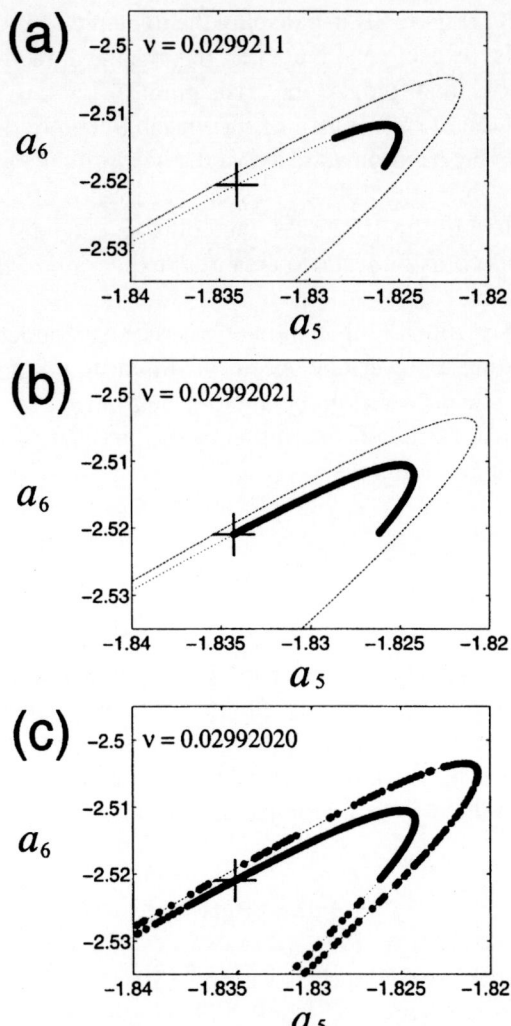

Figure 10. The plots of the strange attractor (dark line) and invariant unstable manifolds (light lines) of the saddle before (a), right before (b) and right after (c) crisis. The cross denotes one of the saddle points.

extended to an arbitrary high number ($N < \infty$) of modes for an appropriate choice of v and L. Figure 9 is a plot of the projection of the invariant unstable manifold of the p-3 UPO onto three axes (a_1, a_{10}, a_{16}) at $v = 0.02992020$, right after the interior crisis. The crosses represent the p-3 UPO.

We proceed next with the characterization of the high-dimensional interior crisis at v_{IC} by showing in Figure 10 the collision of the weak strange attractor with the p-3 UPO in the reduced 2-dimensional Poincaré plane (a_5 vs. a_6), in the vicinity of the upper cross in Figure 9. The dark line denotes the strange attractor, and the light line denotes the numerically computed invariant unstable manifold of the saddle periodic orbit. Figures 10a–c display the dynamics before, right before, and right after the crisis, respectively. Note that the strange attractor always 'overlaps' the invariant unstable manifold. At the crisis point $\nu_{IC} \sim 0.02992021$, the chaotic attractor is the closure of one branch of the unstable manifold of the p-3 UPO, as seen in Figure 10b. The head-on collision of the weak strange attractor with the p-3 UPO at ν_{IC}, shown in Figure 10b, proves the occurrence of a crisis. Although we are showing only two of the 16 modes, the collision can be seen in any choice of Fourier modes. This collision leads to an abrupt expansion of the strange attractor, as seen in Figure 10c.

It was shown in Section 2.2 that an interior crisis can induce a temporal Alfvén intermittency, yielding a power-law spectrum similar to the typical power spectra of Alfvénic turbulence observed in the interplanetary medium. The present section suggests that the spatiotemporal intermittency of interplanetary Alfvén waves can also be driven by an interior crisis.

4. Conclusions

In conclusion, we have shown how the low- and high-dimensional dynamical systems analyses can elucidate the temporal and spatiotemporal evolutions of the space environment turbulence. The development of new mathematical and computational techniques based on the complex systems approach will help us to tackle hard problems such as the dynamics of the eruption of solar flares and coronal mass ejections, as well as the origin of geomagnetic storms.

Acknowledgements

This work is supported by CNPq (Brazil), FAPESP (Brazil) and AFOSR. A. C.-L. Chian and E. L. Rempel wish to thank NITP/CSSM of the University of Adelaide for their kind hospitality, and L. Nocera, the Coordinator of WISER/Pisa, for valuable comments.

References

Abraham, R. H. and Stewart, H. B.: 1986, *Physica D* **21**, 394–400.
Aranson, I. S. and Kramer, L.: 2002, *Rev. of Mod. Phys.* **74**, 99–143.
Bak, P. E., Yoshino, R., Akasura, N. and Nakano, T.: 1999, *Phys. Rev. Lett.* **83**, 1339–1342.
Bohr, T., Jensen, M. H., Paladin, G. and Vulpiani, A.: 1998, *Dynamical Systems Approach to Turbulence*, Cambridge University Press, Cambridge.
Chian, A. C.-L., Borotto, F. A. and Gonzalez, W. D.: 1998, *Astrophys. J.* **505**, 993–998.
Chian, A. C.-L., Borotto, F. A. and Rempel, E. L.: 2002a, *Int. J. Bifurcation Chaos* **12**, 1653–1658.
Chian, A. C.-L., Rempel, E. L., Macau, E. E. N., Rosa, R. R. and Christiansen, F.: 2002b, *Phys. Rev. E* **65**, 035203(R).

Ghosh, S. and Papadopoulos, K.: 1987, *Phys. Fluids* **30**, 1371–1387.
Grebogi, C., Ott, E. and Yorke, J. A.: 1982, *Phys. Rev. Lett* **48**, 1507–1510.
Hada, T., Kennel, C. F., Buti, B. and Mjølhus, E.: 1990, *Phys. Fluids B* **2**, 2581–2590.
Lefebvre, B. and Hada, T.: 2000, *EOS Trans.* **81**(48), AGU Fall Meeting Suppl., Abstract SM62A-09.
Macek, W. M. and Redaelli, S.: 2000, *Phys. Rev. E* **62**, 6496–6504.
Marsch E. and Tu, C.-Y.: 1990, *J. Geophys. Res.* **95**, 8211–8229.
Marsch, E. and Liu, S.: 1993, *Ann. Geophysicae* **11**, 227–238.
Patsourakos, S. and Vial, J. C.: 2002, *Astron. Astrophys.* **385**, 1073–1077.
Pinto, R. D. and Sartorelli, J. C.: 2000, *Phys. Rev. E* **61**, 342–347.
Walsh, R. W. and Galtier, S.: 2000, *Solar Phys.* **197**, 57–73.

PHASE COHERENCE OF MHD WAVES IN THE SOLAR WIND

TOHRU HADA[1], DAIKI KOGA[1] and EIKO YAMAMOTO[2]

[1] *E. S. S. T., Kyushu University, Fukuoka, Japan*
[2] *N.A.S.D.A., Ibaraki, Japan*

Abstract. Large amplitude MHD waves are commonly found in the solar wind. Nonlinear interactions between the MHD waves are likely to produce finite correlation among the wave phases. For discussions of various transport processes of energetic particles, it is fundamentally important to determine whether the wave phases are randomly distributed (as assumed in quasi-linear theories) or they have a finite coherence. Using a method based on a surrogate data technique and a fractal analysis, we analyzed Geotail magnetic field data (provided by S. Kokubun and T. Nagai through DARTS at the Institute of Space and Astronautical Science) to evaluate the phase coherence among the MHD waves in the earth's foreshock region. The correlation of wave phases does exist, indicating that the nonlinear interactions between the waves is in progress.

Key words: MHD turbulence, nonlinear phenomena, phase synchronization

1. Introduction

Magnetohydrodynamic (MHD) waves are ubiquitous in space. In particular, those found in the foreshock region of the earth's bowshock typically have order of unity normalized magnetic field amplitude ($\delta B / B_0 \sim 1$), and thus they have been served as an excellent subject for the research of nonlinear wave theories. In fact, various unique waveforms observed by spacecraft experiments, such as the so-called shocklets (Hoppe *et al.*, 1981), SLAM's (Schwartz and Burgess, 1991), among others, have motivated a number of theoretical as well as numerical simulation studies (Hada *et al.*, 1987; de Wit *et al.*, 1999).

In this paper we focus our attention to the distribution of wave phases of the observed MHD turbulence in space. We note that the wave phases depend on the choice of the origin of the coordinate, which is arbitrary. Thus if one plots the distribution of the wave phases versus the wave number, it will almost always appear to be random even if there is a finite phase coherence – this may be the reason why the phase distribution have been overlooked in past studies of MHD turbulence, even though it contains the equal amount of information as the power spectrum.

In order to avoid the influence of the choice of the origin, we first make two surrogate datasets from the original magnetic field time series observed by Geotail spacecraft, and then characterize differences in these datasets in the real space (by

Space Science Reviews **107**: 463–466, 2003.

evaluating the structure function), instead of dealing with the phase distribution in the Fourier space. Details of the method are given in the next section.

2. Method of Analysis

Suppose we have a sequence of the magnetic field data, $\mathbf{B}(t)$. Since it is usually defined in a coordinate not much relevant to the nature of the turbulence, we should rotate the coordinate axes by, for example, using the minimum variance method. Then let us pick one of the transverse components of $\mathbf{B}(t)$ in the new coordinate and write it $B(t)$. In this paper the original dataset is given by spacecraft experiment, but the present method can be applied to any time series.

From a given dataset (OBS), we make two surrogate datasets: first we Fourier decompse the original data into the power spectrum and the phase. We then randomly shuffle the phase, but keep the power spectrum unchanged, and from these two Fourier data, we inverse Fourier transform to create Phase Randomized Surrogate (PRS). In a similar way we make Phase Correlated Surrogate (PCS), in which the phases are all made equal. The three datasets, OBS, PRS, and PCS, share exactly the same power spectrum, but their phase distributions are different.

As mentioned earlier, the distribution of phases of the OBS dataset looks almost as random as that of the PRS, due to the arbitrary choice of the origin. Thus, we characterize the differences in the phase distribution by the differences in the curves in real space, instead of the Fourier space: when the phases are correlated, it turns out that the curve in the real space appears to be smoother than the case where the phases are random (Higuchi, 1988). Also, the path length of the curves is shorter when the phases are correlated. These differences can be most naturally captured by the structure function,

$$S(m, d) = \sum_x |b(x + d) - b(x)|^m$$

where the unit norm, d, is a measure characterizing the magnification level of the curve.

3. Results

The left panel of Figure 1 shows $S(1, d)$ for the OBS, PRS, and the PCS datasets. Since $m = 1$, the structure function gives the path length measured by a unit length, d. Note that the lengths are different, although by a small amount, for the three datasets. We define the phase coherence index,

$$C_\phi = \frac{L_{PRS} - L_{OBS}}{L_{PRS} - L_{PCS}},$$

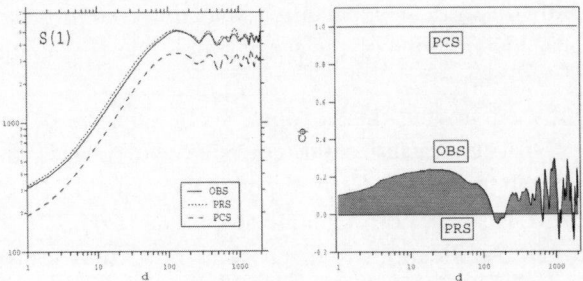

Figure 1. Left: $S(1, d)$ evaluated for the original data (OBS), phase randomized surrogate (PRS), and the phase correlated surrogate (PCS). Right: The phase coherence index as defined in the text.

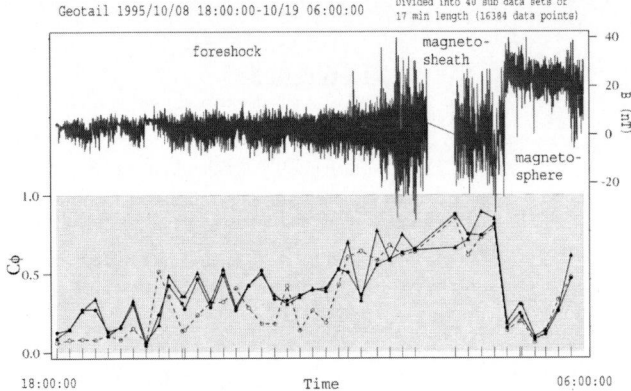

Figure 2. Evolution of C_ϕ as GEOTAIL spacecraft travels from deep foreshock toward the shock, inside of the magnetosheath, and finally into the magnetosphere.

where L_* denotes the value of $S(1, d)$ for dataset *. If the original data is random phase, C_ϕ should be ~ 0, while $C_\phi = 1$ if the data is completely phase correlated. The right panel of Figure 1 shows that the phase coherence does exist up to $d \sim 100$, which roughly corresponds to the time scale of the ion cyclotron period in the plasma rest frame.

In Figure 2, evolution of C_ϕ (bottom panel) is evaluated for a long sequence of data, as the Geotail approaches the bow shock from far upstream. The solid and dashed lines represent C_ϕ of the magnetic field component transverse and parallel to the minimum variance direction, respectively. The field fluctuations (top panel) increase as the spacecraft comes closer to the bowshock, becomes extremely large within the magnetosheath, and decreases again as the spacecraft enters the magnetosphere. The evolution of C_ϕ approximately follows the evolution of the magnetic field turbulence level. This is a natural consequence if the phase coherence is generated via nonlinear interaction between the finite amplitude MHD waves.

We have also examined dependence of C_ϕ to an angle between the average field and the minimum variance directions. Positive correlation among them are found, in agreement with theoretical expectation that the coupling between the MHD waves occurs at a lower order of the wave amplitude for obliquely propagating waves.

In summary, by employing the surrogate technique, we defined the phase coherence index, C_ϕ, to characterize the correlation of phases among waves in a given time series. We found that C_ϕ is almost never equal to zero (typically, $C_\phi = 0.1 \sim 0.5$) when MHD waves of moderate amplitude ($\delta B / B_0 \sim 0.05$) are present. This bears important implications in discussions of various transport processes of charged particles in space, where it is conventional to employ the random phase approximation. Without the approximation the diffusion process can become a qualitatively different one (Kuramitsu and Hada, 2000).

References

Hada, T., Kennel, C. F. and Terasawa, T.: 1987, 'Excitation of Compressional Waves and the Formations of Shocklets in the Earth's Foreshock', *J. Geophys. Res.* **92**, 4423–4435.

Higuchi, T.: 1988, 'Approach to an Irregular Time Series on the Basis of the Fractal', *Physica D* **31**, 277–283.

Hoppe, M. M., Russell, C. T., Frank, L. A., Eastman, T. E. and Greenstadt, E. W.: 1981, 'Upstream Hydromagnetic-waves and their Association with Backstreaming Ion Populations – ISEE-1 and ISEE-2 Observations', *J. Geophys. Res.* **86**, 4471–4492.

Kuramitsu, Y. and Hada, T.: 2000, 'Acceleration of Charged Particles by Large Amplitude MHD Waves: Effect of Wave Spatial Correlation', *Geophys. Res. Lett.* **27**, 629–632.

Schwartz, S. J. and Burgess, D.: 1991, 'Quasi-parallel Shocks – A Patchwork of 3-dimensional Structures', *Geophys. Res. Lett.* **18**, 373–376.

de Wit, T. D., Krasnosel'skikh, V. V., Dunlop, T. and Luhr, H.: 1999, 'Identifying Nonlinear Wave Interactions in Plasmas Using Two-point Measurements: A Case Study of Short Large Amplitude Magnetic Structures', *J. Geophys. Res.* **104**, 17079–17090.

STOCHASTIC ACCELERATION OF CHARGED PARTICLE IN NONLINEAR WAVE FIELD

KAIFEN HE

Institute of Low Energy Nuclear Physics, Beijing Normal University, Beijing, 100875 China

Abstract. Possibility of stochastic acceleration of charged particle by nonlinear waves is investigated. Spatially regular (SR) and spatiotemporal chaotic (STC) wave solutions evolving from saddle steady wave are tested as the fields. In the non-steady SR field the particle is finally trapped by the wave and averagely gains its group velocity, while in the STC field the particle motion displays trapped-free phases with its averaged velocity larger or smaller than the group velocity depending on the charge sign. A simplified model is established to investigate the acceleration mechanism. By analogy with motor protein, it is found that the virtual pattern of saddle steady wave plays a role of asymmetric potential, which and the nonlinear varying perturbation wave are the two sufficient ingredients for the acceleration in our case.

Key words: nonlinear waves, stochastic acceleration

1. Introduction

Stochastic acceleration of charged particles is a fundamental problem in many fields. For example, the origin of high-speed particles in cosmic rays is considered as a result of stochastic acceleration. However, stochastic acceleration should be distinguished with stochastic heating (Sagdeev *et al.*, 1988), to explain stochastic acceleration a key problem is to answer the question why the particle can be accelerated in a prior direction. This problem is especially interesting in some areas. For example, an electron can be accelerated in a wakefield driven by laser (Tajima and Dawson, 1979; Joshi and Corkum, 1995; Bulanov *et al.*, 1999). Another example is motor proteins moving along protein fibers. It is found that Brownian particles can acquire net drift speed in asymmetric potential when subject to a time correlated force (Magnasco, 1991).

In studying this topic the effect of nonlinearity has been taken into account for many years. For example, a particle bouncing between a fixed and an oscillating wall can gain energy if nonlinear dependence of colliding position on particle phase is assumed (Fermi, 1949; Sagdeev *et al.*, 1988). However, in this model the particle is not accelerated in a prior direction. In recent years it has become clear that turbulent waves may arise from intrinsic property of nonlinear systems. A nonlinear wave has certain group velocity, it is therefore reasonable to expect that a particle can gain the velocity by particle-wave interaction. Since nonlinear

Space Science Reviews **107**: 467–474, 2003.

waves may display different space-time patterns, can a particle be accelerated in any type of nonlinear waves, or only in certain types of nonlinear waves? Does particle behavior depend on the wave dynamics? And what are the ingredients necessary for the acceleration? These are our motivations in the present work.

2. Particle Mechanics and Wave Dynamics

Dimensionless Newtonian equations for a single particle in an electric field are

$$\frac{dx}{dt} = v, \tag{1}$$

$$\frac{dv}{dt} = -q\nabla\phi(x,t), \tag{2}$$

here x and v are particle position and velocity respectively, charge number $q = \pm 1$ in the following. In Equation (2) potential $\phi(x,t)$ is chosen as a solution of the driven/damped nonlinear drift-wave equation:

$$\frac{\partial\phi}{\partial t} + a\frac{\partial^3\phi}{\partial t\partial x^2} + c\frac{\partial\phi}{\partial x} + f\phi\frac{\partial\phi}{\partial x} = -\gamma\phi - \varepsilon\sin(x - \Omega t), \tag{3}$$

with a periodic system of length 2π and fixed constants of $a = -0.287$, $c = 1.0$, $f = -6.0$, $\gamma = 0.1$. For the present purpose, we only consider the effect of nonlinear waves of Equation (3) on the particle, while neglect other effect related to the physical system from which the drift-wave equation is derived.

For given (Ω, ϵ) Equation (3) has a steady wave (SW) solution $\phi_0(x - \Omega t)$, satisfying $\partial\phi_0/\partial\tau = 0$ in reference frame $\xi = x - \Omega t, \tau = t$. $\phi_0(\xi)$ can be stable or unstable. By setting $\phi(x,t) = \phi_0(\xi) + \delta\phi(\xi,\tau)$ we obtain an equation for perturbation wave (PW) $\delta\phi(\xi,\tau)$ from Equation (3):

$$\frac{\partial}{\partial\tau}\left[1 + a\frac{\partial^2}{\partial\xi^2}\right]\delta\phi - \Omega\frac{\partial}{\partial\xi}\left[1 + a\frac{\partial^2}{\partial\xi^2}\right]\delta\phi + c\frac{\partial}{\partial\xi}\delta\phi$$
$$+\gamma\delta\phi + f\frac{\partial}{\partial\xi}[\phi_0(\xi)\delta\phi] + f\delta\phi\frac{\partial}{\partial\xi}\delta\phi = 0. \tag{4}$$

If the last term in the left hand side is neglected, Equation (4) becomes linear with respect to $\delta\phi$. By Fourier expanding $\phi_0(\xi)$ and $\delta\phi^l(\xi,\tau)$ the (nonlinear) dispersion relation of $\delta\phi^l$ can be obtained. Here superscript l indicates linear approximation of $\delta\phi$. If $\phi_0(\xi)$ is stable all the mode eigenvalues of $\delta\phi^l$ are complex conjugate with negative real parts. If a real part becomes positive, $\phi_0(\xi)$ is unstable. In particular, if one mode eigenfrequency vanishes, $\phi_0(\xi)$ is unstable due to saddle instability, called as saddle steady wave (SSW) in the following.

Figures 1(a)(b) are contour plots of two dynamically different wave patterns evolving from SSW $\phi_0(x,t)$ respectively. In Figure 1a $\Omega = 0.65, \epsilon = 0.18$, the

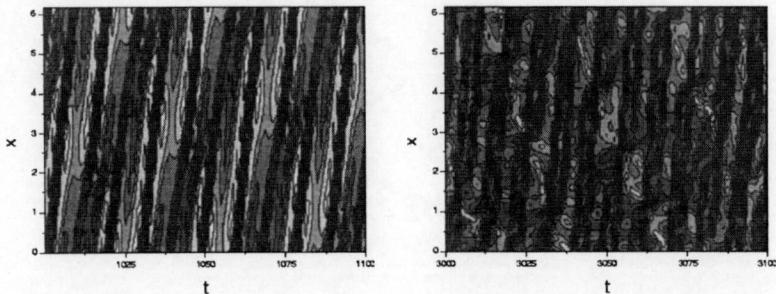

Figure 1. (a) SR wave for $\Omega = 0.65$, $\epsilon = 0.18$, (b) STC wave for $\Omega = 0.65$, $\epsilon = 0.21$.

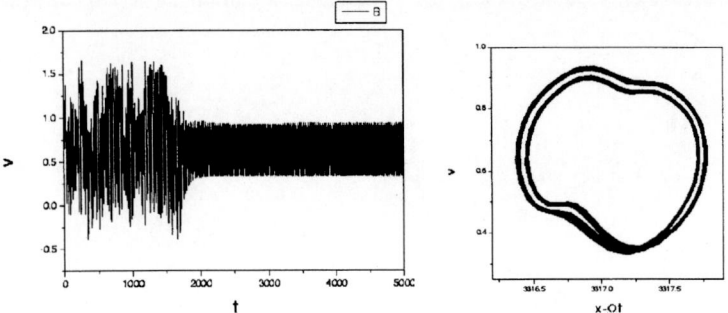

Figure 2. (a) Particle velocity $v(t)$, (b) phase plot $v(\tau)$ vs $\xi(\tau)$ in the SR wave field of $\Omega = 0.65$, $\epsilon = 0.16$.

pattern is spatially regular (SR), but its shape varies with time. In Figure 1b $\Omega = 0.65$, $\epsilon = 0.21$, the wave pattern is chaotic both in time and in space (STC). The spatial power spectra of Figures 1a and b follow different laws respectively. In previous works we have shown that it is a crisis that causes the transition from the SR to STC; and the transition is a critical phenomenon in parameter space, for $\Omega = 0.65$, the critical transition point is $\epsilon_c \approx 0.20$. For the details of the crisis-induced transition to the STC, one can refer to He (1998–2001).

3. Particle Motion in SR and STC Wave Fields

When solving Equation (3) with the pseudospectral method twenty effective modes are used, of which only five modes are used as the field, for characteristic behaviors of the particle are determined only by few key modes.

Figure 2 shows the particle motion in field $\phi(x, t)$ with $\Omega = 0.65$, $\epsilon = 0.16$, here $\phi(x, t)$ is the SR while temporally nearly periodic. In Figure 2a particle velocity v evolves to about periodic oscillations. The averaged particle velocity in the asymptotic periodic motion is $\overline{v} \approx 0.6475$, which changes little when initial values $x(t = 0)$ and $v(t = 0)$ are varied. It is noticed that $\overline{v}$ is very near to the group velocity of SW $\phi_0(x - \Omega t)$, i.e., $\Omega = 0.65$, suggesting that the particle

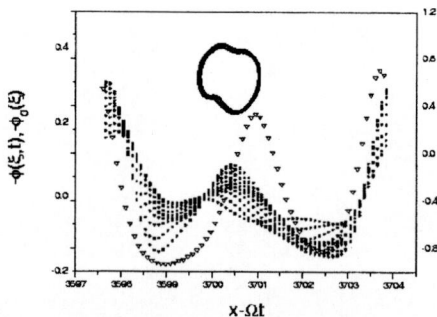

Figure 3. Particle trajectory (solid line) is plotted together with the wave field in the SW frame. Triangle line gives $\phi_0(\xi)$, dotted lines $\phi(\xi, \tau)$, the same case of Figure 2, $q = -1$ is included in the potential.

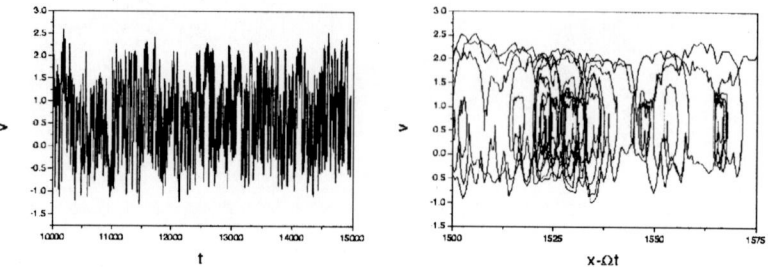

Figure 4. (a) Particle velocity $v(t)$, (b) phase plot $v(\tau)$ vs $\xi(\tau)$ in the STC field of $\Omega = 0.65, \epsilon = 0.21$.

is most probably trapped by the wave. Figure 2b is the phase plot $v(t)$ vs $\xi(t)$, in the SW frame the particle makes a cyclic motion, indicating that the particle is indeed trapped by the wave. If the wave is SR but temporally chaotic, e.g. for $\Omega = 0.65, \epsilon = 0.18$, the particle orbit becomes a little chaotic, however, eventually it is also trapped by a wave trough.

In Figure 3 the particle phase plot is drawn together with field $\phi(x, t)$, again in the SW frame. In the plot, SSW $\phi_0(\xi)$ is shown by a triangle line, which cannot be realized due to the saddle instability; the dotted lines give the realized wave field $\phi(\xi, \tau)$ at several instants τ. One can see that the particle motion is trapped into a wave trough.

Figure 4 is for an STC field with $\Omega = 0.65, \epsilon = 0.21$, the picture is completely different from the case in the SR field. In Figure 4a $v(t)$ is very chaotic, the averaged asymptotic velocities are scattered with variation of initial values. By averaging for seventy runs we get $\bar{v} \approx 0.7513$, which is even larger than the SW velocity, $\Omega = 0.65$. In Figure 4b, $v(t)$ vs $\xi(t)$, one can see that the particle can be trapped or transit freely. Figure 5 is particle position $\xi(\tau)$, corresponding to the free and trapped phases, here one can find jumping and random oscillation phases respectively. On average the particle drifts forward.

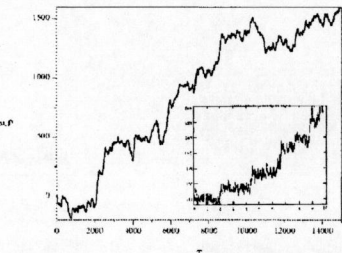

Figure 5. Particle position $\xi(\tau)$, the same case of Figure 4. The inserted is experimental data of motor protein.

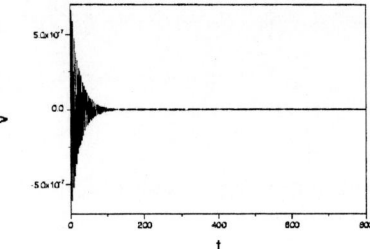

Figure 6. Particle velocity $v(t)$ in $\phi_0(x, t)$ with $\Omega = 0.65$, $\epsilon = 0.16$.

The charge sign has little influence on $\overline{v}$ if the field is SR, but its effect is significant in the STC field. For example, for $\epsilon = 0.21$ with the opposite sign of q we get $\overline{v} \approx 0.4882 < \Omega = 0.65$ in contrast to in Figure 4 where $\overline{v} > \Omega$. In both cases, however, an initially slow particle is accelerated along the SW in the laboratory frame.

In Figure 6 we formally use the virtual pattern of $\phi_0(x - \Omega t)$ for $\Omega = 0.65$, $\epsilon = 0.16$ (see the triangle curve in Figure 3) to compute the particle motion, despite the fact that it can not be realized due to saddle instability. It can be seen that v approaches to zero quickly in coherent structure $\phi_0(x - \Omega t)$. This result indicates that the PW plays a negligible role in the acceleration.

4. Acceleration Mechanism

We notice that the particle motion in the STC field bears similar feature to that of motor protein. The inserted plot in Figure 5 is an experimental result of motor protein displaying jumping and oscillation phases (Svobada *et al.*, 1993). In typical models, a motor protein moves in piecewise linear potential, the second law forbids the particle from displaying any net drift speed, even if the symmetry of potential is broken. But if the particle is subject to an external force having time correlations, detailed balance is lost and the particle can exhibit a nonzero net drift speed. Thus, broken symmetry ('ratchet') and time correlations are sufficient ingredients for particle transport (Magnasco, 1991).

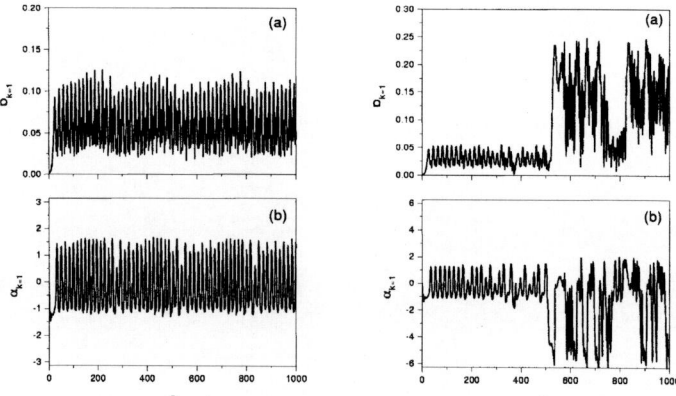

Figure 7. Left: time evolutions of $b_{k=1}(\tau)$ and $\alpha_{k=1}(\tau)$ for SR state($\Omega = 0.65, \epsilon = 0.19$), and right: transition to the STC state ($\Omega = 0.65, \epsilon = 0.22$).

Our situation is very similar to that of motor protein if observed in moving frame (ξ, τ). Actually a nonsteady field $\phi(\xi, \tau)$ includes intrinsically the two ingredients for the acceleration: asymmetric potential and time-correlated force. To demonstrate it let us write $\phi(\xi, \tau) = \phi_0(\xi) + \delta\phi(\xi, \tau)$, where $\phi_0(\xi)$ and $\delta\phi(\xi, \tau)$ can be solved from the mode equations of $\partial\phi_0/\partial\tau = 0$ and Equation (4), respectively. Only few key modes in the results are utilized to construct the wave field:

$$\psi(\xi, \tau) = \sum_{k=1,2} \phi_{0,k}(\xi) + \delta\phi_{k=1}(\xi, \tau) \qquad (5)$$

where $\phi_{0,k}(\xi) = A_k \cos(k\xi + \theta_k)$ and $\delta\phi_k(\xi, \tau) = b_k(\tau) \cos[k\xi + \alpha_k(\tau)]$. In the present case $\phi_{0,k=2}$ is the highest mode, besides, with its inclusion, the potential becomes asymmetric, this is crucial for the acceleration.

For $\Omega = 0.65, \epsilon = 0.19$ and 0.22, the fields thus constructed are denoted as $\psi_{19}(\xi, \tau)$ and $\psi_{22}(\xi, \tau)$, they are the simplified models of the SR and STC fields respectively. Here, for $\epsilon = 0.19$, we have $A_1 = 0.10435, A_2 = 0.38733, \theta_1 = 4.3640, \theta_2 = 0.09039$; for $\epsilon = 0.22$, $A_1 = 0.09165, A_2 = 0.3707, \theta_1 = 4.1359, \theta_2 = 0.1188$. The corresponding data of $b_1(\tau)$ and $\alpha_1(\tau)$ are given in Figures 7a and 7b ($\tau > \sim 600$), respectively. The initial condition is fixed as $\xi(0) = 0, u(0) = 0$, here $u = v - \Omega$.

Figure 8 is asymptotic $u(\tau)$ vs $\xi(\tau)$ of the particle in $\psi_{19}(\xi, \tau)$, it is in a trapped state. Figure 9 shows (a) STC field $\psi_{22}(\xi, \tau)$, (b) $u(\tau)$ vs $\xi(\tau)$, the particle is in a trapped-free state and displays a net drift. These results agree with that in the last section. One can see that ϕ_0 contributions in ψ_{19} and ψ_{22} have little discrepancy, the dramatic different behaviors of particle in them can be easily understood by comparing the PW motion in Figures. 7 (left) and 7 (right).

From the simplified model, we are convinced that the sufficient ingredients for the acceleration are 'ratchet' $\phi_0(\xi)$ and time-correlated force $\delta\phi(\xi, \tau)$. This is in a sense very similar to motor protein. However, there are two important points

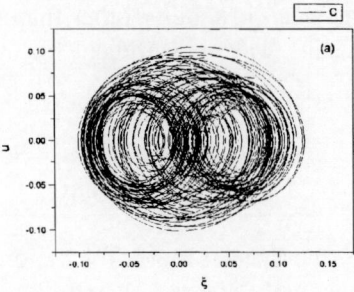

Figure 8. (a) Phase plot $u(\tau)$ vs $\xi(\tau)$ of particle in $\psi_{19}(\xi)$.

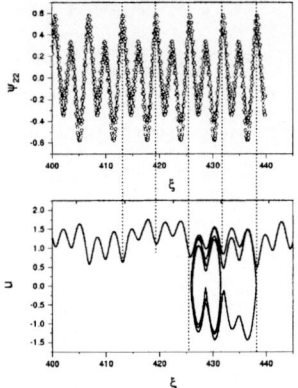

Figure 9. (a) STC potential $\psi_{22}(\xi)$ (the actual fluctuation is much stronger than indicated here), (b) phase plot $u(\tau)$ vs $\xi(\tau)$ in $\psi_{22}(\xi)$.

different from the latter: (1) our fluctuating force, $\delta\phi(\xi, \tau)$, is a result of deterministic nonlinear evolution; (2) 'ratchet' $\phi_0(x, t)$ moves with velocity Ω in laboratory frame, so even in a trapped state the particle can be carried along by the SR wave and accelerated.

5. Conclusion

Charged particle motion in nonlinear wave fields is studied. In all the tested nonsteady wave fields a slow particle can be accelerated in the orientation of the SW. However, in the SR field the particle is finally trapped by a wave trough and eventually acquires the group velocity of the SW; while in the STC field the particle experiences trapped and free phases randomly, depending on the charge sign the averaged velocity can be larger or smaller than the group velocity of the SW. It is shown that the virtual pattern of SSW plays the role of an asymmetric potential, which together with nonzero PW are necessary for the acceleration. It is emphasized that the two ingredients for the acceleration arise from the intrinsic property of the deterministic nonlinear system. Finally we should mention that since in general a SW can be unstable due to an instability other than the saddle type, whether a non-steady nonlinear traveling wave is sufficient for the acceleration needs to be studied further.

Acknowledgements

This work is supported by the Special Funds for Major State Basic Research Projects and by the National Natural Science Foundation of China No. 19975006, and by RFDP No.20010027005.

References

Sagdeev, R. Z., Usikov, D. A. and Zaslavsky, G. M.: 1988, *Nonlinear Physics*, Harwood Academic Publishers, London.
Fermi, E.: 1949, *Phys. Rev.* **75**, 1169.
Tajima, T. and Dawson, J. M.: 1979, *Phys. Rev. Lett* **43**, 267.
Joshi, C. J. and Corkum, P. B.: 1995, *Phys. Today* **48**, 36.
Bulanov, S. V. *et al.*: 1999, *Phys. Rev. Lett.* **82** 3440.
Magnasco, M. O.: 1993, *Phys. Rev. Lett.* **71**, 1477.
Dodd, R. K., Eilbeck, J. C., Gibbon, J. D. and Morris, H. C.: 1982, *Solitons and Nonlinear Wave Equations*, Academic Press, London, p. 596.
He, K.: 1998, *Phys. Rev. Lett.* **80**, 696.
He, K.: 1999, *Phys. Rev. E.* **59**, 5278.
He, K.: 2000, *Phys. Rev. Lett.* **84**, 3290.
He, K.: 2001, *Phys. Rev. E.* **63**, 016218.
Svobada, K. *et al.*: 1993, Nature, **365**, 721.

CRITICAL PHENOMENON, CRISIS AND TRANSITION TO SPATIOTEMPORAL CHAOS IN PLASMAS

KAIFEN HE

Institute of Low Energy Nuclear Physics, Beijing Normal University, Beijing, 100875, China and World Institute for Space Environment Research-WISER, NITP, University of Adelaide, SA 5005, Australia

Abstract. In a driven/damped drift-wave system a steady wave induces nonlinear variation of the dispersion of a perturbation wave (PW). Competition between the nonlinear dispersion with self-nonlinearity of the PW results in rich wave dynamic behaviors. In particular, a steady wave at the negative tangency slope of a hysteresis becomes unstable due to a saddle instability. It is found that such saddle steady wave (SSW) plays an important role in the discontinuous transition from a spatially coherent state to spatiotemporal chaos (STC). The transition is caused by a crisis due to a collision of the PW attactor to an unstable orbit of the SSW. In the time evolution, it is a 'pattern resonance' of the realized wave with the virtual SSW that triggers the crisis. The transition also displays as a critical phenomenon in parameter space, which is related to the change in the symmetry property of the motion of master mode ($k = 1$) of the PW with respect to that of SSW. In the spatially coherent state the former is trapped by the SSW partial wave, while in the STC it can become free from the latter, its trajectory crosses two unstable orbits of the SSW frequently, causing very turbulent behavior.

Key words: crisis, critical phenomenon, pattern resonance, saddle steady wave, spatiotemporal chaos

1. Introduction

There has been a great interest in understanding mechanisms for transition to turbulence. In recent years, many authors studied the problem from the viewpoint of nonlinear dynamics and believe that turbulence is a deterministic spatiotemporal chaos (STC) (Cross and Hohenberg, 1993). However, the terminology of 'STC' is somewhat ambiguous (Sagdeev *et al.*, 1988). In some literatures 'STC' is referred to any chaotic states in space-time dependent systems. We suggest that an 'STC' should be referred to those states where the spatial coherence is destroyed. Therefore, to understand mechanism for onset to STC, a key problem is: what causes collapse of the spatial coherence.

In the last decades a great progress was made in studying nonlinear dynamics of time-dependent systems, different routes of transition to chaos are revealed (Ott, 1993). For explaining onset of turbulence in space-time dependent systems, two pictures have been proposed (Lichtenberg and Lieberman, 1983): in Landau's picture a turbulence is considered as a result of appearances of infinitely many new

Space Science Reviews **107**: 475–494, 2003.
© 2003 *Kluwer Academic Publishers.*

frequencies. This picture, however, is not supported by experiments; In Ruelle-Takens picture, it is shown that the third independent frequency destabilizes the 2d-torus and leads to chaos. This scenario is observed in transition to chaos in low dimensional systems, e.g. in coupled three-wave interactions (Biskamp and He, 1985; Klinger *et al.*, 1997). However, for the onset of fully-developed turbulence appearing in systems with infinitely many dimensions such as nonlinear waves, the mechanism is still open.

It should be stressed that, although many special problems need to answer, nonlinear dynamics of wave systems actually can be closely related to that of time-dependent systems. This problem can be seen clearly when the system has steady wave (SW) solutions. If observed in the reference frame following it, an SW is static with constant mode amplitudes and phases, i.e., it is a fixed point in the Fourier space with infinitely many dimensions. Consequently one can discuss stability property of the fixed point by disturbing its free dimensions, just like what we usually do in time-dependent systems with few dimensions. This procedure is equivalent to studying response of the system when the SW is subject to a perturbation wave (PW).

In general, a stable SW may lose its stability with variation of parameters. In numerical simulations it has been observed that an unstable SW may settle to a state that is chaotic in time, but its spatial behavior remains coherent (Bishop *et al.*, 1983). Therefore, there must be further dynamic process which causes destruction of the spatial coherence. In the present work we will show that a crisis (Grebogi *et al.*, 1982) can be such a process. Crisis is sudden changes in chaotic attractors with parameter variation, e.g. sudden destruction, creation, expansion or shrinking of attractors. In our case, the crisis is caused by a collision of the attractor to an unstable orbit of the saddle steady wave, which spoils the spatial coherence and induces an onset to the STC.

In Section 2 the model equation and its SW solutions are described. In Section 3 we study nonlinear variation of the PW dispersion in the presence of an SW. For the present purpose, the emphasis is on those SW solutions with saddle instability (SSW). In Section 4 linear orbits of PW at an SSW are calculated, and in the following sections the PW self-nonlinearity is included. In Section 5 we find that starting from an unstable SSW the system can settle to a stable gap state. With variation of parameters the gap state may lose stability, if its attractor collides with the saddle point, a crisis occurs, leading to a transition to the STC. This result is given in Section 6. In Section 7 we demonstrate that, it is a 'pattern resonance' with virtual waveform of the SSW that triggers the crisis. In Section 8 a physical reason is discussed to explain the critical phenomenon of transition to the STC in parameter space. In Section 9 and 10, it is shown that the destruction of spatial coherence is associated with a trapped-free state transition of the master PW mode, and unstable orbits of the SSW play an important role in the turbulent behaviors in the STC. Finally Section 11 is a discussion.

2. Model Equation and Hystereses of Steady Wave Solutions

As a model we use the following equation,

$$\frac{\partial \phi}{\partial t} + a\frac{\partial^3 \phi}{\partial t \partial x^2} + c\frac{\partial \phi}{\partial x} + f\phi\frac{\partial \phi}{\partial x} = -\gamma\phi - \epsilon \sin(x - \Omega t) \qquad (1)$$

with periodic boundary condition of length 2π, here $a < 0, c, f, \gamma$ are fixed, and Ω, ϵ are control parameters. If $\epsilon = \gamma = 0$, Equation (1) is the nonlinear drift-wave equation in plasmas (Oraevskii $et~al.$, 1969) or Regularized Long Wave Equation in fluids (Dodd $et~al.$, 1983). In weak nonlinearity Equation (1) approaches to KdV equation, however, it is nonintegrable. With finite (γ, ϵ), by using pseudospectral method we have found rich dynamic behaviors of $\phi(x, t)$ including very turbulent motion (He and Salat, 1989).

Equation (1) has SW solutions in the form of $\phi_0(x - \Omega t)$, in this case, 'wave energy'

$$E(t) = \frac{1}{2\pi}\int_0^{2\pi} \frac{1}{2}[\phi^2(x) - a(\partial\phi/\partial x)^2]dx \qquad (2)$$

approaches to constant, $E_0 \equiv E(\phi_0)$. In reference frame $(\xi \equiv x - \Omega t, \tau = t)$, if expanding

$$\phi_0(\xi) = \sum_{k=1}^{N} A_k \cos(k\xi + \theta_k), \qquad (3)$$

$\{A_k, \theta_k\}$ can be solved from SW equation $\partial\phi_0(\xi)/\partial\tau = 0$.

It is found that for given Ω 'wave energy' E_0 of SW as a function of ϵ constitutes a hysteresis. Figure 1 is an example, a group of hystereses $E_0(\epsilon)$ appears in the range of $0.5 \leq \Omega \leq 0.78$. With decreasing Ω one can find a series of groups of hystereses, they appear in the Ω ranges with smaller and smaller scale respectively.

An interesting phenomenon is, the complicated dynamic behaviors we observed in Equation (1), in particular the very turbulent ones, seem to associate with existence of the hystereses. In the following one can see the reason, i.e., the turbulent motion is related to those SW with a saddle instability, and such saddle SW locate at the negative tangency branches of hystereses.

3. Stability of Steady Wave Under Perturbation

When its free dimensions are disturbed, an SW (a fixed point in Fourier space) can be stable or unstable. To study stability property of an SW $\phi_0(x - \Omega t)$, let us transform Equation (1) into (ξ, τ) frame and set

$$\phi(\xi, \tau) = \phi_0(\xi) + \delta\phi(\xi, \tau), \qquad (4)$$

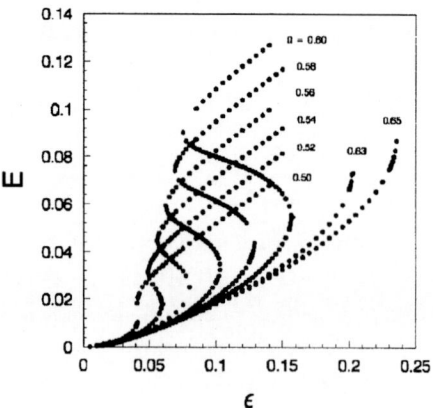

Figure 1. A group of hystereses of E_0 vs ϵ.

then $\delta\phi(\xi, \tau)$ is governed by

$$\frac{\partial}{\partial\tau}[1 + a\frac{\partial^2}{\partial\xi^2}]\delta\phi \; -\Omega\frac{\partial}{\partial\xi}[1 + a\frac{\partial^2}{\partial\xi^2}]\delta\phi + c\frac{\partial}{\partial\xi}\delta\phi + \gamma\delta\phi$$

$$+ f\frac{\partial}{\partial\xi}[\phi_0(\xi)\delta\phi] + f\delta\phi\frac{\partial}{\partial\xi}\delta\phi = 0. \tag{5}$$

It describes response of the system to an initial PW. In Equation (5) the last two terms in the left hand side arise from the system nonlinearity, one of them describes the interaction between $\delta\phi$ and $\phi_0(\xi)$, which gives rise to a nonlinear variation of its dispersion, the other is the self-nonlinearity of $\delta\phi$.

If neglecting the last term (i.e., the nonlinear term with respect to $\delta\phi$) in Equation (5), and writing

$$\delta\phi^l(\xi, \tau) = e^{\lambda\tau}[\sum_{k=1}^{N\to\infty} b_k^l \exp i(k\xi + \alpha_k^l) + c.c.], \tag{6}$$

one can study the linear response of $\delta\phi$ in the presence of $\phi_0(\xi)$, here superscript l indicates the linear approximation. By inserting expansions (3) and (6), an eigen equation for the modes of $\delta\phi^l(\xi)$ can be obtained in the following form

$$|\mathbf{H}(\{A_k, \theta_k\}) - \lambda\mathbf{I}| = 0, \tag{7}$$

Here matrix $\mathbf{H}$ depends on $\phi_0(\xi)$, $\mathbf{I}$ is the Unit Matrix.

Depending on the special form of $\phi_0(\xi)$, eigenvalues λ can be solved from Equation (7). If truncating to N modes, in general one gets N complex conjugate eigenvalues. Two types of instabilities are observed: Hopf instability and saddle instability (He, 1992, 1994, 1995).[1]

---

[1] In Equation (19) of *Phys. Lett.* **A169** a term is missed out in the element of $\mathbf{H}$ for $m < n$, when it is added in the present work a satisfactory agreement in the bifurcation lines with the simulation is obtained.

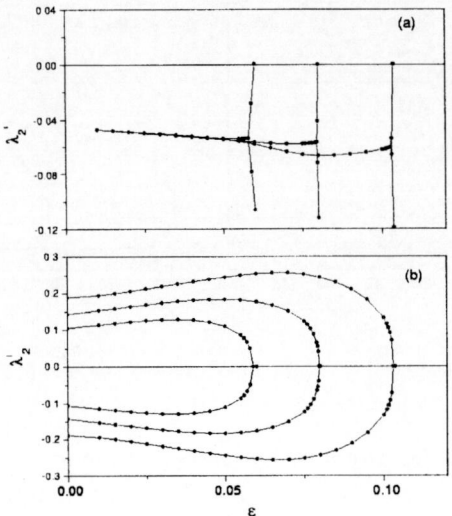

Figure 2. Eigenvalues of mode $k = 2$ when $\phi_0(\xi)$ varying with ϵ along lower branch of the hystereses for $\Omega = 0.52, 0.54, 0.56$ in Figure 1, (a) Real part, (b) imaginary part.

The Hopf instability is caused by nonlinear resonance between two internal modes. In certain parameter regimes, eigenfrequencies of the two modes of $\delta\phi^l$ get locked, of which the growth rate of one mode changes to positive, indicating occurrence of a Hopf instability. In this case the eigenfrequency of resonance modes appears in the spectrum of $E(t)$, or in the spectrum of $\phi(x, t)$ as beat frequencies with Ω (Doppler shift).

On the other hand, if the eigenfrequency of one mode vanishes, the complex conjugate eigenvalues degenerate to two real values. When one of them becomes positive, a saddle instability occurs. A zero eigenfrequency in frame (ξ, τ) indicates that in the laboratory frame the eigenfrequency is Ω, that is, the saddle instability is caused by the resonance between an internal mode and the applied harmonic wave. The resonance is of nonlinear, for the mode eigenfrequency has been varied due to existence of the SW.

For three Ω values Figure 2 gives the variation of $\lambda_{k=2}$ when $\phi_0(\xi)$ varies with ϵ along the lower branch of the respective hysteresis in Figure 1, with (a) real parts $\lambda^r_{k=2}$, (b) imaginary parts $\lambda^i_{k=2}$. For given Ω by comparing Figure 2 with Figure 1 one can see that, when near to the turning point of a hysteresis the two branches of $\lambda^i_{k=2}$ degenerate to zero; in the meantime, $\lambda^r_{k=2}$ bifurcates to two stems, one of which can cross zero and become positive, indicating a saddle instability.

In the whole middle branch of a hysteresis, eigenfrequency $\lambda^i_{k=2}$ keeps as zero, only when going about to the upper branch $\lambda^i_{k=2}$ changes to finite conjugate values. We call a steady wave with saddle instability as SSW, denoted as $\phi_0^*(\xi)$ in the following.

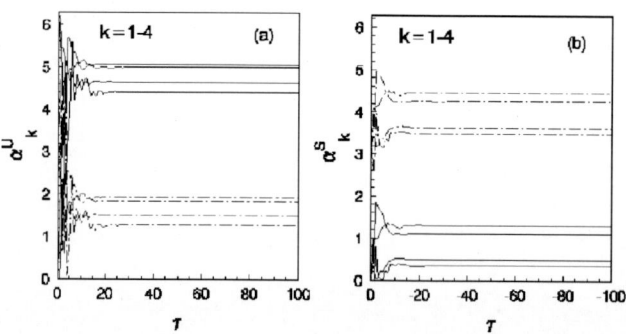

Figure 3. Variations of phase α_k for $k = 1 - 4$ along (a) two unstable orbits, (b) two stable orbits, $\Omega = 0.65$, $\epsilon = 0.22$.

Corresponding to the group of hystereses in Figure 1 the resonance mode is $k = 2$. With decreasing Ω mode $k = 3, 4, 5, \ldots$ can in turn become resonant with the applied wave, which is responsible for the appearance of a series of groups of hystereses mentioned above. Furthermore, there is evidence to show that in each group of hystereses, transition from the lower to upper branch is associated with the resonance mode changing from a (nonlinear) negative-energy mode to a positive-energy mode.

4. Stable and Unstable Orbits of an SSW

To clarify the role of SSW in the transition to STC, let us first see its unstable orbit (UO) and stable orbit (SO). In general one can write

$$\delta\phi(\xi, \tau) = \sum_{k=1}^{\infty} b_k(\tau) \exp i[k\xi + \alpha_k(\tau)] + c.c., \tag{8}$$

then UO/SO of an SSW can be calculated if the last nonlinear term in the left hand side of Equation (5) is neglected. An important feature of the UO/SO is, along them phases $\{\alpha_k\}$ are constants respectively. Take UO as an example: starting from an SSW, with amplitude $\{b_k^l\}$ increasing exponentially phase $\{\alpha_k^l\}$ tend asymptotically to a group of constant values. Since there are two UO's, if denote these two groups of constants as $\{\alpha_k^{U_1}\}$ and $\{\alpha_k^{U_2}\}$ respectively, we find a relation $\{\alpha_k^{U_2}\} = \{\alpha_k^{U_1} + \pi\}$. If reversing the time variable, the SO's can be found, they correspond to another two groups of constant phases $\{\alpha_k^{S_{1,2}}\}$ respectively, with a difference π.

Figure 3 shows an example for $\Omega = 0.65$, $\epsilon = 0.22$ with (a) temporal evolution of $\{\alpha_k^l(\tau)\}$, (b) inversed temporal evolution of $\{\alpha_k^l(\tau)\}$ for $k = 1 - 4$. In both (a) and (b) one can find two groups of curves approaching to constants corresponding to the UO/SO's respectively. This knowledge on the UO/SO is very helpful for understanding the turbulent behavior in the STC, as will be described in Section 10.

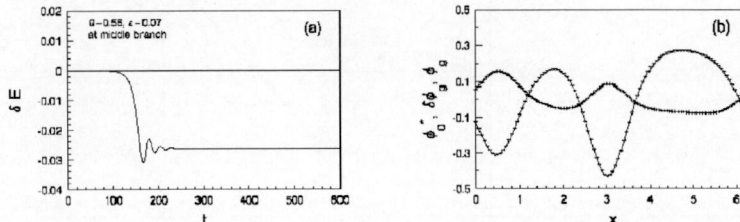

Figure 4. Stable gap solitary wave for $\Omega = 0.56$, $\epsilon = 0.07$. (a) Time evolution of $\delta E(\tau)$, (b) SSW $\phi_0(\xi)$ (triangles) and coherent structure $\delta\phi_g(\xi)$ (crosses) in the asymptotic state.

5. Gap Solitary Wave Coexisting with SSW

In the following we include self-nonlinearity of the PW, i.e., the last term in the left hand side of Equation (5). Again expansion (8) is used. Let us start from an SSW with a tiny initial perturbation. Due to the saddle instability, in the linear stage the PW mode amplitudes grow exponentially along an UO of the SSW, however, attributed to the self-nonlinearity, they gradually approach to a saturated level. In particular, in certain parameter regimes, the amplitudes may settle to constants. In this case $\delta E(\tau) \equiv E[\phi(\tau)] - E_0$, the difference of 'wave energy' to that of the SSW, tends to a constant, as can be seen in Figure 4a. In the meantime, the PW modes build up a coherent structure, denoted as $\delta\phi_g(\xi)$, with its peaks trapped in the trough of SSW, as shown in Figure 4b.

Combination of the SSW and coherent structure, $\phi_g(\xi) \equiv \phi_0^*(\xi) + \delta\phi_g(\xi)$, is also an SW solution of Equation 1, and in particular in Figure 4 it is a stable SW with its energy lower than that of $\phi_0^*(\xi)$. $\phi_g(\xi)$ can be called as a gap solitary wave (He and Zhou, 1997) for the following reason.

As well known, a periodic potential may induce a forbidden gap in the linear spectrum of irradiation, however, when nonlinearity sets in, an irradiation in the gap is allowed to transport through the system with a soliton-like envelope of the amplitudes (Chen and Mills, 1983). In our case, an SSW plays a role of the periodic potential, which induces a forbidden gap in the PW linear spectrum (zero eigenfrequency in the saddle instability indicates that no frequency is allowed to transport through the system for the resonance mode). Similarly, when the PW self-nonlinearity is included, a soliton-like structure $\delta\phi_g(\xi)$ can be built up on the SSW.

Figure 5 gives a hysteresis $E_0(\epsilon)$ (bullets), as well as 'wave energy' of steady gap solitary wave $\phi_g(\xi)$, i.e. $E_g(\epsilon)$ (circle and triangle). These solutions are solved from the SW equation, they can be stable or unstable. One can see that E_g locate about right at the 'forbidden gap', i.e., where the middle branch of the hysteresis is. For a given SSW with different initial PW, one can find two different $\delta\phi_g(\xi)$, furthermore, gap states can also be supported by an SW at the upper branch of hysteresis, that results in three stems of the gap states in Figure 5.

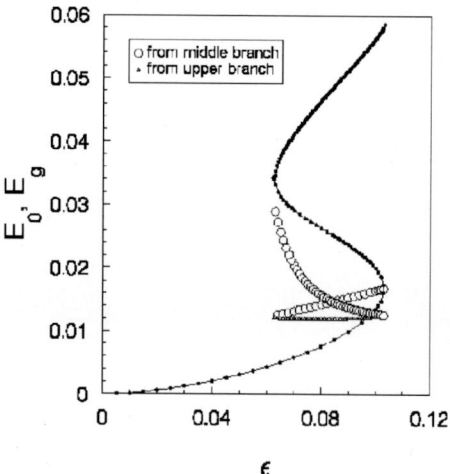

Figure 5. Hysteresis $E_0(\epsilon)$ and gap states E_g associated with the middle branch (circle) and with the upper branch (triangle), $\Omega = 0.56$.

A stable soliton-like structure $\delta\phi_g(\xi)$ as in Figure 4 can be understood as a result of balance between the dispersion and nonlinearity of PW as can be seen in Equation (5), just like the normal solitons, however, here the dispersion experiences a nonlinear variation due to an interaction with the SW.

6. Crisis Due to Collision to UO of SSW and Destruction of Spatial Coherence

For parameters in the concerned regime, now we have two fixed points: one is an unstable SSW, the other is a gap solitary wave. The latter may also become unstable with variation of parameters and be attracted to a periodic or chaotic attractor. In the following we can see, if the attractor expands, it may collide with UO of the saddle point, inducing a crisis (He, 1998).

Let us project $\delta E(\tau)$ to its Fourier components and plot $\delta E_{k=1}$ vs $\delta E_{k=2}$ for different parameters (Ω, ϵ). In each plot of Figure 6 one can find a fixed point of SSW with its UO/SO as well as an atttractor of the gap state. With variation of (Ω, ϵ) one can see that the gap state changes from (a) a fixed point ['$+$'] to (b) a limit cycle; in (c) the limit cycle becomes bigger and a little chaotic; finally in (d) the chaotic attractor collides with an UO of the saddle point. In this case, a crisis occurs, after wandering in the old attractor for a while, suddenly the orbit is attracted to a much larger new attractor. In the old attractor the motion is spatially coherent, while in the new attractor the motion is very turbulent, the spatial coherence is destroyed.

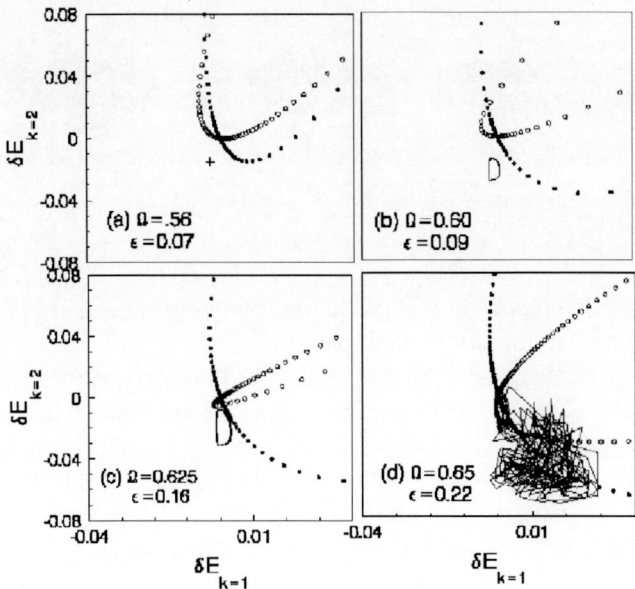

Figure 6. Collision of gap attractor with UO of saddle point. (a) $\Omega = 0.56, \epsilon = 0.07$, (b) $\Omega = 0.60, \epsilon = 0.11$, (c) $\Omega = 0.625, \epsilon = 0.16$, (d) $\Omega = 0.65, \epsilon = 0.22$.

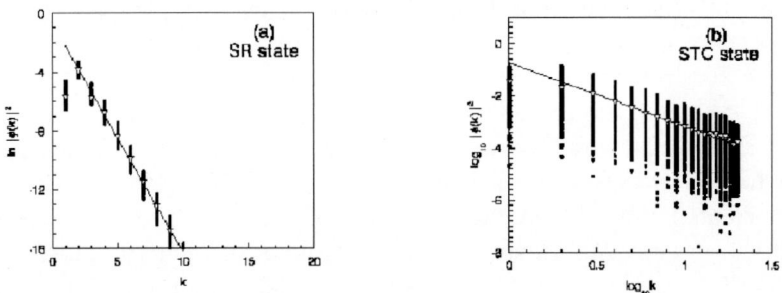

Figure 7. Spatial spectra $\langle \phi^2(k) \rangle$ with (a) exponential law in SR, (b) power law in STC, $\Omega = 0.65, \epsilon = 0.22$.

7. 'Pattern Resonance' and Onset of Crisis

Before the crisis, the wave patterns are of spatially regular (SR), although its temporal behavior can be chaotic. It has a spatial spectrum of an exponential law, $\langle \phi(k)^2 \rangle \sim e^{-\alpha k}$ (Figure 7a); after the crisis, the turbulent wave pattern displays a spatial spectrum of power law, $\langle \phi(k)^2 \rangle \sim k^{-\beta}$ (Figure 7b). In our case, $\alpha \approx 3/2, \beta \approx 5/2$.

Figure 8 shows three wave patterns transitting from an SR to the STC state in time sequence. The transition occurs abruptly, which can be attributed to a collision to the saddle point as we have demonstrated in the above section. However, what

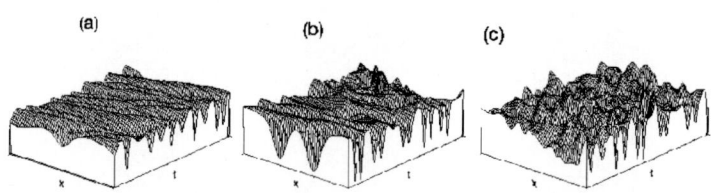

Figure 8. Transition of wave pattern from SR to STC for $\Omega = 0.65$, $\epsilon = 0.22$. (a) Transient SR state, (b) onset of crisis, (c) stationary STC state.

does it mean by 'collision' in reality? In the following we can see that it is actually an occasional realization of the SSW pattern.

Figure 9 gives snapshots of the realized solution $\phi(x, t)$ as well as virtual pattern of SSW $\phi_0^*(x, t)$. In this section, $\phi(x, t)$ are solved from Equation (1). Although normally $\phi_0^*(x, t)$ is not realized due to the saddle instability, it can be solved from the SW equation $\partial \phi_0 / \partial \tau = 0$. In the plots, (a)(b) are of SR before the crisis, (f) is stationary STC after the crisis. The critical transition occurs in (c) $t = t_c \approx 39.0$, one can see that at this critical time the waveform $\phi(x)$ almost coincides with that of $\phi_0^*(x)$, in another words, SSW $\phi_0^*(x)$ is realized for the moment. Right after it the amplitude of $\phi(x, t)$ grows abruptly to a level higher than that of $\phi_0^*(x)$ as given in (d)(e). In this stage the waveform is still very smooth. Following this linear unstable stage the motion finally settles to the STC, as shown in (f).

Figure 10 shows variation of distance between $\phi(x, t)$ and $\phi_0^*(x, t)$, $\Delta(t) = |\phi(x, t) - \phi_0^*(x, t)|$. As can be seen in the plot, right before onset of the transition Δ shows a sharp spike with extremely small value; it is followed by a linear unstable stage where Δ increases exponentially; finally in the STC state, Δ saturates at a level much higher than before the transition.

The (Ω, ϵ) values in Figures 9 and 10 are the same, but with different initial $\phi(x, t = 0)$. Critical transition time t_c sensitively depends on variation of the initial condition. However, in all the examples right before the transition there always appears a very short period during which the realized wave $\phi(x, t)$ evolves to about the same shape of SSW $\phi_0^*(x - \Omega t)$. We call this phenomenon as a 'pattern resonance' (He, 2000). It is the 'pattern resonance' that triggers the onset of transition to the STC, which is the actual meaning of 'collision' to UO of the saddle point.

This phenomenon is novel and significant. An SSW is normally invisible, yet, in certain conditions (in certain parameter regimes), if it gets realized in a very short period, it can steer the system to a dynamically different state.

At the critical moment when the 'pattern resonance' occurs, $\delta\phi(\xi, \tau)$ becomes extremely small, then the linear approximation in Section 3 is valid in reality, accordingly the saddle instability should occur and $\delta\phi(\xi, \tau)$ grows exponentially along UO of the SSW. This explains the linear growing stage in Figures 9 and 10, and the growth rate obtained from Figure 10 is in agreement with that from Equation (7). Furthermore, since in the saddle instability one PW mode becomes reson-

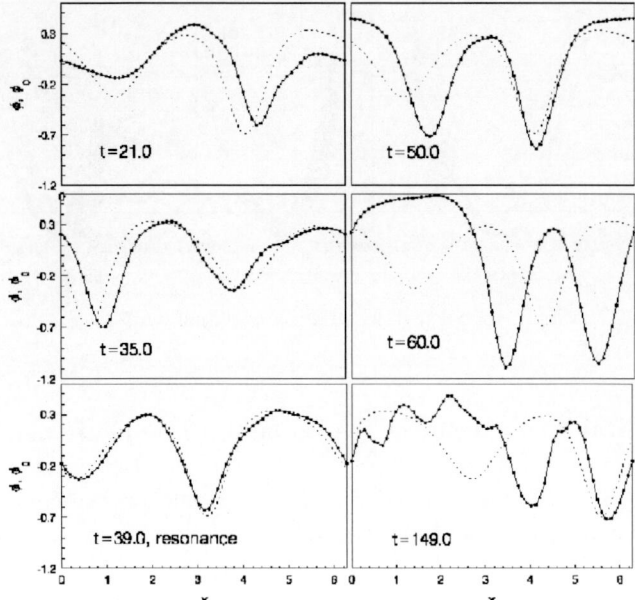

Figure 9. Snapshots of realized wave $\phi(x, t)$ (bullet) and virtual wave $\phi_0^*(x, t)$ (dashed line) for $\Omega = 0.65, \epsilon = 0.22$, with (a)(b) transient SR state($t = 21.0, 35.0$), (c) critical moment for the transition ($t = 39.0$), (d)(e) linear unstable stage ($t = 50.0, 60.0$), (f) stationary STC state ($t = 149.0$).

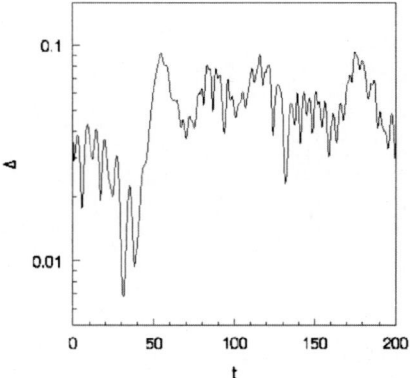

Figure 10. Distance $\Delta(t)$ between the realized wave and SSW, $\Omega = 0.65, \epsilon = 0.22$.

ant with the applied harmonic wave, a 'pattern resonance' is actually a nonlinear frequency resonance, for the mode eigen frequency is altered nonlinearly.

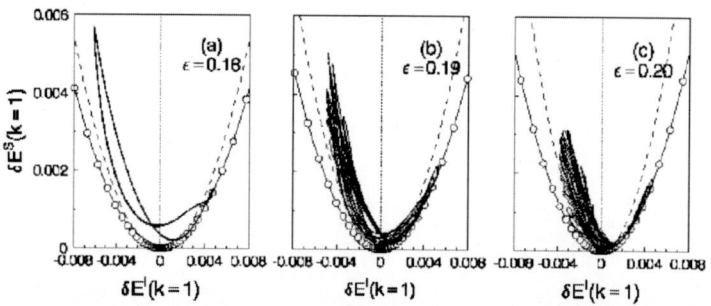

Figure 11. $\delta E_{k=1}^{S}$ vs $\delta E_{k=1}^{I}$ for SR waves before critical point, with (a) $\epsilon = 0.18$, (b) $\epsilon = 0.19$, (c) $\epsilon = 0.20$, $\Omega = 0.65$.

8. Critical Phenomenon of Transition to STC in Parameter Space

Transition to the STC due to crisis displays a critical phenomenon in parameter space. For instance, for $\Omega = 0.65$, we find a critical point $\epsilon_c \approx 0.20$. When $\epsilon < \epsilon_c$ asymptotic solutions are in the SR state with temporally periodic or chaotic behavior; when $\epsilon > \epsilon_c$, in the time evolution one can observe a transition from the SR to STC as shown in the above section.

Let us study this phenomenon in the mode energy representation. A self-energy of an SW mode is given by

$$E_k^S = \frac{1}{4}(1 - ak^2)A_k^2, \tag{9}$$

an interaction energy between PW and SW mode is

$$\delta E_k^I(\tau) = \frac{1}{2}(1 - ak^2)A_k b_k(\tau) \cos[\theta_k - \alpha_k(\tau)], \tag{10}$$

and a self-energy of PW mode is

$$\delta E_k^S(\tau) = \frac{1}{4}(1 - ak^2)b_k^2(\tau). \tag{11}$$

Since E_k^S is constant, we project an orbit to $\delta E_k^I - \delta E_k^S$ plane, and study its behavior when approaching to the critical parameter point.

Figure 11 gives $\delta E_{k=1}^{S}$ vs $\delta E_{k=1}^{I}$ for $\Omega = 0.65$ with (a) $\epsilon = 0.18$, (b) $\epsilon = 0.19$, (c) $\epsilon = 0.20 < \epsilon_c$, respectively. They are all SR waves. Figure 12 is $\delta E_{k=1}^{S}$ vs $\delta E_{k=1}^{I}$ for $\Omega = 0.65$, $\epsilon = 0.22 > \epsilon_c$, with (a) the transient state before onset of the crisis, (b) asymptotic STC state after the crisis. In these plots, dashed line gives the UO's, circles give the SO's. We notice that the PW orbits are constrained by the SO's as if they form an infinitely high barrier.

In general $\delta E_{k=1}^{I}(\tau)$ can be positive or negative. An interesting phenomenon is that, when $\epsilon < \epsilon_c$ in Figure 11 negative values of $\delta E_{k=1}^{I}(\tau)$ are dominant, however, with approaching to critical $\epsilon_c \approx 0.20$, the orbit of $k = 1$ mode looks more and

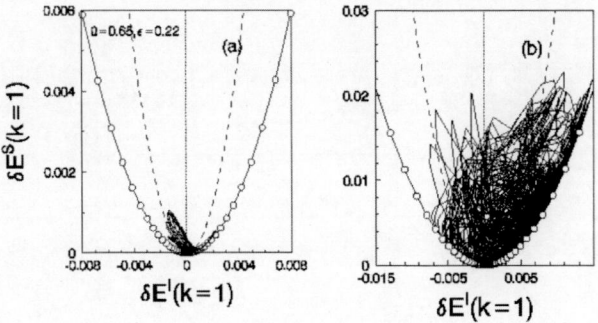

Figure 12. $\delta E_{k=1}^{S}$ vs $\delta E_{k=1}^{I}$ for $\Omega = 0.65$, $\epsilon = 0.22$ beyond critical point. (a) Transient state before the onset of crisis, (b) asymptotic state after the crisis.

more symmetric on the positive and negative sides; When $\epsilon > \epsilon_c$ in Figure 12a, in transient state before the crisis, $\delta E_{k=1}^{I}(\tau)$ becomes very small and symmetric, however, after the crisis in Figure 12b suddenly positive values of $\delta E_{k=1}^{I}(\tau)$ become dominant. This phenomenon suggests that we can define an order parameter according to the symmetry property of PW motion so as to describe the critical phenomenon (He, 2001).

Let us define the following quantity

$$S_k \equiv \langle \delta E_k^{I}(\tau) \rangle^{P} + \langle \delta E_k^{I}(\tau) \rangle^{N} \tag{12}$$

with

$$\langle \delta E_k^{I}(\tau) \rangle^{P} = \frac{1}{T_P} \sum_{\delta E_k^{I} > 0} \delta E_k^{I}(\tau) \Delta \tau, \tag{13}$$

$$\langle \delta E_k^{I}(\tau) \rangle^{N} = \frac{1}{T_N} \sum_{\delta E_k^{I} < 0} \delta E_k^{I}(\tau) \Delta \tau, \tag{14}$$

here $\Delta \tau$ is time step and T_P / T_N are the total time duration when δE_k^{I} takes positive/negative values. That is, $\langle \delta E_k^{I}(\tau) \rangle^{P}$ and $\langle \delta E_k^{I}(\tau) \rangle^{N}$ are averaged positive and negative mode interaction energy respectively.

Figure 13 gives the variation of $S_{k=1}$ with ϵ. In the plot one can see that, if $\epsilon < \epsilon_c$, $S_{k=1}$ is negative, which tends to zero when approaching to ϵ_c; then after crossing the critical point, $\epsilon > \epsilon_c$, $S_{k=1}$ transits discontinuously to a finite positive value. In the plot averaged mode self-energy $\langle \delta E_{k=1}^{S}(\tau) \rangle$ is also shown, when $\epsilon \to \epsilon_c$ it decreases as well but does not reach to zero.

In Figure 14a we draw $|\langle \delta E_{k=1}^{I}(\tau) \rangle^{P} + \langle \delta E_{k=1}^{I}(\tau) \rangle^{N}|$ and $|\langle \delta E_{k=1}^{I}(\tau) \rangle|$ varying with ϵ in the semi-logarithmic plot, here $\langle \delta E_{k=1}^{I}(\tau) \rangle$ is a simple time average of the mode interaction energy. One can see that it is the former but not the latter that has a tendency of reaching zero when approaching to critical point $\epsilon = \epsilon_c$. In Figure 14b it is shown $|S_{k=1}| \sim |\epsilon - \epsilon_c|^{\alpha}$ with $\alpha \approx 1$. Right at the critical parameter

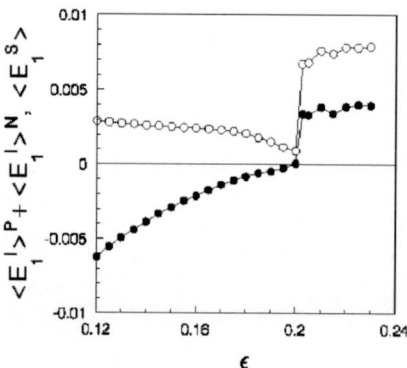

Figure 13. $\langle\delta E_{k=1}^{I}(\tau)\rangle^{P} + \langle\delta E_{k=1}^{I}(\tau)\rangle^{N}$ (bullet) and $\langle\delta E_{k=1}^{S}(\tau)\rangle$ (circle) as functions of ϵ, for $\Omega = 0.65$.

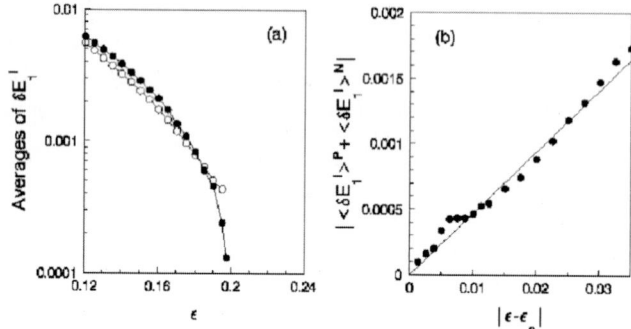

Figure 14. (a) Behaviors of $|\langle\delta E_{k=1}^{I}(\tau)\rangle^{P} + \langle\delta E_{k=1}^{I}(\tau)\rangle^{N}|$ and $|\langle\delta E_{k=1}^{I}(\tau)\rangle|$ vs ϵ, (b) power law of $S_{k=1}$ when approaching to ϵ_{c}, the same case as in Figure 13.

ϵ_{c}, for $k = 1$ mode averaged intensity of the negative interaction energy equals to that of the positive one.

In our examples, as a result of mode couplings $k = 1$ becomes the master mode, it slaves other modes. It can be seen that, in the transition from the SR to STC, only $k = 1$ PW mode shows a symmetry change. In contrast, either before or after the transition, for $k > 1$ modes negative δE_{k}^{I} are dominant. Figure 15 shows $\delta E_{k=2}^{S}$ vs $\delta E_{k=2}^{I}$ with (a) the SR state before the crisis, (b) the STC state after the crisis. Obviously in both cases the orbits are dominant at the negative side, although in the STC state it looks more chaotic.

Figures 16 and 17 are distributions of $\delta E_{k}^{I}(\tau)$ for $k = 1 - 4$ in the SR and STC state respectively. One can see that in Figure 16 all the modes are negative-dominant; in Figure 17, the distributions become broader, however, only $k = 1$ mode displays a symmetry change.

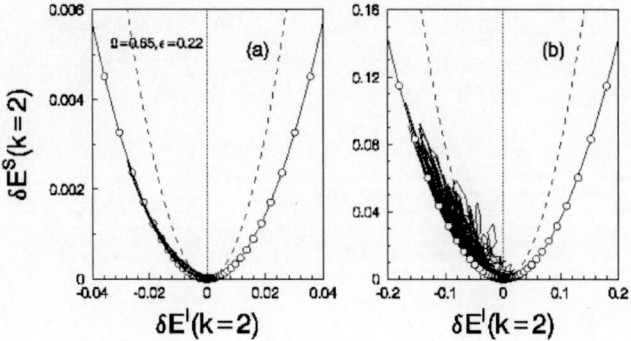

Figure 15. $\delta E^S_{k=2}$ vs $\delta E^I_{k=2}$ for $\Omega = 0.65$, $\epsilon = 0.22$, with (a) before crisis, (b) after crisis.

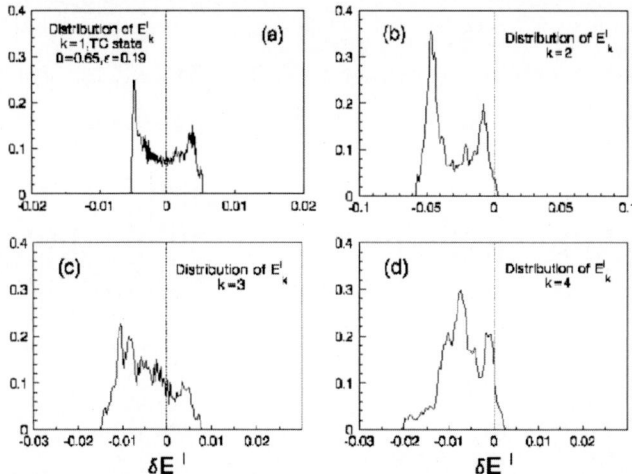

Figure 16. Distribution of $\delta E^I_k(\tau)$ for SR wave with $\Omega = 0.65$, $\epsilon = 0.19$, for (a) $k = 1$, (b) $k = 2$, (c) $k = 3$, (d) $k = 4$.

9. Trapped to Free of $k = 1$ Mode in the Transition

Let us further study the behaviors of PW and SSW partial waves before and after the crisis. In Figure 18 bullets give $\delta\phi_k(\xi, \tau) \equiv b_k(\tau) \cos[k\xi + \alpha_k(\tau)]$ at several instants τ, hollow triangles $\phi^*_{0,k}(\xi) \equiv A_k \cos[k\xi + \theta_k]$ for $k = 1$ mode. One can see that, in (a) the SR state, the peaks of $\delta\phi_{k=1}(\xi, \tau)$ are trapped by $\phi^*_{0,k=1}(\xi)$ as if the latter is a potential well; on the contrary, in (b) the STC state $b_{k=1}(\tau)$ may become larger than $A_{k=1}$. Another dramatic change occurs in the mode phase. In the STC variation of $\alpha_{k=1}(\tau)$ may surpass 2π, in contrast to in the SR where $\alpha_{k=1}(\tau)$ is confined in a range much smaller than 2π.

Figure 19 gives time evolution of (a) $b_{k=1}(\tau)$ and (b) $\alpha_{k=1}(\tau)$ with a crisis to the STC at $\tau \approx 600$, where one can see a sharp increase in the mode amplitude as well as a state transition in the mode phase.

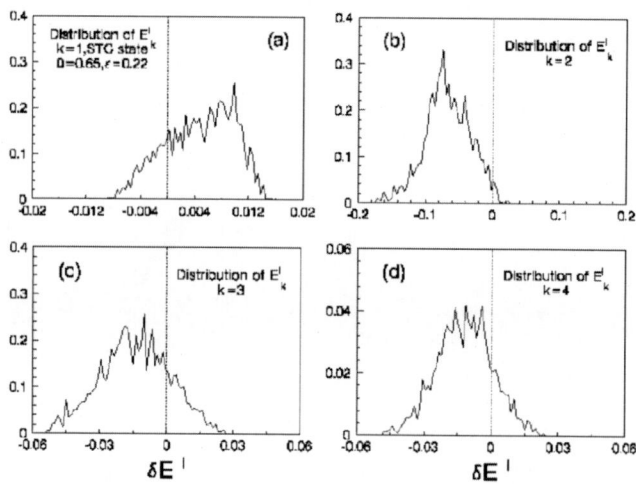

Figure 17. Distribution of $\delta E_k^I(\tau)$ for STC wave with $\Omega = 0.65$, $\epsilon = 0.22$, for (a) $k = 1$, (b) $k = 2$, (c) $k = 3$, (d) $k = 4$.

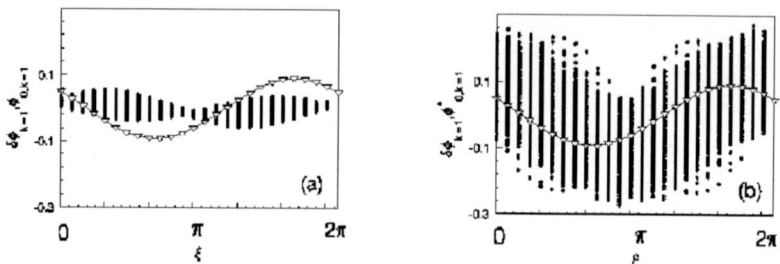

Figure 18. Partial waves $\delta\phi_{k=1}(\xi, \tau)$ at different times (bullet) and $\phi_{0,k=1}^*(\xi)$ (triangle) for $\Omega = 0.65$, $\epsilon = 0.22$ (a) SR wave before crisis, (b) STC wave after crisis.

From the above results one can see that, before the transition $k = 1$ PW mode behaves like a trapped particle vibrating in the potential well $\phi_{0,k=1}^*(\xi)$, while after the transition it makes whirling as well as vibrating, during the whirling phase it can become free from the trapping, just like a transitting particle (He, 1999).

In contrast, $k \neq 1$ modes do not show qualitatively different behavior in the SR and STC states. Figure 20 is an example for $k = 2$ partial wave, either in (a) the SR and (b) STC $\delta\phi_{k=2}(\xi, \tau)$ is trapped by $\phi_{0,k=2}^*(\xi)$, although after the transition amplitude $b_{k=2}(\tau)$ can be comparable with $A_{k=2}$.

10. Effect of Unstable Orbits of SSW in the STC

In Section 6 we have shown that UO of $\phi_0^*(\xi)$ plays a crucial role in the onset of crisis, in the present section we will show that in the asymptotic turbulent motion they still play very important role. To this end let us remember that along UO/SO of the SSW, the PW mode phases are constants. Since an SSW also has constant

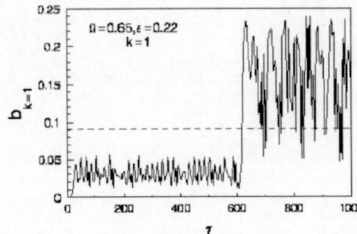 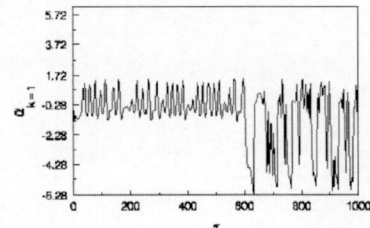

Figure 19. Time evolution of (a) $b_{k=1}(\tau)$ and (b) $\alpha_{k=1}(\tau)$ for $\Omega = 0.65$, $\epsilon = 0.22$, an onset of crisis occurs at $\tau \approx 600$. The dashed line in (a) gives $A_{k=1}$.

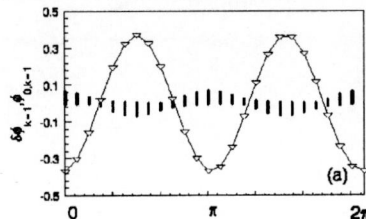 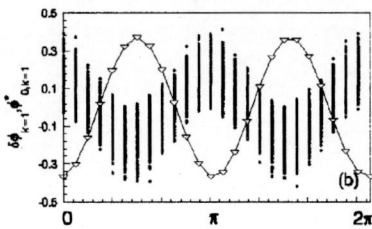

Figure 20. Partial waves $\delta\phi_{k=2}(\xi, \tau)$ (bullet) and $\phi^*_{0,k=2}(\xi)$ for $\Omega = 0.65$, $\epsilon = 0.22$, (a) SR wave before crisis, (b) STC wave after crisis.

phases, it indicates that in linear approximation the relative mode phases between the PW and SSW are very important, depending on their their different values the mode amplitudes can be destabilized or stabilized. For example, there are two groups of relative phases, $\{\Delta\alpha_k^{U_{1,2}}\} \equiv \{\alpha_k^{U_{1,2}} - \theta_k\}$, at which $\{b_k\}$ can be excited exponentially along the UO, respectively.

When the PW self-nonlinearity is included one can imagine that the UO's may still play an important role. If the relative phases between the PW and SSW evolve near to $\{\Delta\alpha_k^{U_{1,2}}\}$, small amplitude $\{b_k\}$ would be strongly excited by the SSW. In the following we test this idea and compare the effects of UO's of the SSW in the SR and STC states.

Let us plot growth rate $db_{k=1}/d\tau$ as a function of phase difference $\Delta\alpha_{k=1} \equiv \alpha_{k=1}(\tau) - \theta_{k=1}$. Figure 21 is for the SR states, in the plot the positions of two UO/SO's are denoted by solid/dashed lines. One can see that in this case PW orbit is confined in between of the two UO's, it never crosses them. Nevertheless one can notice the excitation of amplitude near the two UO.

Figure 22 is $db_{k=1}(\tau)/d\tau$ vs $\Delta\alpha_{k=1}(\tau)$ for the STC case, (a) and (b) are for two sequent time intervals, (c) and (d) are the time averages of (a) and (b), respectively. Since $\alpha_{k=1}(\tau)$ can become whirling, the orbit crosses the UO's frequently. A remarkable phenomenon is that growth rate $db_{k=1}/d\tau$ strongly depends on relative phase $\Delta\alpha_{k=1}$ between the PW and SSW. One can find a peak respectively in (c) and (d), they locate about at the UO's respectively, and around them the data in (a) and (b) are highly scattered.

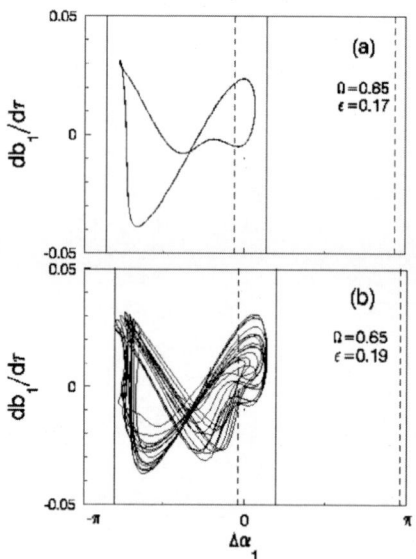

Figure 21. $db_{k=1}/d\tau$ vs $\Delta\alpha_{k=1}(\tau)$ for SR wave, $\Omega = 0.65$, (a) $\epsilon = 0.17$, (b) $\epsilon = 0.19$.

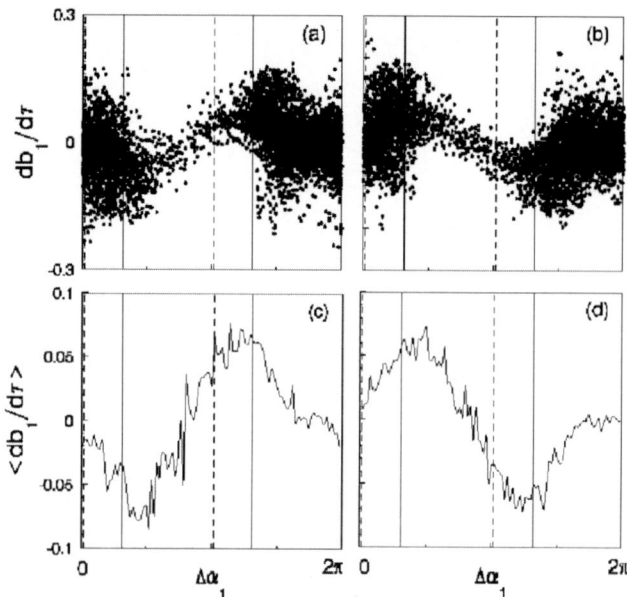

Figure 22. $db_{k=1}(\tau)/d\tau$ vs $\Delta\alpha_{k=1}(\tau)$ for STC wave, $\Omega = 0.65$, $\epsilon = 0.22$, (a) and (b) are for two sequent time intervals, (c) and (d) are the time averages of (a) and (b).

Both the SR and STC in these plots are evolved from SSW respectively, however, depending on whether $k = 1$ PW mode crosses UO's of SSW, their time-space patterns are remarkably different.

11. Conclusion and Discussion

In our nonlinear wave system the complicated dynamic behaviors can be well understood when they are considered as the results of evolution of PW in the existence of an SW. In a sense, the effect of SW is like a periodic potential. The PW dispersion is nonlinearly varied by the interaction with the SW, in particular a saddle instability occurs when the SW is at the negative tangency slope of a hysteresis. An SSW solution is found to play an important role in the discontinuous transition from the SR to STC, where the spatial coherence is destroyed. It is a 'pattern resonance' that triggers a crisis of transition to the STC. As a result of mode couplings, a master mode slaves the other modes, it transits from a trapped state in the SR to a trapped-free state in the STC. The change in time-averaged symmetry property of its motion relative to the SSW gives rise to the critical phenomenon of the transition in parameter space.

The transition discussed in the present work is of discontinuous, and the STC is the most turbulent ones observed in Equation (1), its behavior, e.g. its spatial spectrum of power law, is very much like so-called fully-developed turbulence in fluids. There is evidence that a spatial coherence may also be spoiled through some other route. This problem needs to be studied further.

Acknowledgements

This work is supported by the Special Funds for Major State Basic Research Projects of China and by the National Natural Science Foundation of China No. 19975006, and by RFDP No.20010027005. The author wishes also to thank the NITP/CSSW of the University of Adelaide for their hospitality and AFOSR for support.

References

Ott, E.: 1993, *Chaos in Dynamic Systems*, Cambridge, New York.

Cross, M. C. and Hohenberg, P. C.: 1993, *Rev. Mod. Phys.* **65** 851.

Sagdeev, R. Z. *et al.*: 1988, *Nonlinear Physics From the Pendulum to Turbulence and Chaos*, Harwood Academic Publishers, London, Chapter 11.

Lichtenberg, A. and Lieberman, : 1983, *Regular and Stochastic Motion*, Springer, New York.

Biskamp, D. and He, K.: 1985, *Phys. Fluids* **28** 2172.

Klinger, T., Latten, A., Piel, A., Bonhomme, G., Pierre, T. and de Wit, T. D.: 1997, *Phys. Rev. Lett.* **79**, 3913.

Biskamp, A. R. *et al.*: 1983, *Phys. Rev. Lett.* **50** 1095.

Gregobi, C. *et al.*: 1982, *Phys. Rev. Lett.* **48** 1507.

Oraevskii, V., Tasso, H. and Wobig, H.: 1969, *Proc. 3rd Inter. Conf. on Plasma Physics and Controlled Nuclear Fusion Research*, Novosibirsk, IAEA, Vienna, Vol. I 67.

Dodd, R. K. *et al.*: 1982, *Solitons and Nonlinear Wave Equations*, Academic Press, London.

Chen, W. and Mills, D. L.: 1987, *Phys. Rev. Lett.* **58** 160.

He, K. and Salat, A.: 1989, *Plasma Phys. Contr. Fusion* **31** 123.

He, K.: 1992, *Phys. Lett.* **A169**, 341.

He, K.: 1994, *Phys. Lett.* **A190**, 38.

He, K.: 1995, *Phys. Lett.* **A202**, 369.

He, K. and Zhou, L.: 1997, *Phys. Lett.* **A231**, 65.

He, K.: 1998, *Phys. Rev. Lett.* **80**, 696.

He, K.: 1999, *Phys. Lett. E* **59**, 5278.

He, K.: 2000, *Phys. Rev. Lett.* **84**, 3290.

He, K.: 2001, *Phys. Rev. E.* **63**, 016218.

PHASE COHERENCE OF FORESHOCK MHD WAVES: WAVELET ANALYSIS

DAIKI KOGA and TOHRU HADA

E. S. S. T., Kyushu University, Fukuoka, Japan

Abstract. The earth's foreshock is a region where particularly large amplitude MHD waves are commonly observed. They exhibit various waveforms, suggesting that nonlinear interaction between the waves is in progress. In a previous paper (Hada *et al.*, 2003) we have introduced a method to quantitatively evaluate the strength of phase coherence among the waves from a given time series data. Here we further develop our method by applying wavelet filtering technique. From the analysis it was found that, although the turbulence is consisted of waves with a wide range of plasma rest frame frequencies, only those frequencies lower than the ion gyrofrequency are responsible for generating the phase coherence.

Key words: MHD turbulence, phase coherence, wavelet analysis

1. Introduction

Large amplitude MHD waves (typically, normalized magnetic field amplitude $\delta B / B_0 \sim 1$ or greater) are common in the earth's foreshock. They are thought to be first generated via electromagnetic instabilities driven by ion beams of the bowshock origin, and are continuously amplified to large amplitude as they are convected by the super-Alfvenic solar wind flow. At the same time, the waves nonlinearly interact with each other, resulting in various peculiar waveforms.

The nonlinear wave interaction not only contributes to exchange of energy among the wave modes, but also to synchronization of the wave phases (self-organization of the phase distribution). In a previous paper (Hada *et al.*, 2003), we have developped a method to quantitatively evaluate the phase coherence for a given time series of the wave data, and applied it to magnetic field measurement obtained by Geotail spacecraft (provided by S. Kokubun and T. Nagai through DARTS at the Institute of Space and Astronautical Sciencein Japan). The basic idea was to compare structure function of the original data to those of the phase-shuffled and the phase-coherent surrogate datasets.

Figure 1 shows an example. The original data, $b(t)$, (OBS, the upper left panel) is Fourier decomposed into the power spectrum (center) and the phase (right). Keeping the power spectrum unchanged, the phase distribution is shuffled randomly (middle row) or made completely equal (bottom row). These are then inverse Fourier transformed to construct the Phase Randomized Surrogate (PRS) and the

Space Science Reviews **107**: 495–498, 2003.
© 2003 *Kluwer Academic Publishers*.

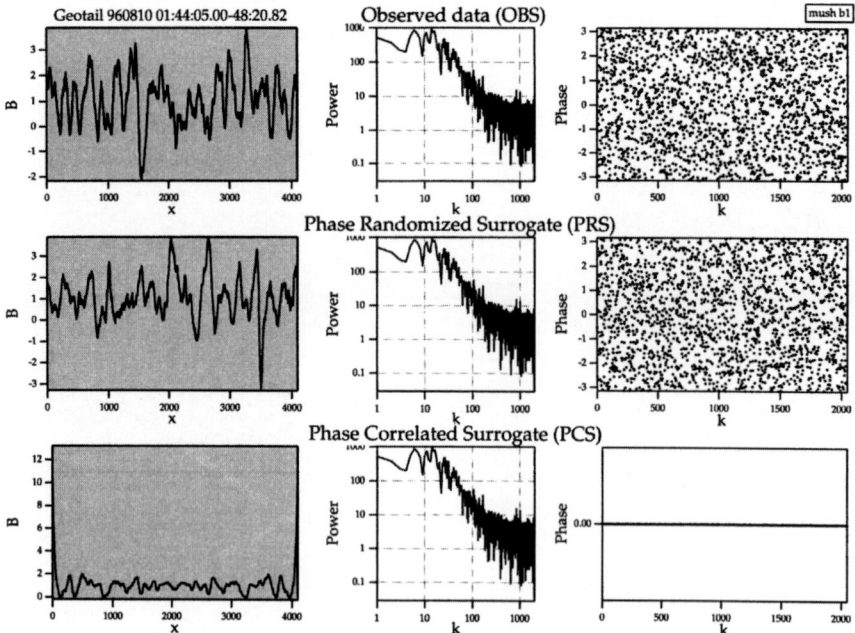

Figure 1. The original time series (OBS, top row), the phase randomized surrogate (PRS, middle row) and the phase correlated surrogates (PCS, bottom row).

Phase Correlated Surrogate (PCS). Note that the phase distribution of the OBS and PRS datasets look both random and almost indistinguishable: this is because the distribution of the phases sensitively depends on the choice of the origin of the coordinate system, which is arbitrary, and so the phase coherence, even if it exists, is easily obscured. Therefore, instead of dealing with the variables in the Fourier space, we compute structure functions, $S(m, d) = \sum_t |b(t+d) - b(t)|^m$, where the unit norm d is a measure charactrizing the coarse-graining of the curve, and them evaluate the 'phase coherence index', $C_\phi = (L_{PRS} - L_{OBS})/(L_{PRS} - L_{PCS})$, where L_* denotes the value of $S(1, d)$ for dataset*. We found that the MHD turbulence in the earth's foreshock does have finite phase coherence, with a range of C_ϕ from 0.1 to 0.5, typically. Furthermore, the value of C_ϕ was found to be positively correlated with the turbulence energy level, and also with propagation angle (with respect to the d.c. magnetic field) of main MHD wave mode in the turbulence.

2. Method of Analysis and Results

In the magnetic field data of the observed MHD turbulence, sharp discontinuities or isolated pulsations are often recognized. Such locally isloated structures may be attributed to the identified phase correlation (superposition of phase correlated

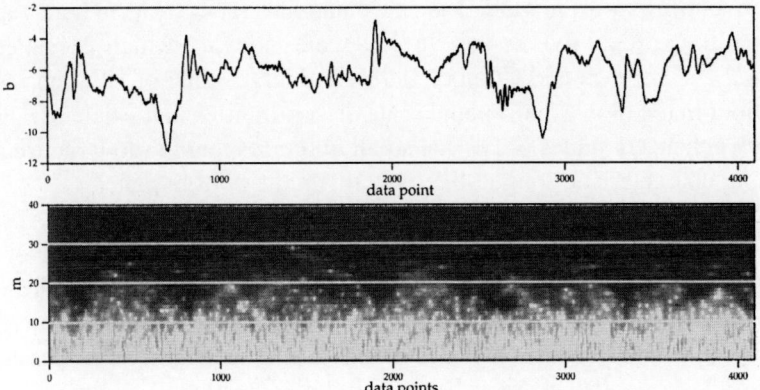

Figure 2. Magnetic field data (top) and its wavelet transform amplitude (bottom).

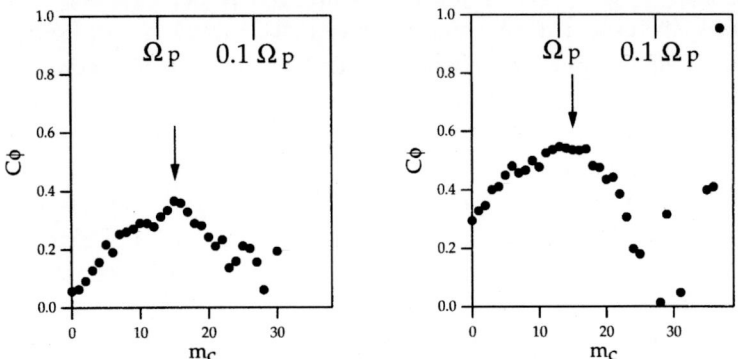

Figure 3. C_ϕ versus m_c (see text for definitions) for the dataset shown in Figure 2.

waves leads to small scale discontinuities: for example, Dirac's delta function is a white noise with completely correlated phases).

In order to discuss the origin of the phase coherence, we have evaluated the phase coherence index of data, processed by wavelet filtering technique, as explained below. The top panel of Figure 2 shows (one component of) the magnetic field data recorded by the Geotail on 8 October 1995 with a sampling interval of $t_0 = 0.0625$ sec (16Hz). The bottom panel is the density plot of the associated wavelet amplitude, $|W_f|$, where

$$W_f(a, t) = \frac{1}{\sqrt{a}} \int f(\tau)\psi^*(\frac{\tau - t}{a})d\tau,$$

the wavelet scale, a, which is related to an instantaneous frequency, $f = 1/a$, the asterisk indicates the complex conjugate, and $\psi(\tau)$ is the mother wavelet (Farge, 1992). In the present analysis we used the Morlet wavelet

$$\psi(\tau) = \pi^{-1/4}e^{-2\pi m\tau}e^{-r^2/2}.$$

The horizontal axis of Figure 2 is the datapoint, $j = t/t_0$, where t is the actual time, the vertical axis is the wavelet scale indes m, which is related to a by $a = 2t_0 2^{m/4}$ (Torrence and Compo, 1998), and large value of $|W_f|$ corresponds to dark colour in the plot. Small spatial scale discontinuities in the data of b are associated with enhanced values of $|W_f|$ at small m (corresponding to large frequencies), producing characteristic tongue-like patterns in the bottom panel.

Now we specify a threshold wavelet scale index, m_c, and let W_f be zero for all $m < m_c$. By inversely wavelet transforming W_f, we obtain wavelet low-pass filtered time series with an instantaneous (logarithmic) frequency threshold, m_c. When $m_c = 0$, the obtained dataset is the same as the original, while increasing m_c gradually eliminates small scale fluctuations from the original time series.

Figure 3 shows the phase coherence index, C_ϕ, plotted versus m_c, using the same interval of data shown in Figure 2. The two panels correspond to the results obtained using normal (right) and one of the transverse (left) components of the field data with respect to the minimum variance directions. In both panels it is clear that C_ϕ increases as m_c is increased to 15, which roughly corresponds to the ion gyro-frequency in the plasma rest frame (the solar wind speed of 400 km s^{-1} has been assumed). This is a rather surprising result since this implies that removal of small scale structure enhances the phase coherence. On the other hand, when m_c is further increased, C_ϕ starts to decrease. From this analysis we infer that, although the turbulence consists of waves with a wide range of plasma rest frame frequencies, only those frequencies lower than the ion gyrofrequency are phase coherent, and the phases of the waves above the ion gyrofrequency are randomly distributed. Thus removal of the latter enhances C_ϕ, while further removing the waves in the former group reduces C_ϕ. More detailed discussion will be given in subsequent publications.

References

Hada, T., Koga, D. and Yamamoto, E.: 2003, 'Phase Coherence of MHD Waves in the Solar Wind', *Space Sci. Rev.* this issue.

Farge, M.: 1992, 'Wavelet Transforms and Their Application to Turbulence', *Ann. Rev. Fluid Mech.* **24**, 395.

Torrence, C. and Compo, G. P.: 1998, 'A Practical Guide to Wavelet Analysis', *Bull. Amer. Meteor. Soc.* **79**, 61.

CROSS FIELD DIFFUSION OF COSMIC RAYS IN A
TWO-DIMENSIONAL MAGNETIC FIELD TURBULENCE

FUMIKO OTSUKA and TOHRU HADA

E.S.S.T., Kyushu University, Fukuoka, Japan

Abstract. Cross field diffusion of energetic particles (cosmic rays) in a two-dimensional static magnetic field turbulence is studied performing test particle simulations. Qualitatively different diffusion processes are observed depending on the ratio of Larmor radius (ρ) to the correlation length (λ) of the magnetic field fluctuations. The diffusion is found to be composed of several regimes with distinct statistical properties, which can be characterized using Levy statistics.

Key words: anomalous diffusion, cosmic rays, Levy process

1. Introduction

Anomalous diffusion is observed in many branches of science, e.g., anomalous diffusion in rotating flows (Solomon *et al.*, 1993), particle motion in nonlinear dynamical systems (Klafter *et al.*, 1993), chaotic phase diffusion in Josephson junctions (Geisel *et al.*, 1985), field line diffusion in solar wind magnetic turbulence (Pommois *et al.*, 2001), and transport in turbulent plasmas (Balescu, 1995). Here the term 'anomalous' is used to emphasize deviation from classical (normal) diffusion, in which the mean squared displacement of particles increases proportional with time. Namely, if we define the diffusion coefficient, D, as

$$D \equiv \frac{< \Delta r^2 >}{\tau} \sim \tau^{\beta}, \tag{1}$$

where Δr is the particle displacement within the time scale τ, and the bracket denotes an ensemble average, then $\beta = 0$ for the classical diffusion. This is a consequence of the well-known central limit theorem, which states that in the long time limit the p.d.f. of Δr approaches a normal (Gaussian) distribution with its variance $\sim \tau$. On the other hand, when a particle can travel long distances ballistically, the so-called Levy flights or Levy walks can arise, and the resultant diffusion process of the ensemble of particles becomes super-diffusive ($\beta > 0$). When a particle can be trapped within a certain bounded region for a long time, then the sub-diffusion ($\beta < 0$) emerges.

In this paper we compute numerically orbits of energetic particles (cosmic rays) in a two-dimensional static magnetic field turbulence, and show that the anomalous diffusion can appear in general. The result may have an important implication

Space Science Reviews **107**: 499–502, 2003.
© 2003 *Kluwer Academic Publishers.*

to plasma astrophysics, since, up to now, various diffusion processes (including the cross-field diffusion) are almost always discussed within the framework of the quasi-linear theory, which in principle is a combination of the classical diffusion equation for particles and an evolution equation for the turbulence energy, which in turn determines the diffusion coefficient. The spatial diffusion problem, in particular, is important for the shock acceleration of cosmic rays.

2. Numerical Model and Results

Although the cross-field diffusion in reality is a three-dimensional problem, we limit our discussion to the case where all the physical variables depend only on two spatial coordinates (x and y), and the magnetic field lines are perpendicular to the $x–y$ plane. By taking such a geometry, we exclude effective cross-field diffusion resulting from parallel motion along the twisted field lines (field-line random walk (Jokipii, 1966)). We note also that, in general, in a model with only two spatial dimensions, particles are tied to the magnetic field lines since the canonical momentum associated with the ignorable coordinate becomes an invariant of the motion (Jokipii et al., 1993). The system we deal with is the only exception to this argument, in which orientations of the field lines and the ignorable coordinate degenerate.

Since we consider energetic particles whose velocities are much larger than the MHD velocities, we assume the field turbulence to be time stationary (fossil turbulence). Then the particle energy is conserved, and the position $\mathbf{r} = (x, y)$ and the velocity $\mathbf{v} = (v_x, v_y)$ obey equations of motion,

$$\dot{\mathbf{v}} = \mathbf{v} \times (1 + b)\mathbf{z} \quad ; \quad \dot{\mathbf{r}} = \mathbf{v}, \tag{2}$$

where $\mathbf{z}$ is a unit vector in the z direction, b is the fluctuation part of the normalized magnetic field, and time is normalized to the reciprocal of the average ion gyrofrequency. The turbulence field is given by,

$$b(x, y) = \sum_m \sum_n A(k) \cos(mx + ny + \phi(m, n)), \tag{3}$$

where $A(k) \sim k^{-\gamma}$ for $k_{min} \leq k \leq k_{max}$, $A(k) \sim k_{min}^{-\gamma}$ for $k_{sys} \leq k < k_{min}$, $k = (m^2 + n^2)^{1/2}$, $k_{sys} = 2\pi/L$, $k_{min} = \pi/L$, $k_{max} = \sqrt{2}\pi$, and L is the system size. Boundary conditions are periodic, and phases $\phi(m, n)$ are random. We define the magnetic field correlation length λ as $\lambda^2 = < (2\pi/k)^2 A(k) > / < A(k) >$. Typically we choose $\gamma = 1.5$, $L = 512$, and $< b^2 >^{1/2} = 0.01$, then $\lambda \sim 61$. By giving different velocities to the particles, we have runs with different ρ/λ, where ρ is the typical Larmor radius.

The results are summarized in Figure 1. Upper panels show D defined in (1) versus τ in logarithmic scales, for three different regimes of ρ/λ, and the lower

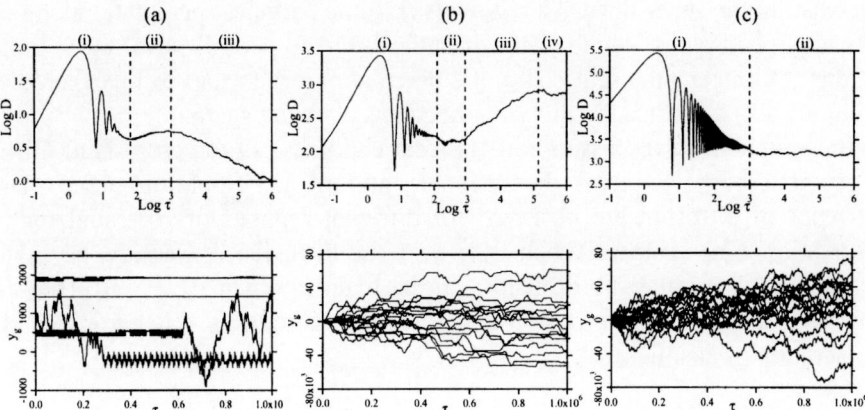

Figure 1. Diffusion coefficients (upper panels) and typical guiding center trajectories (lower panels) for (a) $\rho/\lambda = 0.1$, (b) $\rho/\lambda = 1$, and (c) $\rho/\lambda = 10$.

panels are y component of the guiding center, $\mathbf{r}_g = \mathbf{r} + \mathbf{v} \times \mathbf{z}$ versus time, plotted for some particles. The orbits look quite different for the three runs. When $\rho/\lambda = 10$, the orbits look more or less similar to a Brownian motion, while for $\rho/\lambda = 1$ and $\rho/\lambda = 0.1$, they are composed of segments with different characters – sometimes almost ballistic, sometimes trapped at a certain location, and sometimes like a Brownian motion. This diversity of the types of the orbits is a reflection of the presence of multi-scales in the field turbulence. Namely, if a particle is guided by a large scale island for a longer time period than the 'observation' time scale τ, then its orbit will appear to be almost ballistic, while a particle trapped by a small scale island will appear as trapped if it makes many rotations around the island within τ.

The diffusion coefficients represent the different characteristics of the orbits. Let us first look at the case (c), $\rho/\lambda = 10$. When (i) $\tau < 10^3$, the value of D is still influenced by Larmor rotations (large amplitude oscillations in the figure), and so we do not discuss statistics in this regime. For longer time scale (ii) we find D to be almost constant, suggesting that the diffusion is almost classical and the orbits are essentially Brownian. This is reasonable since, when $\rho/\lambda = 10$, a particle traverses many magnetic field islands during one gyration, and the force acting on the particle, which will be a sum of many fluctuations, will be random and incoherent. When ρ/λ is (much) less than unity at 0.1, the gyration regime (i) is followed by two distinct regimes (ii) and (iii) as τ is increased. In (ii) the process is slightly super-diffusive ($\beta > 0$ in (1)), since within this time scale the majority of the particles gradient-B drift around the field islands without making a complete rotation around the islands. For longer time scales, many particles are trapped (as seen in the lower panel), resulting in the sub-diffusion ($\beta < 0$). The values of β and the transition time scale which separates regimes (ii) and (iii) depend on the parameters for the turbulence, and will be discussed in more detail elsewhere. The case (b) $\rho/\lambda = 1$ illustrates the possibility that even more distinct types of orbits can exist. In the super-diffusive regime (iii), some particles 'percolate' along infinitely long open paths, which result from the assumed periodicity of the simulation system. At longer time scales (iv), the percolation orbits start to mix (percolation random walk), and thus the diffusion becomes classical again.

In summary, we discussed non-classical characters of the cross-field diffusion of energetic particles in two-dimensional static magnetic field turbulence. Different types of diffusion are observed for different regimes of ρ/λ, and for finite observation time scale τ. When $\rho/\lambda > 1$ the diffusion is classical asymptotically ($\tau \to \infty$), while sub-diffusion can be realized when $\rho/\lambda < 1$. The results may have an essential implication to high energy astrophysics, in particular, shock acceleration of cosmic rays.

References

Balescu, R.: 1995, 'Anomalous Transport in Turbulent Plasmas and Continuous-time Random Walks', *Phys. Rev. E* **51**, 4807–4822.

Geisel, T., Nierwetberg, J. and Zacherl, J.: 1985, 'Accelerated diffusion in Josephson Junctions and Related Chaotic Systems', *Phys. Rev. Lett.* **54**, 616.

Jokipii, J. R.: 1966, 'Cosmic-ray Propagation, 1: Charged Particles in a Random Magnetic Field', *Astrophys. J.* **146**, 480.

Jokipii, J. R., Kota, J. and Giacalone, J.: 1993, 'Perpendicular Transport in 1- and 2-dimensional Shock Simulations', *Geophys. Res. Lett.* **20**, 1759–1761.

Klafter, J., Zumofen, G. and Shlesinger, M. F.: 1993, 'Levy Walks in Dynamical-systems', *Physica A.* **200**, 222–230.

Pommois, P., Veltri, P. and Zimbardo, G.: 2001, 'Field Line Diffusion in Solar Wind Magnetic Turbulence and Energetic Particle Propagation Across Heliographic Latitudes', *J. Geophys. Res.* **106**, 24965–24978.

Solomon, T. H., Weeks, E. R. and Swinney, H. L.: 1993, 'Observation of Anomalous Diffusion and Levy Flights in a 2-dimensional Rotating Flow', *Phys. Rev. Lett.* **71**, 3975–3978.

CHAOTIC TEMPORAL VARIABILITY OF MAGNETOSPHERIC RADIO EMISSIONS

E. L. REMPEL[1,2], A. C.-L. CHIAN[1,2] and F. A. BOROTTO[3]

[1] World Institute for Space Environment Research – WISER, NITP/WISER, University of Adelaide, Adelaide, SA 5005, Australia

[2] National Institute for Space Research – INPE, P. O. Box 515, São José dos Campos - SP, CEP 12227-010, Brazil

[3] Universidad de Concepción, Departamento de Física, Concepción, Chile

Abstract. Nonthermal magnetospheric radio emissions provide the radio signatures of solar-terrestrial connection and may be used for space weather forecasting. A three-wave model of auroral radio emissions at the fundamental plasma frequency was proposed by Chian et al. (1994) involving resonant interactions of Langmuir, whistler and Alfvén waves. Chaos can appear in the nonlinear evolution of this three-wave process in the magnetosphere. We discuss two types of intermittency in radio signals driven by temporal chaos: the type-I Pomeau-Manneville intermittency and the interior crisis-induced intermittency. Examples of time series for both types of intermittency are presented.

Key words: chaos, intermittency, magnetosphere, radio emissions, three-wave interaction

1. Introduction

It has been shown that the nonlinear coupling of Langmuir waves with low-frequency magnetic field fluctuations such as Alfvén waves, may provide an efficient mechanism for generating magnetospheric radio waves (Chian et al., 1994; Chian et al., 2002). In this paper, we present a nonlinear dynamical theory of three-wave interactions involving Langmuir, Alfvén and whistler waves in the planetary magnetospheres. This wave triplet is shown to evolve from order to chaos via a number of different routes. The relevance of this theory for detecting chaos in the time series of magnetospheric radio waves is discussed.

2. Nonlinear Coupled Wave Equations

Consider the nonlinear parametric interaction of Langmuir (L), whistler (W) and Alfvén (A) waves, all propagating along the ambient magnetic field $\mathbf{B} = B_0\hat{\mathbf{z}}$. We assume the following phase-matching condition, $\omega_L \approx \omega_W + \omega_A$, $\mathbf{k}_L = \mathbf{k}_W + \mathbf{k}_A$.

The set of coupled wave equations governing the nonlinear interaction $L \rightleftharpoons W + A$ is given by

$$\dot{A}_L = \nu_L A_L + A_W A_A, \tag{1}$$

Space Science Reviews **107**: 503–506, 2003.
© 2003 *Kluwer Academic Publishers.*

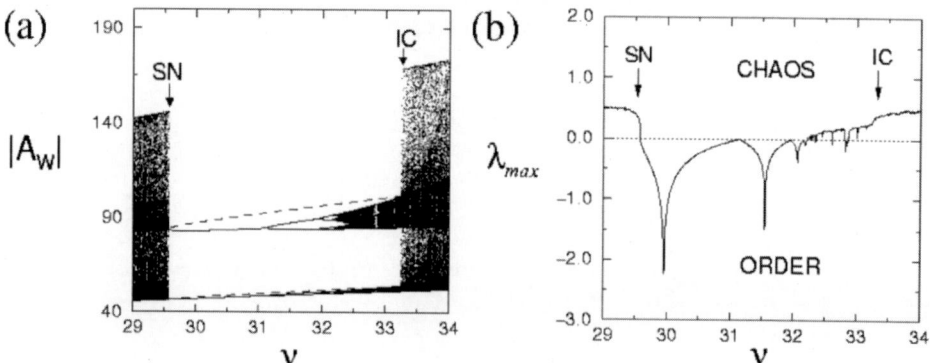

Figure 1. (a) Bifurcation diagram for $|A_W|$ as a function of ν. The dashed lines denote the period-2 unstable periodic orbit. (b) Maximum Lyapunov exponent λ_{max} as a function of ν.

$$\dot{A}_W = \nu_W A_W - A_L A_A^*, \tag{2}$$

$$\dot{A}_A = i\delta A_A + \nu_A A_A - A_L A_W^*, \tag{3}$$

where A_α is the normalized slowly varying complex envelope of the wave electric field, and $\alpha = (L, W, A)$; δ is the frequency mismatch parameter and ν_α are the normalized growth/damping parameters (Chian *et al.*, 2002). The dot denotes derivative with respect to $\tau = k(z - \nu t)$, ν and k are arbitrary wave velocity and wave vector, respectively. We set $\nu_L \equiv 1$ (linearly unstable L wave) and $\nu_W = \nu_A \equiv -\nu$ (linearly damped W and A waves).

3. Nonlinear Dynamics Analysis

The nonlinear dynamics of the system described by Equations (1)–(3) can be studied by constructing a bifurcation diagram. We fix $\delta = 2$ arbitrarily and vary the parameter ν in order to see how the system dynamics changes. For each value of the control parameter ν, we drop the initial transient and plot 400 Poincaré points. The Poincaré points are the projections of the 3-dimensional trajectory onto the 2-dimensional Poincaré plane $((|A_L|, |A_W|))$. A bifurcation diagram is shown in Figure 1a, where the Poincaré points refer to the maxima of A_W. The corresponding behavior of the maximum Lyapunov exponent λ_{max} is plotted in Figure 1b. Chaos (aperiodic solutions) occurs when $\lambda_{max} > 0$, and order (periodic solutions) occurs when $\lambda_{max} < 0$.

Figure 1a indicates that a saddle-node bifurcation (SN) takes place at $29.56 < \nu_{SN} < 29.57$, where a pair of period-2 stable and unstable periodic orbits is created. Figure 2a shows a stable periodic solution with period-two when $\nu > \nu_{SN}$; Figure 2b reveals that just to the left of ν_{SN}, where chaos develops, laminar phases of nearly periodic oscillations are suddenly interrupted by chaotic bursts. This type

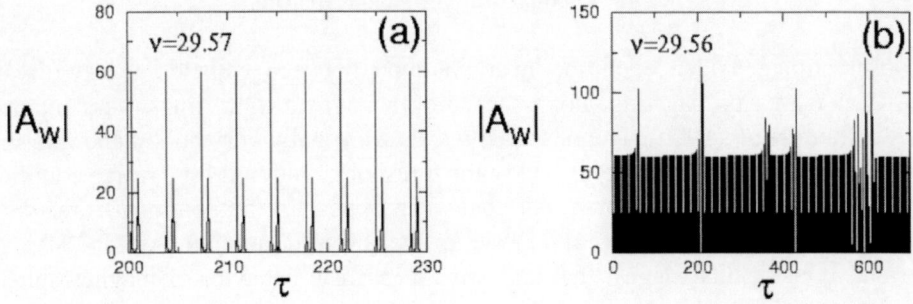

Figure 2. Time series of $|A_W|$ as a function of τ for the type-I Pomeau–Manneville intermittency route to chaos for (a) $v = 29.57$ and (b) $v = 29.56$.

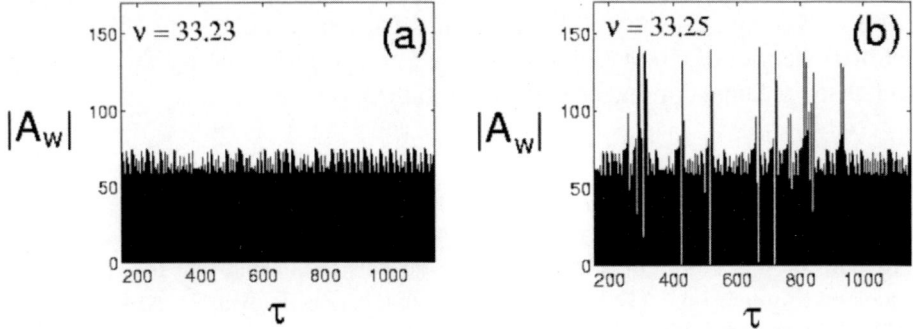

Figure 3. Time series of $|A_W|$ as a function of τ for the crisis-induced intermittency for (a) $v = 33.23$ and (b) $v = 33.25$.

of intermittent behavior is known as type-I Pomeau–Manneville intermittency. Figure 1b shows that, as v decreases past v_{SN}, λ_{max} jumps from negative to positive, indicating a transition from order to chaos.

To the right of v_{SN} the stable periodic orbit created at the saddle-node bifurcation undergoes a cascade of period-doubling bifurcations, leading to chaos, as seen in Figure 1a. At a critical parameter $v = v_{IC} \sim 33.23$, the chaotic attractor collides with the period-2 unstable periodic orbit evolving from the saddle-node bifurcation, leading to an interior crisis (IC), characterized by an abrupt expansion of the size of the attractor, forming a strong chaotic attractor. Figure 3a shows that just before the interior crisis, at $v = 33.23$, the time series displays only weak chaotic fluctuations. After IC, at $v = 33.25$, the 'laminar' phases of weak chaotic fluctuations are randomly interrupted by strong chaotic bursts (Figure 3b). This behavior is known as crisis-induced intermittency. We can see from Figure 1b that for $v > v_{IC}$ the value of λ_{max} increases abruptly, indicating a sudden increase in the system chaoticity.

4. Discussions and Conclusions

Observational evidence of nonlinear coupling between whistler, Langmuir and Alfvén waves propagating along the auroral magnetic field lines were reported by Bohem et al. (1990). The observed waves are usually very bursty and intermittent, and of large-amplitude to satisfy the threshold condition for 3-wave coupling. In this paper, we have shown that chaos can appear in the nonlinear three-wave model proposed by Chian et al. (1994) for the auroral Langmuir–Alfvén–Whistler events. Our results suggest that chaos is an intrinsic behavior of magnetospheric radio emissions. Our theory shows that as the physical parameters, such as the growth/damping parameters, vary, the nonlinear dynamical behavior of radio emissions can change from periodic to chaotic temporal patterns via two different intermittency processes. A systematic analysis of the time series of magnetospheric radio waves, e.g., calculating the Lyapunov exponent (vide Figure 1b) of the data set of Bohem et al. (1990), will enable the observers to identify the intermittent and chaotic features of magnetospheric radio emissions.

Acknowledgements

The authors wish to thank the support of the Australian Institute for Nuclear Science and Engineering (AINSE), CNPq (Brazil), FAPESP (Brazil), AFOSR, and the NITP/CSSM of the University of Adelaide for their hospitality.

References

Boehm, M. H., Carlson, C. W., McFadden, J. P., Clemmons, J. H. and Mozer, F. S.: 1990, 'High-resolution Sounding Rocket Observations of Large-amplitude Alfvén Waves', *J. Geophys. Res.* **95**, 12157–12171.

Chian, A. C.-L., Lopes, S. R. and Alves, M. V.: 1994, 'Generation of Auroral Whistler-Mode Radiation Via Nonlinear Coupling of Langmuir Waves and Alfvén Waves', *Astron. Astrophys.* **290**, L13–L16.

Chian, A. C.-L., Rempel, E. L. and Borotto, F. A.: 2002, 'Chaos in Magnetospheric Radio Emissions', *Nonlinear Proc. Geophys.* **9**, 1–7.

LANGMUIR TURBULENCE AND SOLAR RADIO BURSTS

F. B. RIZZATO[1,2], A. C.-L. CHIAN[2,3], M. V. ALVES[3], R. ERICHSEN[1], S. R. LOPES[4],
G. I. DE OLIVEIRA[5], R. PAKTER[1] and E. L. REMPEL[2,3]

[1] *Instituto de Física, Universidade Federal do Rio Grande do Sul, Caixa Postal 15051,*
91501–970 Porto Alegre, Rio Grande do Sul, Brazil
[2] *World Institute for Space Environment Research (WISER), NITP,*
University of Adelaide, SA 5005, Australia
[3] *National Institute for Space Research (INPE), P. O. Box 515,*
12221-970 São José dos Campos, São Paulo, Brazil
[4] *Instituto de Física, Universidade Federal do Paraná, Caixa Postal 19081,*
81531-990 Curitiba, Paraná, Brazil
[5] *Universidade Federal do Mato Grosso do Sul, Caixa Postal 135,*
79200-000 Aquidauana, Mato Grosso do Sul, Brazil

Abstract. Langmuir waves and turbulence resulting from an electron beam-plasma instability play a fundamental role in the generation of solar radio bursts. We report recent theoretical advances in nonlinear dynamics of Langmuir waves. First, starting from the generalized Zakharov equations, we study the parametric excitation of solar radio bursts at the fundamental plasma frequency driven by a pair of oppositely propagating Langmuir waves with different wave amplitudes. Next, we briefly discuss the emergence of chaos in the Zakharov equations. We point out that chaos can lead to turbulence in the source regions of solar radio emissions.

Key words: chaos, instabilities, radio radiation, Sun, turbulence

1. Introduction

Solar radio bursts provide a powerful tool for predicting the space weather and monitoring the space environment. For example, type-III solar radio bursts are produced by energetic electron beams accelerated in the solar active regions. As the electron beams propagate out of the solar corona and across the solar wind, they interact with the background plasma, resulting in a beam-plasma instability. The growth of this instability leads to large-amplitude Langmuir waves and turbulence, which can generate electromagnetic radiation via nonlinear wave-wave coupling. Since Langmuir waves oscillate at frequencies close to the local plasma frequencies, type-III solar radio bursts may serve as a tracer of the space environment and as a remote sensing technique of Langmuir turbulence in the solar active regions and the interplanetary medium.

A sound interpretation of the observational data of solar radio bursts, from ground radio telescopes and interplanetary spacecrafts, requires a thorough theoretical study of Langmuir waves, instabilities and turbulence. We review in this

Space Science Reviews **107**: 507–514, 2003.
© 2003 *Kluwer Academic Publishers.*

paper a novel theory of the fundamental plasma emission of type-III solar radio bursts, and Langmuir turbulence driven by chaos in the source regions of solar radio emissions.

A theory of type-III solar radio bursts was formulated by Chian and Alves (1988) for the case of two counterpropagating Langmuir pump with the same wave amplitudes. Rizzato and Chian (1992) improved their model by self-consistently including a second grating that assures the symmetry of the wave kinematics and investigated the simultaneous generation of electromagnetic and Langmuir daughter waves. We generalize the model of Rizzato and Chian (1992) to the case of two pump waves with distinct wave amplitudes (Alves *et al.*, 2002). This theory will provide a better framework for understanding the observed features of type-III solar radio bursts which often indicate the signature of two populations of Lagmuir waves traveling in opposite directions with different amplitudes.

2. A Theory of the Fundamental Plasma Emission of Type-III Solar Radio Bursts

The nonlinear coupling of Langmuir waves (L), electromagnetic waves (T) and ion-acoustic waves (S) is governed by the generalized Zakharov equations (Rizzato and Chian, 1992; Alves *et al.*, 2002)

$$
\left(\partial_t^2 - \nu_e \partial_t + c^2 \nabla \times (\nabla \times) - \gamma_e v_{th}^2 \nabla (\nabla \cdot) + \omega_p^2 \right) \vec{E} =
$$
$$
-\frac{\omega_p^2}{n_0} n \vec{E} , \tag{1}
$$

$$
\left(\partial_t^2 - \nu_i \partial_t - v_s^2 \nabla^2 \right) n = \frac{\varepsilon_0}{2m_i} \nabla^2 < \vec{E}^2 > , \tag{2}
$$

where $\vec{E}$ is the high-frequency electric field, n is the ion density fluctuation, $\omega_p^2 = n_0 e^2 / (m_e \varepsilon_0)$ is the electron plasma frequency, c is the velocity of light, $v_{th} = (k_B T_e / m_e)^{1/2}$ is the electron thermal velocity, $v_s = (k_B (\gamma_e T_e + \gamma_i T_i) / m_i)^{1/2}$ is the ion-acoustic velocity, $\nu_{e(i)}$ is the damping frequency for electrons (ions), $\gamma_{e(i)}$ is the ratio of the specific heats for electrons (ions), and the angle brackets denote the fast time average.

In order to derive a dispersion relation from Equations (1)–(2) we assume that the electric field of the Langmuir pump wave is given in the form

$$
\vec{E}_0 = \frac{1}{2} \left(\vec{E}_0^+ exp(i(\vec{k}_0 \cdot \vec{r} - \omega_0 t)) \right.
$$
$$
\left. + \vec{E}_0^- exp(i(-\vec{k}_0 \cdot \vec{r} - \omega_0 t)) \right) + c.c.,
$$

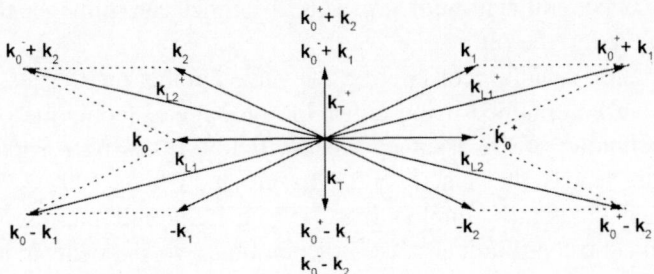

Figure 1. Wave-vector kinematics for our model: $k_0^{+(-)}$ are related to the pump Langmuir waves, $k_{1(2)}$ to ion acoustic waves, $k_0^{+(-)} - k_{1(2)}$ are the electrostatic (oblique to $k_0^{+(-)}$) or electromagnetic ($\perp$ to $k_0^{+(-)}$) Stokes modes and $k_0^{+(-)} + k_{1(2)}$ are the electrostatic (oblique to $k_0^{+(-)}$) or electromagnetic ($\perp$ to $k_0^{+(-)}$) anti-Stokes modes.

which represents two oppositely propagating waves of equal frequencies and opposite wave vectors, the amplitude of two pump waves can be different, $\left|\overrightarrow{E}_0^-\right|^2 = r\left|\overrightarrow{E}_0^+\right|^2$, with $0 \leq r \leq 1$. Each of the two pump waves can generate the Stokes $(\omega_0 - \omega^*)$ and the anti-Stokes $(\omega_0 + \omega)$ modes, either electrostatic or electromagnetic, as known in three-wave processes. We consider the coupling of four triplets. Each two triplets have in common the Langmuir pump wave (forward or backward wave) and a pair of independent density gratings, $n_{1(2)}$, given by

$$n = \frac{1}{2}n_{1(2)}exp\left(i(\overrightarrow{k}_{1(2)} \cdot \overrightarrow{r} - \omega t)\right) + c.c.,$$

with $\left|\overrightarrow{k}_1\right| = \left|\overrightarrow{k}_2\right|$. The electric fields of the daughter waves obey the relation

$$\overrightarrow{E} = \frac{1}{2}\left(\overrightarrow{E}_w^{+(-)}exp(i(\overrightarrow{k}_w^{+(-)} \cdot \overrightarrow{r} - \omega_w^{+(-)}t))\right) + c.c.,$$

where the subscript w represents either electromagnetic (T) or Langmuir (L) daughter wave, respectively; the superscript $+(-)$ represents the anti-Stokes (Stokes) mode, respectively. Figure 1 illustrates this coupling and also shows the wave-vector matching conditions we have assumed.

The assumption of two oppositely propagating pumps implies the need of some process to cause the E_0^+-waves to be scattered into the backward direction. High time resolution electric field waveform measurements from the Galileo spacecraft have shown that Langmuir waves associated with type-III solar radio bursts have a characteristic beat pattern (Gurnett et al., 1993). This beat pattern is produced by two closely spaced narrowband components. One of these components is believed to be the beam-driven Langmuir wave, E_0^+, and the other is believed to be the oppositely propagating Langmuir wave generated by the parametric decay of E_0^+. For the sake of simplicity, we assume in this paper the two pump waves are antiparallel. It is worth pointing out that the Langmuir decay process can generate a spectrum of $\vec{k}$ not collinear with $\vec{k}_0$. This difference can influence the results and should be studied in the future.

The dispersion relation, considering the wave number and frequency matching conditions, can be obtained using either the propagator technique or the Fourier transform technique of the assumed electric fields. It has been shown that these two techniques lead to the same results (Rizzato and Chian, 1992).

The chosen kinematics implies that the two Langmuir pumps propagate oppositely along the longitudinal x-axis, generating two opposite induced electromagnetic modes that primarily propagate along transverse y-axis, plus induced Langmuir modes that mainly propagate along the x-axis. Writing the total high-frequency fluctuating field in terms of its transverse and longitudinal component, $\vec{E} = \vec{E}_L + \vec{E}_T$, imposing perfect $\vec{k}$-matching but allowing for frequency mismatches between the interacting waves and using the kinematic conditions presented in Figure 1 in Equation (1), we can write the variations of the induced waves as follows

$$\mathcal{D}_{L1}^- E_{L1}^- = \frac{\omega_p^2}{n_0} n_1^* E_0^- \, , \ \ \mathcal{D}_{L1}^+ E_{L1}^+ = \frac{\omega_p^2}{n_0} n_1 E_0^+ , \tag{3}$$

$$\mathcal{D}_{L2}^- E_{L2}^- = \frac{\omega_p^2}{n_0} n_2^* E_0^+ , \ \ \mathcal{D}_{L2}^+ E_{L2}^+ = \frac{\omega_p^2}{n_0} n_2 E_0^- , \tag{4}$$

$$\mathcal{D}_T^+ E_T^+ = \frac{\omega_p^2}{n_0} \left(n_1 E_0^- + n_2 E_0^+ \right) ,$$

$$\mathcal{D}_T^- E_T^- = \frac{\omega_p^2}{n_0} \left(n_1^* E_0^+ + n_2^* E_0^- \right) . \tag{5}$$

In Equations (3–5) sub-indexes $L1(2)$ refer to the Langmuir daughter wave due do the first (second) grating.

Introducing Equations (3–5) in Equation (2), we obtain the following equations for the density fluctuations

$$\mathcal{D}_{s1} n_1 = \frac{\varepsilon_0 k_1^2 \omega_p^2}{2 m_i n_0} \left(\frac{n_1 \left| E_0^+ \right|^2}{\mathcal{D}_{L1}^+} + \frac{n_1 \left| E_0^- \right|^2}{\mathcal{D}_{L1}^{-*}} + \frac{n_1 \left| E_0^- \right|^2}{\mathcal{D}_T^+} \right.$$

$$\left. + \frac{n_2 E_0^+ E_0^{-*}}{\mathcal{D}_T^+} + \frac{n_1 \left| E_0^+ \right|^2}{\mathcal{D}_T^{-*}} + \frac{n_2 E_0^+ E_0^{-*}}{\mathcal{D}_T^{-*}} \right) , \tag{6}$$

$$\mathcal{D}_{s2} n_2 = \frac{\varepsilon_0 k_2^2 \omega_p^2}{2 m_i n_0} \left(\frac{n_2 \left| E_0^+ \right|^2}{\mathcal{D}_{L2}^{-*}} + \frac{n_2 \left| E_0^- \right|^2}{\mathcal{D}_{L2}^+} + \frac{n_2 \left| E_0^+ \right|^2}{\mathcal{D}_T^+} \right.$$

$$\left. + \frac{n_1 E_0^{+*} E_0^-}{\mathcal{D}_T^+} + \frac{n_2 \left| E_0^- \right|^2}{\mathcal{D}_T^{-*}} + \frac{n_1 E_0^{+*} E_0^-}{\mathcal{D}_T^{-*}} \right) . \tag{7}$$

In Equations (6) and (7) we have:

$$\mathcal{D}_{s1(2)} = \omega^2 + i\nu_i\omega - v_s^2 k_{1(2)}^2,$$

$$\mathcal{D}_T^{\pm} = \left((\omega_o \pm \omega)^2 + i\nu_T(\omega_o \pm \omega) - c^2 k_T^{\pm 2} - \omega_p^2\right),$$

$$\mathcal{D}_{L1(2)}^{\pm} = \left((\omega_o \pm \omega)^2 + i\nu_L(\omega_o \pm \omega) - v_{th}^2 k_{L1(2)}^{\pm 2} - \omega_p^2\right).$$

Observe that $\left|\vec{k}_{L1}^{\pm}\right| = \left|\vec{k}_{L2}^{\pm}\right|$, and $\left|\vec{k}_T^{+}\right| = \left|\vec{k}_T^{-}\right|$, as shown in Figure 1.

Using the high-frequency approximation

$$\mathcal{D}_T^{\pm} \cong \pm 2\omega_p \left(\omega \pm (\omega_0 - \omega_T) + i\nu_T/2\right),$$

$$\mathcal{D}_{L1(2)}^{\pm} \cong \pm 2\omega_p \left(\omega \pm (\omega_0 - \omega_L) + i\nu_L/2\right),$$

with $\omega_{L(T)}$ representing the linear relation for the Langmuir (electromagnetic) wave, $\left|\vec{k}_1\right| = \left|\vec{k}_2\right|$, and normalizing ω by ω_s and k by $k\lambda_D$ we introduce

$$D_s = \left(\omega^2 - 1 + i\nu_i/2\right)$$

$$D_T^{\pm} = \omega \pm \frac{3}{2}\frac{k_0}{(\mu\tau)^{1/2}} \mp \frac{1}{2}\frac{c^2}{v_{th}^2}\frac{k_T^2}{(\mu\tau)^{1/2} k_0}$$

$$D_L^{\pm} = \omega \mp \frac{9}{2}\frac{k_0}{(\mu\tau)^{1/2}}$$

where $\mu = m_e/m_i$, and $\tau = (\gamma_e T_e + \gamma_i T_i)/T_e$. Finally, by writing $\left|\vec{E}_0^{-}\right|^2 = r\left|\vec{E}_0^{+}\right|^2$, $W_0 = \varepsilon_0\left|\vec{E}_0^{+}\right|^2/(2n_0 k_B T_e)$, and $W_{T0} = (1+r)W_0$, we obtain the general dispersion relation

$$\begin{aligned}
D_s^2 &- \frac{W_{T0}}{4\tau(\mu\tau)^{1/2}k_0}D_s\left(\frac{1}{D_L^+} - \frac{1}{D_L^-} + \frac{1}{D_T^+} - \frac{1}{D_T^-}\right) \\
&+ \frac{W_{T0}^2}{16\mu\tau^3 k_0^2(1+r)^2}\left[\frac{-(1-r)^2}{D_T^+ D_T^-}\right. \\
&+ (1+r^2)\left(\frac{1}{D_T^+ D_L^+} - \frac{1}{D_L^+ D_L^-} + \frac{1}{D_T^- D_L^-}\right) \\
&\left. - 2r\left(\frac{1}{D_L^+ D_T^-} + \frac{1}{D_T^+ D_L^-}\right) + r\left(\frac{1}{D_L^{+2}} + \frac{1}{D_L^{-2}}\right)\right] = 0.
\end{aligned}$$

(8)

In order to verify the relative importance of the second wave pump amplitude, we solve Equation (8) for different values of r and typical physical parameters measured in type-III events in the solar wind ($W_0 = 10^{-5}$, $k_0 = 0.0451$, $\mu = 1/1836$, $v_{th} = 2.2 \times 10^6 m/s$, $T_e = 1.6 \times 10^5 K$, $T_i = 5 \times 10^4 K$). Figure 2 shows

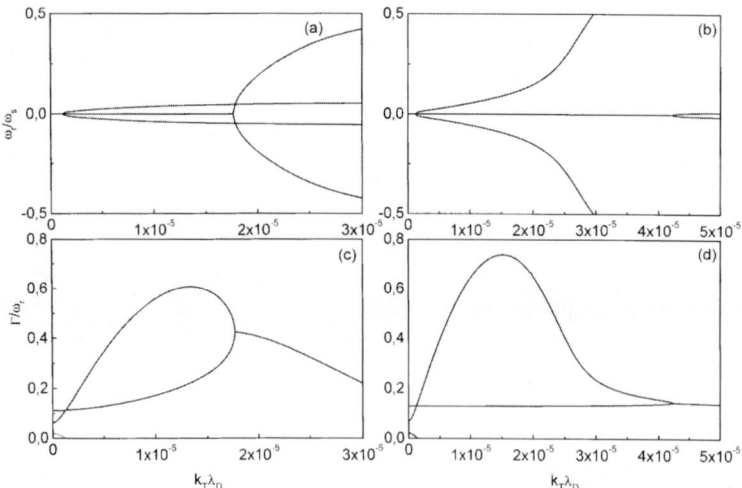

Figure 2. Numerical solutions for the present model with different values of r; (a) and (b) show the real part of the solution and (b) and (d) the growth rates, with $k_0 = 10^{-4}$, within the limit $k_0 < (1/3)W_0^{1/2}$; (a) and (c) refers to $r = 0.5$, and (b) and (d) to $r = 0.95$.

the results for $r = 0.5$ ((a) and (c)), and for $r = 0.95$ (b) and (d)). The presence of a second pump wave with different amplitude from the first one introduces a region of convective instability ($\omega_r \neq 0$), not present when $r = 1$. For small values of $k_T \lambda_D$, the instability is purely growing, $\omega_r = 0$. For a critical value of $k_T \lambda_D$ the real part of the frequency bifurcates off the line $\omega_r = 0$, while still presenting an imaginary part $\Gamma \neq 0$. The critical value of $k_T \lambda_D$ decreases as r decreases.

3. Langmuir Turbulence driven by Chaos in the Source Regions of Solar Radio Emissions

The Zakharov equations, Equations (1)–(2), can model the Langmuir turbulence driven by chaos in the emission regions of solar radio bursts.

De Oliveira, Rizzato and Chian (1995) analysed regular and chaotic dynamics of one-dimensional weakly relativistic Zakharov equations. They first locate a certain region of parameter space where the nonlinear dynamics of Langmuir waves is low-dimensional and regular. They then study the transition to chaos as a function of the system length scale, which follows initial inverse pitchfork bifurcations. The transition includes resonant and quasiperiodic features, as well as separatrix crossing phenomena. Rizzato, de Oliveira and Erichesen (1998) investigated the process of energy transfer in the Zakharov equations. Starting with a low-dimensional quasi-integrable regime where solitons are formed via a modulational instability, they showed that if the largest length scale of the linearly excited modes is much longer than the most unstable one, the interaction of these solitons with ion-acoustic waves leads to energy transfer to smaller length scales due to the stochastic dynamics.

The transition from order to chaos may occur in 3-wave interaction and triplet-triplet interaction in the Zakharov equations. De Oliveira, de Oliveira and Rizzato (1997) found that as the wave amplitude increases a triplet undergoes a transition from a quasi-integrable regular regime to a nonintegrable chaotic regime. Lopes and Rizzato (1999) showed that the coupling of two triplets destroys integrability, leading to spatiotemporal chaos.

The aforementioned works have demonstrated that as the beam-excited Langmuir waves evolve nonlinearly, a variety of dynamical processes may lead to Langmuir turbulence via chaos. Since the electrostatic Langmuir waves are in general coupled to electromagnetic waves, as seen in Equations (1)–(2), solar radio emissions provide an interesting means to probe the turbulent state of the plasma regions traversed by the energetic electron beams emanated from the solar active regions. Further details on this subject can be obtained in the reviews by Chian (1999) and Chian *et al.* (2000).

4. Conclusions

In conclusion, we derived a general dispersion relation for the parametric generation of fundamental plasma emissions due to two counterstreaming Langmuir waves with different wave amplitudes. The presence of a second pump wave with different amplitude from the first one, excites a region of convective instability, not present for equal wave amplitudes. In addition, we discussed the transition from order to chaos which can explain the onset of Langmuir turbulence. There are a number of observational reports of a close correlation of Langmuir and ion-acoustic waves in connection with type-III radio bursts (Lin *et al.*, 1986; Gurnett *et al.*, 1993). Recently, Thejappa and MacDowall (1998) presented experimental results for type-III radio bursts with clear evidence of occurrence of ion-acoustic waves in association with Langmuir waves. Hence, the theory presented in this paper provides a detailed picture of nonlinear wave-wave interaction processes which may be responsible for the excitation of type-III solar radio emissions.

Acknowledgements

F. B. Rizzato, A. C.-L. Chian and E. L. Rempel wish to express their gratitude to NITP/CSSM of the University of Adelaide for their hospitality. This work is supported by CNPq, FAPESP and AFOSR.

References

Alves, M. V., Chian, A. C.-L., de Moraes, M. A. E., Abalde, J. R. and Rizzato, F. B.: 2002, *Astron. Astrophys.* **390**, 351–357.

Chian, A. C. -L.: 1999, *Plasma Phys. Contr. Fusion* **41**, A437–A443.

Chian, A. C.-L. and Alves, M. V.: 1988, *Astrophys. J.* **330**, L77–L80.

Chian, A. C.-L., Abalde, J. R., Borotto, F. A., Lopes, S. R. and Rizzato, F. B.: 2000, *Progr. Theor. Phys. Suppl.* **139**, 34–45.

De Oliveira, G. I., Rizzato, F. B. and Chian, A. C.-L.: 1995, *Phys. Rev. E* **52**, 2025–2036.

De Oliveira, G. I., de Oliveira, L. P. L. and Rizzato. F. B.: 1997, *Physica D* **104**, 119–126.

Gurnett, D. A., Hospodarsky, G. B., Kurth, W. S., Williams, D. J. and Bolton, S. J.: 1993, *J. Geophys. Res.* **98**, 5631.

Lin, R. P., Levedahal, W. K., Lotko, W., Gurnett, D. A and Scarf, F. L.: 1986, *Astrophys. J.* **308**, 854.

Lopes, S. R. and Rizzato, F. B.: 1999, *Phys. Rev. E* **60**, 5375–5384.

Rizzato, F. B. and Chian, A. C.-L.: 1992, *J. Plasma Phys.* **48**, 71–84.

Rizzato, F. B., de Oliveira, G. I. and Erichsen, R.: 1998, *Phys. Rev. E* **57**, 2776–2786.

Thejappa, G. and MacDowall, R. J.: 1998, *Astrophys. J.* **498**, 465–478.

SELF-ORGANIZATION IN A CURRENT SHEET MODEL

J. A. VALDIVIA[1,2], A. KLIMAS[3], D. VASSILIADIS[4], V. URITSKY[4] and J. TAKALO[5]

[1]*Departamento de Fisica, Univerisidad de Chile, Santiago, Chile*
[2]*World Institute for Space Environment Research (WISER) University of Adelaide, Australia*
[3]*NASA Goddard Space Flight Center, Greenbelt, Maryland, U.S.A.*
[4]*Universities Space Research Association, NASA, Greenbelt, Maryland, U.S.A.*
[5]*Departament of Physical Sciences, University of Oulu, Oulu, Findland*

Abstract. Self-organization is a possible solution to the seemingly contradicting observation of the repeatable and coherent substorm phenomena with underlying complex behavior in the plasma sheet. Self-organization, through spatio-temporal chaos, emerges naturally in a plasma physics model with sporadic dissipation.

Key words: self-organizations, space plasmas, substorms

1. Introduction

The magnetotail plasma sheet seems to be a dynamic and turbulent region (Borovsky *et al.*, 1997) at all activity levels. Ohtani *et al.* (1998) found that magnetic fluctuations, associated with substorm onset, were due to a system of chaotic filamentary electric currents. The results are in sharp contrast to the standard picture of plasma sheet transport with laminar earthward flow in a well ordered magnetic field. Instead these turbulent events are probably related to the high-speed, bursty bulk flows (BBF) of the central plasma sheet (Baumjohann *et al.*, 1990), and their intermittent behavior (Angelopoulos *et al.*, 1999). Sergeev *et al.* (1996) postulated impulsive localized dissipation events which are the manifestations of sporadic tail reconnection. Nagai *et al.* (1998) observed that magnetic reconnection stops even while the substorm expansion phase proceeds, implying that reconnection occurs at localized sites that turn on and off over the course of a substorm.

The complex behavior in the plasma sheet is at first sight difficult to reconcile with the predictability of the geomagnetic indices (Vassiliadis *et al.*, 1995; Valdivia *et al.*, 1996; Valdivia *et al.*, 1999) and the coherence and repeatability of the substorm cycle that has led to the identification of its distinct phases (Baker *et al.*, 1996). We suggest that these seemingly contradicting statements may be reconciled by proposing that the plasma sheet is driven into a non-equilibrium self-organized (SO) 'global' state (Chang 1999). Here we will use the concept of SO state, as oppose to self-organized criticality (SOC), in an effort to distantiate our work with the more traditional discrete sandpile-like models (Bak *et al.*, 1987),

Space Science Reviews **107**: 515–522, 2003.
© 2003 *Kluwer Academic Publishers.*

even though we may end up with similar critical behavior belonging to the same universality class. The SO state is characterized by critical behavior with power-law power spectra and scale invariant events with self-similar spatial structure and fractal topology (Milovanov *et al.*, 1996).

There is mounting indirect evidence that a non-equilibrium complex state with self-similar event statistics, as expected of a SO system, really exists in the magnetosphere. Consoliny (1997) found a power law distribution (PLD) of 'burst strength' events in the AL index. Tsurutani *et al.* (1990) showed that the power spectrum for AL consists of two power-law spectral regions. Lui *et al.* (2000) found a PLD of events, except for large ones, of spatial ionospheric energy dissipation from Polar UVI images. And Uritsky *et al.* (2002) showed that this PLD extends even to large events when the spatiotemporal event statistics, integration over time, are considered. Lu (1995) postulated a set of criteria that would be required for a system to reach a SO state, which seem to be satisfied in the plasma sheet. If the plasma sheet is in a SO state, then understanding self-organization may be the key to understanding the substorm evolution. Even though the SO state is a dynamical state in nature with a superimposed unpredictable behavior, its 'global' behavior is inevitable and repeatable (this is true of SOC systems as well). We think that this is the basis for the global coherence and repeatability of the substorm phenomenon in the turbulent plasma sheet. Indeed, Chang (1999) has shown that a system at criticality is expected to exhibit apparent low dimensional 'global' behavior. Thus, we are led to study substorm phenomenon as an ensemble of many dissipation events in the turbulent plasma sheet under the assumption that it is in, or near, a global SO state. We now turn to the description of sporadic current sheet dissipation.

2. Description

Some of the statistical properties of the magnetospheric observations can be reproduced by discrete sandpile-like models (Chapman *et al.*, 1998; Takalo *et al.*, 1999; Takalo *et al.*, 2001). In this paper we study the properties of continuous plasma physics models that evolve naturally into a SO state. We start from the following neutral plasma equations

$$\begin{aligned}
(\partial \mu / \partial t + U_j \partial \mu / \partial x_j) &= -\mu \nabla \cdot \mathbf{U} \\
\mu(\partial \mathbf{U} / \partial t + U_j \partial \mathbf{U} / \partial x_j) &= \mathbf{J} \times \mathbf{B} - \nabla P + \nu \nabla^2 \mathbf{U} \\
(\partial P / \partial t + U_j \partial P / \partial x_j) &= -\gamma P \nabla \cdot \mathbf{U} + (\gamma - 1) \mathbf{J} \cdot (\mathbf{E} + \mathbf{U} \times \mathbf{B}) - \nabla \mathbf{Q} \\
\partial \mathbf{B} / \partial t &= -\nabla \times \mathbf{E}
\end{aligned} \tag{1}$$

and Ohm's law

$$\mathbf{E} + \mathbf{U} \times \mathbf{B} = \eta \mathbf{J} + \alpha_1 \mathbf{J} \times \mathbf{B} + \alpha_2 \frac{\partial \mathbf{J}}{\partial t} + \alpha_3 \nabla \mathbf{P_e} + \dots \tag{2}$$

For this paper, we assume $\alpha_1 = \alpha_2 = \alpha_3 = 0$, $\nabla \mathbf{Q} = 0$, and $B = \nabla \times A$. A 1-D field annihilation model can be obtained from Faradays' law

$$\frac{\partial A_y(z,t)}{\partial t} = \eta \frac{\partial^2 A_y(z,t)}{\partial z^2} + S(z,t) \tag{3}$$

where the current density $J = -\partial^2 A_y / \partial z^2$ and S corresponds to $(\mathbf{U} \times \mathbf{B})_y$. This deceivingly simple diffusion equation will provide the starting point to understand the complex behavior required to establish a global SO state. Klimas et $al.$ (2000) introduced (see Lu, 1995) sporadic dissipation through a spatio-temporal evolution for η as

$$\frac{d\eta}{dt} = \frac{(q(J) - \eta)}{\tau} \qquad q(J) = \begin{cases} \eta_{max} & |J| > J_c \\ \eta_{min} & |J| < \beta J_c \end{cases} \tag{4}$$

The trigger function q, at a given position, is defined on a hysteretic loop. q will remain in the low state $q = \eta_{min}$, until $|J| > J_c$ where it undergoes a transition to the high state $q = \eta_{max}$. Similarly, when in the high state, q will not make the transition to the low state $q = \eta_{min}$ until $J < \beta J_c$ ($\beta < 1$). For this paper, we take $-L \leq z \leq L$, with $L = 20$, $\Delta x = 0.1$, $S(z) = S_o Cos(\pi z/2L)$, $J_c = 0.04$, $\eta_{max} = 5$ (normalized to $c^2 L V_a/4\pi$), V_a a reference Alfven's speed, $\tau = 1$, $\eta_{min} = 10^{-10}$, $S_o = 3 \times 10^{-5}$, and $J = 0$ at the boundaries. In this approach we intend to study the event statistics of the collective effects of many such interacting instability sites, derived from observations and data analysis, in a complementary manner to microphysics and particle kinetics. Hence we use this nonlinear resistivity in an attempt to characterize some of the complex microphysics behavior, with q acting like a physical current driven instability (Papadopoulos et $al.$, 1985) with a threshold J_c that is higher than the value required to maintain the instability.

We start with $A_y(z) = 0$ and we let the system evolve. If we take $\eta = const$, then $J(x,t) = -(S_o/\eta)Cos(\pi z/2L)$ is the solution to Equation 3. A steady solution is permitted as long as $S_o < J_c \eta_{min}$ or $S_o > J_c \eta_{max}$, with some restrictions imposed by the boundary conditions. B_x and η are displayed in Figure 1a and Figure 1b respectively for S_o in the intermediate range. Even though the system can display complex behavior with strong underlying spatio-temporal variability, its global state is robust, coherent and repetitive. This may provide a hint about the coherent and repetitive behavior of the magnetosphere during substorms. If we define the energy as $E(t) = \int B(x,t)^2 dx/2$, then $\dot{E} = -F + I$, with the essentially constant input rate $I(t) = \int S(x)B(x,t)dx$. The sporadic dissipation rate $F(t) = \int \eta(x,t)J(x,t)^2 dx$ is shown in Figure 1c, due to the spatio-temporal dissipation ηJ^2 given in Figure 1d. This complexity emerges from purely deterministic spatio-temporal chaotic dynamics, with well defined loading-unloading cycles. Using a long time series of F (Figure 2a) we compute the event statistics of energy dissipated (Figure 2b) and duration (Figure 2c). The statistics seem to be insensitive to $F > F_{min} = 10^{-5}$ used to define an event. The power spectrum

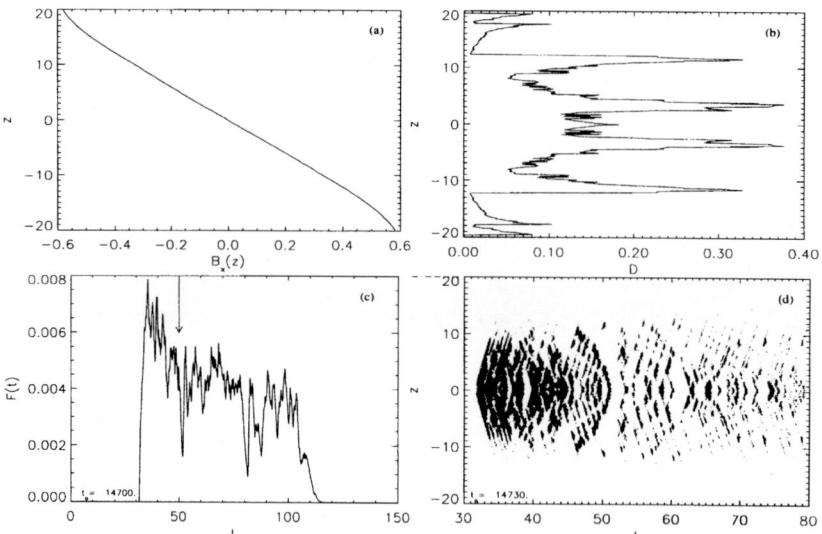

Figure 1. Spatial profile of (a) B_x and (b) η for $\beta = 0.9$ (at time defined by arrow in (c)). (c) F(t). (d) The spatio-temporal evolution of $\eta J^2 > 10^{-2} \eta_{max} J_c^2$.

shows a characteristic $1/f$ spectrum but is not shown for space reasons. The event duration distribution seems to agree with the statistics of the duration of BBFs in the plasma sheet (Angelopoulos *et al.*, 1999), but here we cannot have BBFs since we don't have a plasma yet (see below).

In practice the robust critical behavior occurs only for $J_c \eta_{max} < S_o < S_c < J_c \eta_{max}$ at which point the duration of the loading cycle is about the same as the duration of the unloading cycle. Uritsky *et al.* (2002) found that $S_c = 2 \times 10^{-2} J_c \eta_{max}$ for $\beta = 0.9$ and $\eta_{min} = 0$. Therefore, this model clearly departs from standard sandpile models in which critical behavior is only realized in the limit $S_o \rightarrow 0$. It is important not to over extrapolate from this model, but the transition at $S_o \sim J_c \eta_{max}$ is similar to a first order phase transition, and may provide an explanation for the observation of Sitnov *et al.* (2000). S_c, depends strongly on β, as can be seen from the periodic dissipations shown in Figure 2d for $\beta = 0.5$. If we take $\beta = 1$, then η relaxes very quickly after q is turned on, hence we move away from criticality again. It turns out that the hysteresis is present in virtually all SO models, including sandpiles, and it is the reason behind the existence of the loading-unloading cycle necessary for the intermittent behavior. This hysteresis is also present in deterministic and stochastic systems that evolve into intermittent behavior without threshold conditions. These issues are very important and the details will be presented elsewhere.

Chang (1999) postulated that the multiscale turbulence in the magnetotail evolves into a SO state where the locally, $J \times B$, accelerated plasma interacts with the turbulent field and plasma around it generating bursts remeniscent of BBFs. The SO state evolves precisely when the turbulent field and bursty plasma gen-

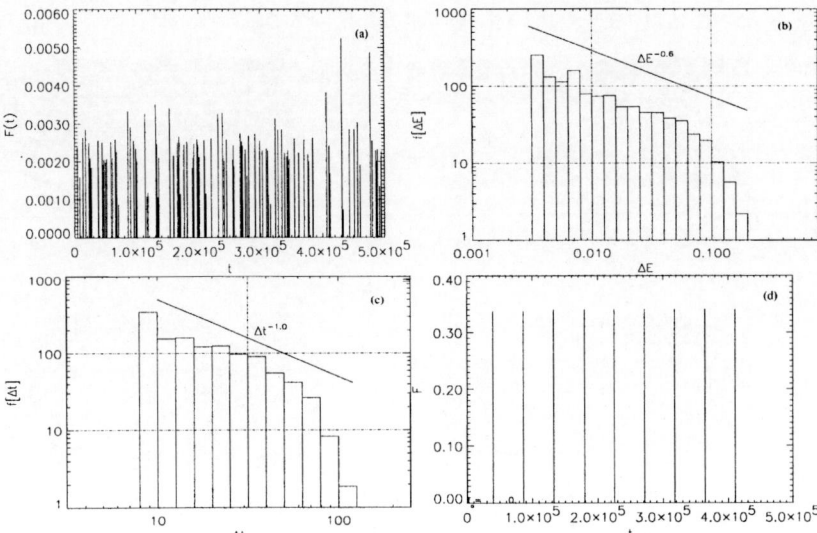

Figure 2. (a) F(t) for a long integration time using $\beta = 0.9$. The event distribution of (b) energy dissipation and (c) duration. The event statistics are constructed using a logarithmic histogram with a log bin size of 0.1. This is then normalized by the linear size of each bin. (d) F(t) for a long integration time using $\beta = 0.5$.

erate coherent structures. The results from our 1-D model suggest the following question: Can the intermittent dissipation be used as the core of a 2-D plasma physics model of the plasma sheet, that incorporate self-consistently the turbulent flow velocity and the magnetic field, and evolves into a SO state? Since we do not examine the kinetic details, we model the plasma sheet as a magnetofluid (This does not mean that MHD can account for the microphysics of plasma sheet) and solve Equation (1) in two dimensions, but with the dissipation Equation (4). The system is similar to Ugai and Tsuda (1977) in which we have a symmetric system at both $x = 0$ and $z = 0$, with an imposed constant inflow z-velocity $U_{z,o}$ and a constant magnetic field B_o at $z = L_z$. We have outgoing conditions at $x = L_x$.

If we first take $\beta \rightarrow 0$, hence a constant $\eta = \eta_{max}$, the incoming plasma gets transported out of the system with the generation of a thin current sheet (see Biskamp, 2000), as shown in Figure 3a. If we now include the hysteretic dissipation term, $\beta = 0.9$, we can study how the plasma turbulence enhances or quenches the transport. Figures 3b–d show the turbulent current density, at different times, with underlying spatio-temporal diffusion. The details of the characterization of this system will be published elsewhere, but we can already see that even though we have strong underlying turbulence, there is a well defined global state that permits the dissipation and transport of energy through the system, but in a bursty fashion.

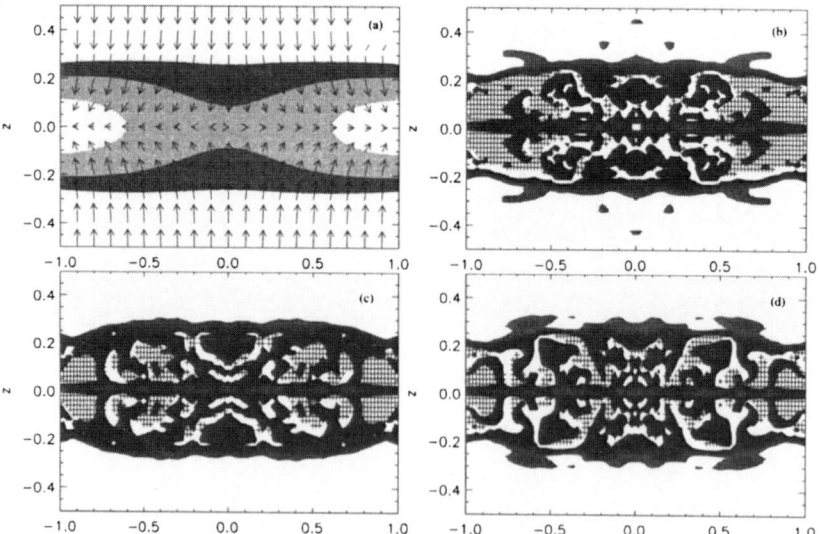

Figure 3. (a) The evolution of the current sheet model (including flow direction) when $\eta = const.$ The current density J is also shown at three different times in (b), (c), (d) respectively. The + signs correspond to places where the instability is active. Darker corresponds to stronger J.

3. Conclusions

Recent observations seem to suggest that space plasmas, under certain conditions, demonstrate very complex behavior, from self-organization to spatio-temporal chaos to phase transitions, etc. While deterministic chaotic models are providing a new tool for studying turbulent behavior in space plasmas, the concept of a SO state allows for the possibility of a repeatable coherent global state, with underlying complex dynamics at all scales. In this needed framework, it is necessary to clarify the emergence of turbulent behavior from stochastic and deterministic spatio-temporal dynamics. In the magnetosphere the robust SO state is a possible solution to the seemingly contradicting observation of the repeatable and coherent substorm phenomena with underlying turbulent and complex behavior in the plasma sheet. In this work, we assumed a simple parameterization of the dissipation η since we intended to study the collective effect of many interacting instability sites in a complementary manner to microphysics and particle kinetics. Even though the exact details of the microphysics (ballooning, cross field current, variant of tearing, etc.) cannot be accounted by such simple parameterizations, it appears that the statistical behavior of many complex distributed systems is more a property of their SO state, if it is achieved, than the details of the physical processes that allow such state. It is probable that the statistics of substorms, pseudobreakups, and even the evolutions of the growth and expansion phases, are unrelated to the details of the dissipation process (Shay *et al.*, 1998) other than that dissipation allows for the establishment of a SO state. The very simple 1-D model already has some of the properties observed in the magnetospheric dynamics, and it is expected that the sporadic 2-D plasma model may help us bring some light about the relationship between the energy dissipation with the self-similar complex behavior observed in the magnetosphere (Biskamp, 2000).

Acknowledgements

We acknowledge support by FONDECYT project N° 1000808, FONDECYT project N° 10307. FONDECYT project N° 1020152, and AFOSR. We thank the hospitality of NITP/CSSM at the University of Adelaide.

References

Angelopoulos V., Mukai T. and Kokubun S.: 1999, 'Evidence for Intermittency in Earth's Plasma Sheet and Implications for Self-organized Criticality', *Phys. Plasmas* **6**, 4161.

Bak, P., Tang, C. and Wiesenfeld, K.: 1987, 'Self-organized Criticality: An Explanation of 1/f Noise', *Phys. Rev. Lett.* **59**, 381.

Baker, D. *et al.*: 1996, 'Neutral Line Model of Substorms: Past Results and Present View', *J. Geophys. Res.* **101**, 12975.

Baumjohann, W., Paschmann, G. and Luhr, H.: 1990, 'Characteristics of High Speed Ion Flows in the Plasma Sheet', *J. Geophys. Res.* **95**, 3801.

Biskamp, D.: 2000, *Magnetic Reconnection in Plasmas*, Cambridge University Press.

Borovsky, J. E., Elphic, R. C., Funsten, H. O. and Thomsen, M. F.: 1997, 'The Earth's Plasma Sheet as a Laboratory for Flow Turbulence in High-beta MHD', *J. Plasma Phys.* **57**, 1.

Chang, T.: 1999, 'Self-organized Criticality, Multi-fractal Spectra, Sporadic Localized Reconnections and Intermittent Turbulence in the Magnetotail', *Phys. Plas.* **6**, 4137.

Chapman, S. *et al.*: 1998, 'A Simple Avalanche Model as an Analogue for Geomagnetic Activity', *Geophys. Res. Lett.* **25**, 2397.

Consolini, G.: 1997, 'Sandpile Cellular Automata and Magnetospheric Dynamics', in S. Aiello, N. Lucci, G. Sironi, A. Treves and U. Villante (eds), Cosmic Physics in the Year 2000, Proc. of VIII GIFCO Conference, p. 123, SIF, Bologna.

Klimas, A., Valdivia, J. A., Vassiliadis, D., Takalo, J. and Baker, D.: 2000, 'Self-organized Criticality in the Substorm Phenomenon and its Relation to Localized Reconnection in the Magnetospheric Plasma Sheet', *J. Geophys. Res.* **105**, 18765.

Lu, E. T.: 1995, 'Avalanches in Continuum Driven Dissipative Systems', *Phys. Rev. Lett.* **74**, 2511.

Lui, A. T. Y., Chapman, S. C., Liou, K., Newell, P. T., Meng, C. I., Brittnacher, M. and Parks, G. K.: 2000, 'Is the Dynamic Magnetosphere an Avalanching System?' *Geophys. Res. Lett.* **27**, 911.

Milovanov, A. V., Zelenyi, L. M. and Zimbardo, G.: 1996, 'Fractal Structures and Power Law Spectra in the Distant Earth's Magnetotail', *J. Geophys. Res.* **101**, 19903.

Nagai, T. *et al.*: 1998, 'Structure and Dynamics of Magnetic Reconnection for Substorm Onset with Geotail Observations', *J. Geophys. Res.* **103**, 4419.

Ohtani, S. *et al.*: 1998, 'AMPTE/CCE-SCATHA Simultaneous Observation of Substorm Associated Magnetic Fluctuations', *J. Geophys. Res.* **103**, 4671.

Papadopoulos, K.: 1985, 'Microinstabilities and Anomalous Transport, R. G. Stone and B. T. Tsurutani (eds), *in Collisionless Shocks in the Heliosphere: A Tutorial Review*, Vol 34, Washington DC, American Geophysical Union, p. 59.

Sergeev, V. A., Pulkkinen, T. I. and Pellinen, R. J.: 'Coupled-mode Scenario for the Magnetospheric Dynamics', *J. Geophys. Res.* **101**, 13047.

Shay, M. A. *et al.*: 1998, 'Structure for the Dissipation Region During Collisionless Magnetic Reconnection', *J. Geophys. Res.* **103**, 9165.

Sitnov, M. I., Sharma, A. S., Papadopoulos, K., Vassiliadis, C., Valdivia, J. A. and Klimas, A. J.: 2000, 'Phase Transition-like Behavior of the Magnetosphere During Substorms', *J. Geophys. Res.* **105**, 12955.

Takalo, J., Timonen, J., Klimas, A., Valdivia, J. and Vassiliadis, D.: 1999, 'A Coupled Map Model for the Magnetotail Current Sheet', *Geophys. Res. Lett.* **26**, 2913.

Takalo, J., Timonen, J., Klimas, A., Valdivia, J. A. and Vassiliadis, D.: 2001, 'A Coupled Map as a Model of the Dynamics of the Magnetotail Current Sheet', *J. Atmos. Solar-Terrestrial Phys.* **63**, 1407.

Tsurutani, B. T. *et al.*: 1990, 'The Nonlinear Response of AE to the IMF Bz: A Spectral Break at 5 hours', *Geophys. Res. Lett.* **17**, 279.

Ugai, M. and Tsuda, T.: 1977, *J. Plasma Phys.* **17**, 337.

Uritsky, V. M., Klimas, A. J. and Vassiliadis, D.: 2002, 'Multiscale Dynamics and Robust Scaling in a Continuum Current Sheet Model', *Phys. Rev. E* **65**, 046113-1.

Uritsky, V. M., Klimas, A. J. and Vassiliadis, D.: 2002, *Geophys. Res. Lett.* (in press).

Valdivia, J. A., Sharma, A. and Papadopoulos, K.: 1996, 'Prediction of Magnetic Storms by Nonlinear Dynamical Methods', *Geophys. Res. Lett.* **23**, 2899.

Valdivia, J. A., Vassiliadis, D. and Klimas, A.: 1999, 'Modeling the Spatial Structure of the High Latitude Magnetic Perturbation and the Related Current System', *Phys. of Plasmas* **6**, 4185.

Vassiliadis, D. *et al.*: 1995, 'A Description of Solar Wind-magnetosphere Coupling Based on Nonlinear Filters', *J. Geophys. Res.* **100**, 3495.

DATA GRIDS AND HIGH ENERGY PHYSICS: A MELBOURNE PERSPECTIVE

LYLE WINTON

School of Physics, The University of Melbourne, Melbourne, Australia
(e-mail: winton@physics.unimelb.edu.au)

Abstract. The University of Melbourne, Experimental Particle Physics group recognises that the future of computing is an important issue for the scientific community. It is in the nature of research for the questions posed to become more complex, requiring larger computing resources for each generation of experiment. As institutes and universities around the world increasingly pool their resources and work together to solve these questions, the need arises for more sophisticated computing techniques. One such technique, grid computing, is under investigation by many institutes across many disciplines and is the focus of much development in the computing community. 'The Grid', as it is commonly named, is heralded as the future of computing for research, education, and industry alike. This paper will introduce the basic concepts of grid technologies including the Globus toolkit and data grids as of July 2002. It will highlight the challenges faced in developing appropriate resource brokers and schedulers, and will look at the future of grids within high energy physics.

Key words: E-Science, experimental high energy physics, grid, high performance computing

1. Introduction

1.1. THE GRID

'The Grid' (Foster and Kesselman, 1978) is the name often given to a future world-wide infrastructure of distributed computing resources. These resources will be owned and managed by various organisations, however, the grid infrastructure will enable collaborative use of these resources globally. The development of the grid infrastructure, both hardware and software, has recently become the focus of a large community of researchers and developers in both academia and industry. This has lead to the development of smaller grid testbeds, often termed Grids, Application Grids, or Analysis Grids.

Perhaps the best way to describe the grid is in the words of the grid pioneers:

> *The Grid refers to an infrastructure that enables the integrated, collaborative use of high-end computers, networks, databases, and scientific instruments owned and managed by multiple organizations.* – Globus

The Globus organisation are developers of the most commonly used foundation software within grid testbeds (Globus, 2002).

Space Science Reviews **107**: 523–540, 2003.
© 2003 *Kluwer Academic Publishers.*

Tomorrow, the grid will be a single, sustained engine for scientific invention. It will link petaflops of computing power, petabytes of data, simulation and modeling codes of every stripe, sensors and instruments around the globe, and tools for discovering and managing these resources. At your desktop and at your whim, you'll have access to the world and its computing assets. – NCSA (National Center for Supercomputing Applications)

NCSA are a US based grid pioneer providing software and computing resources. (NCSA, 2002)

The grid …consists of physical resources (computers/clusters, disks and networks) and 'middleware' (software) that ensures the access and the co-ordinated use of such resources. – EDG (European Data Grid)

The EDG work closely with research organisations and industry to provide software and infrastructure for the grid in Europe (EDG, 2002).

A common misconception is that The Grid will be the next internet. This is not true. The grid will use the existing internet infrastructure, however, there are several grid projects aimed at extending the existing internet infrastructure in preparation.

The grid is not designed to replace Cluster or Parallel computing and does not attempt to solve the problems and limitations with these technologies. It is designed to help utilise these technologies, and share computing resources. In the grid a cluster or multiple processor machine becomes a single grid node or computation resource. Detailed information about these resources and their status are made available to the grid, and jobs may be submitted to these nodes/resources based on this information and the job requirements.

A common way to describe the grid is with the analogy of an electrical power grid. In the grid there are nodes that supply power (which are resources that provide computing power) and nodes that use power (users accessing the grid). Standards are required to connect these together seamlessly. Users will access and consume power (computing resources) as required without the need to know where the resources are coming from. The end goal is to make the use of computing power as simple as the use of electricity today, at the flick of a switch.

1.2. DATA GRIDS

Data Grids are a specific application of grids where the access to data is as important, or more important, than access to computational resource. These grids occur in most collaborative research areas where access to common data is required or data intensive research areas.

Experimental high energy physics is a good example as it is a data focused research area where access to experimental and simulated data is required by large collaborations of institutes spread around the world. To give an indication the ATLAS experiment, situated on the Large Hadron Collider (LHC) at CERN in Geneva, will need to produce around 3.5 PB of data per year. The ATLAS collaboration consists of over 1800 people in 150 institutes around the world (ATLAS Proposal, 2002).

Another example of a data grid is the Earth Systems Grid in the united states (ESG, 2002). This will produce 3 PB (petabytes) of data per year which will require 3 TFLOPS (tera FLOPS – floating point operations per second) of processing power to analyse.

Data grids may also be of interest for commercial applications. Teleimmersive Applications which include engineering design environments require world-wide access to parts description databases, and assembly information. Computing resources are used for finite element analysis and the rendering of parts and products. The entertainment industry could benefit from the use of grids in the storage of high definition digital footage, scene overlays, and computer graphics. Processing power is required for image manipulation and rendering. Work loads could be split across multiple production centres in different countries, each having access to the film data and sharing computing resources.

1.3. DATA GRIDS IN PRACTICE

To understand the enthusiasm about grids we need to understand the problems that they will solve. The problem with existing computing methods is they will become more cumbersome due to the increasing size of collaborations, experiments, and their data. From an administration point of view much time is already spent in large organisations managing software, accounts, and security. From the point of view of data management, the storage and transfer of large amounts of data can be expensive. Commercially, data transfer can cost of the order of cents per MB (megabyte), disk storage around a cent per MB, and tape storage around a dollar per GB (gigabyte). Backup of this data, particular using traditional methods, can be expensive or impossible. Rapid changes in software versions together with many users and their private analyses can lead to the proliferation of data. The software and data need to be tracked and managed. From the point of view of computing power, the CPU required for peak periods can be expensive, however idle CPU will exist if loads vary. Service and task replication exists between current high energy physics experiments, leading to the reinvention of code to perform these common tasks.

It is the aim of data grid development to solve these problems by providing simple centralised administration tools, providing tools for world wide access to data and computing facilities, and by providing standard software services and APIs (application programmer interfaces). Tools for data manipulations will including intelligent replication and caching for faster cheaper access to data. The grid may also provide better ways of filing data with well formed descriptions and versioning information (metadata). Access to more computing power will help spread CPU load meaning less CPU is required at each facility to meet peak loads. Services for job management will be developed providing intelligent job location

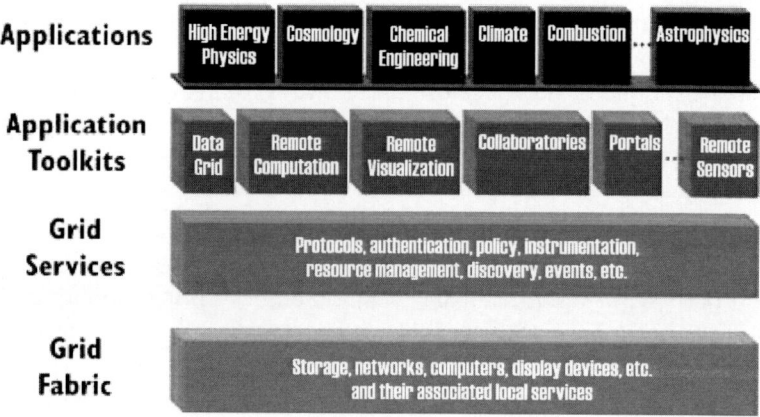

Figure 1. Grid services and software architecture (courtesy of Globus).

(e.g. moving the code to the data or vice-versa) for faster processing times and reduced cost. Standard APIs will be provided for parallel computing, network access, and for many other common tasks.

These development tasks are quite difficult and raise many new problems. The grid will have to provide for administration of tens of thousands of computers, thousands of users, access privileges between these and the data produced, monitoring facilities, descriptions of user jobs and computing resources, and job management which will determine where and when the job will run using information such as required CPU time or number of CPUs for parallel jobs, required memory or hardware, location of the data, current loads on resources, the time or cost of data transfer, user permissions, and resource restrictions. Many of these problems have not been solved and software packages are still in the development and testing phases.

2. Overview of Products

The term middleware is used to describe grid services and software. Figure 1 illustrates the relationship between the various components within a grid. The grid services sit on top of the hardware (or fabric) and provide the means of communication with the hardware and include the security for the grid. The application toolkits, of which data grid specific services are a part, provide tools and interfaces for higher level grid usage, such as job scheduling and data management. The applications themselves are the tools and analysis code that we use today.

2.1. PACKAGES

Of the main grid packages that are available the Globus Toolkit is at the core of most grid middleware (Globus, 2002). This provides the low level grid services as displayed in Figure 1. The European Data Grid (EDG or EU-DataGrid), the Virtual Data Toolkit (VDT – GriPhyN project), and the Particle Physics Data Grid (PPDG) are examples of data grid packages that provide the higher level services (EDG, 2002; GriPhyN, 2002; PPDG, 2002). These are all built on the Globus Toolkit. Nile and the Sun Grid Engine are a notable products as they are one of the few which are not built on a Globus core (Nile, 2002; SGE, 2002). Many of these packages can be broken down into separate services or products.

2.2. TOOLS AND SERVICES

The following is a non-exhaustive list of available tools and services. Some are Grid specific while others are independant and commonly used within Grid environments. You will notice that the Globus Toolkit provides many of the services listed.

- Security
  - GSI – Grid Security Infrastructure (globus)
- Service/Resource Information (data/machine)
  - MDS – Metacomputing Directory Service (globus)
  - GRIS – Grid Resource Information Service (globus; MDS subsystem)
  - GIIS - Grid Index Information Service (globus; MDS subsystem)
- Resource Management
  - RSL – Resource Specification Language (globus)
    method to exchange resource info & requirements
  - GRAM – Globus Resource Allocation Manager (globus)
    standard interface to computation resources like PBS/Condor
  - DUROC – Dynamically-Updated Request Online Coallocator (globus)
  - WMS – Workload Management System (EDG)
- Data Management
  - GSIFTP – high performance, secure FTP uses GSI (globus)
  - Replica Catalog - data filing and tracking system (globus)
  - GASS – Globus Access to Secondary Storage (globus)
    access data stored in any remote file system by URL; Unix like calls fopen(), fclose()
  - GDMP – Grid Data Mirroring Package (EDG,GriPhyN,PPDG)
  - Magda – distributed data manager (PPDG)
  - Spitfire – grid enabled access to any RDBMS (EDG)
  - RLS – Replica Location Service (EDG)
- Mass Storage
  - HPSS – high performance storage system

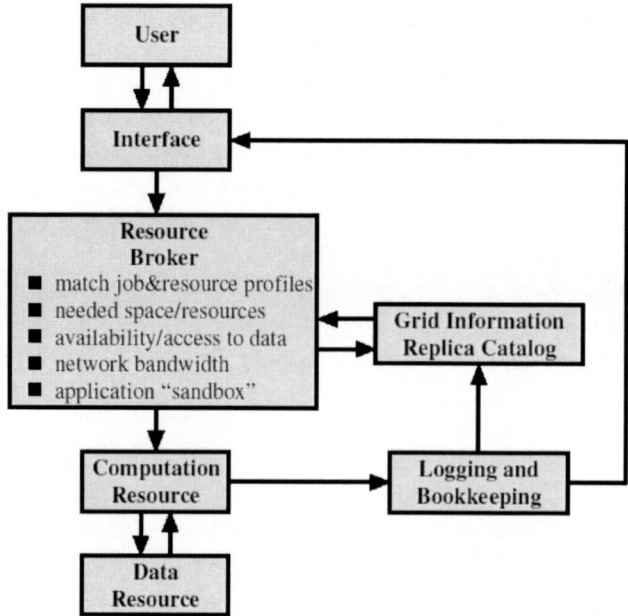

Figure 2. Simplified EDG workload manager flowchart.

- SRB – Storage Resource Broker
- Communication
  - Nexus and MPICH-G (globus)
- Monitoring
  - HBM – Heartbeat Monitor (globus)
- Job Managers
  - PBS – portable batch system
  - Condor – distributed computing environment
- Fabric Management
  - Cfengine – configuration manager

Some of the central services are GSI which provides the security, GRIS which provides the information about grid computing resources, GRAM which runs tasks on computing resources, GASS and GSIFTP which allow access to remote files, and the Replica Catalog which provides the filing system for the data.

2.3. FUNCTIONALITY

Figure 2 shows a simplified flowchart for the workload manager in the European Data Grid package. The general functionality is to simply accept a job and execute it on one or more computing resources.

A user will submit a job via an interface. This is passed on to the resource broker which will match the job's profile with the computing resource profiles

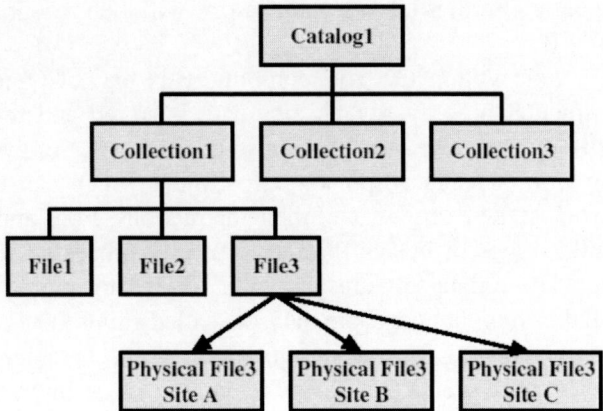

Figure 3. Basic replica catalog structure.

that are available. Checks are performed for required disk space, memory, CPU, and network bandwidth. The location, availability, and access permissions to the required data are determined. This is done by accessing the grid information services and replica catalog. The presence of the 'application sandbox' is ensured. The application sandbox or environment may include application code and all necessary auxiliary files to run the users application. If required, the resource broker can move the data to an appropriate location before the job is started and the replica catalog is updated accordingly. Finally, the resource broker actually runs the job on one or more computers. The status and output of the job are monitored via the logging and bookkeeping services so the grid information can be kept up to date and to inform the user of the status of their job.

An important issue within data grids is that of file management and replication. Data files can be organised as Logical File Names within a Virtual Directory Structure by using a Replica Catalog (Figure 3). Each logical file name maps to one or many Physical Files located somewhere on grid, usually specified via URLs. A virtual directory structure within the replica catalog is organised into Catalogs and Collections. In a typical job the user will specify the logical file that is to be processed and the Resource Broker will resolve this to the most appropriate physical file location. The job may require the file to be staged locally before execution or the application will stream the file from the current physical location via grid protocols.

The GDMP tool may be used to replicate files from one system to another (GDMP, 2002). A transfer of a file or group of files from site A to site B can be triggered in two ways, by site A publishing a list of files that need to go to other sites, or by site B requesting a list of files. In the later case, site B may be configured by an administrator to request certain files that will be used commonly, or a user application running at site B may request files for processing. Once the file has been transferred the Replica Catalog, which holds the list of physical file locations, is updated. Further jobs or applications will then recognise the presence of the file on site B.

A subject of note within the grid community is that of Object Replication. Existing experimental high energy physics data is stored and replicated as files containing many independent measurements called events. An event will include all information from sensors within the experiment which can be correlated to reconstruct particles, their origins, and their interactions. Files and groups of files containing similar types of events that are typically processed together and are called data-sets. These data-sets are generally filtered to extract events of most interest into smaller more manageable datasets called skims. The computing community in general is moving towards treating data as objects rather than streams of information. At first this seems the most efficient way of dealing with data. In high energy physics each event is already an independent object which could be stored separately. If this approach was taken, data-sets would become a collection of event IDs. Skims, for example, would take the form of a list of references to events and would be very small. As events in data-sets and skims may potentially overlap with events in other data-sets, the duplication of events would be eliminated by storing references and not actual events. Data processing could then be performed on the nearest event within the grid. Events may be cached locally, at nearby institutions, or overseas. With the file replication approach, processing will generally occur only in locations that have the whole file or the file must be transferred.

The problem with object replication is most current storage systems (databases and file systems) cannot handle the quantity of objects required, which may be greater than 10^9 in number. File replication problems are currently solve by replicating smaller skim files shared by many people rather than large data-sets. A compromise solution may develop where skim files can be constructed or reconstructed at process time by extracting events from the nearest data file or skim. However, this still requires the tracking of large numbers of events.

2.4. AN EXAMPLE

The following are examples of how a user might use various grid software to perform a simple task. The first example is the traditional method of running a job and does not involve grid software. The example uses secure shell to connect to remote resources (*ssh* command). It is assumed the user has access to at least one remote resource, a cluster or multiple CPU machine, via secure shell. A user must first choose a location or resource to run their job (*qstat* is used to check the status of some common queue systems), then copy their auxiliary files such as libraries, scripts, and configuration files to the remote site (*scp* secure copy is used), then either run the job or submit it to an existing queue or batch system (*qsub* is used to submit a job on some common queue systems). This is repeated for as many job as are needed, potentially across multiple remote resources.

```
>   ssh remote qstat
```

```
>  scp auxfiles remote:dir
>  ssh remote qsub < myrun.csh
```

Using the Globus Toolkit we may access grid computation resources, usually clusters or multiple CPU machines. A user signs on to the grid by creating a proxy, they query grid resources to choose an appropriate location to run, then copy auxiliary files, then submit the job to the resource. This has the advantage of a single point of authentication and potentially greater access to remote resources. The disadvantage is that this is essentially the same effort from a user point of view as the traditional method. However, this is not surprising as, by using Globus, we are accessing only the low level Grid services.

```
>  grid-proxy-init
>  grid-info-search remote
>  globus-url-copy auxfiles remote
>  globus-job-run remote myrun.csh
```

To gain more from using the Grid higher level application tools must be used, such as resource brokers or schedulers. The following example is with the use of an economic scheduler called Nimrod/G as an interface to grid resources (Abramson *et al.*, 2000). This involves the previously mentioned sign-on to the grid by creating a proxy, then registration of a predefined plan for a group of jobs. You may then specify a budget or deadline for this run. The advantages are that file transfer, the choice of grid resource, and cost and time considerations are all handled by Nimrod/G. A single run can easily be split over multiple resources. The disadvantage is that data location and cost of transfer is not build into the scheduler, but is currently under development by Melbourne GridBus group (GridBus, 2002).

```
>  grid-proxy-init
>  nimrod myrun.plan
   <Specify budget/deadline>
```

The final example uses the Resource Broker within the European Data Grid Workload Manager which is currently under development. A job profile is submitted to the resource broker, security sign-on and all decisions are handled. The advantages are the very simple interface, file transfer and choice of resource is handled by the broker. The disadvantage is that cost, time, and feasibility are not yet considered by resource brokers.

```
>  dg-job-submit myrun.profile
```

The ideal future method for submitting jobs to the grid will consist of a simple interface, it will handle all file transferring and access transparently, the choice of grid resource will be handled taking into account the requirements of the job, and also the cost, quotas, time, and job feasibility with respect to computing resources and data resources.

3. High Energy Physics Grid Projects

3.1. RHIC, HENP GRAND CHALLENGE

The first project that should be mentioned was led by the High Energy and Nuclear Physics groups associated with the RHIC facility in the US (HENP, 2000). Several experiments were involved in the 'HENP Data Access Grand Challenge' which was to ensure access to 50 TB (tera bytes) of data for hundreds of simultaneous users across many institutes. As part of the project 'Mock Data Challenges' were constructed to test their systems for readiness. The challenges were completed and successful and RHIC has been up and running since mid 2000. The network of computing resources of this project were not viewed as a data grid, however, there are many similarities. Many of the people from this project are now working on the LHC grid, and are bringing their experiences from RHIC to this.

3.2. LHC COMPUTING GRID

The LHC computing grid will be constructed to service the experiments on the Large Hadron Collider facility in CERN and their collaborations (LCG, 2002). These are ATLAS, LHCb, ALICE, CMS. The data grid is currently in the design and testing phase and each experiment will be performing their own 'Data Challenges' to test the readiness of the grid systems for physics. The ATLAS data challenges will be mentioned in a later section.

The infrastructure of the LHC grid will follow the 'MONARC Model' (Models Of Networked Analysis at Regional Centers) (Table I). This model arose from the recommendations of the CERN computing division following simulations of grid infrastructures. It consists of a hierarchy of Grid nodes or resources, each existing on one of a number of Tiers or levels. The tier denotes the separation from the central CERN node (Tier 0).

Tier 0 will consist of a single node located at CERN and will be connected to the experiments and store all raw data. This tier will do all of the first pass reconstruction of raw data to offline data. Detector or sensor information for each event or measurement are correlated and reconstructed into particle tracks which are more palatable for physics analysis. There will be one Tier 1 per LHC experiment situated at CERN plus regional centres situated in other countries. These may contain a little raw data and all of the offline data. They will be used for additional reconstruction and simulation. There will be many Tier 2 nodes situated at various laboratories and cities around the world. These will be accessible to all users within the experiments and will contain part of the offline data for reconstruction and simulation support. Tier 3 nodes will be at institution levels

TABLE I

LHC grid infrastructure MONARC model

Tier 0	CERN (1 node)	Raw data
	connected to experiments	First pass reconstruction
	limited access	
Tier 1	One per experiment	Some raw data, offline data
	plus regional centres	Additional reconstruction and simulation
	accessible to all	
Tier 2	Cities or labs	Partial offline data
	accessible to all	Reconstruction and simulation support
Tier 3	Institute level	Partial offline data, private data
	local access	Personal analysis
Tier 4	Desktop PC level	Private data
	personal or local access	
Tier 5	Laptops or transients	Private data
	personal access	

and will not be accessible to the whole grid. They will contain some offline data and users private data. Tier 4 nodes consist of personal desktop machines that are permanently connected to the grid. They will generally be used as portals to the grid and for normal desktop operations. Tier 5 are the transient machines such as laptops which are not guaranteed to be on the network.

ATLAS expect to have around 10 PB of storage in total over all of tier 1 in about 10 nodes. They expect to have between 400 and 1000 TB of storage within Tier 2 amongst 12 to 25 nodes.

One of the most important specifications within the grid hierarchy structure is the network connections between tiers. The network must be such that raw and re-constructed data can be transferred to the correct level within the tiers, and the data can be access by users and application from lower levels within the tiers. The CMS collaboration estimate that a 100 MB/s (mega bytes per second) connection will be required between their experiment and tier 0 at CERN. A 2.5 Gb s^{-1} (giga bits per second) connection will be required between tier 0 and tier 1, 0.6–2.5 Gb s^{-1} will be required between tiers 1 and 2, and a 622 Mb s^{-1} connection is recommended between tiers 2 and 3.

3.3. ATLAS AND LHC COMPUTING GRID

The ATLAS experiment expects to record around 10^9 events per year and the collaboration will generate 3.5 PB of data including simulation and experimental data (ATLAS Proposal, 2002). They estimate there will be 1000–1500 users, typically 150 of whom will be performing analysis jobs simultaneously, and that up to 20 separate analyses may be undertaken. Over the period of the experiment they expect they will require 20 PB of tape storage, 2.6 PB of disk storage and 2000 kSI95 in CPU resources (2000 kSI95 $\approx$ 20 000 Intel P4 2GHz).

To investigate the problems that may occur with implementing such a large computing infrastructure, the LHC experiments, including ATLAS, have proposed several data challenges (ATLAS DC, 2002). These will test the infrastructure, network, hardware, calculated estimates, and push the grid software towards a scalable solution. The ATLAS collaboration have specified 3 different data challenges which build on each other over time. These are called DC0, DC1, and DC2.

Data Challenge 0 was completed in March 2002 and was designed to process a sample of 10^5 events in 1 month. This was not a test of processing power of the grid as the power required for this was trivial for existing hardware. The main aim was to test the existing software and interoperability of packages.

Data Challenge 1 which is scheduled for completion in 2002 is split into 3 main phases. Phase 0 involved preparation of hardware and operating system including analysis and comparison of differing systems. Phase 1 was to ensure the generation of 10^7 events in 10–20 days. The collaboration will also use this to test some physics of the detector and analysis techniques. Together with the Melbourne Advanced Research Computing Centre a total of 16 CPUs were utilised at the University of Melbourne to participate in this challenge. Phase 2 will test different software solutions and the infrastructure of the grid.

Data Challenge 2 will occur in 2003 and will construct and process a sample of 10^8 events in 3 months. They expect to perform this data challenge on an analysis grid equivalent to 50% of the expected ATLAS grid that will be built by 2006–2007. This is mainly to test the infrastructure and hardware that will be used.

3.4. THE UNIVERSITY OF MELBOURNE

At the University of Melbourne in the Experimental Particle Physics group we are undertaking a number of grid activities (University of Melbourne, EPP, 2002). We have obtained funding from the Victorian Partnership for Advanced Computing to build expertise and resources in High Performance Computing and Data Grid technologies within high energy physics (VPAC, 2002). In particular we have identified two major projects. The first is to investigate the advantages of using the grid for analysis within the Belle experiment (Belle, 2002). The Belle experiment is situated at the KEK B factory in Japan and is researching CP violation in the standard model. We are implementing a grid architecture to enable collaborative access to resources and data for an existing physics application. This is an important contribution as it will provide a real-life testbed for the use of the grid within high energy physics. The second major project is to gain software expertise and to develop software for the grid and high energy physics. Expertise in grid software will help us take advantage of collaborative computing and is important for the future of high energy physics research in Australia as most facilities are situated overseas. Contributing to software development enables us to develop user driven applications that will aid us in our research.

We are also participating in a number of collaborative grid activities. Working together with the MARCC group (the Melbourne Advanced Research Computing Centre) we are building a data grid infrastructure for use within high energy physics and to take part in the ATLAS data challenges mentioned in the previous section (MARCC, 2002).

We are working with members of Computer Science department at the University of Melbourne in developing an Ontological framework for experimental high energy physics analysis (UM CS, 2002). This is a framework for high-level descriptions for use by analysis code or agents.

We are working with another group within the Computer Science Department, called GridBus, in developing tools for resource brokering and economic job scheduling for use within high energy physics (GridBus, 2002). The effectively means the ensuring the best and cheapest use of available grid resources.

We are also working with members of the Computer Science department of the Royal Melbourne Institute of Technology on developing agent based technologies for high energy physics analysis (RMIT CS, 2002). These will be intelligent services able to dissect, suggest, and perform analysis based on high level descriptions of the analysis.

3.4.1. *The Belle Project*

The aim of our first major project is to attempt to use the Belle collaboration's analysis code in a working grid environment. If the concept is proven locally and found useful it may be adopted by the international Belle collaboration. On a wider level, using the grid infrastructure for existing physics applications will be an important test for future communities investigating the grid.

The first step was to construct a grid node then initial tests were performed using the Belle software without modification. A sample of Belle data was then made accessible via grid protocols and the Belle analysis code was modified to access this data via the grid. The analysis code was then tested via submission to the grid node together with grid enabled data access. The results of these tests were successful and proved the principle of basic grid functionality.

The next step is the construction of additional grid resources. A similar node has been constructed at the University of Sydney and tests have been carried out between the University of Melbourne and Sydney. Jobs were successfully submitted to both locations from either site. Preliminary results are encouraging but further quantitative tests need to be performed.

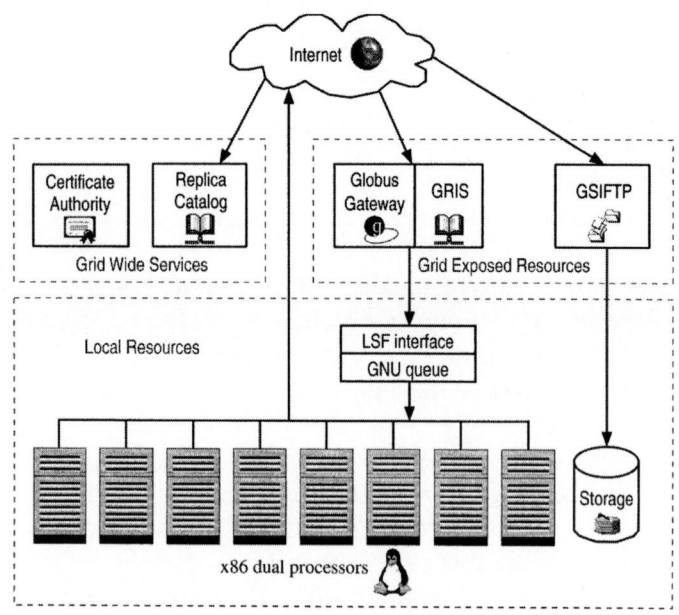

Figure 4. The University of Melbourne experimental particle physics grid node.

The next step will be to implement or build a resource broker to help better utilise multiple grid resources. This will also enable users to specify complex job requirements and utilise new resources as they become available.

Figure 4 is an overview of the grid node at Melbourne University. The existing resources consist of 8 AMD dual processor machines running Debian Linux accessible on a PBS or GNU queue system, and around 200 GB of disk storage. These have been exposed as grid resources by installing a Globus Gateway and Grid Resource Information System, and a GSIFTP server for data access. Globus GRAM software for the GNU queue system is not provided so an LSF like interface to the GNU queue system was built to solve the incompatibility. Grid wide services have been implemented such as a Certificate Authority to approve user and host certificates, and a Replica Catalog for the filing and tracking of data. Our local resources are now accessible to any authorised users on the internet as grid resources. Users and applications on our local resources can access our grid resources and other grid resources via the internet.

Figure 5 illustrates the future Belle analysis grid with other grid resources incorporated. Through collaboration with the MARCC group (Melbourne Advanced Research Computing Centre) some of their resource will be made available to the grid for our use. Through our collaboration with the Belle experiment at KEK, Japan, we hope to gain access to other resources for our mutual use. In the future Belle analysis grid there may be several cloned grid wide resources such as replica catalogs at various locations. A grid user is identified by a certificate signed by a trusted certificate authority. The user signs on to the grid by creating a grid-proxy

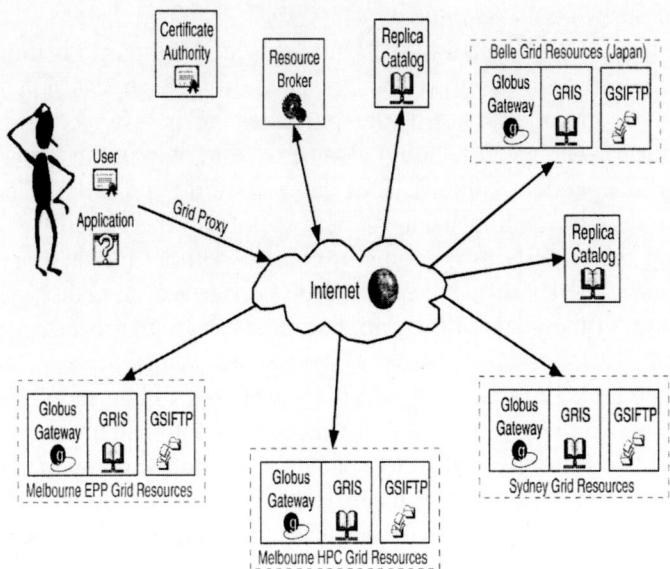

Figure 5. Future Belle analysis grid.

which enables the submission of jobs via the internet to grid resources. When the job application is running on these resources it inherits the grid proxy of the user, allowing it to further utilise grid resources via the internet.

As an analysis test case the research of Rohan Dowd was chosen, a Ph.D. student at the University of Melbourne. His analysis is the investigation of charmless B meson decays to two vector mesons. This is used to determine two angles of the CKM (Cabibbo–Kobayashi–Maskawa) unitarity triangle. These types of decays have not been measure previously. A number of preliminary tests were performed using this analysis on the grid. The first test was processing of 10 sample files, a total of 2 GB. When processed serially all files took 95 minutes. When processed using our facilities as a grid resource the files took 35 minutes to process. This is not surprising as the files are processed in parallel and the time is limited by the longest file. Next the two available data access protocols GASS and GSIFTP were tested across our 100 Mb network. For comparison NFS access to files is around 8.5 MB s^{-1}. File access via GASS averaged 4.8 MB s^{-1} and GSIFTP averaged 9.1 MB s^{-1}. Based on this initial test GSIFTP will be used for further grid file access as it appears to have less overhead. The GASS protocol is built on secure HTTP inheriting the larger overhead from data encryption. GSIFTP provides encrypted authentication but the transferred data is not encrypted. Combining the two test, the Belle analysis was performed using grid data access via GSIFTP and found very little difference in performance compared to running over NFS. This indicates that this particular analysis is CPU and not I/O limited and the grid does not degrade the performance.

3.4.2. *Grid Software Development*

A software project undertaken at the University of Melbourne is the development of a generic interface to the grid. There is a need for a high level job control which is currently not provided by software packages, or is provided for specific grid implementations. An interface could also be used to help in the automatic division of jobs into independent subjobs. This can also add another layer between users and the systems, decoupling what the user sees from the underlying environment and allowing the underlying systems to change without effect. Users can access different grid or cluster software without the need to learn another system. In fact, these are some of the goals of the grid, but most efforts in this direction are still in development.

There are a number of existing efforts looking at grid interfaces. Firstly, Globus provides access to the grid via its Resource Specification Language RSL (Globus RSL, 2002). The European Data Grid has provided tools on top of this that access the grid via a job description language (EDG WP1, 2002). Jefferson Laboratories are developing a web interface to the grid communicating via XML, however, physicists are used to using command line environments and web interfaces can be restrictive and cumbersome (JLAB, 2002). The CMS collaboration have specified they will need an extra layer between the physicist and the grid. The development of this will probably commence at a later stage. This extra layer could reduce the complexity of the interface from the users point of view. The CMS layer may control the automated creation of other subjobs for data control or output colla-tion. It may ensure access to auxiliary files needed within the application. It may automatically create what are called 'job hints' which estimate the memory size, system load, and output size of a given job.

From inspecting these efforts the design of the generic grid interface must be both graphical and command-line. The basis for this software development project will be a job description that is stored and communicates via XML (extensible markup language). XML is a well structured document format used for storing and passing information. The principle is that generic job descriptions in XML can undergo a grid/network specific transformation to produce a job description that incorporates knowledge about the type of network it will run on. This will be done using XSLT, an industry standard for the transformation of XML docu-ments into other documents. For example, a description of a job can be created and used to perform a PBS-Transformation resulting in PBS commands, scripts, and directives that enable the job to be run on a PBS system. The same job de-scription can be used performing a LHC-Data-Grid-Transformation resulting in European Data Grid commands and grid specific information for running the job on the LHC grid. With the CMS specification in mind, this same job description can be used performing a CMS transformation which will create any transparent subjobs, include auxiliary files, and job hint calculations. This transformation will incorporate the LHC-Data-Grid-Transformation resulting in commands and grid specific information for running the job on the LHC grid with the CMS extras.

In addition to this development project replica catalog tools have been developed for simple navigation and manipulation of replica catalog files and collections. A further project is underway to develop a resource broker that is aware of system loads, data availability in the replica catalog, complex job requirements, and job scheduling taking into account data location.

4. Summary and Conclusion

In summary, the grid will play an important role in collaborative research and, in particular, the future of high energy physics. It must be noted, however, that the technology is still in development. The ATLAS Data Challenges are underway at the University of Melbourne which will increase our understanding of the future of grid computing in physics, and the results will aid in the development of grid software and the LHC Grid infrastructure.

The construction of a grid infrastructure for Belle experiment analysis is continuing and further testing is planned. We have successfully used grid resources in an analysis test.

The generic grid interface development is still to be completed. Contributing to the wider community in this project and in others will help develop our experience with grid computing.

Acknowledgements

We at the University of Melbourne experimental particle physics group would like to acknowledge the help of the Melbourne Advanced Research Computing Centre (MARCC), Rajkumar Buyya and the GridBus team, and the Victorian Partnership for Advanced Computing (VPAC) for their generous funding.

References

Foster, I. and Kesselman, C.: 1999, *The Grid: Blueprint for a New Computing Infrastructure,* Morgan Kaufmann Publishers, USA.

The University of Melbourne, Experimental Particle Physics Group
    http://www.ph.unimelb.edu.au/epp/

Victorian Partnership for Advanced Computing (VPAC)
    http://www.vpac.org/

Melbourne Advanced Research Computing Centre (MARCC)
    http://www.hpc.unimelb.edu.au/

The University of Melbourne, Department of Computer Science & Software Engineering
    http://www.cs.mu.oz.au/

The GridBus project (Grid Computing and Business)
    http://www.gridbus.org/

Abramson, D., Giddy, J., Foster, I. and Kotler, L.: 2000, *High Performance Parametric Modeling with Nimrod/G: Killer Application for the Global Grid?* International Parallel and Distributed Processing Symposium, Cancun, Mexico.
http://www.csse.monash.edu.au/~davida/nimrod.html

Royal Melbourne Institute of Technology RMIT, Computer Science
http://www.cs.rmit.edu.au/

The Belle Collaboration, K. Abe *et al.*: 2002, *Nucl. Inst. Meth. A* **479**, 117.
http://belle.kek.jp/

ATLAS Technical Proposal, *Printed at CERN*, CERN/LHCC/94-43 LHCC/P2, 1994.
http://atlas.web.cern.ch/

The ATLAS Data Challenges (ATLAS DC)
http://atlas.web.cern.ch/Atlas/GROUPS/SOFTWARE/DC

The LHC Computing Grid project (LCG)
http://lhcgrid.web.cern.ch/

HENP Data Access Grand Challenge
http://www-rnc.lbl.gov/GC/default.htm

The Globus Project
http://www.globus.org/

The Globus Resource Specification Language RSL v1.0
http://www.globus.org/gram/rsl_spec1.html

The National Center for Supercomputing Applications (NCSA)
http://www.ncsa.uiuc.edu/

European Data Grid project (EDG)
http://www.eu-datagrid.org/

European Data Grid Work Package 1 (Workload Management). JDL Attributes *Internal Document DataGrid-01-NOT-0101_07*
http://www.eu-datagrid.org/

Jefferson Lab Data Grid Activities
http://www.jlab.org/hpc/datagrid/

Grid Data Mirroring Package (GDMP)
http://project-gdmp.web.cern.ch/projectgdmp/

GriPhyN – Grid Physics Network
http://www.griphyn.org/

Particle Physics Data Grid (PPDG)
http://www.ppdg.net/

Nile – National Challenge Computing
http://www.nile.cornell.edu/

Sun Grid Engine (SGE)
http://wwws.sun.com/software/gridware

The Earth Systems Grid (ESG)
http://www.earthsystemgrid.org/

Printed by Books on Demand, Germany